Maneuverable Formation Control in Constrained Space

Inspired by the community behaviors of animals and humans, cooperative control has been intensively studied by numerous researchers in recent years. Cooperative control aims to build a network system collectively driven by a global objective function in a distributed or centralized communication network and shows great application potential in a wide domain. From the perspective of cybernetics in network system cooperation, one of the main tasks is to design the formation control scheme for multiple intelligent unmanned systems, facilitating the achievements of hazardous missions – e.g., deep space exploration, cooperative military operation, and collaborative transportation. Various challenges in such real-world applications are driving the proposal of advanced formation control design, which is to be addressed to bring academic achievements into real industrial scenarios. This book extends the performance of formation control beyond classical dynamic or stationary geometric configurations, focusing on formation maneuverability that enables cooperative systems to keep suitable spacial configurations during agile maneuvers. This book embarks on an adventurous journey of maneuverable formation control in constrained space with limited resources, to accomplish the exploration of an unknown environment. The investigation of the real-world challenges, including model uncertainties, measurement inaccuracy, input saturation, output constraints, and spatial collision avoidance, brings the value of this book into the practical industry, rather than being limited to academics.

Automation and Control Engineering
Series Editors - Frank L. Lewis, Shuzhi Sam Ge, and Stjepan Bogdan

Synchronization and Control of Multiagent Systems
Dong Sun

System Modeling and Control with Resource-Oriented Petri Nets
MengChu Zhou, Naiqi Wu

Deterministic Learning Theory for Identification, Recognition, and Control
Cong Wang and David J. Hill

Optimal and Robust Scheduling for Networked Control Systems
Stefano Longo, Tingli Su, Guido Herrmann, and Phil Barber

Electric and Plug-in Hybrid Vehicle Networks
Optimization and Control
Emanuele Crisostomi, Robert Shorten, Sonja Stüdli, and Fabian Wirth

Adaptive and Fault-Tolerant Control of Underactuated Nonlinear Systems
Jiangshuai Huang, Yong-Duan Song

Discrete-Time Recurrent Neural Control
Analysis and Application
Edgar N. Sánchez

Control of Nonlinear Systems via PI, PD and PID
Stability and Performance
Yong-Duan Song

Multi-Agent Systems
Platoon Control and Non-Fragile Quantized Consensus
Xiang-Gui Guo, Jian-Liang Wang, Fang Liao, Rodney Swee Huat Teo

Classical Feedback Control with Nonlinear Multi-Loop Systems
With MATLAB® and Simulink®, Third Edition
Boris J. Lurie, Paul Enright

Motion Control of Functionally Related Systems
Tarik Uzunović and Asif Sabanović

Intelligent Fault Diagnosis and Accommodation Control
Sunan Huang, Kok Kiong Tan, Poi Voon Er, Tong Heng Lee

Nonlinear Pinning Control of Complex Dynamical Networks
Edgar N. Sanchez, Carlos J. Vega, Oscar J. Suarez and Guanrong Chen

Adaptive Control of Dynamic Systems with Uncertainty and Quantization
Jing Zhou, Lantao Xing and Changyun Wen

Robust Formation Control for Multiple Unmanned Aerial Vehicles
Hao Liu, Deyuan Liu, Yan Wan, Frank L. Lewis, Kimon P. Valavanis

Variable Gain Control and Its Applications in Energy Conversion
Chenghui Zhang, Le Chang, Cheng Fu

Maneuverable Formation Control in Constrained Space
Dongyu Li, Xiaomei Liu, Qinglei Hu, and Shuzhi Sam Ge

For more information about this series, please visit: https://www.crcpress.com/Automation-and-Control-Engineering/book-series/CRCAUTCONENG

Maneuverable Formation Control in Constrained Space

Dongyu Li, Xiaomei Liu, Qinglei Hu, and
Shuzhi Sam Ge

CRC Press
Taylor & Francis Group
Boca Raton London New York

CRC Press is an imprint of the
Taylor & Francis Group, an **informa** business

First edition published 2024
by CRC Press
2385 NW Executive Center Drive, Suite 320, Boca Raton FL 33431

and by CRC Press
4 Park Square, Milton Park, Abingdon, Oxon, OX14 4RN

CRC Press is an imprint of Taylor & Francis Group, LLC

© 2024 Dongyu Li, Xiaomei Liu, Qinglei Hu and Shuzhi Sam Ge

Reasonable efforts have been made to publish reliable data and information, but the author and publisher cannot assume responsibility for the validity of all materials or the consequences of their use. The authors and publishers have attempted to trace the copyright holders of all material reproduced in this publication and apologize to copyright holders if permission to publish in this form has not been obtained. If any copyright material has not been acknowledged please write and let us know so we may rectify in any future reprint.

Except as permitted under U.S. Copyright Law, no part of this book may be reprinted, reproduced, transmitted, or utilized in any form by any electronic, mechanical, or other means, now known or hereafter invented, including photocopying, microfilming, and recording, or in any information storage or retrieval system, without written permission from the publishers.

For permission to photocopy or use material electronically from this work, access www.copyright.com or contact the Copyright Clearance Center, Inc. (CCC), 222 Rosewood Drive, Danvers, MA 01923, 978-750-8400. For works that are not available on CCC please contact mpkbookspermissions@tandf.co.uk

Trademark notice: Product or corporate names may be trademarks or registered trademarks and are used only for identification and explanation without intent to infringe.

ISBN: 978-1-032-27722-6 (hbk)
ISBN: 978-1-032-28813-0 (pbk)
ISBN: 978-1-003-29861-8 (ebk)

DOI: 10.1201/9781003298618

Typeset in CMR10 font
by KnowledgeWorks Global Ltd.

Publisher's note: This book has been prepared from camera-ready copy provided by the authors.

Contents

PART I Formation Control with Maneuverability

Preface

Over the past few decades, the rapid development of intelligence on the individual agent has brought many novel opportunities to research and engineering. Motivated by the concept of collective behaviors from the human community, cooperative control has been intensively studied by numerous researchers aiming to build a multi-agent system collectively driven by a global objective function in a distributed or centralized communication network. It would be widely used in the domain of the military, civilian, and aerospace, with the utilization of surface robots, flapping-wing vehicles, unmanned aerial vehicles, and satellites. The interest in research diverges widely from the area of computer science to autonomous control.

The intelligence of a single agent has been incredibly boosted by the development of edge computing power, multiple sensor networks, and wireless communication modules. Nevertheless, considering the expense of building a single super-smart autonomous agent, it is costly to grant the whole group the same level of sensory and computation capability. That naturally motivates the generation of the leader-follower relationship, which is also inspired by the community behaviors of animals and humans. The leaders own the main resource and play the key role as decision-makers in the group. The decision from the leaders is spread to the remaining followers in the group through a communication network, while followers are attempting to infer the action or the optimal increment to update their state by information exchanged among the networks. The ultimate objective is to drive collective behavior in a designed manner.

From the perspective of the control plant, the main task is to design the control scheme for subsystems in the group based on the own knowledge and perception output, and the information spread from neighbor agents. Various challenges in real-world applications are driving the proposal of diverse controller design. The constraints are coming from both the internal and the external. The internal constraints result mainly from hardware limitations, including communication range, measurement, actuator output range, or energy consumption. Depending on the connectivity of the communication network, centralized, decentralized, or distributed control is studied by researchers. The uncertainties in system dynamics, sensor measurement noise, and communication delays impose on us the challenges to be addressed to bring academic achievements into real industrial scenarios. The external constraint we are talking mainly about in this book is the constrained space regarding mobility control. They may be known constraints to the multi-agent system or the unknown and the time-varying obstacles for which real-time reaction is needed to accomplish the exploration of an unknown environment.

This book is mainly inspired by the maneuverable formation control problem, with the consideration of realistic challenges in industrial applications. A few highlights of this are worth mentioning. First, though the focus of the problem in this book is on formation control, the mathematical tools introduced by authors on system dynamic modeling, the acquisition and the proof of the system convergence and stability, and the controller design could be generally extended into other cooperative control domains. Second, the investigation of the real-world challenges, including model uncertainties, measurement inaccuracy, input saturation, output constraints, and spatial collision avoidance, brings the value of this book into the practical industry, rather than being limited to academics.

Acknowledgment

We would like to express our sincere appreciation to numerous valuable suggestions, comments, and multi-faceted support to this monograph – in particular, from (in alphabetical order of their last names)

Rodrigo Aldana of University of Zaragoza,
Linda Bushnell of University of Washington,
Hui Cao of Beihang University,
Hui Cao of Xi'an Jiaotong University,
Kun Cao from Nanyang Technological University,
Yanning Guo from Harbin Institute of Technology,
Yufeng Gao of Harbin Institute of Technology,
Qinglei Hu of Beihang University,
Wei He of University of Science and Technology Beijing,
Ruihang Ji of National University of Singapore,
Wanyue Jiang of Qingdao University,
Vijay Kumar of University of Pennsylvania,
Tong Heng Lee of National University of Singapore,
Frank L. Lewis of University of Texas at Arlington,
Shuai Liu of Shandong University,
Xiaoming Liu of Beihang University,
Chuanjiang Li of Harbin Institute of Technology,
Yueyong Lv from Harbin Institute of Technology,
Xiaoling Liang of National University of Singapore,
Xing Liu of Northwestern Polytechnical University,
Guangfu Ma of Harbin Institute of Technology,
Jun Ma of Hong Kong University of Science and Technology,
Lorenzo Marconi of University of Bologna,
Yongduan Song of Chongqing University,
Xiaodong Shao of Beihang University,
Keng Peng Tee of Agency for Science, Technology and Research (SG),
Yan Wu of Agency for Science, Technology and Research (SG),
Lihua Xie of Nanyang Technological University,
Keyou You of Tsinghua University,
Haoyong Yu of National University of Singapore,
Wei Zhang of Shanghai Institute of Satellite Engineering,
Yuxiang Zhang of National University of Singapore,
and Jianying Zheng of Beihang University, among others.

Special gratitude towards Nora Konopka, Paul Boyd, and Swapnil Joshi from CRC Press, for their consistent and reliable support in the process of publishing the book, and towards IEEE and Elsevier for granting us the permission to revisit materials from our publications.

Acronyms and Notation

Acronyms

AI	Artificial Intelligence
APF	Artificial Potential Field
BLF	Barrier Lyapunov Function
CCN	Cooperative Circumnavigation
CW	Clohessy-Wiltshire
CMCA	Cohesion Maintenance and Collision Avoidance
DSC	Dynamic Surface Control
DTW	Dynamic Time Warping
EA	Evolutionary Algorithm
ELS	Euler-Lagrange System
FWV	Flapping-Wing Vehicle
ISS	Input to State Stable/Stability
LAF	Layered Affine Formation
LDFE	Layered Distributed Finite-time Estimator
LOS	Line-of-Sight
LVLH	Local-Vertical-Local-Horizontal
MAS	Multi-Agent System
MELS	Multiple Euler-Lagrange System
MFCC	Mel-Frequency Cepstral Coefficient
MIMO	Multiple-Input Multiple-Output

MLFC	Multi-Layer Formation Control
NN	Neural Network
RBF	Radial Basis function
RF	Random Forest
SGUUB	Semi-Globally Uniformly Ultimately Bounded
SISO	Single-Input Single-Output
UAV	Unmanned Aerial Vehicle
UGV	Unmanned Ground Vehicle
USV	Unmanned Surface Vehicle
VS	Virtual Structure

Notation

$\mathbb{R}$	Field of real numbers
$\mathbb{R}^+$	Field of nonnegative real numbers
$\mathbb{Z}^+$	Field of positive integers
Σ	Summation
max	Maximum
min	Minimum
$\forall$	For all
$\in$	Belongs to
$> (<)$	Greater (less) than
$\geq (\leq)$	Greater (less) than or equal to
$t \to T$	t approaches T
$t \uparrow T$	t approaches T from below (in an increasing manner)
$t \downarrow T$	t approaches T from above (in a decreasing manner)

$\otimes$	Kronecker product
w.r.t.	With respect to
s.t.	Such That or Subject To
i.e.	The Latin phrase id est, meaning "that is"
e.g.	The Latin phrase exempli gratia, meaning "for example"
$\mathbb{R}^n$	The n-dimensional Euclidean space
I_n	The n-dimensional identity matrix
$\mathbf{1}_n$	The n-dimensional vector with all entries being 1
$\lvert \bullet \rvert$	The absolute of a scalar $\bullet$
$\lVert \bullet \rVert$	The norm of a vector $\bullet$
$\dot{f}$	The first derivative of f with respect to time
$\ddot{f}$	The second derivative of f with respect to time
$\mathbf{D}^+ f$	The right-hand upper Dini derivative of f w.r.t time
$\text{diag}\{x_1 \cdots x_n\}$	The diagonal matrix with diagonal elements x_1 to x_n
$H > 0$	A positive definite matrix H
$H \geq 0$	A positive semi-definite matrix H
X^T	The transpose of matrix/vector X
$\lambda_{\max}(A)$	The maximum eigenvalue of matrix A
$\lambda_{\min}(A)$	The minimum eigenvalue of matrix A
$\text{vec}(A)$	The vectorization form of matrix A
$\text{tr}(A)$	The trace of matrix A
$A^\dagger$	The Moore-Penrose pseudo-inverse of matrix A

1 Introduction

This chapter begins by introducing the multi-agent system (MAS) control and outlining research results on the multi-agent formation control. Multi-agent formation control, in general, contains two parts: 1) formation shape control which drives a collection of agents to a specific spacial configuration; and 2) formation maneuver control which continuously maneuvers an entire group to realize certain spacial transformations based on mission requirements. In Section 1.1, multi-agent cooperative control is introduced and categorized. In Section 1.2, affine formation control is introduced which is capable of achieving formation with flexible geometric maneuverability. In Section 1.3, formation control in constrained space is discussed since agents usually move in a complex environment with irregular obstacles in practical applications. Finally, in Section 1.4, a robust preview of the following chapters is provided.

1.1 MULTI-AGENT COOPERATIVE CONTROL

We first introduce a group of cooperative agents known as MASs. Distributed artificial intelligence (AI) is likewise given as a superclass of MASs [121]. And several different classifications of MASs are given.

In general, an agent is defined as an object in a space that perceives several factors before making a choice based on its own objective. And based on this choice, the object takes some essential actions on the environment [121]. MASs are explored by dividing a complex problem into many individual subproblems and assigning them to multiple agents. Compared to a single-agent system, MASs have special advantages because of their cooperation such as higher expandability, flexibility, robustness, and energy efficiency.

An MAS can be categorized as a kind of distributed AI; other categories of distributed AI include parallel AI and distributed problem-solving [121]. Parallel AI usually focuses on investigating parallel algorithms, architectures, and languages to increase the efficiency of AI algorithms. Distributed problem-solving focuses on dividing a task into subtasks and allocating each subtask to a node of a group of cooperative nodes (called computing entities). Just like distributed problem-solving, MAS solves problems distributively with a set of nodes called agents. First of all, any agent can learn the environment and make autonomous decisions proactively instead of passively dealing with assigned tasks. Second, agents communicate with their neighbors or perceive the environment to learn new actions and contexts. Eventually, agents utilize the acquired knowledge to determine and complete the following actions in the environment, guaranteeing to complete their own task, while gradually completing an overall task or global optimization.

DOI: 10.1201/9781003298618-1

During the past few decades, the MAS research has sparked widespread interest in a variety of subjects, for instance, mathematics, physics, biology, engineering science, and sociology [193, 268, 372, 397, 541]. There are many interesting topics for multi-agent research, including distributed consensus, formation, flocking, computing, and learning [48, 88, 140, 271, 430, 442].

Investigations on multi-agent control could have many classifications based on different focuses. An agent is called a leader who is responsible for raising goals for other agents according to the global objective. For whether there exists at least a leader or not, we can classify multi-agent control as leader-follower [138, 158] and leaderless [6, 270]. It can also be classified into homogeneous [122], heterogeneous [234, 238], linear [536, 554], non-linear [124, 282], first-order [341], second-order [493], and higher-order [493] according to the model of an agent. Based on other practical constraints, researchers also explore cooperative control of MASs on time-delay [143, 295, 493], actuator failures or network-induced packet loss [313, 493], switching mechanism [295, 364], event/time-triggered [143, 162, 265] (depending on the triggered manner of data sharing and environment sensing of an agent) and circular (or other shapes) formation [472, 473].

Although elegant MAS results have emerged over the last decades, there are still open challenges in cooperative control [325]. The following subsections provide more in-depth introductions to cooperative control including MAS consensus, synchronization, controllability, connection, and formation, among other interesting topics.

1.1.1 CONSENSUS CONTROL

Consensus control refers to the design of distributed controllers that enable all agents to concur on specific quantities of interest through local contacts [178, 391, 543].

The consensus schemes have been used in multi-agent coordination in a number of applications like vehicle formations [289, 330], attitude alignment [22, 249, 400], rendezvous problem [288, 403], coordinated decision-making [4, 23], flocking [363, 459, 459], coupled oscillators [205, 424], and robot position synchronization [410].

In the context of MASs, cooperative control refers to the ability of multiple agents to work together toward a common objective by sharing information and coordinating their actions. As MASs are often used in networked systems, where agents are connected through a communication network, the study of cooperative control in such systems is known as networked MASs control.

One of the most important performance metrics for networked MAS cooperative control is convergence rate, which measures how quickly agents can reach an agreement or consensus on a particular variable of interest. In many applications of networked MASs, such as vehicle formations, attitude alignment, and coordinated decision-making, rapid convergence is critical to achieving the desired performance. Furthermore, in the case of networked MASs subject to disturbances or uncertainties, a fast convergence rate can enhance the robustness of the system by reducing the impact of these factors on the control performance.

Overall, the convergence rate is an essential performance metric for networked MAS control because it reflects how quickly the system can achieve a desired objective while maintaining coordination among the agents. It is broadly classified into groups [436]: convergence in infinite time [1, 84, 214, 219, 530, 531] and convergence in finite time [114, 270, 297, 435, 486]. In the case of convergence in infinite time, the convergence rate can only be exponential with infinite settling time. Convergence in finite time is to realize the control goals in finite time. It also has higher robustness and disturbance rejection.

1.1.2 SYNCHRONIZATION

Synchronization refers to the time of one agent's actions in relation to those of other agents [75, 257, 281, 308, 344, 420, 495, 514]. The consensus issue and synchronization are closely related: consensus is the state of all agents coming to the same value on a specific amount of interest, whereas synchronization describes the time-correlated behavior of various agents [443]. However, increasing agent heterogeneity makes synchronization more difficult. For heterogeneous agents, synchronization must be accomplished across a number of different features, but for homogeneous agents, synchronization may be conducted above a single shared feature.

As MASs become more heterogeneous, achieving synchronization becomes increasingly challenging since agents may differ in various characteristics, such as their dynamics, communication capabilities, or objectives. This heterogeneity can affect the agents' ability to coordinate their actions in time, which is essential for achieving consensus on a common value of interest. To address this issue, different synchronization methods have been proposed in the literature, such as output or partial-state synchronization [234], which aim to synchronize only a portion of the agents' state variables that are physically comparable. These methods can help to mitigate the effects of heterogeneity on synchronization and enable agents to cooperate effectively despite their differences.

1.1.3 CONTROLLABILITY

Controllability is usually emphasized and analyzed for MASs under a switching topology. It describes the circumstance of whether or not MASs can be guided by particular rules from an initial point to a desired condition [211, 212, 294–296, 303, 319, 393, 432]. Controlling robots, automobiles, and airplanes are just a few of the many uses for controllability [295]. The degrees of topological dynamism and environmental determinism are two important variables that have an impact on the MAS controllability. In a dynamic MAS, the topology frequently changes, affecting the connections between agents and thus impacting their ability to cooperate. Because the outcomes of actions in nondeterministic environments are unpredictable, agents must wait to take new actions until they have seen the effects of their prior ones, which adds latency and reduces agent pro-activity. Controllability is generally attained through a centralized process in which leaders give followers instructions on how to accomplish a specific goal.

1.1.4 CONNECTIVITY

In some cases, agents need to be continuously connected to each other. For example, agents must be linked to the leaders under a fixed or dynamics topology in the leader-follower MAS structure. The connection is complicated to analyze because of the issues listed below: 1) noisy environments: the link between two agents might be disrupted by any environmental intervention that needs to reestablish these connections; 2) mobility: due to the high-speed movement and maneuvering of agents, connections between them frequently break down and must then be reestablished; and 3) constrained sensing ability: positioning the agent to gain maximum connection is likewise challenging since agents have restricted views (e.g., line-of-sight).

MASs are divided into connectionless [312] or connected [120, 128, 446] according to their connectivity. In the connected model, stable and permanent connectedness between agents is ensured while in the connectionless model connectedness is not guaranteed. Connectivity is a commonly considered aspect in flocking and vehicular systems [481]. Connectivity in flocking and vehicular systems refers to the ability of agents to establish and maintain communication links with one another. This is crucial for the agents to coordinate their behaviors and achieve their collective goals. In flocking systems, agents may use communication to share information about their local velocity and position, allowing them to align their movement and maintain a cohesive group. In vehicular systems, connectivity can facilitate the exchange of information about traffic and road conditions, enabling vehicles to adapt to changes in their environment and avoid collisions.

1.1.5 FORMATION

For MASs, we need to organize a set of agents to form a specific configuration and maintain the spatial shape for a while. For instance, unmanned aerial vehicles (UAVs) are required to construct a specific formation to detect multiple environmental parameters or locate a specific object in the environment. For a specific MAS mission, finding the structure that will work best for all agents, grouping agents around that structure, and preserving such structure for a predetermined time period are the key phases in formation. Multi-agent formation control consists of formation shape control and formation maneuver control: 1) control a group of agents to achieve a desired geometric pattern;

and 2) maneuver the whole group to realize translation, rotation, scale, shear, and other geometric transformations continuously.

Numerous methodologies have been employed to investigate the multi-agent formation control problem. The early techniques, such as behavior-based methods, can tackle challenging formation tasks. However, it is challenging to mathematically demonstrate whether or not these approaches can reach system convergence. But convergence is critical since it ensures that the system will act as predicted. Due to the successful application of the consensus theory in formation control [290,402], plenty of convergence-guaranteed formation control approaches have been developed [41, 360].

1.2 AFFINE FORMATION MANEUVER CONTROL

In this section, we introduce the recent development, major problems, and practical applications of MAS affine formation. Affine formation grants MAS maneuverability by affine transformations, e.g., scaling, rotating, shearing, and collinear.

1.2.1 BACKGROUND OF AFFINE FORMATION

The multi-agent formation control problem has garnered plenty of interest, and many delicate studies have been conducted.

Early approaches like behavior-based ones (to name a few [16,92,102,250]) can handle complicated formation tasks, but their convergence is difficult to prove mathematically. Since convergence is vital to guarantee the MAS behavior as anticipated, tremendous research has been done to ensure convergence during establishing formation control strategies (see [41, 360] for recent surveys). Formation control approaches can be divided into several different classes in line with how the desired formation is articulated: distance-based [11, 74, 336], displacement-based [10], bearing-based [200, 457, 542, 548, 549], angle-based [58, 62, 218], etc. The maneuverability of the target formation is strongly influenced by the invariance of the constant restrictions. In certain cases, it limits the above-mentioned approaches to realizing translation transformation, rotation transformation, and scaling transformation together. For example, inter-agent displacement constraints keep the formation translation invariant but variant to rotation transformation and scaling transformation. As a result, displacement-based formation control can only implement translation transformation. Thus, in order to achieve the necessary formation movements, certain modified approaches are proposed. For instance, the strategy in [83] modified the displacement-based formation control approach by adding a formation scale estimation mechanism. It allows translation transformation and scaling transformation to be implemented simultaneously. The approach proposed in [198] modifies the distance-based formation control method by introducing a new variable representing the scale of the formation and by incorporating a control law to regulate the scale of the formation. This allows the formation to converge to a desired shape, while the scale of the formation is not pre-specified but instead emerges from the interactions among the agents.

However, while these methods can be effective in achieving the desired formation size and shape, they often introduce additional parameters which can complicate the estimation and control processes. This may lead to problems such as non-convergence, instability, or reduced performance. In order to address these issues, it may be necessary to augment the system with extra sensing or communication capabilities, which can help ensure that the agents are able to adjust their behavior as needed and maintain the desired formation.

There are several recent approaches utilizing different constant limits to form the nominal target, for instance, local bearings [535], complex Laplacians [169,292], barycentric coordinates [167], and stress matrices [290]. These approaches enhance the invariance of the constraints, which means more variations of formation configuration can be done at the same time. The complex Laplacian is invariant over the formation's translation, rotation, and scaling changes. But, it is only suitable for two-dimensional formation control. The stress matrix can be viewed as a generalized graph Laplacian [290]. It is a symmetric matrix that characterizes the internal forces in a network and reflects the connectivity and geometry of the network. But as opposed to traditional graph Laplacian matrices, in a stress matrix, an edge's weight can be positive, negative, or zero. The stress matrix-based one, called affine formation control [546], appears to have a better chance of achieving the extensive objective of being applicable in any number of dimensions and invariant to any affine transformation.

The affine formation is a set of conditions for a group of agents that associate with the desired formation through an affine transformation. It maintains collinearity and distance ratios. In other words, the initial state of agents laying in a line remains in a line and keeps the distance ratio in relation to the desired formation [290]. Some advantages of affine formation are listed as follows: 1) by manipulating just a few agents (usually referred to as the network's leaders), the affine formation of agents can be changed into a rigid formation or translational formation [290]. Additionally, compared to the traditional Laplacian matrix, the stress matrix has a lower rank if it meets the stability requirement. This leads to higher free degrees when taking the stress matrix's null space into account to construct different transformations [497]. It is beneficial for MAS formation to be more capable of adapting to changing environments while fulfilling other requirements like collision avoidance; 2) cooperative localization in sensor networks is related to affine formation control. It just takes a few agents in a sensor network called anchors to know their absolute coordinates, another sensor agents can determine their absolute coordinates using affine formation control principles [509, 513].

1.2.2 MAJOR PROBLEMS IN AFFINE FORMATION CONTROL

Two major problems in affine formation control will be introduced: the leader selection and the graphical design.

1.2.2.1 Leader Selection

The leader-follower strategy is mostly employed in affine formation control. For example, in [546], only a small subset of agents named as leaders are aware of the desired formation maneuvers, while the remaining agents known as followers merely have to follow the leaders through proposed specific control protocols for different complex assigns. Some other examples using the leader-follower formation strategy of MASs can be found in [60, 266, 476, 505, 507–510, 512, 522, 546, 561].

In the leader-follower formation strategy, the leader selection problem has attracted considerable focus in the fields of MAS control. Different control strategies have been investigated for various practical uses. In [428], it proposes a general graphical method for ensuring the controllability of the signed network. In [429], the energy-related controllability is studied using the controllability Gramian. When agents are exposed to stochastic disturbances, the coherence is enhanced [374]. For reducing the overall mean-square error of the followers' states when connection noise is present, a supermodular optimization technique is suggested [79]. An essential indicator of network effectiveness is the convergence rate of consensus network, which is introduced in [77, 78, 375, 523]. In [546], it solved the leader selection problem in affine formation control and put forward the concept of "affine formation localizability", which can judge if the chosen leaders have the ability to control the whole formation to produce the appropriate affine transformations. In [497], the proposed leader selection algorithm is designed to achieve two objectives: control energy and convergence speed, which are aimed at two mixed integer semi-definite programming optimization problems. The proposed framework can provide less communication cost, higher tolerance for time delay, and faster convergence speed, all of which can be freely regulated by changing the corresponding bounds in the mixed integer semi-definite programming formulation.

Rather than the conventional leader-follower formation model, the authors in [97] proposed a leaderless strategy by altering the original weights that construct the Laplacian matrix so that via the local interactions of the agents, the designed steady-state motion with the desired shape can emerge.

1.2.2.2 Graph Design

The topology of affine formation can be either directed or undirected. Affine formation control with undirected graphs is considered in [477, 510, 520, 561]. Elegant studies also consider the affine formation control with directed interaction graphs [49, 263, 505, 507–509, 512, 513, 520, 558].

The topology of interaction graphs is hard to keep fixed because of the uncertainty in the environment as well as some constraints on agent dynamics. An affine formation tracking problem of nonholonomic systems on SE(3) was solved under fixed and switching topologies in [520].

Topology design is another problem that should be importantly considered in affine formation control. The construction of the stress matrix is the foremost concern in the affine formation control problem. According to consensus theory, the stress matrix may be considered as a more generalized version of the Laplacian matrix. Additionally, the underlying communication topology among agents is implicitly captured by the stress matrix. Keeping the connections between agents in order to enhance network performance is the main focus of the topology design challenge. To achieve the stress matrix with high tolerance to time delay and communication cost, a topology design approach using mixed integer semi-definite programming is given in [497].

1.2.3 AFFINE FORMATION CONTROL IN PRACTICAL APPLICATIONS

Affine formation control problems in practical applications will be introduced, such as affine maneuverability, model uncertainties, collision avoidance, time delay, path length, fuel consumption, and time.

Stress matrices are used to stabilize the stationary target formation in [290]. And in [546], it studies to use stress matrices to solve formation maneuver control problems. The ability of the stress matrix to cope with the formation maneuver control problem is discovered by utilizing a leader-follower control structure. Additionally, the control algorithms developed in [546] can perform numerous formation maneuvers in two and three dimensions, including translation, rotation, scaling, and shearing. Due to the fact that the affine image can represent configuration alterations in any dimension, the preceding works are soon popularized, and subsequent works investigate the general linear system [365], the second-order system [263, 560], and the high-order system [60].

The model uncertainty is another concerned issue in practical applications. Reference [365] studies the problem of distributed adaptive affine formation maneuver control with parameter uncertainties for general linear dynamics. Considered systems having dynamic and static coupling gains, robust control laws to uncertainty are suggested. For uncertain Euler-Lagrange MASs, reference [505] presents a distributed hybrids affine formation control mechanism. The performance of the Euler-Lagrange systems is improved by using the adaptive neural networks approach to account for the uncertainty. Other interesting results such as [61, 507] are also capable of tackling model uncertainties.

Avoiding collisions is essential in practical conditions, e.g., multiple agents pass through a narrow passage. Particularly, avoiding collisions with other spacecraft and other suddenly emerging risks is crucial for controlling several spacecraft in formation. The work in [507] combines backstepping controllers with specific artificial potential functions and adaptive neural networks, which address the issues of collision avoidance, formation reconfiguration and of Euler-Lagrange models at the same time.

Time delay in affine formation control is another particular concern issue in practical applications, e.g., the inevitable time delays during information and signal exchanges. In [477], it investigated the affine formation control of general linear MASs over an undirected graph with time delays. Different delays circumstances have been taken into account, and predictive observers are created so that followers can forecast future situations. The projected states are used to develop affine formation control laws, and adequate conditions are obtained to guarantee the convergence of the formation errors. The work in [513] deals with the affine formation maneuvers when there is a non-uniform or time-varying time delay. Compared to time-domain Lyapunov-type functional methods proposed in previous works such as [117], the approach presented in this paper proves the upper bound of delays in the frequency domain, making it suitable for scenarios where the delays can be non-uniform and time-varying.

The paper [562] studies how to optimize the different performances of affine formation control, such as path length, fuel consumption, and time, which can make different shapes with stress matrices. The paper uses a leader-follower method, where some agents are controlled and others follow them. The paper has two ways to find the best control laws for the agents. The first way uses the inverse optimality strategy, which makes a control law that minimizes a performance index and makes the followers reach their targets. The second way uses a cooperative estimator, which guesses the targets of the followers and minimizes another performance index. The paper shows that the control laws work well and are optimal with simulations.

1.3 FORMATION CONTROL IN CONSTRAINED SPACE

In recent years, a variety of applications for formation control are proposed, including mobile robots [8, 153], vehicles [478, 517], aircraft [15, 161], vessels [32, 115], and spacecraft [25, 345]. Many challenges are to be overcome to guarantee the successful utilization of formation control in the real world, such as multiple constraints which include complex environmental constraints and system dynamics constraints [333, 394], the rapid collision avoidance [407,427,439], and the information exchange among MASs [131]. In these problems, MAS control tasks in constrained spaces have received increased interest in recent research from various disciplines. Constrained spaces refers to a space with boundaries or obstacles such as UAVs performing missions in canyons, forests, etc. The effects of boundaries and obstacles will ruin the formation and cause agent positions to be uncertain. Hence, designing a strategy that takes into account both formation maintenance and obstacle avoidance is challenging, especially when the environment is unknown, and it is important to adapt to complex environments in time.

We first review the most commonly used formation control strategies. Then we will review the cooperative formation path planning while it is acting as a command generator of a formation control system that accepts an environment description and outputs sequences of waypoints as trajectories.

1.3.1 CONTROL STRATEGIES

There are three primary types of formation maintenance [311]: 1) formation generation and maintenance which refers to forming a formation shape from a situation where the agents' location and headings are arbitrary and maintaining the formation shape to complete the mission; 2) formation maintenance during trajectory tracking refers to maintaining a formation shape when the formation is moving along a predetermined path; and 3) formation shape variation and regeneration which not only needs to maintain the formation shape but also requires adjustment and regeneration of the formation shape when the mission objective changes or MASs are dynamically adapting the environment (e.g, collision avoidance in constrained space). Type 1 and Type 2 are commonly used in simple environments with few obstacles. While in extremely complex environments existing multiple moving obstacles or cluttered obstacles, Type 3 is mainly used. So in the following introduction, the capability to perform these three types of formation maintenance is regarded as an indicator to show the ability of collision avoidance in constrained space.

A variety of methods, including behavior-based [16, 494], leader-follower [82, 175], and Virtual Structure (VS) [25, 252, 456], have been proposed to accomplish formation maintenance. While these three methods have their own feature, their ability for collision avoidance and accomplishing tasks in constrained space is different. We now introduce and analyze the three methods.

1.3.1.1 Leader-Follower Formation Control

At least one agent is selected to be the leader of the group, having complete access to global information, and acts as the formation's reference agent in the leader-follower control strategy. The designated virtual leader takes the place of the actual agent in the formation [67]. Other agents in the formation, aside from the leader(s), are considered as followers. According to the leader's static or dynamic position, orientation, the followers' main task is to maintain the formation shape by keeping the attitude angle to the leader(s) and keeping the desired distance from the leader(s). The main advantage of leader-follower control strategies is simplicity since the entire formation mainly depends on the leader(s) trajectory. As a result, the formation problem can be reduced to a trajectory-tracking problem since the formation stability is provided by the control laws of the individual agents. The main drawbacks of this strategy include the lack of explicit feedback from the followers to the leader(s) and formation failure due to leader(s) errors.

In [102], two types of leader-follower control strategies using polar coordinates are proposed: separation-bearing control and separation-separation control. In the separation-bearing control, followers maintain the formation by keeping the desired formation distance and desired formation angle to the leader(s). In the separation-separation control, followers maintain the formation by keeping a fixed distance to two neighbor agents. The authors in [273] present a kinematic model for the leader-follower formation control using Cartesian coordinates. Employing the Cartesian coordinates instead of the polar coordinates, the singularity in the approach [102] can be avoided. When formation control is confronted with complex environment, the authors in [103] propose a collision avoidance formation control strategy in order to realize the leader-follower task. The obstacle is avoided by allowing the agent to keep a redefined distance from the other agents and the obstacle. Its shape may be modified flexibly while avoiding the obstacle, and once when the risk of collision is eliminated, it can back to the desired shape. The approach proposed in [469] highly depends on omni-directional vision. The approach estimates the position and velocity of the leader by using the motion segmentation technique and omni-directional visual is used for tracking and obstacle avoidance.

1.3.1.2 Virtual Structure Formation Control

A method called Virtual Structure (VS) [3,111,153,307,527], which is a crucial formation control strategy, is developed in [456]. The VS refers to a group of agents that are rigidly geometrically related to another and to a reference frame [252]. The basic idea of VS is to consider the formation shape as a rigid body or a VS and keep the formation by minimizing the position error between the actual formation position and the VS. To accomplish this goal, a bidirectional control strategy is presented where the VS locations are decided by the formation locations, and the agents are driven by the force that is applied virtually to the VS [311].

A comparison between the VS method and the leader-follower approach reveals that while the former has superior fault-tolerant ability, it has certain limitations in formation modification. A malfunctioning agent will not be identified by other agents due to the absence of feedback on the locations of each agent in the formation under leader-follower control, leading the formation to disintegrate. However, it may be avoided by using the VS method. Because the formation is structured to match the rigid body VS, the VS approach gives higher formation maintenance performance. However, such high formation maintenance performance becomes troublesome for formation modification. The formation alterations involve redesigns of the VS, which may possibly increase the formation's computational cost. The rigidity of the formation restricts the VS's ability to deal with obstacles and avoid collisions, therefore it is inappropriate for formation shape generation and formation reconfiguration.

1.3.1.3 Behavior-based Formation Control

The behavior-based formation control approach is inspired by natural formation behaviors like schooling and flocking. Possible behaviors include formation keeping, collision avoidance, and goal seeking. In contrast to the VS approach, the benefit of the behavior-based approach is decentralization with fewer communications. However, the theoretical analysis of system stability is quite hard because the proposed controller is not based on the dynamic/kinematic features of the agents. Therefore, guaranteeing the convergence of the formation is still an open issue [250]. The original proposal for behavior-based formation control appears in [16]. Through the use of a hybrid vector-weighted control function that can provide control commands based on various sorts of formation missions, which can address the issue of formation control. Implementing behavior-based formation control enables to achieve collision avoidance, formation generation, and formation maintenance simultaneously.

Overall, behavior-based formation control and the genetic algorithm are two promising approaches to address the challenges of formation control. While the behavior-based approach allows for decentralized control and simultaneous achievement of multiple tasks, such as collision avoidance and goal seeking, it faces difficulties in ensuring system stability due to the lack of consideration for the kinematic features of agents. On the other hand, the genetic algorithm provides a stochastic optimization solution to determine the optimal weighted gains for each behavior, making the system more adaptable to different formation tasks and environments. The combination of behavior-based formation control and the genetic algorithm has been applied to various scenarios, from formation maintenance to formation control in unfamiliar and dynamic environments with moving obstacles.

The genetic algorithm is a stochastic optimization algorithm. It is inspired by biological evolution and adopts the principle of natural selection and natural genetics. The genetic algorithm, together with behavior-based formation control, is used in [42] to estimate the weighted gains for each behavior. In this study, each agent can maintain formation while moving as swiftly as possible toward the predetermined target without any collisions with static objects or other agents. Work in [43] investigates formation control in an unfamiliar environment with dynamic obstacles. To predict the positions of moving obstacles in order to avoid collisions, this work uses a recurrent least squares algorithm-based prediction model. Then, considering the task and the environment, combined with the traditional behaviors, a behavior called random behavior is designed.

1.3.2 COOPERATIVE PATH PLANNING

In general, path planning algorithms may be grouped into two main methods based on their searching characteristics [455]: heuristic path planning and deterministic path planning. Due to the fact that it is carried out by following a series of predefined processes, the deterministic searching strategy has the advantages of consistency and completeness. As long as the result exists, a search result can always be found; additionally, if the searching environment does not change, then the output is the same every time. The exact algorithm is another name for the deterministic algorithm. The optimization method [537], the roadmap-based algorithm [355], and the artificial potential field (APF) [13, 93, 322] are deterministic approaches. The heuristic searching strategy [215, 361, 366] is presented to precisely solve the issue that a deterministic approach cannot efficiently handle. When exact answers are difficult, it may also offer an approximate solution, hence the heuristic approach is usually known as an approximation algorithm [243]. But it cannot ensure the global optimality of the findings, because it only searches a subset of the search area, implying that only near-optimal solutions can be found. Furthermore, since the algorithm searches inside the space stochastically, the consistency of the findings is lower than that of the deterministic method [454]. Heuristic algorithms for example evolutionary algorithm (EA) including ant colony optimization, particle swarm optimization, and genetic algorithm [455]. For formation path planning which means maintaining the predetermined formation along the path and also retaining the time for reaching the desired target, common approaches are artificial potential field (APF) [368,524], evolutionary algorithms [411], optimal control method [526], neural networks [274], etc.

1.3.2.1 Artificial Potential Field Methods

The APF approach is first proposed in [232] to operate a robot manipulator. The APF approach uses the potential field to describe the configuration space, which is made up of repulsive fields around obstacles and an attracting field around the target point. The repulsive force is only effective in specific areas around the obstacles and is inversely linked to the distance to them. In contrast, the attractive field exists throughout the whole space and is proportional to the distance to the target point. The route is computed by the total force, including the sum of fields gradient, at each site.

In order to accomplish formation path planning with the APF, additional fields are necessary to describe cooperative formation behaviors. To preserve the formation shape, an inner attractive potential field must be built first. When an agent departs from its formation place, the force may drag it back to keep the formation shape. To keep two agents from colliding, internal repulsive fields are also necessary. In [475], it created potential fields by drawing on electric field concepts. Each agent is regarded as an electric field point source with varying electrical polarity. In [376], an attractive potential field related to the deviation between the intended and actual distances is developed to increase control precision, allowing any deviation from the desired location to be promptly adjusted and rectified.

Based on the APF, the authors in [524] publish a study on autonomous underwater vehicle formation motion planning in an obstacle-filled environment. The autonomous underwater vehicle formation is treated by the algorithm as a multi-body system. Each agent is simulated as a point mass with complete actuation. And this work concentrates on overall mission objectives rather than constructing a control law for each individual agent.

The fundamental disadvantage of utilizing the APF should be noted: the problem of local minima. Many researchers addressed this issue by creating different types of fields, for example, the harmonic potential fields [81, 91, 236, 332]. In [144, 156], an approach for solving the local minima problem is presented, which builds the potential field using the Fast Marching Method (FMM).

1.3.2.2 Evolutionary Algorithm

The EA technique is a heuristic search algorithm based on biological evolution [94, 222, 392, 502, 556]. The EA method includes ant colony optimization, particle swarm optimization, genetic algorithm, etc. While using EA in path planning, EA simulates the process of natural evolution to obtain the plan. The EA treats paths as individuals and allows individuals to produce new individuals by reproducing, recombining, and mutating. For example, mutations can be generated by randomly fine-tuning the waypoints of an existing path. Then EA filters out the desired individuals by a function. This process is called natural selection. Each individual's quality is determined by the function, which is a weighted function made up of various optimization factors. Only the one having the highest fitness score is chosen as the final path. When an EA is used for MAS formation path planning, a two-layer evolution process is normally used. First, each agent executes an EA to obtain a trajectory subject to each individual agent's planning conditions. Then, the master EA algorithm compares each individual path to a master fitness function to produce cooperative behavior. The master fitness function considers factors such as the distance between each agent, the target point arrival time, the internal collisions, etc. However, the results produced by EA have strong randomness which makes it difficult to obey the strict requirement of formation shape preservation. One solution to address the issue of randomness is to use hybrid optimization algorithms, which can combine the strengths of different optimization techniques.

1.3.2.3 Optimal Control Method

Another common technique for multi-agent cooperative path planning is the optimal control method [173, 315, 567]. This method is regarded path planning as a numerical optimization problem with restrictions [226]. Multi-agent cooperation is accomplished by following a predetermined collection of cooperation constraints, which transforms multi-agent path planning into several single path planning procedures.

Single-agent optimization constraints included state and control input boundary conditions, as well as position constraints to avoid fixed and dynamic obstacles. Therefore, in this case, a cooperation constraint is created to keep two agents at a safe distance. Mixed integer linear programming has been applied to identify the best control input for every agent by submitting it to all constraint conditions, which could then be utilized to construct a viable trajectory by replacing that with the system's dynamic functions. The authors in [526] expand this method to a larger scale multiple vehicle cooperation. To achieve this cooperation, constraints are devised to maintain sufficient distances for robust communication. However, the mixed integer linear programming method requires a large amount of computation. For this reason, it is not suitable for achieving real-time planning. To improve the mixed integer linear programming method, the authors in [26] use receding horizon control to address optimization problems for UAV formations. Unlike traditional methods that seek the global optimal solution, the receding horizon control only seeks the optimal solution in each time step. This significantly reduces the computation time. By integrating the receding horizon control with the APF approaches, the work in [69] creates a formation hybrid formation path planner for UAVs. An extra control force provided by APF is added to the system control input to increase the formation's collision avoidance capacity.

1.4 ROBUST PREVIEW OF CHAPTERS

The following chapters are mainly divided into two parts. The first is Layered Affine Formation Control and the second Formation Control in Constrained Space. For the first part, we first introduce mathematical preliminaries in Chap. 2. In Chap. 3, distributed formation control based on layered finite-time estimators are presented. Chap. 4 proposes a two-layer formation control method combined with adaptive neural networks and attitude estimation for flapping-wing vehicles under model uncertainties and measurement errors. In Chap. 5, a distributed formation controller is disposed of for networked Euler-Lagrange systems under a directed interaction topology to ensure that closed-loop errors converge to a small region of the origin in a finite time without any global information. Chap. 6 presents a control scheme, the networked microsatellites can achieve elliptical motion around a host spacecraft with the desired spaced formation in 3-D space. In Chap. 7, a cooperative circumnavigation control problem for networked unmanned aerial vehicles is defined and explored for circumnavigating a moving or static target. For the second part, while all the chapters are in Chaps. 8–15 Consider collision avoidance. Chap. 8 considers the all-to-all communication scheme and creates a measure-theoretic methodology for producing the flocking behavior of planar autonomous vehicles. Chap. 9 presents a leader-follower formation control method for a group of waterjet unmanned marine surface vehicles. Chap. 10 proposes a velocity-free platoon formation control strategy for unmanned surface vehicles under model uncertainties and output constraints. In Chap. 11, a formation potential field method is proposed for MASs, which achieves a specific formation with collision avoidance in spatial constraints. Chap. 12 extends Chap. 11 to the formation tracking control issue. Chap. 13 considers formation tracking for MASs with system uncertainties to track a predefined trajectory with switching formation in non-omniscient constrained space. Chap. 14 investigates a role-switching strategy with communication limitations. Eventually, vision-based formation tracking is investigated for MASs under visibility restrictions without communication among agents in Chap. 15.

1.4.1 PART I: LAYERED AFFINE FORMATION CONTROL

In Chap. 3, we explore a multilayer formation control strategy of multiple Euler-Lagrange systems. First, consider that the first layer forms a primary formation configuration described by the position-based interaction, other layers form and preserve dependent formation configurations by the bearing constraints with their prior layers. Agents enable instinctively receive and transmit information among the layers. Next, layered distributed finite-time estimators are designed for each layer, and the precise estimations of the bearing-based velocity and position of the target formation in finite time, respectively. Furthermore, a model-based multilayer formation controller for multiple Euler-Lagrange systems is investigated to realize multilayer formation, where the formation configurations can be constant or time-variant. Eventually, simulation results showcase the validity of the given approaches with a class of satellites in a time-variant formation condition. In Chaps. 4 and 5, we present two practical cases respectively. In Chap. 4, the formation issues for flapping-wing vehicles are subjected to system uncertainty and measurement inaccuracy. A two-layer formation control protocol together with adaptive neural networks and attitude estimation is proposed for the flapping-wing vehicle formation. And tracking errors and estimation errors are semi-globally uniformly ultimately bounded. We especially, utilize the coupling characteristics between translational motion and rotational motion of the flapping-wing vehicles to design a coupling-based attitude angle estimation method to solve the repeated integration of the measurement error accumulation of attitude angular velocity. Simulation results verify the effectiveness of the proposed strategy. In Chap 5, a distributed layered affine formation control method based on an adaptive neural network is presented for networked Euler-Lagrange systems under a directed graph. Based on the property of the affine transformation, the necessary and sufficient condition for the affine maneuverability is given. The proposed control protocols ensure that closed-loop system errors are semi-globally practical and finite-time stable. Simulation results showcase the effectiveness of the proposed method. Chap. 6 defines and investigates the cooperative circumnavigation problem for networked microsatellites, which drives multiple microsatellites to circumnavigate a host spacecraft on predefined planar ellipse while maintaining a geometric formation configuration. First, several potential functions are constructed to guide the microsatellites to the preset planar elliptical orbit with a proper radius and a circulating rate. Next, for the sake of characterizing the desired geometric formation shape of microsatellites in line with the orbit, the affine Laplacian matrix is introduced. Furthermore, found on the potential functions and the Laplacian matrix, a cooperative circumnavigation controller is presented, which ensures that the closed-loop errors converge to the origin exponentially. Simulation results of multiple microsatellites with Earth-orbiting mission scenarios are showcased, where the natural trajectory motion is incorporated which consumes nearly zero fuel. In Chap. 7, the cooperative circumnavigation control problem is explored for multiple networked unmanned aerial vehicles (UAVs) under a directed interaction topology. The control protocol presents two main advantages. First, in light of affine shape, the designed controller realizes an elliptical travel corresponding to a moving target in 3-D space with arbitrarily spaced formation. Second, the given protocol alleviates the follower UAVs' requirements on global information, which is more general and nontrivial as only the displacements with neighbors are required for follower UAVs.

1.4.2 PART II: FORMATION CONTROL IN CONSTRAINED SPACE

In Chap. 8, a measure-theoretic approach to derive the collective behavior of flocking of planar autonomous vehicles is proposed. The presented control method also achieves two important performances of cohesion maintenance and collision avoidance. In line with a measure-valued transition equation from the exact multi-body particle model of the collective system and the auxiliary continuum model, the control protocol is formulated. Simulation results demonstrate the effectiveness of the theoretical method with practical mobile robots examples. In Chap. 9, a novel formation control strategy for marine surface vehicles based on a bioinspired neurodynamics method subjected to formation tracking errors constraints is designed. First, three bioinspired neurodynamic shunting models are derived to diminish the complex differential calculations and simplify the control protocol. Second, a tan-type barrier Lyapunov function is given to guarantee communication connectivity and collision avoidance. Furthermore, the command filter method is incorporated with the controller to alleviate the computational burden in the control algorithm. Simulations prove that the line-of-sight range and angle errors converge into an arbitrarily small neighborhood of origin under the proposed method. Chap. 10 defines and presents a velocity-free formation control method for autonomous unmanned surface vehicles with model uncertainties and output constraints. First, a barrier Lyapunov functions method is established to tackle the constraint problems of multiple unmanned surface vehicles, meanwhile which can guarantee satisfactory formation performance with collision avoidance and connectivity maintenance. Second, an adaptive neural networks control together with a symmetric barrier Lyapunov functions is investigated for autonomous unmanned surface vehicles to deal with the unmodeled dynamics and parameter uncertainties caused by hydrodynamic damping terms.

The proposed controller ensures that the system is asymptotically stable and that the line-of-sight range and angle errors constraint is never violated and. achieves the desired formation pattern. The simulation results verify the performance of the presented algorithm. Chap. 11 proposes a formation potential field method to control multi-agents to automatically generate and maintain a desired formation configuration while avoiding collisions among agents. Each agent will automatically be attracted to the optimal desired formation position. Meanwhile, the attractive potential functions presented in the exponential form will significantly improve the time cost. In addition, a global potential field is employed to cover the shortage of local formation potential fields. In this way, the requirement of agents' initial positions can be relaxed and the robustness can be enhanced. Furthermore, the proposed two controllers take into account the input saturation to achieve a stable dynamic formation. Simulation results demonstrate the validity and optimality of the proposed approaches. In Chap. 12, formation tracking control in non-omniscient constrained space is investigated in this chapter, a novel control approach based on artificial potential field method for formation tracking control of multi-agents in constrained space is presented. A novel modeling approach for spatial constraints is introduced and a Dirac delta function is used to solve the situation of multiple complex spatial constraints, which enables to analyze of the effect of suddenly detected spatial constraints on system stability. Formation tracking problems in a bounded space and rapid obstacle avoidance have been both solved by the proposed protocol. Besides, an optimization algorithm is designed to minimize the formation time. Simulation results demonstrate that all agents can successfully track the predefined trajectory in a desired formation with collision avoidance. In Chap. 13, a novel tracking control with switching formation of MASs with system uncertainties in a non-omniscient constrained environment is explored. First, the established potential functions are utilized to track the preset trajectory and avoid collisions. Second, a local path replanning method is proposed to act the potential field generated by an inevitable disturbance on the originally desired trajectory rather than the agent's direct motion. Moreover, a neural networks-based controller is designed for tracking problems in restricted spaces. Especially, the passability of a restricted path is mathematically given, while the dissatisfaction will trigger the formation switching to achieve rapid collision avoidance with spatial constraints. Simulations showcase the effectiveness of the proposed approaches. In Chap. 14, we propose a novel role-switching formation tracking strategy for multi-agent systems in constrained space with limited communications. First, a novel role "coordinator" is presented as the coordination mechanism between the leader and followers to achieve collision avoidance and maintain formation configuration. Second, we design an adaptation law for the formation scaling factor, which ensures that the size of the specific formation configuration enables it to be scaled up or scaled down in line with various spatial constraints. Furthermore, the proposed controller makes MAS asymptotic stable, meanwhile, the formation tracking task should be relinquished when the maximum duration that the conditions cannot be satisfied. Simulation results illustrate the performance of the proposed approaches. Chap. 15 showcases a Kinect-based formation tracking control for MASs under visibility constraints without communication among agents. First, to estimate the relative orientation and relative positions of the local leader, a pose estimation algorithm based on RGBD images are designed. Second, a dipolar vector field is constructed to form and maintain the desired formation based on the preset position displacement between agents. The visibility constraints, generated from the limited vision range of the Kinect used as the sole sensor for formation, are transformed into input constraints. By utilizing the constructed auxiliary system, the control protocol achieves the formation tasks while maintaining the visibility constraints. Simulation results demonstrate the performance under the proposed control strategy by using two robot operating systems.

2 Preliminaries

2.1 INTRODUCTION

In this chapter, we provide mathematical preliminaries of stability theorems, sign functions and differential inclusions for control design and analysis. Graph description and neural networks are then introduced to facilitate the articulation of multi-agent control problems and approximation of unknown system dynamics.

2.2 STABILITY THEOREMS

We consider the following system dynamics,

$$\dot{x} = f(x), \; x(0) = x_0, \tag{2.1}$$

where $x = [x_1, \ldots, x_n]^T \in \mathbb{R}^n$, and with respect to an open neighborhood $\mathcal{D}_0 \subseteq \mathbb{R}^n$ of the origin, $f : \mathcal{D}_0 \to \mathbb{R}^n$ is continuous. We assume that the system (2.1) has a unique solution for all initial conditions in forward time.

Remark 1. *Stability refers to a specific equilibrium with regard to the target system. In this book, we mainly focus on the equilibrium at the origin of a given system. We refer to the stability of a system under certain cases without ambiguity.*

Definition 1. *(Stability in the sense of Lyapunov [434]) The system* (2.1) *is stable in the sense of Lyapunov, if for any* $\varepsilon > 0$ *and initial time* $t_0 \geq 0$*, there is* $\delta = \delta(\varepsilon, t_0) > 0$*, such that*

$$\|x(t_0)\| < \delta \Rightarrow \|x(t)\| < \varepsilon, \quad \forall t \geq t_0. \tag{2.2}$$

Definition 2. *(Class $\mathcal{K}$-functions [231]) A continuous function* $\alpha : [0, a) \to [0, \infty)$ *belongs to class $\mathcal{K}$-functions if the following conditions hold:*

i. α *is strictly increasing;*
ii. $\alpha(0) = 0$.

Definition 3. *(Class $\mathcal{K}_\infty$-functions [231]) A class $\mathcal{K}$-function* $\alpha : [0, a) \to [0, \infty)$ *belongs to class $\mathcal{K}_\infty$-functions if the following conditions hold:*

i. $a = \infty$*;*
ii. $\lim_{r \to \infty} \alpha(r) = \infty$.

DOI: 10.1201/9781003298618-2

Definition 4. *(Lyapunov stability [419]) The system* (2.1) *is*

uniformly stable iff there exist a class $\mathcal{K}$ *function* $\gamma(\cdot)$ *and a positive constant* c *independent of* t_0*, such that*

$$\|x(t)\| \leq \gamma\left(\left\|x\left(t_0\right)\right\|\right), \quad \forall \geq t_0 \geq 0, \quad \left\|x\left(t_0\right)\right\| \leq c \tag{2.3}$$

uniformly asymptotically stable, iff there exists a class $\mathcal{K}$ L *function* $\beta(\cdot,\cdot)$ *and a positive constant* c *independent of* t_0*, such that*

$$\|x(t)\| \leq \beta\left(\left\|x\left(t_0\right)\right\|, t-t_0\right), \quad \forall t \geq t_0 \geq 0, \quad \left\|x\left(t_0\right)\right\| \leq c \tag{2.4}$$

globally uniformly asymptotically stable iff (2.4) *is satisfied for any initial state* $x\left(t_0\right)$.

exponentially stable if there exists positive constants c*,* k *and* λ *such that*

$$\beta(r, s)=k\left\|x\left(t_0\right)\right\| e^{-\lambda\left(t-t_0\right)}, \quad \forall\left\|x\left(t_0\right)\right\| \leq c \tag{2.5}$$

globally exponentially stable if the foregoing inequality holds $\forall x\left(t_0\right)$.

The following criterion of stability is called the First (or indirect) method of Lyapunov. Let

$$Df(x)=\left[\frac{\partial f}{\partial x}\right]_{x=0} \tag{2.6}$$

be the Jacobian matrix of system (2.1) evaluated at the zero equilibrium.

Theorem 2.1

(*First method of Lyapunov — autonomous systems [228]*) If all the eigenvalues of $Df(x)$ have a negative real part, then the equilibrium of system (2.1) is asymptotically stable. ■

Theorem 2.2

(*Second method of Lyapunov — autonomous systems [228]*) The system (2.1) is globally (over the entire domain $\mathcal{D}_0$) and asymptotically stable about its zero equilibrium, if there exist a scalar-valued function, $V(x)$, defined on $\mathcal{D}_0$, such that

1. $V(0)=0$;

2. $V(x) > 0$ for all $x \neq 0$ in $\mathcal{D}_0$;

3. $\dot{V}(x) < 0$ for all $x \neq 0$ in $\mathcal{D}_0$.

■

In Theorem 2.2, function V is called a Lyapunov function. The method of constructing a Lyapunov function for stability determination is called the second (or direct) method of Lyapunov.

Note that if Condition 3 in Theorem 2.2 is replaced by

$$\dot{V}(x) \leq 0, \ \forall x \neq 0 \in \mathcal{D}_0 \tag{2.7}$$

then the resulting stability is only in the sense of Lyapunov but may not be asymptotic.

Consider a general non-autonomous or time-variant nonlinear system

$$\dot{x}(t) = f(x,t), \quad x(t) \in \mathbb{R}^n \tag{2.8}$$

where $f : [0, \infty) \times \mathcal{D}_0 \rightarrow \mathbb{R}^n$ is a piecewise continuous function in x and $\mathcal{D}_0 \subset \mathbb{R}^n$ is a domain that contains the origin $x = 0$.

Definition 5. *The origin is an equilibrium point of* (2.8) *if*

$$f(t,0) = 0, \ \forall t \geq 0. \tag{2.9}$$

Theorem 2.3

(*Second method of Lyapunov — non-autonomous systems [228]*) The system (2.8) is globally (over the entire domain $\mathcal{D}_0$), uniformly (with respect to the initial time over the entire time interval $[t_0, \infty)$), and asymptotically stable about its zero equilibrium, if there exist a scalar-valued function, $V(x,t)$, defined on $\mathcal{D}_0 \times [t_0, \infty)$, and three functions $\alpha(\cdot), \beta(\cdot), \gamma(\cdot) \in \mathcal{K}$, such that

1. $V(0, t_0) = 0$;

2. $V(x,t) > 0$ for all $x \neq 0$ in $\mathcal{D}_0$ and all $t \geq t_0$;

3. $\alpha(\|x\|) \leq V(x,t) \leq \beta(\|x\|)$ for all $t \geq t_0$;

4. $\dot{V}(x,t) \leq -\gamma(\|x\|)$ for all $t \geq t_0$.

■

Theorem 2.4

(*LaSalle theorem [228]*) Let $\Omega \subset \mathcal{D}_0$ be a compact set that is positively invariant with respect to (2.1). Let $V : \mathbb{R}^n \rightarrow \mathbb{R}$ be a continuously differentiable function such that $\dot{V}(x) \leq 0$ in Ω. Let $\mathcal{E}$ be the set of all points in Ω

where $\dot{V}(x) = 0$. Let $\mathcal{M}$ be the largest invariant set in $\mathcal{E}$. Then, every solution starting in Ω approaches $\mathcal{M}$ as $t \to \infty$. ■

Corollary 1. *[228] Let $x = 0$ be an equilibrium point for (2.1). Let $V : \mathbb{R}^n \to \mathbb{R}$ be a continuously differentiable, positive definite function on domain $\mathcal{D}_0$ containing the origin $x = 0$, such that $\dot{V}(x) \leq 0$ in $\mathcal{D}_0$. Let $S = \left\{x \in \mathbb{R}^n \mid \dot{V}(x) = 0\right\}$ and suppose that no solution can stay permanently in $\mathcal{S}$, other than the trivial solution $x \equiv 0$. Then, the origin is asymptotically stable.*

Corollary 2. *[228] Let $x = 0$ be an equilibrium point for (2.1). Let $V : \mathbb{R}^n \to \mathbb{R}$ be a continuously differentiable, radially unbounded, positive definite function such that $\dot{V}(x) \leq 0$ for all $x \in \mathbb{R}^n$. Let $\mathcal{S} = \left\{x \in \mathbb{R}^n \mid \dot{V}(x) = 0\right\}$ and suppose that no solution can stay permanently in $\mathcal{S}$, other than the trivial solution $x \equiv 0$. Then, the origin is globally asymptotically stable.*

Lemma 2.1

(Barbalat's lemma [228]) Let $\phi : \mathbb{R}^n \to \mathbb{R}$ be a uniformly continuous function on $[0, \infty)$. Suppose that

$$\lim_{t\to\infty} \int_0^t \phi(\tau) d\tau \tag{2.10}$$

exists and is finite. Then,

$$\phi \to 0 \text{ as } t \to \infty. \tag{2.11}$$

■

Definition 6. *(Lie derivative [228]) Let $h : \mathcal{D}_0 \to \mathbb{R}$ be a scalar function and $f : \mathcal{D}_0 \to \mathbb{R}^n$ be a sufficiently smooth vector field in a domain $\mathcal{D}_0 \subset \mathbb{R}^n$. The Lie derivative of h with respect to f, or along f, written as $L_f h$, is defined by*

$$L_f h(x) = \frac{\partial h}{\partial x} f(x) \tag{2.12}$$

Results on input-to-state stability are recalled and introduced.

Definition 7. *(Input-to-state stability [438]) The system (2.1) is input-to-state stable (ISS) if there exist a $\mathcal{KL}$-function γ_1 and a $\mathcal{K}$-function γ_2 such that, for each input $u \in L_\infty$ and each $x_0 \in \mathbb{R}^n$, it holds that*

$$|x(t, x_0, u)| \leq \gamma_1(|x_0|, t) + \gamma_2(\|u\|), \ \forall t \geq 0. \tag{2.13}$$

Lemma 2.2

(*ISS Lyapunov function [438]*) A smooth function $V : \mathbb{R}^n \to \mathbb{R}_+$ is an ISS Lyapunov function if there exist functions $\alpha_1, \alpha_2 \in \mathcal{K}_\infty$ and $\alpha_3, \ell \in \mathcal{K}$, such that for $\forall x \in \mathbb{R}^n, u \in \mathbb{R}^m$, the following conditions hold

$$\alpha_1(|x|) \leqslant V(x) \leqslant \alpha_2(|x|), \tag{2.14}$$

$$|x| \geqslant \ell(|u|) \Rightarrow L_{f(x,u)}V(x) \leqslant -\alpha_3(|x|), \tag{2.15}$$

where ℓ is the Lyapunov gain. ■

Lemma 2.3

(*Conditions of ISS [438]*) The system (2.1) is ISS iff there exists an ISS Lyapunov function for it. ■

Some well-known concepts and results on finite-time stability are next introduced.

Definition 8. *(Finite-time stability) The equilibrium of system* (2.1) *is finite-time stable if*

i. Finite-Time Convergence: For an open neighborhood $\mathcal{N}_0 \subseteq \mathcal{D}_0$ *of the origin,* $\forall x_0 \in \mathcal{N}_0 \backslash \{0\}$, $\lim_{t \uparrow T(x_0)} x\,(t) = 0$, *where* $T\,(x_0)$ *is the settling time.*
ii. Lyapunov stability: Given any open neighborhood $\mathcal{U}_\varepsilon$ *of 0, there exits an open subset* $\mathcal{U}_\delta$ *of* $\mathcal{N}_0$ *containing 0 such that, for* $\forall x_0 \in \mathcal{U}_\delta \backslash \{0\}$, $x\,(t) \in \mathcal{U}_\varepsilon$, *for* $\forall t \in [0, T(x_0))$.

The equilibrium is globally finite-time stable if it is finite-time stable with $\mathcal{N}_0 = \mathcal{D}_0 = \mathbb{R}^n$.

Definition 8 is motivated by [367, 563], which differs slightly from the definition of finite-time stability in [28]. The definition in [28] requires $\forall x_0 \in \mathcal{N}_0 \backslash \{0\}$, $x\,(t) \in \mathcal{N}_0 \backslash \{0\}$, $\lim_{t \to T(x_0)} x\,(t) = 0$ while Definition 8 only requires $\forall x_0 \in \mathcal{N}_0 \backslash \{0\}$, $\lim_{t \uparrow T(x_0)} x\,(t) = 0$. This is essential when considering finite-time control for non-autonomous systems.

Definition 9. *(Fixed-time stability [384]) The equilibrium of the system* (2.1) *is fixed-time stable if*

i. it is globally finite-time stable in the sense of Definition 8;
ii. the upper bound of the settling-time $T\,(x_0)$ *is independent of any initial system state, i.e., for* $\forall x_0$, *we have* $T\,(x_0) < T_m$, *where* T_m *is a positive constant.*

Lemma 2.4

(*Sufficient conditions of finite-time stability [28]*) Considering a Lyapunov function $V(x)$, for any real numbers $c > 0$ and $\alpha \in (0,1)$, the origin of the system (2.1) is finite-time stable if

$$\dot{V}(x) + cV^{\alpha}(x) \leq 0, \tag{2.16}$$

where the settling time is estimated by

$$T(x_0) \leq \frac{V^{1-\alpha}(x_0)}{c(1-\alpha)}. \tag{2.17}$$

■

Lemma 2.5

(*Sufficient conditions of fixed-time stability [385]*) The equilibrium of the system (2.1) is fixed-time stable, if there exists a Lyapunov function $V(x)$ such that, along the state trajectory of (2.1), we have

$$\dot{V}(x) \leq -\left(\alpha V^{p}(x) + \beta V^{g}(x)\right)^{\chi}, \tag{2.18}$$

where α, β, p, g and χ are positive constants, with $p\chi < 1$ and $g\chi > 1$. The settling time $T(x_0)$ is bounded as

$$T(x_0) \leq \frac{1}{\alpha^{\chi}(1-p\chi)} + \frac{1}{\beta^{\chi}(g\chi - 1)}. \tag{2.19}$$

■

The concepts of semi-globally uniform ultimate boundedness and practical finite-time stability, and their sufficient Lyapunov conditions are introduced together in the following lemmas. They show that the system state is driven to a small bounded neighborhood around the origin when encountering disturbances and uncertainties. Stability of semi-globally uniform ultimate boundedness is first proposed in [146], and then practical finite-time stability is introduced in [566]. As shown below, they essentially hold the same definitions.

Definition 10. *(Semi-global uniform ultimate boundedness [146]) The equilibrium of the system* (2.1) *is semi-globally uniformly ultimately bounded, if for any open neighborhood* $\mathcal{N}_0 \subseteq \mathcal{D}_0$ *of the origin, a compact subset of* $\mathbb{R}^n$ *and all* $x_0 \in \mathcal{N}_0$*, there exist a constant* $\mu > 0$ *and a settling time* $T(\mu, x_0)$*, s.t.,* $\forall t > t_0 + T$*,* $\|x(t)\| < \mu$*.*

Lemma 2.6

(*Sufficient conditions of semi-global uniform ultimate boundedness [146]*) The system (2.1) is semi-globally uniformly ultimately bounded, if there exists a Lyapunov function $V(x)$ such that,

$$\dot{V} \leq -\rho V + c, \tag{2.20}$$

where ρ and c are positive constants.

Proof. Multiplying (2.20) by $e^{-\rho t}$, we have the following by integrating both sides,

$$\begin{aligned} V(t) \leq & V(x_0)e^{-\rho_i t} + \frac{c}{\rho}\left(1 - e^{-\rho_i t}\right) \\ \leq & V(x_0) + \frac{c}{\rho}. \end{aligned} \tag{2.21}$$

□

■

Definition 11. *(Practical finite-time stability [566]) The equilibrium of the system* (2.1) *is practical finite-time stable, if for any open neighborhood $\mathcal{N}_0 \subseteq \mathcal{D}_0$ of the origin, a compact subset of $\mathbb{R}^n$ and all $x_0 \in \mathcal{N}_0$, there exist a constant $\mu > 0$ and a settling time $T(\mu, x_0)$, s.t., $\forall t > t_0 + T$, $\|x(t)\| < \mu$.*

Lemma 2.7

(*Sufficient conditions of practical finite-time stability [566]*) For the system (2.1), if there exists a continuous function $V(x)$, s.t.,

$$\dot{V}(x) \leq -\lambda V^{\ell}(x) + \eta, \tag{2.22}$$

where λ, ℓ, and η are scalars satisfying that $\lambda > 0$, $0 < \ell < 1$, and $0 < \eta < \infty$, the equilibrium of the system (2.1) is practical finite-time stable, i.e., the trajectories are bounded in finite time T_m as

$$\lim_{t \to T_m} x \in \left\{ x \in \mathbb{R}^d | V^{\ell}(x) \leq \frac{\eta}{(1-\theta)\lambda} \right\}, \tag{2.23}$$

$$T_m \leq \frac{V^{1-\ell}(x_0)}{\lambda\theta(1-\ell)}, \tag{2.24}$$

where θ is a scalar $\in (0, 1)$. ■

It is clear that they refer to the same characterization of system stability from Definitions 10 and 11. However, they seem to have different Lyapunov-like conditions in Lemmas 2.6 and 2.7. In the following lemma, we prove that they are actually the same formulation.

Lemma 2.8

(*Relationship between semi-globally uniform ultimate boundedness and practical finite-Time stability [259]*) If the system (2.1) meets the condition in Lemma 2.7, it is also semi-globally uniformly ultimately bounded. If the system (2.1) meets the condition in Lemma 2.6, it is also practical finite-time stable.

Proof. First, if the system (2.1) meets the condition in Lemma 2.7, it is clearly practical finite-time stable according to Lemma 2.7. Accordingly, we have

$$\lim_{t \to T_m} x \in \left\{ x \in \mathbb{R}^d | V^{\ell}(x) \le \frac{\eta}{(1-\theta)\lambda} \right\}, \tag{2.25}$$

$$T_m \le \frac{V^{1-\ell}(x_0)}{\lambda\theta(1-\ell)}. \tag{2.26}$$

The above formula satisfies the definition of semi-global uniform ultimate boundedness in Definition 10.

Next, if the system (2.1) meets the condition in Lemma 2.6, we have $\dot{V} \le -\rho V + c$. Further, after some manipulations, it reads

$$\begin{aligned}
\dot{V} &\le -\rho V + c \\
&\le -\rho\left(\frac{1}{2}V + \frac{1}{2}\left(V^{\frac{1}{2}} - V^{\frac{1}{4}}\right)^2 - \frac{1}{2}V^{\frac{1}{2}} + V^{\frac{3}{4}}\right) + c \\
&\le -\rho\left(\frac{1}{2}V - \frac{1}{2}V^{\frac{1}{2}} + V^{\frac{3}{4}}\right) + c \\
&\le -\rho\left(\frac{1}{2}V - \frac{1}{2}\left(\frac{1}{2}V + \frac{1}{2}\right) + V^{\frac{3}{4}}\right) + c \\
&\le -\rho V^{\frac{3}{4}} + c + \frac{\rho}{4}.
\end{aligned} \tag{2.27}$$

The inequality (2.27) satisfies the conditions in Lemma 2.7. Thus, the system (2.1) is also practical finite-time stable. □

■

2.3 SIGN FUNCTIONS

The classical sign function is defined as

$$\operatorname{sign_c}(x_i) \triangleq \begin{cases} +1, & x_i > 0, \\ 0, & x_i = 0, \\ -1, & x_i < 0, \end{cases} \tag{2.28}$$

where its vector form and exponential form are defined as

$$\operatorname{sign_c}(x) = [\operatorname{sign_c}(x_1), \ldots, \operatorname{sign_c}(x_n)]^T, \tag{2.29}$$

$$\operatorname{sig}_{\mathrm{c}}^{\alpha}(x) = [\operatorname{sign_c}(x_1)|x_1|^{\alpha}, \ldots, \operatorname{sign_c}(x_n)|x_n|^{\alpha}]^T. \tag{2.30}$$

Remark 2. *The classical sign function is not necessarily equal to zero when $x = 0$. Especially, when differential inclusions are involved and considered under the Filippov sense [135], the classical sign function can be treated as a set-valued function,*

$$\operatorname{sign_c}(x_i) = \begin{cases} 1, & \text{if } x_i > 0, \\ [-1, 1], & \text{if } x_i = 0, \\ -1, & \text{if } x_i < 0. \end{cases} \tag{2.31}$$

Another sign function (actually a unit vector function) called norm-normalized sign function is defined in the following.

Definition 12. *(Norm-normalized sign functions [258]). For a vector $x = [x_1, x_2, \ldots, x_n]^T \in \mathbb{R}^n$, a norm-normalized sign function and its exponential form are defined as*

$$\operatorname{sign_n}(x) \triangleq \begin{cases} \frac{x}{\|x\|}, & x \neq 0, \\ 0, & x = 0, \end{cases} \tag{2.32}$$

$$\operatorname{sig}_{\mathrm{n}}^{\alpha}(x) \triangleq \|x\|^{\alpha} \operatorname{sign_n}(x). \tag{2.33}$$

Remark 3. *In Definition 12, the norm-normalized sign function is introduced. It incorporates an input vector and its norm by normalization computation, which also serves as a unit direction vector. It is recalled as a sign function since* $\operatorname{sign_n}$ *holds properties similar to those of the classical sign function, e.g., they both show the general direction of an input vector. For example, the function* $\operatorname{sign_c}(x)$ *defines 3^n vectors evenly distributed in the n-dimensional space (including $0 \in \mathbb{R}^n$ for $x = 0$), while* $\operatorname{sign_n}(x)$ *defines an n-dimensional unit ball. The norm-normalized sign function has wide applications in the literature, and is also referred as the unit vector [126] (Chapter 3.6, The Unit Vector Approach), the unit controller [465] (Chapter 3.5, Unit Control), the scaled unit vector [165], the vector-valued function [418], to name a few.*

2.4 DIFFERENTIAL INCLUSIONS

We introduce the concept of differential inclusions and some relevant preliminaries. The following definitions are given.

Definition 13. *(Open cube) A product of open intervals*

$$U=(a_1,b_1)\times(a_2,b_2)\times\cdots\times(a_n,b_n), \tag{2.34}$$

is called an open cube, and its volume is defined as

$$\mathrm{vol}(U)=(b_1-a_1)(b_2-a_2)\cdots(b_n-a_n). \tag{2.35}$$

Definition 14. *(Measure zero [209]) A set $X\in R^n$ has measure zero if $\forall\epsilon>0$, $\exists U_1,U_2,\cdots$ (open cubes), s.t. $X\subseteq\bigcup_{i=1}^{\infty}U_i$ and $\sum_{i=1}^{\infty}\mathrm{vol}(U_i)<\epsilon$.*

Definition 15. *(Lipschitz continuous [240]) A function $f(x,t)$ is Lipschitz continuous in x with Lipschitz constant $L_C^f>0$ if*

$$\|f(x,t)-f(y,t)\|\geq L_C^f\|x-y\| \tag{2.36}$$

for all x,y,t.

Definition 16. *(Quasi-continuity [353]) Let X be a topological space. A real-valued function $f:X\to\mathbb{R}$ is quasi-continuous at a point $x\in X$ if for any $\epsilon>0$ and any open neighborhood U of x there is a non-empty open set $G\subset U$ such that $|f(x)-f(y)|<\epsilon$, $\forall y\in G$.*

Note that in the above definition, it is not necessary that $x\in G$.

Consider system dynamics with discontinuous right-hand sides

$$\dot{x}=f(t,x),\ t\in[0,\infty),\ x\in\mathbb{R}^n, \tag{2.37}$$

where the function $f:\mathbb{R}\times\mathbb{R}^n\to\mathbb{R}^n$ is quasi-continuous and the set of discontinuous points has *measure zero*.

The concept of *differential inclusions* is then introduced, where the quasi-continuous system (2.37) is handled and transformed into dynamics (called differential inclusion) with *multi-value right-hand sides*,

$$\dot{x}\in F(t,x), \tag{2.38}$$

where $t\in\mathcal{D}_t\subset\mathbb{R}, x\in\mathcal{D}_x\subset\mathbb{R}^n$, and $F(t,x)$ is a multi-value function, mapping points (t,x) of the region $\mathcal{D}=\mathcal{D}_t\times\mathcal{D}_x$ to the set $F(t,x)$ of vectors $\in\mathbb{R}^n$. Clearly, the essential behind differential inclusions involves a set-valued map, mapping a point to a set of points.

Definition 17. *(Absolutely continuous functions [386]) A mapping* φ : $[a, b] \to \mathbb{R}$ *is an absolutely continuous function, if* $\forall \varepsilon \in (0, \infty)$, $\exists \delta \in (0, \infty)$, *s.t., given an arbitrary finite disjoint open intervals* $\{(a_1, b_1), \ldots, (a_n, b_n)\} \in [a, b]$ *with*

$$\sum_{i=1}^{n} (b_i - a_i) < \delta, \tag{2.39}$$

the following inequality holds

$$\sum_{i=1}^{n} |\gamma(b_i) - \gamma(a_i)| < \varepsilon. \tag{2.40}$$

Definition 18. *(Solution of differential inclusion [80]) A absolutely continuous function* $x(t) \in \mathbb{R}^n$ *is a solution of differential inclusion* (2.38) *if when* $\dot{x}(t)$ *exists,*

$$\dot{x}(t) \in f(t, x(t)). \tag{2.41}$$

If the solution of differential inclusions is said to be understood in the *Filippov sense* [135], it meets the following *Filippov set-valued map.* The core of *Filippov solution* lies in observing the vector field within the neighborhood of each point.

Definition 19. *(Filippov solution [136]) For a function* $f : \mathbb{R}^n \to \mathbb{R}^n$, *the Filippov Set-Valued Map is defined as* $F[f] : \mathbb{R}^n \to \mathfrak{B}(\mathbb{R}^n)$ *by*

$$F[f](x) \triangleq \bigcap_{\delta>0} \bigcap_{\mu(S)=0} \overline{\mathrm{co}}\{f(\mathfrak{B}(x, \delta) \backslash S)\}, \quad x \in \mathbb{R}^n, \tag{2.42}$$

where $\overline{\mathrm{CO}}$ *represents convex closure,* μ *Lebesgue measure, and*

$$\mathfrak{B}(x, \delta) = \{y \in R^n \| \|y - x\| < \delta\}. \tag{2.43}$$

If $x(t)$ *is a solution of the following differential inclusion*

$$\dot{x} \in F[f](x), \tag{2.44}$$

then $x(t)$ *is a Filippov solution of the system* (2.37).

From Definition 19, in terms of a point x, we know that $F[f](x)$ is independent of the actual value of $f(x)$. Except uniqueness, the Filippov solution involves most of the well-known standard properties.

Simple first-order affine dynamics in $\mathbb{R}^n$ is considered,

$$\dot{x} = f(x) + b(x) u, \tag{2.45}$$

where $x = [x_1, x_2, \ldots, x_n]^T \in \mathbb{R}^n$ denotes the state vector, $u \in \mathbb{R}^n$ the control input, and $f(x) \in \mathbb{R}^n$ and $b(x) \in \mathbb{R}^{n \times n}$ $(\det(b(x)) \neq 0)$ the known parts.

We propose two control laws u_c and u_n based on classical and norm-normalized sign functions,

$$u_c = - b^{-1}(x)\left(k\mathrm{sign_c}(x) + f(x)\right), \tag{2.46}$$
$$u_n = - b^{-1}(x)\left(k\mathrm{sign_n}(x) + f(x)\right). \tag{2.47}$$

Taking the controllers (2.46) and (2.47) respectively into the system (2.45), we have

$$\dot{x} = - k\mathrm{sign_c}(x), \tag{2.48}$$
$$\dot{x} = - k\mathrm{sign_n}(x). \tag{2.49}$$

Note that the closed-loop systems (2.48) and (2.49) have discontinuous right-hand sides. They should be understood in the Filippov sense. Then, the closed-loop dynamics (2.48) and (2.49) become

$$\dot{x} \in F\left[-k\mathrm{sign_c}(x)\right](x), \tag{2.50}$$
$$\dot{x} \in F\left[-k\mathrm{sign_n}(x)\right](x). \tag{2.51}$$

Since the Filippov set-valued maps

$$F\left[-k\mathrm{sign_n}(x)\right](x) \text{ or } F\left[-k\mathrm{sign_c}(x)\right](x), \tag{2.52}$$

at the point x_0 ($x_0 = 0$ or any element of x_0 is equal to zero) are independent of the actual value of $\mathrm{sign_n}(x_0)$ or $\mathrm{sign_c}(x_0)$, one can thus redefine/generalize the multi-variable classical sign function $\mathrm{sign_c}$ and the norm-normalized sign function $\mathrm{sign_n}$ as follows,

$$\mathrm{sign_c}(x) = \left[\mathrm{sign_c}(x_1), ..., \mathrm{sign_c}(x_n)\right]^T, \tag{2.53}$$
$$\mathrm{sign_n}(x) = \frac{x}{\|x\|}, \tag{2.54}$$

where

$$\mathrm{sign_c}(x_i) = \begin{cases} 1, & \text{if } x_i > 0, \\ [-1, 1], & \text{if } x_i = 0, \\ -1, & \text{if } x_i < 0. \end{cases} \tag{2.55}$$

Remark 4. *Please note that a Lyapunov function $V(t)$ is not necessarily differentiable along the absolutely continuous Filippov solutions of* (2.48) *and* (2.49)*. In [99], Proposition 14.1, the generalized Lyapunov theorem actually requires only continuity of $V(t)$, which can be applied to system dynamics involving differential inclusions. In contrast, the classical Lyapunov theorem requires the continuous right-hand side of the differential equation and/or a differentiable (or at least locally Lipschitz continuous) Lyapunov function candidate. This issue has also been discussed in [157, 351].*

2.5 GRAPH DESCRIPTION

A graph is used to describe the network topology of a multi-agent system. A graph $\mathcal{G} = (V, E)$ consists of the vertex set V and the edge set E.

For example, there are N leaders and M followers. Let $V = [1, 2, \cdots, N + M]$ and $E \subseteq V \times V$ describe the sized of agents and the connection between vertexes. $(i, j) \in E$ indicates that agent j can achieve information from agent i. The adjacency matrix $\mathcal{A} = [a_{ij}] \in \mathbb{R}^{(N+M)\times(N+M)}$ is the nonnegative weight matrix of the edges. If $(i, j) \in E$, $a_{ij} > 0$, otherwise $a_{ij} = 0$. The Laplacian matrix is defined as $\mathcal{L} = [l_{ij}]$, where $l_{ii} = \sum_{j\neq i} a_{ij}$ and $l_{ij} = -a_{ij}$ for $i \neq j$. It is noteworthy that the matrix weights can also be negative $a_{ij} < 0$ in certain cases such as the stress matrix.

A graph can be directed or undirected. In the undirected cases, the agents have bidirectional measurement and the edges $a_{ij} = a_{ji}$. In the directed cases, the Laplacian matrix L is not necessarily symmetric as $a_{ij} \neq a_{ji}$.

The above is merely simple demonstration of how a underlying graph is constructed. Other modified types of graphs associated with different Laplacian matrices with be detailed in subsequent chapters when they are properly invoked and applied, such as displacement-based, distance-based, bearing-based, angle-based graphs.

Kronecker product is recalled to better facilitate the analysis of multi-agent systems associated with graphs.

Definition 20. *(Kronecker product [314]) The Kronecker product of matrices A and B is defined as*

$$A \otimes B = \begin{bmatrix} a_{11}B & \cdots & a_{1m}B \\ \vdots & \ddots & \vdots \\ a_{n1}B & \cdots & a_{nm}B \end{bmatrix} \tag{2.56}$$

where if $A \in \mathbb{R}^{n\times m}$ matrix and $B \in \mathbb{R}^{p\times q}$ matrix, then $A \otimes B \in \mathbb{R}^{np\times mq}$ matrix.

Definition 21. *Vector $A \otimes f(x_i, \mathbf{t})$ is defined as*

$$A \otimes f(x_i, t) = \begin{bmatrix} a_{11}f(x_1, t) + a_{12}f(x_2, t) + \cdots + a_{1m}f(x_m, t) \\ \vdots \\ a_{n1}f(x_1, t) + a_{n2}f(x_2, t) + \cdots + a_{nm}f(x_m, t) \end{bmatrix} \tag{2.57}$$

where if $A \in \mathbb{R}^{n\times m}$ matrix and $f \in \mathbb{R}^p$ is a function, then $A \otimes f(x_i, t) \in \mathbb{R}^{np}$ is a vector.

2.6 NEURAL NETWORK APPROXIMATION

Let the function $f(x)$ be an unknown component in system dynamics. In practice, some controllers cannot be implemented due to the uncertainty in

$f(x)$. Some existing results employ the radial basis function neural network (RBFNN) [174] to approximate $f(x) : \mathbb{R}^m \rightarrow \mathbb{R}^m$ as follows

$$f(x) = W^{\mathrm{T}} S(x) \tag{2.58}$$

where weight matrix $W \in \mathbb{R}^{l \times m}$, l denotes the number of neurons, and $S(x) = [s_1(x), ..., s_l(x)]^{\mathrm{T}}$ with

$$s_j(x) = \exp[\frac{-(x-\mu_j)^{\mathrm{T}}(x-\mu_j)}{\sigma_j^2}], \quad j = 1, ..., l \tag{2.59}$$

and $\mu_j \in \mathbb{R}^m$ is the center of receptive field and in general, $(\sqrt{2}\sigma_j)/2$ is the width of the Gaussian function.

For any given positive constant d_0, there exists the weight matrix W^* such that

$$f(x) = W^{*\mathrm{T}} S(x_i) + d_i \tag{2.60}$$

giving a sufficiently large l, where d is the bounded function approximation error satisfying $\|d\| < d_0$ in Ω_Z.

The optimal matrix W^* can be theoretically derived from

$$W^* = \arg \min_{W \in \mathbb{R}^{l \times m}} \{ \sup_{x_i \in \Omega_Z} \|f_i(x) - W^{\mathrm{T}} S(x_i)\| \} \tag{2.61}$$

For real application, its estimation $\hat{W}$ is used for the practical function approximation. The estimation of $f(x)$ is given by

$$\hat{f}(x) = \hat{W}^{\mathrm{T}} S(x). \tag{2.62}$$

The updating law for the RBFNN weight matrix $\hat{W}$ is to be designed and integrated into the control law.

2.7 SUMMARY

This chapter has presented mathematical preliminaries. They are useful in the subsequent chapters where a series of maneuvrable formation control problems are formally introduced and various control strategies are to be designed.

Part I

Formation Control with Maneuverability

3 Multilayer Formation Control of Multi-Agent Systems

This chapter first defines a multilayer formation control problem, where the agents can receive and transmit information among the layers. A layered distributed finite-time estimator is proposed for agents in each layer to obtain their target positions and velocities based on the information of agents in their prior layers. A model-based control law is then proposed to achieve multilayer formation, where the formation configurations can be constant or time-varying. This chapter also gives a clue on multilayer formation control design in the presence of practical issues, such as model uncertainties and loss of velocity measurements. Simulation results are given to show the effectiveness of the proposed approaches with a group of satellites in a time-varying formation case.

3.1 INTRODUCTION

Multi-agent system (MAS) control has been an active area with existing works focusing on interesting topics, such as the consensus problem, the tracking problem, the formation problem, and the containment problem. As for the consensus problem, the state of each agent aims to converge to a common value via information flow exchanged with its neighbors [203, 362, 401, 515]. Considering a moving leader, the multi-agent tracking problem has been studied in [39, 189, 283, 377]. The leader-follower formation control problem has been investigated in [50, 110, 168, 275], where a group of agents track a spatial configuration. In [210, 357], the containment control problem has been addressed, where all followers are driven to a specific region spanned by the leaders.

Integrating the containment problem into the formation, the formation-containment problem defined by a leader-follower structure has attracted much interest [118, 488–490]. In fact, in a leader-follower problem, the agent dynamics and the distributed communication of the leaders should be considered to design an implementable control algorithm more in line with practical tasks. Unfortunately, most existing formation [50,110,168,275] and formation-containment formulations [118, 488–490] assume that there is no communication among leaders or use target trajectories to describe the leaders, which should be defined as a single-layer formation problem [255]. Using a leading layer and a following layer to characterize agents' behavior from groups, a

DOI: 10.1201/9781003298618-3

two-layer formation control problem is next addressed in [255]. Note that in practical tasks with a large number of agents (multi-robot exploration, networked drone cruise, and satellite formation), based on the agents' levels of intelligence, equipped payloads, and task areas, they can be divided into different layers. For each individual layer, the agents within are able to receive information from the prior layers, interact within the current layer, and transmit information to the subsequent layers. Inspired by the discussion above, we aim to define a multilayer formation control (MLFC) problem. This framework contains an arbitrary number of layers to design the formation shape and the communication topology naturally.

We give attention to the MLFC of multiple Euler-Lagrange Systems (MELSs), as the Euler-Lagrange Systems (ELSs) can describe a class of practical systems such as underwater robots, autonomous vehicles, and spacecraft [180]. In this chapter, the two-layer approach in [255] is extended to an MLFC problem. The extension is non-trivial, and the following tasks should be solved. First, an appropriate mathematical definition should be proposed to characterize the problem with multilayer formation networks. Second, the relationship among the multilayer framework, formation configurations, and communication topologies is intrinsic and highly coupled, which should also be investigated. Note that in the second layer of our previous approach [255], we deal with a containment problem rather than a formation problem. Moreover, the finite-time estimators introduced in [255] are confined in an undirected graph and a constant formation configuration. Thus, the results in [255] cannot be directly employed to design a desired formation configuration and to build the subsequent layers in this approach. Consequently, it is important to design a new structure of the multilayer framework, which can naturally characterize the communication topology and the target formation shape of each layer.

In this chapter, a unified framework of multilayer formation is articulated. Different with the above mentioned formation-containment results [118, 255, 488–490], this framework can instinctively represent the formation shape and the structure of the information flow of practical multi-agent missions with a great number of agents or with a relatively large distance among agent groups. With the proposed framework, the topology constraints (e.g., bearing and affine Laplacian matrices) can be easily constructed layer by layer in advance. Due to the layered structure, this approach holds advantages when characterizing heterogeneous multi-agent systems under heterogeneous communication graph [172]. Next, we propose a bearing-based MLFC law for MELSs, driving agents in each layer to form a proper formation shape based on the information from their current layer and prior layers, where formation shape is preserved during translational or scaling motions. Since the localization and stability of prior layers will not be influenced by subsequent layers, the networked systems are still stable when the subsequent layers suffer an unpredictable attack or a system failure. Analogously, agents in prior layers can achieve a faster formation convergence speed. In addition, numerical evidence in [31] approves that robustness and synchronizability of the multi-agent systems can be enhanced by a multilayer framework. Finally, the exponential convergence of all closed-loop errors is proved via Lyapunov direct method, and a clue on how to design an MLFC law in the presence of practical issues is also given.

3.2 LAYERED FORMATION PROBLEM FORMULATIONS

A multilayer framework is considered, where the first layer characterized by the position-based interaction forms a primary formation configuration, and the other layers form and maintain dependent formation configurations by the bearing constraints with their prior layers. The MLFC (whose mathematical definition of MLFC is formally given in Problem 1, Section 3.3) can be applied to not only the formation-containment configuration in [118,255,488–490] but also more complex formation shapes.

3.2.1 AGENT DYNAMICS

Suppose that the multilayer formation consists of l layers with N_n agents in the nth layer, $n = 1, 2, \ldots, l$. Let $m = N_1 + N_2 + \ldots + N_l$. The ith agent is modeled by the following Euler-Lagrange (EL) equation

$$M_i(q_i)\ddot{q}_i + C_i(q_i, \dot{q}_i)\dot{q}_i + g_i(q_i) = \tau_i(t), \tag{3.1}$$

where $q_i \in \mathbb{R}^p$, $i = 1, 2, \ldots, m$ denotes the position vector of agent i, $M_i(q_i) \in \mathbb{R}^{p\times p}$ is the symmetric positive definite inertia matrix, $C_i(q_i, \dot{q}_i) \in \mathbb{R}^{p\times p}$ is the vector of Coriolis and centrifugal matrix, $g_i(q_i) \in \mathbb{R}^p$ denotes the gravitational force.

3.2.2 GRAPH THEORY

Suppose that the agents are ordered by layers. We utilize a communication graph $\mathcal{G}_n = (V_{Ln}, E_n)$ to characterize the underlying information flow among the agents in the nth layer, $n = 1, 2, \ldots, l$, where $V_{Ln}=\{\sum_{j=1}^{n-1} N_j + 1, \sum_{j=1}^{n-1} N_j + 2, \ldots, \sum_{j=1}^{n} N_j\}$ and $E_n \subseteq V_{Ln} \times \{V_{Ln} \cup V_{L(n-1)} \cup \ldots \cup V_{L1}\}$ denote the edge set and the vertex set of the nth layer, respectively. Denote $\mathcal{N}_i$ as the in-neighbor set of agent i, where $\mathcal{N}_i = \{j \cdot (j, i) \in E_n\}$. For convenience, let $V_{LM} = V_{L1} \cup V_{L2} \cup \ldots \cup V_{Ll}$.

We next introduce an ordinary Laplacian matrix for the first layer and a bearing Laplacian matrix for the other $l-1$ layers. For convenience, define an additional virtual agent labeled as 0. Based on agent 0, the target formation of the first layer is constructed by different offsets, where agent 0 can be seen as the *virtual center* of the target formation for the first layer.

An ordinary Laplacian is designed as $\mathcal{L}_1 = [l_{ij}] \in \mathbb{R}^{N_1+1\times N_1+1}$ (including the virtual agent 0), where $l_{ii} = \sum_{j=1, j\neq i}^{N_1+1} a_{ij}$ and $l_{ij} = -a_{ij}$, $i \neq j$. In addition, $a_{ij} > 0$ if $(j,i) \in E_1$, and $a_{ij} = 0$ otherwise, where the graph has no self-loop, i.e., $a_{ii} = 0$.

We use a bearing-based approach to describe the interaction in the nth layer, $n = 2, 3, \ldots, l$, and the bearing of agent j relative to agent i is given as [552]

$$g_{ij} = \frac{q_i - q_j}{\|q_i - q_j\|}, \tag{3.2}$$

$$P_{g_{ij}} = I_p - g_{ij} g_{ij}^T, \; i \in V_{Ln}, \; j \in V_{LM}, \tag{3.3}$$

where the desired inter-neighbor bearing is denoted as g_{ij}^*, $P_{g_{ij}}$ is positive semidefinite, and $P_{g_{ij}} = 0_{p \times p}$ if there is no communication between agent j and agent i. The Laplacian matrix $\mathcal{B}_{L_n v}$, $n = 2, 3, \ldots, l$, $v = 2, 3, \ldots, n$, is defined with the ijth block as

$$[\mathcal{B}_{L_n v}]_{ij} = \begin{cases} -P_{g_{ij}^*}, & i \neq j, \; (i,j) \in E_n, \\ \sum_{k \in \mathcal{N}_n} P_{g_{ik}^*}, & i = j, \; i \in V_{Ln}, \end{cases} \tag{3.4}$$

where $P_{g_{ij}^*} = I_d - g_{ij}^* g_{ij}^{*T}$.

We next formally define a multilayer Laplacian matrix $\mathcal{L} \in \mathbb{R}^{p(m+1) \times p(m+1)}$,

$$\mathcal{L} = \begin{bmatrix} 0 \otimes I_p & 0_{1 \times N_1} \otimes I_p & 0_{1 \times N_2} \otimes I_p & \cdots & \cdots & 0_{1 \times N_l} \otimes I_p \\ \mathcal{L}_{L_1 2} \otimes I_p & \mathcal{L}_{L_1 1} \otimes I_p & 0_{N_1 \times N_2} \otimes I_p & \cdots & \cdots & 0_{N_2 \times N_l} \otimes I_p \\ 0_{N_2 \times 1} \otimes I_p & \mathcal{B}_{L_2 2} & \mathcal{B}_{L_2 1} & \cdots & \cdots & 0_{N_3 \times N_l} \otimes I_p \\ \vdots & \vdots & \vdots & \ddots & & \vdots \\ \vdots & \vdots & \vdots & & \mathcal{B}_{L_{l-1} 1} & 0_{N_{l-1} \times N_l} \otimes I_p \\ 0_{N_l \times 1} \otimes I_p & \mathcal{B}_{L_l l} & \mathcal{B}_{L_l l-1} & \cdots & \mathcal{B}_{L_l 2} & \mathcal{B}_{L_l 1} \end{bmatrix} \tag{3.5}$$

where $\mathcal{L}_{L_1 1}$ represents the directed communication topology among the agents in the first layer, $\mathcal{L}_{L_1 2}$ represents the topology between agents in the first layer and virtual agent 0, $\mathcal{B}_{L_n 1}$ represents the bidirectional communication topology among the agents in the nth layer, and $\mathcal{B}_{L_n v}$, $v = 2, 3, \ldots, n-1$, represents the directed communication interaction between the agents in the vth layer and the agents in the nth layer. Let $\mathcal{G}$ denote the directed graph corresponding to $\mathcal{L}$. In this chapter, we assume the internal information flow in the nth layer, $n = 2, 3, \ldots, l$, is bidirectional, while the internal information flow in the first layer and the communication topologies between different layers are all directed.

3.3 LAYERED DISTRIBUTED FINITE-TIME ESTIMATOR AND CONTROL DESIGN OF EULER-LAGRANGE SYSTEMS

In this section, a layered distributed finite-time estimator (LDFE) is first designed for agents in each layer. For the first layer, an LDFE is introduced for each agent to estimate the target position and velocity of the time-varying formation configuration in finite time. For other layers, a similar LDFE is given to estimate the bearing-based positions and velocities of the target formation in finite time. Then, an MLFC law is designed to steer all the agents to the target formation based on the estimate from LDFEs. The MLFC problem to be solved in this section is formally stated below.

Problem 1. *Given the initial formation $\mathcal{G}(q_i(0))$, $i \in V_{LM}$, and feasible bearing constraints $P_{g_{ij}^*}$, $i \in V_{LM} \backslash V_{L1}$, $j \in V_{LM}$, design an LDFE with $\hat{v}_i$ and $\hat{q}_i$ for agent $i \in \mathcal{V}_{L_1}$ such that $\hat{v}_i(t) \to \dot{q}_0(t) + \dot{r}_i(t)$, and $\hat{q}_i(t) \to q_0(t) + r_i(t)$ in finite time, where $r_i(t)$ is the auxiliary function decomposed from the target formation configuration of the first layer, i.e., the target relative position between agent 0 and agent i. Similarly, design an LDFE for agent $i \in V_{Ln}$, $n = 2, 3, \ldots, l$, such that $\hat{v}_i(t) \to \dot{c}_i(t)$, and $\hat{q}_i(t) \to c_i(t)$, where c_i denotes the target formation position characterized by the desired constant bearings for agent i (c_i will be formally introduced in Section 3.3.2). Next, based on the estimates $\hat{v}_i$ and $\hat{q}_i$ from the LDFE, design the control input $\tau_i(t)$ for agent $i \in \mathcal{V}_{LM}$ such that $q_i(t) \to \hat{q}_i(t)$ and $\dot{q}_i(t) \to \hat{v}_i(t)$ as $t \to \infty$.*

In fact, the LDFE can be viewed as a second-order sliding mode estimator to generate accurate target position and velocity with finite-time convergence. Using the LDFEs, we can straightforward decouple the layered formation problem into two subtasks, i.e., target position and velocity estimation (in Sections 3.1 and 3.2) and formation tracking control (in Section 3.3) as defined in Problem 1. We next propose the LDFE protocol (see (3.6a)–(3.7b)), where $\beta_1, \beta_2, \ldots, \beta_{2n}$, $n = 2, 3, \ldots, l$ are positive constants, ϑ_i is an auxiliary variable, and the sign function has been defined in (2.29). As shown in (3.6a)-(3.6c), only a subset of agents whose parameter $a_{i0} \neq 0$ can directly access information from agent 0, while all the agents in the first layer can acquire their own desired displacements r_i associated with agent 0.

$$
i \in V_{L1} \begin{cases} \dot{\hat{q}}_i = \hat{v}_i - \beta_1 \text{sign}_c \left[\sum\limits_{j \in V_{L1}} a_{ij} (\hat{q}_i - r_i - \hat{q}_j + r_j) + a_{i0} (\hat{q}_i - r_i - q_0) \right] & \text{(3.6a)} \\ \hat{v}_i = \vartheta_i + \dot{r}_i & \text{(3.6b)} \\ \dot{\vartheta}_i = -\beta_2 \text{sign}_c \left[\sum\limits_{j \in V_{L1}} a_{ij} (\vartheta_i - \vartheta_j) + a_{i0} (\vartheta_i - \dot{q}_0) \right] & \text{(3.6c)} \end{cases}
$$

$$
i \in V_{Ln} \begin{cases} \dot{\hat{q}}_i = \hat{v}_i - \beta_{2n-1} \text{sign}_c \left[\sum\limits_{j \in V_{Ln}} P_{g_{ij}^*} (\hat{q}_i - \hat{q}_j) + \sum\limits_{j \in V_{L(n-1)} \cdots V_{L1}} P_{g_{ij}^*} (\hat{q}_i - q_j) \right] & \text{(3.7a)} \\ \dot{\hat{v}}_i = -\beta_{2n} \text{sign}_c \left[\sum\limits_{j \in V_{Ln}} P_{g_{ij}^*} (\hat{v}_i - \hat{v}_j) + \sum\limits_{j \in V_{L(n-1)} \cdots V_{L1}} P_{g_{ij}^*} (\hat{v}_i - \dot{q}_j) \right] & \text{(3.7b)} \end{cases}
$$

3.3.1 ESTIMATOR DESIGN FOR THE FIRST LAYER

We made the following assumptions on the acceleration of agent 0 and the directed graph in the first layer before giving the main result.

Assumption 1. *For the N_1 agents ($N_1 \geq 2$) in the first layer, there exists a directed spanning tree in $\mathcal{G}_{L_1}$, where the root of $\mathcal{G}_{L_1}$ is the virtual agent 0.*

Assumption 2. *Suppose that the acceleration of the virtual agent 0 is bounded by a nonnegative constant δ_{L_1}, such that $\|\ddot{q}_0\| \leq \delta_{L_1} < \infty$.*

Theorem 3.1

Let $\beta_1 > 0$ and $\beta_2 > \delta_{L_1}$. If Assumptions 1 and 2 hold, under the LDFEs (3.6a) and (3.6b), the agents in the first layer can acquire the precise estimations of the target formation velocity and position in finite time, respectively, which means

$$\hat{v}_i(t) = \dot{q}_0(t) + \dot{r}_i(t),\ t \geq T_1, \tag{3.8}$$

$$\hat{q}_i(t) = q_0(t) + r_i(t),\ t \geq T_2,\ i \in V_{L1}, \tag{3.9}$$

where T_1 and T_2 are positive constants which will be introduced later. ■

Proof. Consider agent i in the first layer, $i \in V_{L1}$. Define $\tilde{v}_i = \hat{v}_i - \dot{q}_0 - \dot{r}_i$, $\tilde{v}_L^+(t) \triangleq \max_{i \in V_{L1}} \tilde{v}_i(t)$, and $\tilde{v}_L^-(t) \triangleq \min_{i \in V_{L1}} \tilde{v}_i(t)$. We can then get

$$\begin{aligned} \dot{\tilde{v}}_i &= -\beta_2 \mathrm{sign_c} \sum_{j=0}^{N} a_{ij}\left(\hat{v}_i - \dot{r}_i - \dot{q}_0 - \hat{v}_j + \dot{r}_j + \dot{q}_0\right) - \ddot{q}_0 \\ &= -\beta_2 \mathrm{sign_c} \sum_{j=0}^{N} a_{ij}\left(\tilde{v}_i - \tilde{v}_j\right) - \ddot{q}_0,\ i \in V_{L1}. \end{aligned} \tag{3.10}$$

We can get the convergence time for $\hat{v}_i(t) \to \dot{q}_0(t) + \dot{r}_i(t)$ following the proof of Theorem 3.1 in [40]. For the integrity, the proof is briefly described by the following three cases:

Case 1 $\tilde{v}_L^+(0) \leq 0$ and $\tilde{v}_L^-(0) \leq 0$. It is trivial to derive $\tilde{v}_L^+(t) = \tilde{v}_L^-(t) = 0$, $\forall t > 0$, if $\tilde{v}_L^+(0) = 0$ and $\tilde{v}_L^-(0) = 0$ [40]. Next, consider the case with $\tilde{v}_L^+(0) \leq 0$ and $\tilde{v}_L^-(0) < 0$, where $\tilde{v}_L^+(t) \leq 0$, $\forall t > 0$ because $\dot{\tilde{v}}_L^+(t) \leq -\beta_2 - \ddot{q}_0 \leq -\beta_2 + \delta_{L_1} < 0$ when $\tilde{v}_L^+(t) > 0$. Similarly, we know $\tilde{v}^-(t)$ is nondecreasing. Since either $\dot{\tilde{v}}_L^-(t) = -\ddot{q}_0$ or $\dot{\tilde{v}}_L^-(t) > \beta_2 - \delta_{L_1} > 0$, $\dot{\tilde{v}}_L^-(t) = -\ddot{q}_0$ only exists at isolated time instants when $\tilde{v}_L^-(t) < 0$. Assume that $\tilde{v}_L^-(t) < 0$ for $T > 0$, and

$\dot{\tilde{v}}_L^- (t) = -\ddot{q}_0$ for $t \in [t_1, t_2]$ with $t_1 < t_2 < T$. Therefore, there exists leader j satisfactory $\dot{\tilde{v}}_j (t) = -\ddot{q}_0$ for $t \in [t_1, t_3]$, where $t_1 < t_3 \le t_2$. Then, according to (3.10) and Assumption 1, it follows that $\tilde{v}_i (t) = \tilde{v}_j (t) = \tilde{v}_L^- (t) < 0$, $\forall i \in \{0, 1, \ldots, N_1\}$, for $t \in [t_1, t_3]$, which is contradicted by $\tilde{v}_0 (t) = 0$. Thus, except for several isolated instants, we know that $\tilde{v}_L^- (t)$ will keep increasing when $t_1 < T$ with a speed larger than $\beta_2 - \delta_{L_1}$. The convergence time is given by $t \le \left|\tilde{v}_L^- (0)\right| / (\beta_2 - \delta_{L_1})$.

Case 2 $\tilde{v}_L^+ (0) \ge 0$ and $\tilde{v}_L^- (0) \ge 0$. Utilizing a similar analysis given in Case 1, the convergence time is given by $t \le \left|\tilde{v}_L^+ (0)\right| / (\beta_2 - \delta_{L_1})$.

Case 3 $\tilde{v}_L^+ (0) \ge 0$ and $\tilde{v}_L^- (0) \le 0$. Combine Case 1 and Case 2, and we can derive the convergence time as $t \le \max_{i \in V_{L1}} \{|\tilde{v}_i (0)|\} / (\beta_2 - \delta_{L_1})$.

We next study the convergence time for $\hat{q}_i (t) \to q_0 (t) + r_i (t)$: Define $\tilde{q}_i = \hat{q}_i - q_0 - r_i$ and $T_1 = \max_{i \in V_{L1}} \{|\tilde{v}_i (0)|\} / (\beta_2 - \delta_{L_1})$. Since $\hat{v}_i (t) = \dot{q}_0 (t) + \dot{r}_i$, $\forall t \ge T_1$, we have

$$\dot{\tilde{q}}_i = -\beta_1 \mathrm{sign_c} \sum\nolimits_{j=0}^{N} a_{ij} (\tilde{q}_i - \tilde{q}_j), \; i \in V_{L1}. \tag{3.11}$$

Similar to the above-mentioned three cases, we can derive $\hat{q}_i (t) = q_0 (t) + r_i (t)$, for $T_2 = T_1 + \max_{i \in V_{L1}} \{|\tilde{q}_i (T_1)|\} / \beta_1$, which completes the proof. Therefore, $\hat{q}_i$ and $\hat{v}_i$, $i \in V_{L1}$ can be used to replace the position and velocity of the target formation when $t \ge T_2$. □

3.3.2 ESTIMATOR DESIGN FOR THE N^{TH} LAYER

Next, consider the nth layer with $n = 2, 3, \ldots, l$. We first define the following compact vectors for convenience of the analysis

$$\begin{aligned}
Q_{L_1} &= \left[q_1^T, q_2^T, \ldots, q_{N_1}^T\right]^T, \\
C_{L_2} &= \left[c_{N_1+1}^T, c_{N_1+2}^T, \ldots, c_{N_1+N_2}^T\right]^T \\
&= -\left(\mathcal{B}_{L_2 1}^{-1} \mathcal{B}_{L_2 2} \otimes I_p\right) Q_{L_1}, \\
&\vdots \\
Q_{L_{n-1}} &= \left[q_1^T, q_2^T, \ldots, q_{N_1+\ldots+N_{n-2}+N_{n-1}}^T\right]^T, \\
C_{L_n} &= \left[c_{N_1+\ldots+N_{n-1}+1}^T, c_{N_1+\ldots+N_{n-1}+2}^T, \right. \\
&\quad \left. \ldots, c_{N_1+\ldots+N_{n-1}+N_n}^T\right]^T \\
&= -\left(\mathcal{B}_{L_n 1}^{-1} \left[\mathcal{B}_{L_n l}, \mathcal{B}_{L_n l-1}, \ldots, \mathcal{B}_{L_n 2}\right]\right) Q_{L_{n-1}}, \\
n &= 2, 3, \ldots, l.
\end{aligned}$$

More specifically, c_i, $i \in V_{LM} \backslash V_{L1}$ denotes the target formation position for agent i, which is governed by the predefined bearings. It is worth mentioning that the invertibility of $\mathcal{B}_{L_2 1}, \ldots, \mathcal{B}_{L_n 1}$ can be derived by Assumptions 3 [552].

Before giving the main result, we next introduce a well-known lemma on the finite-time stability.

Lemma 3.1

[28] For any real numbers $c > 0$ and $\alpha \in (0,1)$, the origin of a Lyapunov function $V(x)$ is finite-time stable if $\dot{V}(x) + cV(x)^{\alpha} \leq 0$, where the settling time is estimated by $T(x_0) \leq \frac{V(x_0)^{1-\alpha}}{c(1-\alpha)}$. ■

Next, the definition of infinitesimal bearing rigidity and an assumption on the target formation are made.

Definition 22. *(Infinitesimal bearing rigidity [550]). A framework is infinitesimally bearing rigid if all the infinitesimal bearing motions are trivial.*

Assumption 3. *The target formation associated with the topology graph in the* n^{th} *layer* $(n = 2, 3, \ldots, l)$ *is infinitesimal bearing rigid, respectively, which correspondingly means the formation is unique and* $\mathcal{B}_{L_n 1}$ *is positive semidefinite and nonsingular.*

According to Corollary 3 in [551], Assumption 3 together with Assumption 1 imposes a sufficient condition for the uniqueness of the target formation, i.e., the target formation is infinitesimally bearing rigid and there exist at least two agents in the first layer.

Theorem 3.2

Let $\beta_{2n-1} > 0$ and $\beta_{2n} > \frac{\delta_{Ln} \| B_{L_n 1}^{-1} B_{L_n S} \|}{\lambda_{\min}(B_{L_n 1})}$. If Assumption 3 holds, under the DFSEs (3.7a) and (3.7b), the agents can acquire the precise estimations of the bearing-based velocity and position of the target formation in finite time, respectively, which means

$$\hat{v}_i(t) = \dot{c}_i(t), \ t \geq T_{2n-1}, \tag{3.12}$$

$$\hat{q}_i(t) = c_i(t), \ t \geq T_{2n}, \ i \in V_{Ln}, \tag{3.13}$$

where T_{2n-1} and T_{2n} are positive constants which will be introduced later. ■

Proof. Let $\tilde{v}_i = \hat{v}_i - \dot{c}_i$ and $\tilde{q}_i = \hat{q}_i - c_i$, $i \in V_{Ln}$. Define

$$\tilde{v}_{L_n} = \Big[\tilde{v}^T_{N_1+\ldots+N_{n-1}+1}, \tilde{v}^T_{N_1+\ldots+N_{n-1}+2}, \ldots, \tilde{v}^T_{N_1+\ldots+N_{n-1}+N_n}\Big]^T, \tag{3.14}$$

$$\tilde{q}_{L_n} = \Big[\tilde{q}^T_{N_1+\ldots+N_{n-1}+1}, \tilde{q}^T_{N_1+\ldots+N_{n-1}+2}, \ldots, \tilde{q}^T_{N_1+\ldots+N_{n-1}+N_n}\Big]^T, \tag{3.15}$$

$$\mathcal{B}_{L_nS} = [\mathcal{B}_{L_nl}, \mathcal{B}_{L_nl-1}, \ldots, \mathcal{B}_{L_n2}]. \tag{3.16}$$

Consider the Lyapunov function $V_{2n-1} = \frac{1}{2}\tilde{v}^T_{L_n}\mathcal{B}_{L_n1}\tilde{v}_{L_n}$. According to Assumption 3 and the bidirectional information flow among the nth layer, we know that $\mathcal{B}_{L_n1}$ is positive definite. Taking the derivative of V_{2n-1} yields

$$\begin{aligned}\dot{V}_{2n-1} &= \tilde{v}^T_{L_n}\mathcal{B}_{L_n1}\left[-\beta_{2n}\text{sign}_c\left(\mathcal{B}_{L_n1}\tilde{v}_{L_n}\right) - \ddot{C}_{L_n}\right]\\ &\leq -\left(\beta_{2n}\|\mathcal{B}_{L_n1}\|_1 - \delta_{Ln}\|\mathcal{B}_{L_nS}\|_1\right)\|\tilde{v}_{L_n}\|_1\\ &\leq -\sqrt{2}\frac{\beta_{2n}\|\mathcal{B}_{L_n1}\|_1 - \delta_{Ln}\|\mathcal{B}_{L_nS}\|_1}{\lambda_{\max}\left(\mathcal{B}_{L_n1}\right)}V^{\frac{1}{2}}_{2n-1},\end{aligned} \tag{3.17}$$

where δ_{Ln} is defined as $\|\ddot{Q}_{L_{n-1}}\| \leq \delta_{Ln} < \infty$. This boundedness property will be discussed later in Remark 7. According to Lemma 3.1, we know $\hat{v}_i \to \dot{c}_i$, $i \in V_{Ln}$ in a settling time $T_{2n-1} = \frac{\sqrt{2}\lambda_{\max}(\mathcal{B}_{L_n1})}{\left(\beta_{2n}\|\mathcal{B}_{L_n1}\|_1 - \delta_{Ln}\|\mathcal{B}_{L_nS}\|_1\right)}V^{\frac{1}{2}}_{2n-1}(0)$. When $t \geq T_{2n-1}$, consider the Lyapunov function $V_{2n} = \frac{1}{2}\tilde{q}^T_{Ln}\mathcal{B}_{L_n1}\tilde{q}_{Ln}$. Taking the derivative of V_{2n} and following the proof of V_{2n-1}, we can get the settling time for $q_i \to c_i$, $i \in V_{Ln}$ as $T_{2n} = \frac{\sqrt{2}\lambda_{\max}(\mathcal{B}_{L_n1})}{\beta_{2n-1}\|\mathcal{B}_{L_n1}\|_1}\sqrt{2}V^{\frac{1}{2}}_{2n}(T_{2n-1})$. Then, we can use $\hat{q}_i$ and $\hat{v}_i$, $i \in V_{Ln}$ to substitute the bearing-based weighted average of the positions and velocities of the agents in the $n-1^{\text{th}}$ layer when $t \geq T_{2n}$. □

Combining Theorem 3.1 and Theorem 3.2, we know that the agents in all layers can obtain target position and velocity estimation in settling time $T_M = \max\{T_2, T_4, \ldots, T_{2n}\}$.

Remark 5. *In all layers, the estimation error $\tilde{q}_i$ for agent i is bounded for $0 \leq t \leq T_M$. Consider the first layer with $i \in V_{L1}$. It follows from* (3.10) *that $|\tilde{v}_i(t)| \leq |\tilde{v}_i(0)| + (\delta_{L_1} + \beta_2)t$ for $0 \leq t \leq T_1$. From* (3.11), *we know $|\tilde{q}_i(t)| \leq |\tilde{q}_i(0)| + (|\tilde{v}_i(0)| + \beta_1)t + (\delta_{L_1} + \beta_2)t^2/2$ for $0 \leq t \leq T_1$. Thus, $\tilde{q}_i$, $i \in V_{L1}$ is bounded for $0 \leq t \leq T_1$. For the other layers, we can also prove $\tilde{q}_i$, $i \in V_{Ln}$, $n = 2, 3, \ldots, n$ is bounded for $0 \leq t \leq T_M$.*

Remark 6. *We extend the two-layer distributed estimator in [255] to LDFEs with multilayers. Compared with the distributed estimator dealing with undirected graphs in [255], the first layer of the proposed LDFE can be applied to* *a directed graph. The LDFE tackles a more general task with the multilayer agents and the bearing-based topology. It is also noteworthy that the LDFEs are independent of control design and system dynamics.*

3.3.3 CONTROL DESIGN OF EULER-LAGRANGE SYSTEMS

Based on the proposed LDFE (3.6a)–(3.7b), we next design the MLFC of MELSs. First, define the following variables:

$$z_{1i} = q_i - \hat{q}_i, \tag{3.18}$$

$$z_{2i} = \dot{q}_i - \alpha_{1i},\ i \in V_{LM}, \tag{3.19}$$

where the virtual control $\alpha_{1i} = \hat{v}_i - K_{1i} z_{1i}$ and the gain matrix $K_{1i} = K_{1i}^T > 0$. Propose a model-based control algorithm as

$$\begin{aligned} \tau_i =& - z_{1i} - K_{2i} z_{2i} \\ &+ M_i\left(q_i\right)\dot{\alpha}_{1i} + C_i\left(q_i, \dot{q}_i\right)\alpha_{1i} + g_i\left(q_i\right), \end{aligned} \tag{3.20}$$

where K_{2i} is a symmetric positive definite matrix.

Lemma 3.2

In each layer, considering the network of MELSs governed by (3.1), under the MLFC law (3.20), the multilayer formation can be achieved, and the closed-loop errors z_1 and z_2 exponentially converge to the origin.

Proof. Consider the Lyapunov function $V_i = (1/2)z_{1i}^T z_{1i} + (1/2)z_{2i}^T M_i\left(q_i\right) z_{2i}$. Substituting (3.20) into the derivative of V_i, we have $\dot{V}_i = -z_{1i}^T K_{1i} z_{1i} - z_{2i}^T K_{2i} z_{2i} \leq -\rho_i V_i$, where $\rho_i = \min\{2K_{1i}, 2K_{2i}\}$. By using Lyapunov stability theory [230], it can be concluded that z_{1i} and z_{2i} converge to the origin exponentially. According to Theorems 3.1 and 3.2, the proposed LDFEs can, respectively, yield accurate target formation position and velocity estimations for agents in each layer when $t \geq T_M$. Hence, the layered formation control objective is achieved. Next, the stability analysis when $t < T_M$ is discussed. Under the MLFC law (3.20), the error signals z_{1i} and z_{2i} in all layers are bounded when $t \leq T_M$. According to Remark 5, the estimation error $\tilde{q}_i$, $i \in V_{LM}$ is bounded for $0 \leq t \leq T_M$. Utilizing manipulations similar in Remark 5, we can get $\tilde{v}_i$, $i \in V_{LM}$ are bounded for $0 \leq t \leq T_M$. Therefore, we know the error signals z_{1i} and z_{2i} in each layer are bounded when $t \leq T_M$. □

■

Remark 7. *From the above analysis, we know the accelerations of the agents in the prior $n-1$ layers are bounded if we want to analyze the stability of*

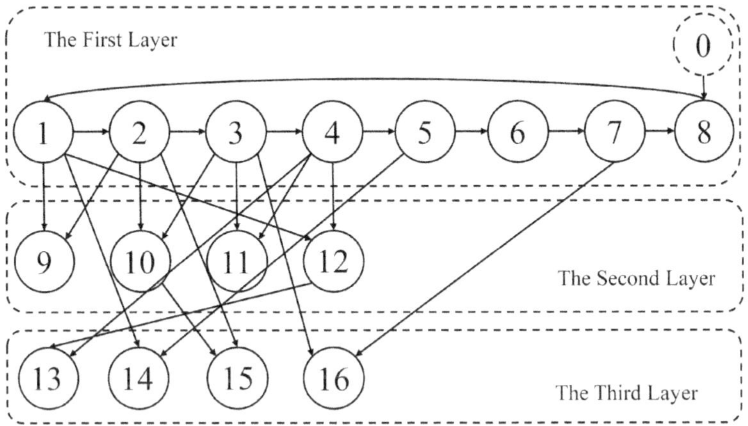

Figure 3.1 The layered communication graph among 16 satellites.

the nth layer. In other words, this boundedness property can be utilized as a precondition in the proof of Theorem 3.2 for the nth layer, $n = 2, 3, \ldots, l$.

Remark 8. *For control with practical issues, such as the problems of system uncertainties and loss of velocity information, we can apply the control techniques addressed in [255] where the stability analysis can also be found.*

3.4 SIMULATION

To illustrate the full power of proposed MLFC algorithms, we consider a group of 16 networked satellites with 8 satellites labeled as $1, \ldots, 8$ in the first layer, 4 satellites labeled as $9, \ldots, 12$ in the second layer, and 4 satellites labeled as $13, \ldots, 16$ in the third layer. In the Local-Vertical-Local-Horizontal (LVLH) rotating frame, the relative orbital motion of the ith satellite and the initial elements of the near-circular orbit are the same as those in [255]. We evaluate the LDFE (3.6a)–(3.7b) and the MLFC law (3.20) for networked satellites, where the target formation configuration is time-varying. The layered interaction graph among 16 satellites is shown in Fig. 3.1. The control parameters are chosen as $K_{1i} = 0.1I_3$ and $K_{2i} = 5I_3$. The continuous and smooth formation functions $r_i(t)$ are constructed by proper designed sigmoid functions. The detailed Laplacian matrix $\mathcal{L}$, $r_i(t)$, and $q_i(0)$ in this simulation are omitted due to the limited pages. The trajectory of the virtual agent 0 is given as $q_0 = [400\sin(\frac{\pi t}{500}),\ 400\cos(\frac{\pi t}{500}),\ 400\sin(\frac{\pi t}{500})]^T$.

The trajectories of satellites are given in Fig. 3.2, where it can be observed that the MLFC algorithms can guarantee satisfactory performance. Compared with the formation-containment configuration in [118, 255, 488–490], 3 different formation configurations are achieved and transformed flexibly. Fig. 3.3 shows that the formation errors of 16 satellites converge to the origin within

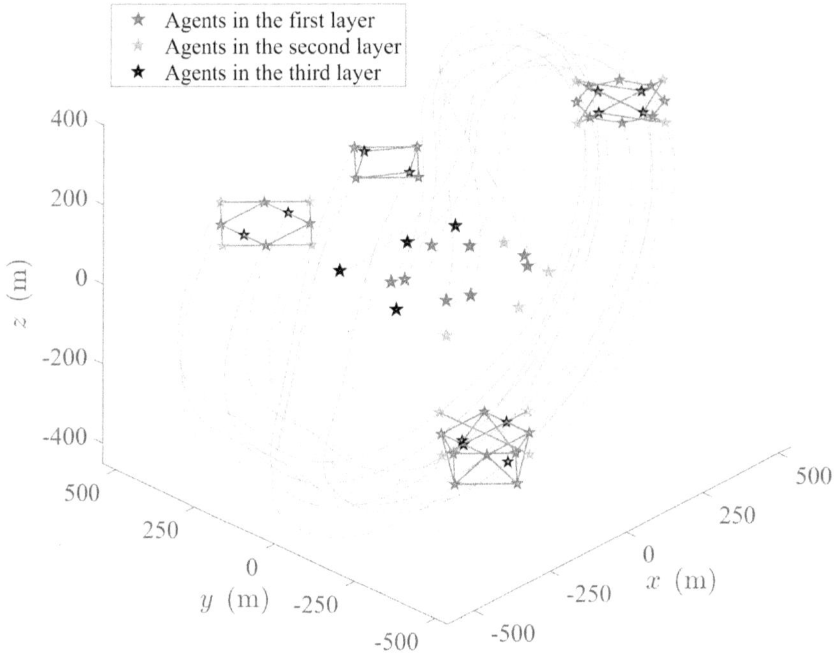

Figure 3.2 Trajectories of 16 satellites (3D view).

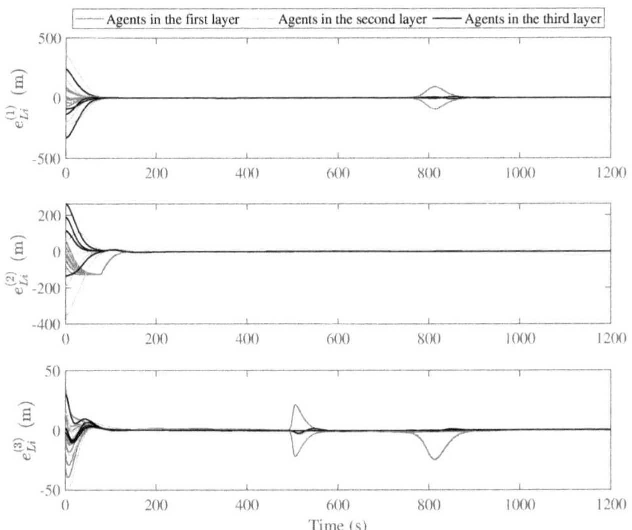

Figure 3.3 Formation errors of 16 satellites in three layers.

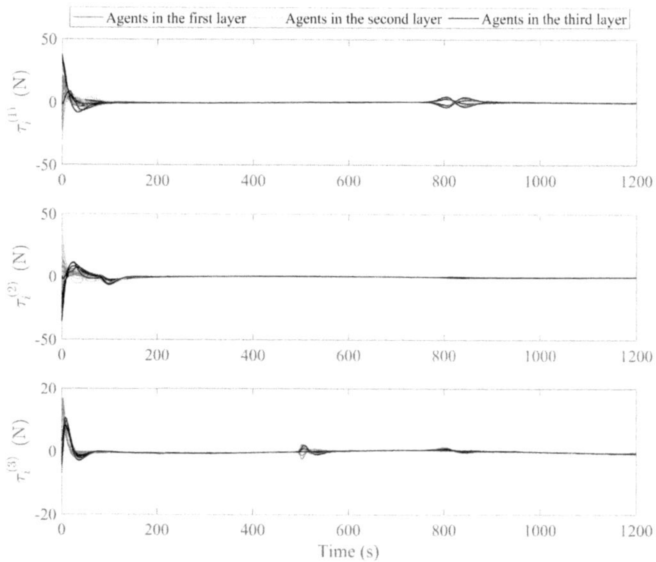

Figure 3.4 Control forces of 16 satellites in three layers.

200 seconds, where the formation errors in the three layers increase when the agents are transforming the formation configuration. The corresponding control inputs of satellites are given in Fig. 3.4, which are acceptable for satellites in practice. The above numerical results have shown that the proposed LDFE and MLFC are effective and feasible.

3.5 CONCLUSION

This chapter proposed a multilayer framework to describe and extend the leader-follower problem with an arbitrary number of layers. We designed an LDFE and an MLFC law, where the finite-time stability and the exponential stability are guaranteed, respectively. The approaches in this chapter provide a new insight into the multi-agent formation problems with an arbitrary number of formation layers and flexible formation configurations, where the information flow among agents can also be explored and designed naturally.

4 Micro Flapping-Wing Vehicles Formation Control with Attitude Estimation

This chapter addresses the formation control problem of flapping-wing vehicles (FWVs) under the model uncertainty and the measurement inaccuracy. A two-layer formation strategy is adopted, which consists of a formation control layer for the leaders, and a containment control layer for the followers. In both layers, attitudes and positions are required by the formation geometry. A formation state estimation algorithm is designed to achieve desired formation states from local neighborhoods. In FWVs, attitude angles are usually achieved from angular velocities, whose measurement error accumulates during integration and leads to divergence of the system. In order to solve this problem, we explore the coupling property between the translational motion and the rotational motion of FWVs, and design a coupling-based estimation method for attitude angles. To compensate for the model uncertainty, the measurement error, and the estimation error, adaptive neural networks are developed together with the control algorithm. The stability of both the control algorithm and the estimation algorithm are guaranteed based on the Lypanov stability theory. Simulations are conducted to validate our method, and the results illustrate its effectiveness.

4.1 INTRODUCTION

Inspired by the characteristics of insects and birds, flapping-wing vehicles (FWVs) are developed by researchers [326]. Compared with fixed-wing aerial vehicles and the rotorcraft, flapping-wing structure is more fitted to the micro aircraft. FWVs combine the advantages of the two previous categories, such as high-speed forward flight, hovering, vertical take-off and landing, and good maneuverability. Moreover, they can naturally conceal themselves or pretend to be an insect to conduct special tasks.

As early as 1993, the mechanism of the FWV is studied [101] and its aerodynamic model is constructed [100]. The dynamics and control mechanism is further investigated by many researchers [187,417]. In 2011, a Nano Hummingbird is built by the Aero Vironment team after over four years of work [225]. An attitude and position control method for this flapping-wing model is designed on the basis of the adaptive sliding mode technique [17]. The model uncertainty is considered and an adaptive neural network (NN) method is adopted to control the FWV [183]. Disturbance rejection schemes of FWVs

DOI: 10.1201/9781003298618-4

are also explored for the outdoor wind condition [72]. The manual control and autonomous flight can be achieved by these methods, however, they are designed for a single FWV. Cooperation of multiple FWVs can be more advantageous for complicated tasks.

The formation problem is one of the most fundamental and important tasks for multi-agent cooperation. The generalized formation control is defined as driving multiple agents to achieve prescribed constraints on their states [256, 359], whereas the narrow formation control emphasizes on driving multiple agents to form a desired geometry pattern [68]. Despite from the work on formation geometry, there are also efforts made on the cooperative control of multiple agents. The consensus control aims to make the states of agents reach a common value [324, 564], whereas the containment control requires that a team of followers are guided by multiple leaders [38, 299].

The rotorcraft, the fixed-wing aerial vehicle, and the spacecraft are three typical types of agents in the aerial formation control. In [116], the formation tracking problem is studied subjected to switching topologies with application to the quadrotor, only translational motion is considered for the quadrotor and it is modeled as a second-order system. In [251], a translational tracking control method for the spacecraft is designed to construct a formation. Despite from the translational motion, formation problem with respect to the rotational motion is also studied [197, 261, 544]. In some works, the translational motion and the rotational motion are regarded as a coupled system, for example, the formation control problem of fixed-wing aerial vehicles or ground vehicles [98, 223, 458]. They are always described as the so-called unicycle model, which actually have three degrees of freedom: the forward velocity, the pitch angle, and the heading angle. The motion of the two-dimensional unicycle agent is always decomposed to the heading vector and its perpendicular vector [130, 287, 545], and the control input is linearly formulated by these intermediate vectors. The polar coordinate is also widely used for unicycle agents [123, 452]. In [484], both the translational motion and the rotational motion are controlled for the formation of quadrotors, and they are considered as decoupled motion and handled separately. In the formation control of the spacecraft, there are also works deal with both the translational motion and the rotational motion, whereas the coupling between them is analyzed for decoupled control [373, 449]. Different from these works, we consider both the translational motion and the rotational motion in FWVs, and the coupling property between them is helpful in the design.

Attitude angles of many small aircraft are achieved mediately from the integration of angular velocities [137]. Measurement errors accumulate over time by integration and lead to drift in long-time flight. Usually, auxiliary sensors are adopted to solve this problem at the price of larger space and weight. In this chapter, both model uncertainties and measurement errors are considered during flight, an adaptive NNs based control algorithm and a state estimation algorithm are designed for the formation control problem.

The contributions of this chapter are as follows:

1. A two-layer formation control method combined with adaptive NNs and attitude estimation is proposed for the FWV formation under model uncertainties and measurement errors. Using the proposed method, both tracking errors and estimation errors will remain within small compact sets. In contrast, most of the existing works for multiple Euler-Lagrange Systems deal with either known systems or parametric uncertainties with the linearity-in-parameters assumption [152, 337].
2. Both the translational motion control and the rotational motion control are considered in the formation, and the coupling property between them is analyzed. Defined as "coupled by coordinate transformation", the coupling property is considered as useful information in our scheme. In contrast, most existing works either consider the translational tracking control [251] and the rotational tracking control [197, 544] separately for the formation, or analyze the coupling property for decoupling [373, 449].
3. Using the specific coupling property, an estimation algorithm for attitude angles is designed to prevent them from drifting away, which is caused by the measurement inaccuracy of angular velocities. In contrast, most of the state estimators consider only the output-feedback situation, and the outer-loop information is needed for unmeasurable states [63,306,474,519].
4. A finite-time sliding-mode state estimator is designed to achieve the desired translational and rotational state information. Using the estimator, the control law proposed in this chapter is fully distributed, whereas global information is commonly required in the control level of formation problems [437, 499].

4.2 PROBLEM FORMULATION AND PRELIMINARIES

4.2.1 GRAPH THEORY

Graphs are commonly used to describe network topologies of multi-agent systems. A graph $\mathcal{G} = (V, E)$ consists of two parts: a vertex set V and an edge set E. The size of V is determined by the number of agents. We have N leaders and M followers in the system, thus $V = [1, 2, \cdots, N + M]$. The edge set $E \subseteq V \times V$ describes the connection between vertexes. $(i, j) \in E$ indicates that the agent j can achieve information from the agent i. The adjacency matrix $\mathcal{A} = [a_{ij}] \in \mathbb{R}^{(N+M)\times(N+M)}$ is the nonnegative weight matrix of the edges. If $(i, j) \in E$, $a_{ij} > 0$, otherwise $a_{ij} = 0$. The Laplacian matrix is defined as $\mathcal{L} = [l_{ij}]$, where $l_{ii} = \sum_{j\neq i} a_{ij}$ and $l_{ij} = -a_{ij}$ for $i \neq j$.

4.2.2 AGENT DYNAMICS

The dynamics of the FWV is introduced [17]. The translational motion equation of the FWV is derived directly from aerodynamic forces. Using the mass matrix $M_{iT} = m_i I_{3\times3}$ and the gravity vector $G_{iT} = [0, 0, -m_i g]^T$, the Euler-Lagrange equation for the translational motion can be written as

$$M_{iT}\ddot{\boldsymbol{q}}_{iT} + G_{iT} = H_{IB}(\boldsymbol{q}_{iR})\boldsymbol{u}_{iT}, \tag{4.1}$$

where $\boldsymbol{u}_{iT}$ is the translational motion control input, $\boldsymbol{q}_{iT} = [x_i, y_i, z_i]^T$ is the translational state vector, subscript i represents the i-th FWV, subscript T stands for the translational motion, and H_{IB} is the rotation matrix from the body frame to the inertial frame,

$$\begin{aligned} H_{IB}(\boldsymbol{q}_{iR}) &= \begin{bmatrix} \cos\psi_i & -\sin\psi_i & 0 \\ \sin\psi_i & \cos\psi_i & 0 \\ 0 & 0 & 1 \end{bmatrix} \\ &\times \begin{bmatrix} \cos\theta_i & 0 & \sin\theta_i \\ 0 & 1 & 0 \\ -\sin\theta_i & 0 & \cos\theta_i \end{bmatrix} \times \begin{bmatrix} 1 & 0 & 0 \\ 0 & \cos\phi_i & -\sin\phi_i \\ 0 & \sin\phi_i & \cos\phi_i \end{bmatrix}. \end{aligned} \tag{4.2}$$

Remark 9. *The rotation matrix H_{IB} is always invertible. Its inverse matrix H_{BI} is the rotation matrix from the inertial frame to the body frame. H_{BI} is also the transposed matrix of H_{IB}, namely $H_{BI} = H_{IB}^{-1} = H_{IB}^T$.*

The rotational motion equation is calculated in the body frame,

$$\begin{aligned} \dot{\boldsymbol{\eta}}_i &= \bar{I}_3\dot{\boldsymbol{w}}_i + \boldsymbol{w}_i \times \bar{I}_3\boldsymbol{w}_i \\ &= \bar{I}_3\frac{d}{dt}(H_i\dot{\boldsymbol{q}}_{iR}) + (H_i\dot{\boldsymbol{q}}_{iR}) \times \bar{I}_3(H_i\dot{\boldsymbol{q}}_{iR}) = \boldsymbol{u}_{iR}, \end{aligned} \tag{4.3}$$

where $\boldsymbol{\eta}_i$ is the angular momentum, $\bar{I}_3$ is the moment of inertia matrix, $\boldsymbol{q}_{iR} = [\phi_i, \theta_i, \psi_i]^T$ is the rotational state vector with ϕ the roll angle, θ the pitch angle, and ψ the yaw angle. Subscript R stands for the rotational motion. We use bars to distinguish the inertia matrix from the identity matrix. H_i is the transfer matrix from rotational velocities to translational velocities,

$$H_i = \begin{bmatrix} 1 & 0 & -\sin\theta_i \\ 0 & \cos\phi_i & \cos\theta_i\sin\phi_i \\ 0 & -\sin\phi_i & \cos\theta_i\cos\phi_i \end{bmatrix}. \tag{4.4}$$

Using a square matrix L_i such that

$$(H_i\dot{\boldsymbol{q}}_{iR}) \times \bar{I}_3(H_i\dot{\boldsymbol{q}}_{iR}) = L_i\dot{\boldsymbol{q}}_{iR}, \tag{4.5}$$

equation (4.3) can be written as

$$M_{iR}\ddot{\boldsymbol{q}}_{iR} + C_{iR}\dot{\boldsymbol{q}}_{iR} = \boldsymbol{u}_{iR}, \tag{4.6}$$

where $M_{iR} = \bar{I}_3 H_i$ and $C_{iR} = \bar{I}_3\dot{H}_i + L_i$.

Bring the translational motion and the rotational motion together, the FWV can be modeled as an Euler-Lagrange system,

$$M_i(\boldsymbol{q}_i)\ddot{\boldsymbol{q}}_i + C_i(\boldsymbol{q}_i, \dot{\boldsymbol{q}}_i)\dot{\boldsymbol{q}}_i + G_i = H(\boldsymbol{q}_i)\boldsymbol{u}_i, \tag{4.7}$$

where $\boldsymbol{q}_i = [\boldsymbol{q}_{iT}^T, \boldsymbol{q}_{iR}^T]^T$, $M_i(\boldsymbol{q}_i) \in \mathbb{R}^{6\times 6}$ is the symmetric positive definite inertia matrix, $C_i(\boldsymbol{q}_i, \dot{\boldsymbol{q}}_i) \in \mathbb{R}^{6\times 6}$ is the matrix of Coriolis forces, G_i denotes the gravitational force, $\boldsymbol{u}_i$ is the control input vector, and $\boldsymbol{q}_i$ denotes the state vector of the i-th FWV. The matrices and vectors in (4.7) is presented as follows:

$$\begin{aligned} M_i &= \begin{bmatrix} M_{iT} & 0 \\ 0 & M_{iR} \end{bmatrix}, \ C_i = \begin{bmatrix} 0 & 0 \\ 0 & C_{iR} \end{bmatrix}, \ G_i = \begin{bmatrix} G_{iT} \\ 0 \end{bmatrix} \\ H(\boldsymbol{q}_i) &= \begin{bmatrix} H_{IB}(\boldsymbol{q}_{iR}) & 0 \\ 0 & I_{3\times 3} \end{bmatrix}, \boldsymbol{u}_i = \begin{bmatrix} \boldsymbol{u}_{iT} \\ \boldsymbol{u}_{iR} \end{bmatrix}. \end{aligned} \tag{4.8}$$

Property 1. *[280] The matrix $\dot{M}_{iR} - 2C_{iR}$ is skew-symmetric.*

Remark 10. *The translational control input $\boldsymbol{u}_{iT}$ is physically the aerodynamic force in the body frame, and the rotational control input $\boldsymbol{u}_{iR}$ is physically the aerodynamic torque in the body frame. The Euler-Lagrange formulation in (4.7) fits for a variety of aerial objects. Since the translational motion is more important in the inertial frame than in the body frame, $H_{IB}(\boldsymbol{q}_{iR})$ in (4.1) is used for coordinate transformation.*

4.2.3 PROBLEM DESCRIPTION AND COUPLING ANALYSIS

Consider that the formation problem contains N leader FWVs labelled as $1, 2, \cdots, N$ and M follower FWVs labelled as $N+1, N+2, \cdots, N+M$. In the system, the Laplacian matrix $\mathcal{L} \in \mathbb{R}^{(N+M)\times(N+M)}$ can be written as

$$\mathcal{L} = \begin{bmatrix} \mathcal{L}_{L1} & 0_{N\times M} \\ \mathcal{L}_{F2} & \mathcal{L}_{F1} \end{bmatrix}, \tag{4.9}$$

where $\mathcal{L}_{L1}$ is the network topology among the leaders, $\mathcal{L}_{F1}$ is the network topology among the followers, and $\mathcal{L}_{F2}$ is the network topology between the leaders and the followers. Let $\mathcal{L}_{L2} = [-a_{10}, -a_{20}, \cdots, -a_{N0}]^T$ denote whether the leader can access the formation trajectory, where $a_{i0} = 1$ if yes, and $a_{i0} = 0$ otherwise. Denote $\mathcal{G}_l = (V_l, E_l)$ the communication network topology graph of the N leaders, with $V_l = [1, 2, \cdots, N]$ and $E_l = V_l \times V_l$. The adjacency matrix of the N leaders is

$$\mathcal{A}_l = \begin{bmatrix} a_{00} & \cdots & a_{0N} \\ \vdots & \ddots & \vdots \\ a_{N0} & \cdots & a_{NN} \end{bmatrix}, \tag{4.10}$$

where a_{ij} is the nonnegative weight of the edges in the graph.

Assumption 4. *[404] For the N leaders, $\mathcal{G}_l$ is connected. For the M followers, there exists at least one path from the leader set to each follower.*

Lemma 4.1

[339] Each entry of $-\mathcal{L}_{F1}^{-1}\mathcal{L}_{F2}, -\mathcal{L}_{L1}^{-1}, -\mathcal{L}_{L1}^{-1}\mathcal{L}_{L1}$ is negative and each row sum of $-\mathcal{L}_{F1}^{-1}\mathcal{L}_{F2}, -\mathcal{L}_{L1}^{-1}, -\mathcal{L}_{L1}^{-1}\mathcal{L}_{L1}$ is equal to one. $\mathcal{L}_{L1} \in \mathbb{R}^{N\times N}$ and $\mathcal{L}_{F1} \in \mathbb{R}^{M\times M}$ are symmetric positive definite. ■

Remark 11. *Note that the concept of the leader in [339] is properly modified here for greater clarity. In this approach, a leader does not necessarily have no neighbors, but it does not have follower neighbors. Therefore, leaders do not get information from followers, whereas followers achieve information from leaders. Since the connectivity is the same in both systems, Lemma 1 works.*

This chapter focuses on the two-layer formation control problem of FWVs, whose formation structure is similar as in [264]. In the first layer, leaders move together to achieve a specified formation with a desired reference. In the second layer, followers converge into the convex hull spanned by leaders. To construct a splendid formation, the configuration is required not only for the translational motion of the FWVs but also for their rotational motion.

We consider the formation control problem in which attitude angles cannot be achieved directly and angular velocities cannot be measured accurately. To address this problem, we design an estimation algorithm for attitude angles. Moreover, the adaptive NNs are employed to compensate for model uncertainties.

Remark 12. *In [264], different estimation and control methods are designed for constant and time-varying configurations. In this chapter, we use a unified method for both constant and time-varying configurations. Both the translational motion and the rotational motion are controlled for the desired configuration.*

Remark 13. *Attitude angles of the aircraft are commonly achieved from the integration of angular velocities, which are measured by gyroscopes. The measurement error accumulates over time and lead to drift in long-time flight. Usually, magnetometer and accelerometer are used as auxiliary sensors to prevent the drift of attitude angles. In micro FWVs, extra sensors cause larger space and weight.*

Remark 14. *Compared with output feedback control methods in [63, 306, 474, 519] which estimate unmeasurable states using output states, the estimation of attitude angles is much more challenging. The reason is that output states are the outer-loop information for their unmeasurable states, whereas angular* *velocities which can be measured in our system are the inner-loop information. Attitude angles cannot be estimated using only angular velocities due to the measurement inaccuracy, therefore coupling information is explored to help.*

In order to estimate attitude angles of the FWV without drift, the coupling information is used. In (4.7), the translational motion of the FWV is coupled with its rotational motion via a matrix $H_{IB}(\boldsymbol{q}_{iR})$, we call it "coupled by coordinate transform".

Definition 23. *In an Euler-Lagrange system, the translational motion is said to be coupled by coordinate transform with the rotational motion if the translational control input has component generated directly in the body frame and the component has full rank.*

The coupling concept has the following features:

The translational control input has components generated in the body frame. For the control of the translational motion in the motion frame (usually inertial frame), coordinate transformation has to be conducted from the body frame to the motion frame. Rotational states are coupled during the coordinate transformation process.
The component of the translational control input in the body frame has full rank. The translational motion can be controlled by the translational control input with any rotational state.

In our case, the control input of the translational motion has the same rank with the translational states, and its coefficient matrix is invertible. Therefore the translational motion is controllable.

The schematic diagram is in Fig. 4.1. In the design of the translational motion control algorithm, estimated attitude angles are used instead of their real values. Since the estimation error leads to the tracking error in the translational motion, we can adjust the estimated attitude angles using translational information and prevent attitude angles from drifting away.

4.3 DISTRIBUTED FORMATION CONTROL METHOD

4.3.1 FINITE-TIME ESTIMATOR FOR THE LEADERS

The following necessary assumption is made.

Assumption 5. *The measured angular velocity $\bar{\boldsymbol{v}}_R$ is in the neighborhood of its real value $\boldsymbol{v}_R$. Mathematically, there exists a positive constant $\epsilon_{\boldsymbol{v}}$ such that*

$$\|\tilde{\boldsymbol{v}}_R\| \leqslant \epsilon_{\boldsymbol{v}}, \tag{4.11}$$

where $\tilde{\boldsymbol{v}}_R = \boldsymbol{v}_R - \bar{\boldsymbol{v}}_R$, and $\boldsymbol{v}_R = \dot{\boldsymbol{q}}_R$.

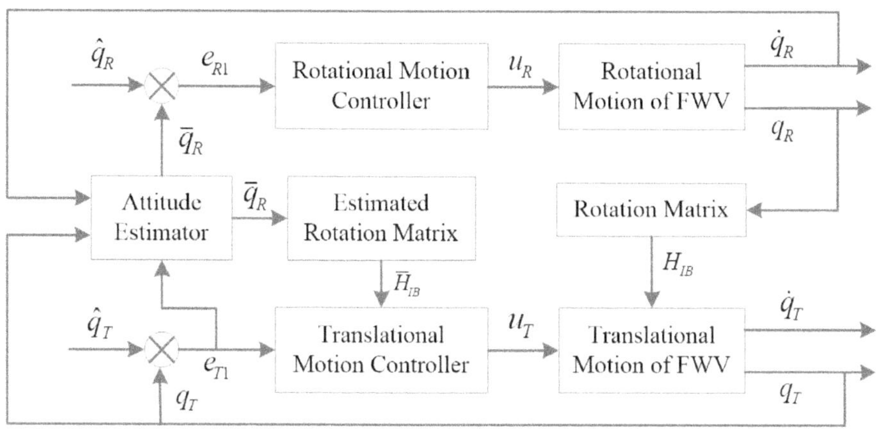

Figure 4.1 Estimate attitude angles using coupling information.

Inspired by [264], we use the following estimation algorithm for the formation states of the leaders,

$$\begin{aligned}\dot{\hat{\boldsymbol{q}}}_i &= \hat{\boldsymbol{v}}_i - \beta_1 \mathrm{sgn} \sum_{j=0}^{N} a_{ij}(\hat{\boldsymbol{q}}_i - \hat{\boldsymbol{q}}_j - \boldsymbol{\delta}_{ij}), \\ \dot{\hat{\boldsymbol{v}}}_i &= -\beta_2 \mathrm{sgn} \sum_{j=0}^{N} a_{ij}(\hat{\boldsymbol{v}}_i - \hat{\boldsymbol{v}}_j - \dot{\boldsymbol{\delta}}_{ij}),\end{aligned} \tag{4.12}$$

where β_1 and β_2 are positive parameters with respect to the convergence rate of the formation state estimation process, $\hat{\boldsymbol{q}}_i$ and $\hat{\boldsymbol{v}}_i$, $i = 1, 2, \cdots, N$ are the desired formation positions and velocities for the translational motion, and the desired attitude angles and angular velocities for the rotational motion, distingushed by subscripts T and R. Denote $\boldsymbol{\delta}_{ij} = \boldsymbol{\delta}_i - \boldsymbol{\delta}_j$ the desired relative distance between FWV i and FWV j, with $\boldsymbol{\delta}_i$ and $\boldsymbol{\delta}_j$ the desired trajectory of them. $\boldsymbol{\delta}_{i0}$ represents the relative distance between FWV i and the desired trajectory of the formation. Initially, $\hat{\boldsymbol{q}}_0 = \boldsymbol{q}_0 = \boldsymbol{\delta}_0$, $\hat{\boldsymbol{v}}_0 = \dot{\boldsymbol{q}}_0 = \dot{\boldsymbol{\delta}}_0$. Subscript 0 stands for the formation trajectory information, and subscript T and R are ignored because the algorithm in (4.12) fits for both translational and rotational motions.

The operation of the function sgn is [147]

$$\mathrm{sgn}(\boldsymbol{x}) \triangleq \frac{\boldsymbol{x}}{\|\boldsymbol{x}\|}, \boldsymbol{x} \neq 0. \tag{4.13}$$

The function defines a unit vector in the direction of $\boldsymbol{x}$. It covers an n-dimensional unit ball.

Remark 15. *$\boldsymbol{\delta}_{ij}$ controls the shape of the formation. In our scheme, $\boldsymbol{\delta}_{ij}$ is known instead of $\boldsymbol{\delta}_i$ and $\boldsymbol{\delta}_j$. In translational motion, $\boldsymbol{\delta}_{ij}$ can be set fixed for* *a specified formation, or vary smoothly for a continuously varying formation geometry. In rotational motion, it is always given as 0 to achieve consensus of rotational states.*

Define

$$\begin{aligned} Q_L &= [\boldsymbol{q}_1^T, \boldsymbol{q}_2^T, \cdots, \boldsymbol{q}_N^T]^T, \\ \boldsymbol{\delta}_L &= [\boldsymbol{\delta}_1^T, \boldsymbol{\delta}_2^T, \cdots, \boldsymbol{\delta}_N^T]^T, \end{aligned} \tag{4.14}$$

and set the following assumption for the desired formation trajectory,

Assumption 6. *There exist a nonnegative constant $\delta_{\max}$ such that for every FWV i*

$$\|\ddot{\boldsymbol{\delta}}_i\| \leqslant \delta_{\max} < \infty. \tag{4.15}$$

We have the following theorem for the estimation of the leaders.

Theorem 4.1

Let $\beta_1 > 0$ and $\beta_2 > \delta_{\max}$. If Assumptions 4 and 6 holds, under protocol (4.12), the leaders can achieve precise estimations of the desired formation states in finite time t_s, which means

$$\begin{aligned} \hat{\boldsymbol{q}}_i &= \boldsymbol{q}_0 + \boldsymbol{\delta}_{i0}, \\ \hat{\boldsymbol{v}}_i &= \dot{\boldsymbol{\delta}}_i, \quad i = 1, 2, \cdots, N. \end{aligned} \tag{4.16}$$

■

Proof. Let $\tilde{\boldsymbol{v}}_i = \hat{\boldsymbol{v}}_i - \dot{\boldsymbol{\delta}}_i$ and $\tilde{\boldsymbol{q}}_i = \hat{\boldsymbol{q}}_i - \boldsymbol{q}_0 - \boldsymbol{\delta}_{i0}, i = 1, 2, \cdots, N$. Define the error vectors $\tilde{\boldsymbol{v}}_L = [\tilde{\boldsymbol{v}}_1^T, \tilde{\boldsymbol{v}}_2^T, \cdots, \tilde{\boldsymbol{v}}_N^T]^T$ and $\tilde{\boldsymbol{q}}_L = [\tilde{\boldsymbol{q}}_1^T, \tilde{\boldsymbol{q}}_2^T, \cdots, \tilde{\boldsymbol{q}}_N^T]^T$ for the velocity estimation and the position estimation, respectively. Construct the following Lyapunov function for the velocity estimation,

$$V_{\boldsymbol{v}} = \frac{1}{2}\tilde{\boldsymbol{v}}_L^T[\mathcal{L}_{L1} \otimes I_p]^T \tilde{\boldsymbol{v}}_L. \tag{4.17}$$

According to [264], taking the derivative of $V_{\boldsymbol{v}}$ yields

$$\begin{aligned} \dot{V}_{\boldsymbol{v}} =& \tilde{\boldsymbol{v}}_L^T[\mathcal{L}_{L1} \otimes I_p]^T \dot{\tilde{\boldsymbol{v}}}_L \\ =& \tilde{\boldsymbol{v}}_L^T[\mathcal{L}_{L1} \otimes I_p]^T \left[-\beta_2 \mathrm{sgn}((\mathcal{L}_{L1} \otimes I_p)\tilde{\boldsymbol{v}}_L) - \ddot{\boldsymbol{\delta}}_L\right] \\ \leqslant& -\sqrt{2}(\beta_2 - \delta_{\max})\frac{\lambda_{\min}(\mathcal{L}_{L1})}{\sqrt{\lambda_{\max}(\mathcal{L}_{L1}))}}\sqrt{V_{\boldsymbol{v}}}, \end{aligned} \tag{4.18}$$

According to Lemma 2.4, $\hat{\boldsymbol{v}}$ approaches its desired value $\dot{\boldsymbol{\delta}}_i$ for $i = 1, 2, \cdots, N$ in a settling time t_1

$$t_1 = \frac{\sqrt{2\lambda_{\max}(\mathcal{L}_{L1})}}{(\beta_2 - \delta_{\max})\lambda_{\min}(\mathcal{L}_{L1})}\sqrt{V_{\boldsymbol{v}}(0))}. \tag{4.19}$$

When $t \geqslant t_1$, construct a Lyapnov function $V_{\boldsymbol{q}}$,

$$V_{\boldsymbol{q}} = \frac{1}{2}\tilde{\boldsymbol{q}}_L^T[\mathcal{L}_{L1} \otimes I_p]^T \tilde{\boldsymbol{q}}_L. \tag{4.20}$$

Similar as in (4.18),

$$\dot{V}_{\boldsymbol{q}} \leqslant -\sqrt{2}\beta_1 \frac{\lambda_{\min}(\mathcal{L}_{L1})}{\sqrt{\lambda_{\max}(\mathcal{L}_{L1}))}} \sqrt{V_{\boldsymbol{q}}}. \tag{4.21}$$

According to Lemma 2.4, $\hat{\boldsymbol{q}}$ approaches its desired value $\boldsymbol{q}_0 + \boldsymbol{\delta}_{i0}$ for $i = 1, 2, \cdots, N$ in a settling time t_2

$$t_2 = \frac{\sqrt{2\lambda_{\max}(\mathcal{L}_{L1})}}{(\beta_1 - \dot{\boldsymbol{\delta}}_{\max})\lambda_{\min}(\mathcal{L}_{L1})} \sqrt{V_{\boldsymbol{q}}(t_1))} + t_1. \tag{4.22}$$

□

Note that

$$\dot{\hat{\boldsymbol{q}}}_i - \hat{\boldsymbol{v}}_i = -\beta_1 \text{sgn} \sum_{j=0}^{N} a_{ij}(\hat{\boldsymbol{q}}_i - \hat{\boldsymbol{q}}_j - \boldsymbol{\delta}_{ij}) \tag{4.23}$$

is bounded, and $\|\dot{\hat{\boldsymbol{q}}}_i - \hat{\boldsymbol{v}}_i\| \leqslant 3\beta_1$ for both the translational motion and the rotational motion.

4.3.2 TRANSLATIONAL MOTION CONTROL

A control law for the translational motion is designed for the formation flight. The NN technique is applied to compensate for the model uncertainty. To simplify the notations, we omit the subscript i hereinafter which is the FWV index.

Consider the translational motion equation in (4.1) and the desired formation position in (4.12), the second-order dynamics can be obtained as

$$\ddot{\boldsymbol{q}}_T = M_T^{-1}(H_{IB}\boldsymbol{u}_T - G_T). \tag{4.24}$$

Define the tracking error as

$$\begin{aligned} \boldsymbol{e}_{T1} &= \boldsymbol{q}_T - \hat{\boldsymbol{q}}_T, \\ \boldsymbol{e}_{T2} &= \dot{\boldsymbol{q}}_T - \boldsymbol{\alpha}_T, \end{aligned} \tag{4.25}$$

where $\boldsymbol{\alpha}_T = \hat{\boldsymbol{v}}_T - A_{T1}$, and $A_{T1} = K_{T1}\boldsymbol{e}_{T1}$. $K_{T1} \subset \mathbb{R}^{3\times 3}$ is a control parameter matrix for the translational motion.

Differentiating (4.25) with respect to time yields

$$\begin{aligned} \dot{\boldsymbol{e}}_{T1} &= \dot{\boldsymbol{q}}_T - \dot{\hat{\boldsymbol{q}}}_T \\ &= \boldsymbol{e}_{T2} - A_{T1} + \hat{\boldsymbol{v}}_T - \dot{\hat{\boldsymbol{q}}}_T, \\ \dot{\boldsymbol{e}}_{T2} &= \ddot{\boldsymbol{q}}_T - \dot{\boldsymbol{\alpha}}_T \\ &= (M_T)^{-1}(H_{IB}\boldsymbol{u}_T - G_T) - \dot{\boldsymbol{\alpha}}_T. \end{aligned} \tag{4.26}$$

The attitude angles cannot be directly measured in this problem. We use its estimation instead. Denote $\bar{\boldsymbol{q}}_T$ the estimation of $\boldsymbol{q}_T$, and $\bar{H}_{IB}$ the estimation of H_{IB}. $\bar{H}_{IB}$ depends on $\bar{\boldsymbol{q}}_R = [\bar{\phi}, \bar{\theta}, \bar{\psi}]$, which is calculated according to the measurement of the angular velocity and the translational control information.

$$\begin{aligned}\bar{H}_{IB}(\bar{\boldsymbol{q}}_R) = &\begin{bmatrix} \cos\bar{\psi} & -\sin\bar{\psi} & 0 \\ \sin\bar{\psi} & \cos\bar{\psi} & 0 \\ 0 & 0 & 1 \end{bmatrix} \\ &\times \begin{bmatrix} \cos\bar{\theta} & 0 & \sin\bar{\theta} \\ 0 & 1 & 0 \\ -\sin\bar{\theta} & 0 & \cos\bar{\theta} \end{bmatrix} \times \begin{bmatrix} 1 & 0 & 0 \\ 0 & \cos\bar{\phi} & -\sin\bar{\phi} \\ 0 & \sin\bar{\phi} & \cos\bar{\phi} \end{bmatrix}.\end{aligned} \tag{4.27}$$

To enhance the control performance, we use the NN technique to compensate for model uncertainties, which is detailed in [245, 503]. Let $\boldsymbol{z}_T = [\boldsymbol{q}_T^T, \bar{\boldsymbol{q}}_T^T, \boldsymbol{\alpha}_T^T, \dot{\boldsymbol{\alpha}}_T^T, c]^T$, where c is a constant value to fit the constant part of the uncertainty. The uncertainty part can be represented by $f_T(\boldsymbol{z}_T) = G_T + M_T\dot{\boldsymbol{\alpha}}_T$. Using a constant weight matrix $W_T^* \in \mathbb{R}^{l\times 3}$, and an activation function $\boldsymbol{\xi}_T = [\xi_{T1}, \xi_{T2}, \cdots, \xi_{Tl}]^T \in \mathbb{R}^l$, $f_i(\boldsymbol{z}_T)$ can be estimated as

$$f_{iT}(\boldsymbol{z}_T) = (W_T^*)^T\boldsymbol{\xi}_T(\boldsymbol{z}_T) + \epsilon, \tag{4.28}$$

where ϵ is the estimation error with an upper bound ϵ_T, and $l > 1$ is the NN node number. The elements of activation functions are constructed using Guassian functions

$$\xi_T^t(\boldsymbol{z}_T) = \exp\left[\frac{-(\boldsymbol{z}_T - \boldsymbol{\mu}_T(j))^T\Xi_T(\boldsymbol{z}_T - \boldsymbol{\mu}_T(j))}{(\eta_T(j))^2}\right], \tag{4.29}$$

where $j = 1, 2, \cdots, l$, η_T is the width of the Gaussian function and μ_T is the receptive field center [145]. Different from [183, 264], we use a diagonal weight matrix Ξ_T here to balance the order of magnitudes in vector $\boldsymbol{z}_T$. Without Ξ_T, dimensions with higher orders of magnitudes tend to dominate the function $\xi_T^t(\boldsymbol{z}_T)$, whereas dimensions with lower orders of magnitudes are usually ignored.

In (4.28), the weight matrix W_T^* is ideal and unknown. We use its estimation $\hat{W}_T$ instead, and update $\hat{W}_T$ according to the tracking error. The error between the estimated value and the ideal value is $\tilde{W}_T$, $\tilde{W}_T = \hat{W}_T - W_T^*$. Denote $\hat{W}_{T,k}$ the k-th column vector of $\hat{W}_T$.

Since the estimation of attitude angles are used, we have to make an assumption about their boundary before we present the theorem. It can be observed that both $\bar{H}_{IB}$ and H_{IB} are bounded, therefore $H_{IB}\bar{H}_{IB}^{-1} - I$ is bounded. We introduce the unity negative-definite lemma for H_{IB},

Lemma 4.2

$H_{IB}\bar{H}_{IB}^{-1} - I$ has the following properties:

1. $\boldsymbol{x}^T(H_{IB}\bar{H}_{IB}^{-1} - I)\boldsymbol{x} \leqslant 0$ for any $\boldsymbol{x} \in \mathbb{R}^3$, and $\boldsymbol{x}^T(H_{IB}\bar{H}_{IB}^{-1} - I)\boldsymbol{x} = 0$ holds only if $\bar{\boldsymbol{q}}_R = \boldsymbol{q}_R$ or $\boldsymbol{x} = \boldsymbol{0}$.
2. There exist a positive constant ϵ_h such that

$$\boldsymbol{x}^T(H_{IB}\bar{H}_{IB}^{-1} - I)\boldsymbol{y} \leqslant \epsilon_h(\boldsymbol{x}^T\boldsymbol{x} + \boldsymbol{y}^T\boldsymbol{y}). \tag{4.30}$$

■

The proof of the lemma is in Appendix A.

Theorem 4.2

Given that the translational states of the desired formation trajectory are bounded in Ω_{T0}, there exist a positive constant σ_T and a parameter matrix K_{T2} such that the tracking errors for the leaders' positions (as modeled in (4.1) and (4.25)) are semi-globally uniformly ultimately bounded (SGUUB) under the control law

$$\begin{aligned} \boldsymbol{u}_T =& \bar{H}_{IB}^{-1}(-\boldsymbol{e}_{T1} - K_{T2}\boldsymbol{e}_{T2} + \hat{W}_T^T\boldsymbol{\xi}_T(\boldsymbol{z}_{iT})), \\ \dot{\hat{W}}_{T,k} =& -\left(\boldsymbol{\xi}_T(\boldsymbol{z}_i)\boldsymbol{e}_{T2} + \sigma_T\hat{W}_{T,k}\right), \end{aligned} \tag{4.31}$$

if the Assumption 2 holds. The error signals will remain within the compact sets respectively,

$$\begin{aligned} \Omega_{\boldsymbol{e}_{T1}} &= \left\{\boldsymbol{e}_{T1} \in \mathbb{R}^{3\times1} \middle| \|\boldsymbol{e}_{T1}\| \leqslant \sqrt{D_T}\right\}, \\ \Omega_{\boldsymbol{e}_{T2}} &= \left\{\boldsymbol{e}_{T2} \in \mathbb{R}^{3\times1} \middle| \|\boldsymbol{e}_{T2}\| \leqslant \sqrt{D_T/\lambda_{\min}(M_T)}\right\}, \\ \Omega_{\tilde{W}_T} &= \left\{\tilde{W}_T \in \mathbb{R}^{l\times3} \middle| \|\tilde{W}_T\| \leqslant \sqrt{D_T}\right\}, \end{aligned} \tag{4.32}$$

where D_T will be specified later. ■

Proof. Consider the following Lyapunov function,

$$V = \frac{1}{2}\boldsymbol{e}_{T1}^T\boldsymbol{e}_{T1} + \frac{1}{2}\boldsymbol{e}_{T2}^T M_T\boldsymbol{e}_{T2} + \frac{1}{2}\sum_{k=1}^{3}\tilde{W}_{T,k}^T\tilde{W}_{T,k}, \tag{4.33}$$

where $\tilde{W}_{T,k}$ is the k-th column of the matrix $\tilde{W}_T$. According to (4.26), the derivative of V is

$$\begin{aligned}\dot{V} = & -\boldsymbol{e}_{T1}^T K_{T1} \boldsymbol{e}_{T1} + \boldsymbol{e}_{T1}^T (\hat{\boldsymbol{v}}_T - \dot{\boldsymbol{q}}_T) \\ & - \boldsymbol{e}_{T2}^T \left[\boldsymbol{e}_{T1} + H_{IB}\boldsymbol{u}_T - G_T - M_T \dot{\boldsymbol{\alpha}}_T\right] \\ & - \sum_{k=1}^{3} \left[\tilde{W}_{T,k}^T \boldsymbol{\xi}_T(\boldsymbol{z}_i) \boldsymbol{e}_{T2}^T(k) + \sigma_T \tilde{W}_{T,k}^T \hat{W}_{T,k}\right]. \end{aligned} \tag{4.34}$$

Substituting $\boldsymbol{u}_T$ in (4.31) and the NN estimation in (4.28) yields

$$\begin{aligned}\dot{V} \leqslant & -\boldsymbol{e}_{T1}^T K_{T1} \boldsymbol{e}_{T1} + \boldsymbol{e}_{T2}^T \left[\boldsymbol{e}_{T1} + \tilde{W}_T^T \boldsymbol{\xi}_T(\boldsymbol{z}) - \hat{W}_T^T \boldsymbol{\xi}_T(\boldsymbol{z})\right] \\ & + \epsilon_T \|\boldsymbol{e}_{T2}\| - \boldsymbol{e}_{T2}^T H_{IB} \bar{H}_{IB}^{-1} (-\boldsymbol{e}_{T1} - K_{T2}\boldsymbol{e}_{T2} + \hat{W}_T^T \boldsymbol{\xi}_T(\boldsymbol{z}_{iT})) \\ & + 3\beta_1 \|\boldsymbol{e}_{T1}\| - \sum_{k=1}^{3} \left[\tilde{W}_{T,k}^T \boldsymbol{\xi}_T(\boldsymbol{z}_i) \boldsymbol{e}_{T2}^T(k) + \sigma_T \tilde{W}_{T,k}^T \hat{W}_{T,k}\right] \\ \leqslant & -\boldsymbol{e}_{T1}^T K_{T1} \boldsymbol{e}_{T1} - \boldsymbol{e}_{T2}^T K_{T2} \boldsymbol{e}_{T2} + \epsilon_T \|\boldsymbol{e}_{T2}\| + 3\beta_1 \|\boldsymbol{e}_{T1}\| \\ & - \boldsymbol{e}_{T2}^T (H_{IB} \bar{H}_{IB}^{-1} - I)(-\boldsymbol{e}_{T1} - K_{T2}\boldsymbol{e}_{T2} + \hat{W}_T^T \boldsymbol{\xi}_T(\boldsymbol{z}_{iT})) \\ & - \sum_{k=1}^{3} \sigma_T \tilde{W}_{T,k}^T \hat{W}_{T,k}, \end{aligned} \tag{4.35}$$

in which

$$\begin{aligned} & -\sigma_T \tilde{W}_{T,k}^T \hat{W}_{T,k} \\ = & -\sigma_T \tilde{W}_{T,k}^T (\tilde{W}_{T,k} + W_T^*(k)) \\ \leqslant & -\sigma_T \|\tilde{W}_{T,k}\|^2 + \frac{\sigma_T}{2} \|\tilde{W}_{T,k}\|^2 + \frac{\sigma_T}{2} \|W_T^*(k)\|^2 \\ \leqslant & \frac{\sigma_T}{2} \left(\|W_{T,k}^*\|^2 - \|\tilde{W}_{T,k}\|^2\right). \end{aligned} \tag{4.36}$$

According to Lemma 4.2,

$$\begin{aligned}\dot{V} \leqslant & -\boldsymbol{e}_{T1}^T K_{T1} \boldsymbol{e}_{T1} - \boldsymbol{e}_{T2}^T K_{T2} \boldsymbol{e}_{T2} + 3\beta_1 \|\boldsymbol{e}_{T1}\| \\ & + \epsilon_h \left(3\boldsymbol{e}_{T2}^T \boldsymbol{e}_{T2} + \boldsymbol{e}_{T1}^T \boldsymbol{e}_{T1} + \|\tilde{W}_T^T \boldsymbol{\xi}_T(\boldsymbol{z}_{iT}))\|^2\right) \\ & + \sum_{k=1}^{3} \frac{\sigma_T}{2} \left(\|W_{T,k}^*\|^2 - \|\tilde{W}_{T,k}\|^2\right) + \epsilon_T \|\boldsymbol{e}_{T2}\| \\ & + \epsilon_h \|(W_T^*)^T \boldsymbol{\xi}_T(\boldsymbol{z}_{iT})\|^2. \end{aligned} \tag{4.37}$$

Since $\boldsymbol{\xi}_T(\boldsymbol{z}_{iT})$ is a Guassian function and stays in the range $(0,1)$, for both W_T^* and $\tilde{W}_T^T$,

$$\begin{aligned}
\|W_T^T\boldsymbol{\xi}_T(\boldsymbol{z}_{iT})\|^2 &\leqslant \sum_{m=1}^{3}\sum_{n=1}^{3}\sum_{k=1}^{3} W_T(m,k)W_T(n,k) \\
&\leqslant \sum_{m=1}^{3}\sum_{n=1}^{3}\sum_{k=1}^{3} \frac{W_T^2(m,k)+W_T^2(n,k)}{2} \\
&\leqslant \sum_{n=1}^{3}\sum_{k=1}^{3} 3W_T^2(n,k) \\
&\leqslant 3\sum_{k=1}^{3}\|W_{T,k}\|^2, \qquad (4.38)
\end{aligned}$$

where $W_T(m,k)$ is the element in the m-th row and k-th column of matrix W_T.

Then the derivative of the Lyapunov function becomes

$$\begin{aligned}
\dot{V} \leqslant& -\boldsymbol{e}_{T1}^T(K_{T1}-\epsilon_h)\boldsymbol{e}_{T1} - \boldsymbol{e}_{T2}^T(K_{T2}-3\epsilon_h)\boldsymbol{e}_{T2} \\
&+\epsilon_T\|\boldsymbol{e}_{T2}\| + 3\beta_1\|\boldsymbol{e}_{T1}\| \\
&+\sum_{k=1}^{3}\frac{\sigma_T+6\epsilon_h}{2}\|W_T^*(k)\|^2 - \sum_{k=1}^{3}\frac{\sigma_T-6\epsilon_h}{2}\|\tilde{W}_{T,k}\|^2 \\
\leqslant& -\rho_T V + \varrho_T\sqrt{V} + c_T, \qquad (4.39)
\end{aligned}$$

where ρ_T, ϱ_T, and c_T are three constants,

$$\begin{aligned}
\rho_T =& \min\left(2\lambda_{\min}(K_{T1}-\epsilon_h), 2\lambda_{\min}(K_{T2}-3\epsilon_h)M_T^{-1}\right., \\
&\left.\sigma_T - 6\epsilon_h\right), \\
\varrho_T =& \max\left(3\sqrt{2}\beta_1, \epsilon_T\sqrt{2M_T^{-1}}\right), \\
c_T =& \sum_{k=1}^{3}\frac{\sigma_T+6\epsilon_h}{2}\|W_{T,k}^*\|^2. \qquad (4.40)
\end{aligned}$$

Notice that

$$\dot{V} \leqslant \begin{cases} -(\rho_T-\varrho_T)V + c_T, & \text{if } V > 1, \\ -\rho_T V + (\varrho_T + c_T), & \text{if } V \leqslant 1, \end{cases} \qquad (4.41)$$

$\dot{V}$ can be further represented as

$$\dot{V} \leqslant -\rho_T' V + c_T', \qquad (4.42)$$

where

$$\begin{aligned}
\rho_T' &= \rho_T - \varrho_T, \\
c_T' &= c_T + \varrho_T.
\end{aligned} \qquad (4.43)$$

Design control gain matrices K_{T1}, K_{T2} and σ_T such that

$$\begin{aligned}\lambda_{\min}(K_{T1}) &> \epsilon_h + 0.5\varrho_T, \\ \lambda_{\min}(K_{T2}) &> 3\epsilon_h + 0.5\varrho_T\lambda_{\max}(M_T), \\ \sigma_T &> 6\epsilon_h + \varrho_T,\end{aligned} \tag{4.44}$$

$\rho'_T > 0$ can be assured. Then we have

$$\begin{aligned}V(t) &\leqslant V(0)\exp(-\rho'_T t) + \frac{c'_T}{\rho'_T}(1-\exp(-\rho'_T t)) \\ &\leqslant V(0) + \frac{c'_T}{\rho'_T} = 2D_T,\end{aligned} \tag{4.45}$$

therefore the tracking errors $\boldsymbol{e}_{T1}$, $\boldsymbol{e}_{T2}$, and $\tilde{W}_T$ are SGUUB as in (4.32). □

4.3.3 ATTITUDE ESTIMATION

For FWVs, gyroscopes are used to measure angular velocities, whereas attitude angles cannot be measured directly by onboard sensors. They are always achieved by integrating the angular velocities, and the integration process leads to error accumulation. To address this problem, we design an estimation algorithm for attitude angles and analyze the stability of the algorithm.

Based on the observation that the rotational states and the translational states are coupled by coordinate transform, we estimate attitude angles according to the variation of the translational states.

Substitute (4.31) to (4.26) yields

$$\begin{aligned}&M_T\dot{\boldsymbol{e}}_{T2} + M_T\dot{\boldsymbol{\alpha}}_T + G_T \\ =&M_T\dot{\boldsymbol{e}}_{T,2} + (W_T^*)^T\boldsymbol{\xi}_T + \epsilon \\ =&H_{IB}\bar{H}_{IB}^{-1}\left[-\boldsymbol{e}_{T1} - K_{T2}\boldsymbol{e}_{T2} + \hat{W}_T^T\boldsymbol{\xi}_T\right].\end{aligned} \tag{4.46}$$

Let $\gamma_1 = M_T\dot{\boldsymbol{e}}_{T2} + \hat{W}_T^T\boldsymbol{\xi}_T$, $\gamma_2 = -\boldsymbol{e}_{T1} - K_{T2}\boldsymbol{e}_{T2} + \hat{W}_T^T\boldsymbol{\xi}_T$, and $\eta = \epsilon - \tilde{W}_T^T\boldsymbol{\xi}_T$, we have

$$\gamma_1 + \eta = H_{IB}\bar{H}_{IB}^{-1}\gamma_2. \tag{4.47}$$

Denote $\tilde{\boldsymbol{q}}_R = \boldsymbol{q}_R - \bar{\boldsymbol{q}}_R$ as the difference between $\boldsymbol{q}_R$ and its estimated value. The second-order approximation to an arbitrary function $f(\boldsymbol{q}_R)$ around $\tilde{\boldsymbol{q}}_R$ is

$$(\boldsymbol{q}_R) = f(\tilde{\boldsymbol{q}}_R) + \tilde{\boldsymbol{q}}_R^T\nabla f(\boldsymbol{q}_R) + 0.5\tilde{\boldsymbol{q}}_R^T\mathcal{H}(f)\tilde{\boldsymbol{q}}_R + O(f(\boldsymbol{q}_R)), \tag{4.48}$$

where $\nabla f(\boldsymbol{q}_R)$ is the gradient matrix, $\mathcal{H}(f)$ is the Hessian matrix at $\bar{\boldsymbol{q}}_R$, and $O(f(\boldsymbol{q}_R))$ contains the higher order terms.

Approximate each element in H_{IB} based on $\bar{H}_{IB}$ as in (4.48), and substitute the approximation to (4.46),

$$\begin{aligned}\gamma_1 - \gamma_2 + O(H_{IB}(\boldsymbol{q}_R)) + \eta = &\sum_m \tilde{\boldsymbol{q}}_R(m)\nabla_{\boldsymbol{q}_R(m)}\bar{H}_{IB}(\boldsymbol{q}_R)\gamma_2 \\ &+ \frac{1}{2}\sum_m\sum_n \tilde{\boldsymbol{q}}_R(m)\tilde{\boldsymbol{q}}_R(n)\mathcal{H}_{m,n}(\boldsymbol{q}_R)\gamma_2. \end{aligned} \tag{4.49}$$

In the above equation,

$$\begin{aligned}\nabla_{\boldsymbol{q}_R(m)}\bar{H}_{IB}(\boldsymbol{q}_R) &= \frac{\partial \bar{H}_{IB}(\boldsymbol{q}_R)}{\partial \boldsymbol{q}_R(m)}, \\ \mathcal{H}_{m,n}(\boldsymbol{q}_R) &= \frac{\partial^2 \bar{H}_{IB}(\boldsymbol{q}_R)}{\partial \boldsymbol{q}_R(m)\partial \boldsymbol{q}_R(n)} = \frac{\partial^2 \bar{H}_{IB}(\boldsymbol{q}_R)}{\partial \boldsymbol{q}_R(n)\partial \boldsymbol{q}_R(m)}.\end{aligned} \tag{4.50}$$

Property 2. *$\nabla_{\boldsymbol{q}_R(m)}\bar{H}_{IB}(\boldsymbol{q}_R)$, $\mathcal{H}_{m,n}(\boldsymbol{q}_R)$, and $O(H_{IB})$ are all bounded.*

Remark 16. *According to (4.27), all the elements in $\bar{H}_{IB}(\boldsymbol{q}_R)$ are the sums of products of sine and cosine functions. Any order of the partial differential term of $\bar{H}_{IB}(\boldsymbol{q}_R)$ is also a combination of sine and cosine functions. Therefore we can easily get Property 2.*

Ignore the high-order terms $O(H_{IB}(\boldsymbol{q}_R))$ and the small value η, we can approximate the estimation error $\tilde{\boldsymbol{q}}_R$ by solving the following equation,

$$\begin{aligned}\gamma_1 - \gamma_2 = &\sum_m \bar{\tilde{\boldsymbol{q}}}_R(m)\nabla_{\boldsymbol{q}_R(m)}\bar{H}_{IB}(\boldsymbol{q}_R)\gamma_2 \\ &+ \frac{1}{2}\sum_m\sum_n \bar{\tilde{\boldsymbol{q}}}_R(m)\bar{\tilde{\boldsymbol{q}}}_R(n)\mathcal{H}_{m,n}(\boldsymbol{q}_R)\gamma_2.\end{aligned} \tag{4.51}$$

Denote $\bar{\tilde{\boldsymbol{q}}}_R$ the solution of (4.51), and $\tilde{\tilde{\boldsymbol{q}}}_R = \tilde{\boldsymbol{q}}_R - \bar{\tilde{\boldsymbol{q}}}_R$ the difference between the estimation error and its calculated solution. The following lemma can be achieved.

Lemma 4.3

There exist a positive constant $\epsilon_{\boldsymbol{q}}$ such that

$$\|\tilde{\tilde{\boldsymbol{q}}}_R\| \leqslant \epsilon_{\boldsymbol{q}}. \tag{4.52}$$

■

Proof. Substituting $\tilde{\boldsymbol{q}}_R = \tilde{\tilde{\boldsymbol{q}}}_R + \bar{\tilde{\boldsymbol{q}}}_R$ to (4.51) yields

$$\begin{aligned}
&\gamma_1 - \gamma_2 + O(H_{IB}(\boldsymbol{q}_R)) + \eta \\
=&\sum_m \bar{\tilde{\boldsymbol{q}}}_R(m) \nabla_{\boldsymbol{q}_R(m)} \bar{H}_{IB}(\boldsymbol{q}_R)\gamma_2 \\
&+\frac{1}{2}\sum_m \sum_n \bar{\tilde{\boldsymbol{q}}}_R(m)\bar{\tilde{\boldsymbol{q}}}_R(n)\mathcal{H}_{m,n}(\boldsymbol{q}_R)\gamma_2 \\
&+\sum_m \tilde{\tilde{\boldsymbol{q}}}_R(m) \nabla_{\boldsymbol{q}_R(m)} \bar{H}_{IB}(\boldsymbol{q}_R)\gamma_2 \\
&+\sum_m \sum_n (\frac{1}{2}\tilde{\tilde{\boldsymbol{q}}}_R^2(m) + \tilde{\tilde{\boldsymbol{q}}}_R(m)\bar{\tilde{\boldsymbol{q}}}_R(n))\mathcal{H}_{m,n}(\boldsymbol{q}_R)\gamma_2 \\
=&\gamma_1 - \gamma_2 + \sum_m \tilde{\tilde{\boldsymbol{q}}}_R(m) \nabla_{\boldsymbol{q}_R(m)} \bar{H}_{IB}(\boldsymbol{q}_R)\gamma_2 \\
&+\sum_m \sum_n (\frac{1}{2}\tilde{\tilde{\boldsymbol{q}}}_R^2(m) + \tilde{\tilde{\boldsymbol{q}}}_R(m)\bar{\tilde{\boldsymbol{q}}}_R(n))\mathcal{H}_{m,n}(\boldsymbol{q}_R)\gamma_2.
\end{aligned} \tag{4.53}$$

Therefore

$$\begin{aligned}
O(H_{IB}) + \eta =& \sum_m \tilde{\tilde{\boldsymbol{q}}}_R(m) \\
&\cdot \left[\nabla_{\boldsymbol{q}_R(m)} \bar{H}_{IB} + \sum_n \left(\frac{1}{2}\tilde{\tilde{\boldsymbol{q}}}_R(m) + \bar{\tilde{\boldsymbol{q}}}_R(n)\right)\mathcal{H}_{m,n}\right]\gamma_2,
\end{aligned} \tag{4.54}$$

where η and γ_2 are bounded according to Theorem 4.2, $\nabla_{\boldsymbol{q}_R(m)} H_{IB}$, $\mathcal{H}_{m,n}$, and $O(H_{IB})$ are bounded according to Property 2.

Suppose that there does not exist a positive constant $\epsilon_{\boldsymbol{q}}$ which satisfies (4.52), there will be a dimension m such that $\tilde{\tilde{\boldsymbol{q}}}_R(m)$ goes to infinity, which contradicts the bounding property of $O(H_{IB}) + \eta$ in (4.54). Therefore Lemma 4.3 holds. □

Remark 17. *In simulations, we choose only the feasible solution of (4.51) as the calculated estimation error. Feasible here means that the solution satisfies the physical constrains of attitude angles.*

Theorem 4.3

Under Property 2 and Lemma 4.3, the estimation error of attitude angles are SGUUB under the estimation law

$$\dot{\hat{\boldsymbol{q}}}_R = \bar{\boldsymbol{v}}_R + \sigma_{\boldsymbol{v}} \bar{\tilde{\boldsymbol{q}}}_R, \tag{4.55}$$

if Assumption 5 holds, the estimation error will remain within the compact set

$$\Omega_{\tilde{\boldsymbol{q}}_R} = \left\{\tilde{\boldsymbol{q}}_R \in \mathbb{R}^{3\times 1} \middle| \|\tilde{\boldsymbol{q}}_R\| \leqslant \epsilon_{\boldsymbol{v}}/\sigma_{\boldsymbol{v}} + \epsilon_{\boldsymbol{q}}\right\}. \tag{4.56}$$

■

Proof. Construct a Lyapnov function with respect to the estimation error $\tilde{\boldsymbol{q}}_R$,

$$V = \frac{1}{2}\tilde{\boldsymbol{q}}_R^T\tilde{\boldsymbol{q}}_R. \tag{4.57}$$

Substituting (4.55) into the derivative of V yields

$$\begin{aligned} \dot{V} =& \tilde{\boldsymbol{q}}_R^T(\dot{\boldsymbol{q}}_R - \dot{\hat{\boldsymbol{q}}}_R) \\ =& \tilde{\boldsymbol{q}}_R^T(\boldsymbol{v}_R - \bar{\boldsymbol{v}}_R - \sigma_{\boldsymbol{v}}\tilde{\tilde{\boldsymbol{q}}}_R) \\ =& \tilde{\boldsymbol{q}}_R^T\left[\tilde{\boldsymbol{v}}_R - \sigma_{\boldsymbol{v}}(\tilde{\boldsymbol{q}}_R - \tilde{\tilde{\boldsymbol{q}}}_R)\right] \\ \leqslant& -\sigma_{\boldsymbol{v}}\tilde{\boldsymbol{q}}_R^T\tilde{\boldsymbol{q}}_R + \|\tilde{\boldsymbol{q}}_R\|(\|\tilde{\boldsymbol{v}}_R\| + \sigma_{\boldsymbol{v}}\|\tilde{\tilde{\boldsymbol{q}}}_R\|). \end{aligned} \tag{4.58}$$

Applying Lemma 4.3 and Assumption 5,

$$\begin{aligned} \dot{V} \leqslant& -\left(\sigma_{\boldsymbol{v}}\|\tilde{\boldsymbol{q}}_R\| - (\epsilon_{\boldsymbol{v}} + \sigma_{\boldsymbol{v}}\epsilon_{\boldsymbol{q}})\right)\sqrt{2V} \\ \leqslant& -f(\boldsymbol{q}_R)\sqrt{2V}. \end{aligned} \tag{4.59}$$

If $\|\tilde{\boldsymbol{q}}_R\| > \epsilon_{\boldsymbol{v}}/\sigma_{\boldsymbol{v}} + \epsilon_{\boldsymbol{v}}$, $f(\boldsymbol{q}_R) > 0$ and $\dot{V} < 0$, the estimation error $\|\tilde{\boldsymbol{q}}_R\|$ keeps decreasing until it goes into the region as in (4.56). Hence, it is straight forward to present that the estimation error $\|\tilde{\boldsymbol{q}}_R\|$ is SGUUB and will finally remain in the region $\Omega_{\tilde{\boldsymbol{q}}_R}$, which completes the proof. □

Remark 18. *According to the compact set (4.56) and Theorem 4.3, the estimation error of the attitude angle is very related to the measuring accuracy of the angular velocity* $\tilde{\boldsymbol{v}}_R$ *and the calculation error* $\tilde{\tilde{\boldsymbol{q}}}_R$. *We can achieve the estimation of attitude angles with higher accuracy by improving the measuring accuracy of* $\bar{\boldsymbol{v}}_R$ *and calculating accuracy of* $\tilde{\boldsymbol{q}}_R$.

4.3.4 ROTATIONAL MOTION CONTROL

Given the desired formation attitude angles and angular velocities as in (4.12), the rotational control algorithm can be designed. Consider the rotational motion equation in (4.6), the second order dynamics can be obtained as

$$\ddot{\boldsymbol{q}}_R = M_{iR}^{-1}(\boldsymbol{u}_R - C_R\dot{\boldsymbol{q}}_R). \tag{4.60}$$

Although attitude angles are achieved by estimation and angular velocities are measured inaccurately, the stability of the system is analyzed using real states,

$$\begin{aligned} \boldsymbol{e}_{R1} &= \boldsymbol{q}_R - \hat{\boldsymbol{q}}_R, \\ \boldsymbol{e}_{R2} &= \boldsymbol{v}_R - \boldsymbol{\alpha}_R. \end{aligned} \tag{4.61}$$

Estimated error signals are also calculated,

$$\begin{aligned}\bar{\boldsymbol{e}}_{R1} &= \bar{\boldsymbol{q}}_R - \hat{\boldsymbol{q}}_R = \boldsymbol{e}_{R1} + \bar{\boldsymbol{q}}_R - \boldsymbol{q}_R,\\ \bar{\boldsymbol{e}}_{R2} &= \bar{\boldsymbol{v}}_R - \bar{\boldsymbol{\alpha}}_R = \boldsymbol{e}_{R2} - \tilde{\boldsymbol{v}}_R + K_{R2}(\boldsymbol{q}_R - \bar{\boldsymbol{q}}_R),\end{aligned} \tag{4.62}$$

where $\boldsymbol{\alpha}_R = \hat{\boldsymbol{v}}_R - A_{R1}$, $A_{R1} = K_{R1}\boldsymbol{e}_{R1}$, $\bar{\boldsymbol{\alpha}}_R = \hat{\boldsymbol{v}}_R - \bar{A}_{R1}$, and $\bar{A}_{R1} = K_{R1}\bar{\boldsymbol{e}}_{R1}$.

Substituting $\dot{\boldsymbol{q}}_R = \boldsymbol{v}_R$ and differentiating (4.61) with respect to time yields

$$\begin{aligned}\dot{\boldsymbol{e}}_{R1} &= \dot{\boldsymbol{q}}_R - \dot{\hat{\boldsymbol{q}}}_R = \boldsymbol{e}_{R2} - A_{R1} + \hat{\boldsymbol{v}}_R - \dot{\hat{\boldsymbol{q}}}_R,\\ \dot{\boldsymbol{e}}_{R2} &= \dot{\boldsymbol{q}}_R - \dot{\boldsymbol{\alpha}}_R = M_R^{-1}(\boldsymbol{u}_R - C_R\dot{\boldsymbol{q}}_R) - \dot{\boldsymbol{\alpha}}_R.\end{aligned} \tag{4.63}$$

The NN technique is also used in the rotational motion to compensate for model uncertainties, estimation error, and measurement inaccuracy. Let $\boldsymbol{z}_R = [\bar{\boldsymbol{q}}_R^T, \boldsymbol{v}_R^T, \boldsymbol{\alpha}_R^T, \dot{\boldsymbol{\alpha}}_R^T]^T$, the NN function is constructed as

$$\begin{aligned}f_R(\boldsymbol{z}_R) =& C_R\dot{\boldsymbol{q}}_R + M_R\dot{\boldsymbol{\alpha}}_R\\ &- (1 + K_{R2}^2)(\boldsymbol{q}_R - \bar{\boldsymbol{q}}_R) - K_{R2}\tilde{\boldsymbol{v}}_R.\end{aligned} \tag{4.64}$$

Equation (4.64) can also be described using a constant weight matrix $W_R^* \in \mathbb{R}^{l\times 3}$ and an activation function $\boldsymbol{\xi}_R = [\xi_{R1}, \xi_{R2}, \cdots, \xi_{Rl}]^T \in \mathbb{R}^l$,

$$f_R(\boldsymbol{z}_T) = (W_R^*)^T\boldsymbol{\xi}_R(\boldsymbol{z}_R) + \epsilon, \tag{4.65}$$

where ϵ has an upper bound ϵ_R. The weight matrix W_R^* is ideal and unknown. We use its estimation $\hat{W}_R$ instead, and update $\hat{W}_R$ according to the tracking error. The error between the estimated value and the ideal value is $\tilde{W}_R$, $\tilde{W}_R = \hat{W}_R - W_R^*$. Denote $\boldsymbol{e}_{R2}(k)$ and $\hat{W}_{R,k}$ the k-th element and the k-th column vector of $\boldsymbol{e}_{R2}$ and $\hat{W}_R$, respectively.

Theorem 4.4

Given that the rotational states of the desired trajectory are bounded in Ω_{R0}, the tracking errors for the leaders' angles (as modeled in (4.6) and (4.61)) are SGUUB under the control law

$$\boldsymbol{u}_R = -(\bar{\boldsymbol{e}}_{R1} + K_{R2}\bar{\boldsymbol{e}}_{R2} - \hat{W}_R^T\boldsymbol{\xi}_R(\boldsymbol{z}_R)), \tag{4.66}$$

$$\dot{\hat{W}}_{R,k} = -\left(\boldsymbol{\xi}_R(\boldsymbol{z})\boldsymbol{e}_{R2}(k) + \sigma_R\hat{W}_{R,k}\right). \tag{4.67}$$

The error signals will remain within the compact sets respectively,

$$\begin{aligned}\Omega_{e_{R1}} &= \left\{\boldsymbol{e}_{R1} \in \mathbb{R}^{3\times 1} | \|\boldsymbol{e}_{R1}\| \leqslant \sqrt{D_R}\right\},\\ \Omega_{e_{R2}} &= \left\{\boldsymbol{e}_{R2} \in \mathbb{R}^{3\times 1} | \|\boldsymbol{e}_{R2}\| \leqslant \sqrt{D_R/\gamma_{\min}(M_R)}\right\},\\ \Omega_{\tilde{W}_R} &= \left\{\tilde{W}_R \in \mathbb{R}^{l\times 3} | \|\tilde{W}_R\| \leqslant \sqrt{D_R}\right\},\end{aligned} \tag{4.68}$$

where D_R will be detailed later. ■

Proof. Consider the following Lyapunov function,

$$V = \frac{1}{2}\boldsymbol{e}_{R1}^T\boldsymbol{e}_{R1} + \frac{1}{2}\boldsymbol{e}_{R2}^T M_R \boldsymbol{e}_{R2} + \frac{1}{2}\sum_{k=1}^{3}\tilde{W}_R^T\tilde{W}_R. \tag{4.69}$$

According to (4.61), (4.62), and (4.66), the derivative of V is

$$\begin{aligned}
\dot{V} =& \boldsymbol{e}_{R1}^T\left(\boldsymbol{e}_{R2} - K_{R1}\boldsymbol{e}_{R1} + \hat{\boldsymbol{v}}_R - \dot{\hat{\boldsymbol{q}}}_R\right) \\
&+ \boldsymbol{e}_{R2}^T\left[-\bar{\boldsymbol{e}}_{R1} - K_{R2}\bar{\boldsymbol{e}}_{R2} + \hat{W}_R^T\boldsymbol{\xi}_R(\boldsymbol{z}_R) - C_R\dot{\boldsymbol{q}}_R - M_R\dot{\boldsymbol{\alpha}}_R\right] \\
&- \sum_{k=1}^{3}\left[\tilde{W}_{R,k}^T\boldsymbol{\xi}_R(\boldsymbol{z})\boldsymbol{e}_{R2}(k) + \sigma_R\tilde{W}_{R,k}^T\hat{W}_{R,k}\right] \\
\leqslant& -\boldsymbol{e}_{R1}^T K_{R1}\boldsymbol{e}_{R1} - \boldsymbol{e}_{R2}^T K_{R2}\boldsymbol{e}_{R2} + 3\beta_1\|\boldsymbol{e}_{R1}\| \\
&+ \boldsymbol{e}_{R2}^T\left[(1+K_{R2}^2)(\boldsymbol{q}_R - \bar{\boldsymbol{q}}_R) + K_{R2}\tilde{\boldsymbol{v}}_R\right] \\
&+ \boldsymbol{e}_{R2}^T\left[\hat{W}_R^T\boldsymbol{\xi}_R(\boldsymbol{z}_R) - C_R\dot{\boldsymbol{q}}_R - M_R\dot{\boldsymbol{\alpha}}_R\right] \\
&- \sum_{k=1}^{3}\left[\tilde{W}_{R,k}^T\boldsymbol{\xi}_R(\boldsymbol{z})\boldsymbol{e}_{R2}(k) + \sigma_R\tilde{W}_{R,k}^T\hat{W}_{R,k}\right]
\end{aligned} \tag{4.70}$$

Substituting (4.64) and (4.31) into the above equation yields

$$\begin{aligned}
\dot{V} \leqslant& -\boldsymbol{e}_{R1}^T K_{R1}\boldsymbol{e}_{R1} - \boldsymbol{e}_{R2}^T K_{R2}\boldsymbol{e}_{R2} + 3\beta_1\|\boldsymbol{e}_{R1}\| + \epsilon_R\|\boldsymbol{e}_{R2}\| \\
&- \sum_{k=1}^{3}\sigma_R\tilde{W}_{R,k}^T\hat{W}_{R,k},
\end{aligned} \tag{4.71}$$

where

$$-\tilde{W}_{R,k}^T\hat{W}_{T,k} \leqslant \frac{1}{2}\left(\|W_R^*(k)\|^2 - \|\tilde{W}_{R,k}\|^2\right). \tag{4.72}$$

Therefore

$$\begin{aligned}
\dot{V} =& -\boldsymbol{e}_{R1}^T K_{R1}\boldsymbol{e}_{R1} - \boldsymbol{e}_{R2}^T K_{R2}\boldsymbol{e}_{R2} - \frac{\sigma_R}{2}\sum_{k=1}^{3}\|\tilde{W}_{R,k}^T\|^2 \\
&+ 3\beta_1\|\boldsymbol{e}_{R1}\| + \epsilon_R\|\boldsymbol{e}_{R2}\| + \frac{\sigma_R}{2}\sum_{k=1}^{3}\|W_R^*(k)\|^2 \\
\leqslant& -\rho_R V + \varrho_R\sqrt{V} + c_R,
\end{aligned} \tag{4.73}$$

where ρ_{iR} and c_{iR} are two constants,

$$\rho_R = \min\left(2\lambda_{\min}(K_{R1}), 2\lambda_{\min}(K_{R2}), \sigma_R\right), \tag{4.74}$$

$$\begin{aligned}
\varrho_R &= \max\left(\sqrt{2M_R^{-1}\epsilon_R}, 3\sqrt{2}\beta_1\right), \\
c_R &= \sum_{k=1}^{3}\frac{\sigma_R}{2}\|W_R^*(k)\|^2.
\end{aligned} \tag{4.75}$$

Notice that

$$\dot{V} \leqslant \begin{cases} -(\rho_R - \varrho_R)V + c_R, & \text{if } V > 1, \\ -\rho_R V + (\varrho_R + c_R), & \text{if } V \leqslant 1, \end{cases} \tag{4.76}$$

$\dot{V}$ can be further represented as

$$\dot{V} \leqslant -\rho_R' V + c_R', \tag{4.77}$$

where

$$\begin{aligned} \rho_R' &= \rho_R - \varrho_R, \\ c_R' &= c_R + \varrho_R. \end{aligned} \tag{4.78}$$

Design control gain matrices K_{R1}, K_{R2} and σ_R such that $\lambda_{\min}(K_{R1}) > 0.5\varrho_R$, $\lambda_{\min}(K_{R2}) > 0.5\varrho_R$, and $\sigma_T > \varrho_R$, $\rho_T' > 0$ is always assured. Then,

$$\begin{aligned} V(t) &\leqslant V(0)\exp(-\rho_R' t) + \frac{c_R'}{\rho_R'}(1 - \exp(-\rho_R' t)) \\ &\leqslant V(0) + \frac{c_R'}{\rho_R'} = 2D_R. \end{aligned} \tag{4.79}$$

Therefore the tracking error $\boldsymbol{e}_{R1}$, $\boldsymbol{e}_{R2}$, and $\tilde{W}_R$ are SGUUB and remain within the compact sets as in (4.68). □

4.3.5 FORMATION CONTROL FOR THE FOLLOWERS

Assume that the leaders and the followers are isomorphic, the follower model is also presented in equation (4.1), (4.6), and (4.7). The control objective of this part is to keep the followers in a convex hull with respect to the leaders. The followers should move with the leaders.

The following estimation algorithm is used to estimate the desired translational and rotational information of the followers,

$$\begin{aligned} \dot{\hat{\boldsymbol{q}}}_i &= \hat{\boldsymbol{v}}_i - \beta_3 \text{sgn} \sum_{j=N+1}^{N+M} a_{ij}(\hat{\boldsymbol{q}}_i - \hat{\boldsymbol{q}}_j), \\ \dot{\hat{\boldsymbol{v}}}_i &= -\beta_4 \text{sgn} \sum_{j=N+1}^{N+M} a_{ij}(\hat{\boldsymbol{v}}_i - \hat{\boldsymbol{v}}_j). \end{aligned} \tag{4.80}$$

The definition of $\hat{\boldsymbol{q}}_i$ and $\hat{\boldsymbol{v}}_i$ are similar as in previous subsections. Note that $i = N+1, N+2, \cdots, N+M$ for the followers. The precise estimation of the desired follower states can be achieved in limited time.

The translational motion and the rotational control problem for the followers can also be handled separately by following the process in the previous subsections.

Table 4.1
Parameters for simulation

Parameters	Values	Parameters	Values
K_{T1}	I_3	K_{R1}	I_3
K_{T2}	$8I_3$	K_{R2}	$8I_3$
β_1	5	σ_T	4
β_2	2.5	σ_R	4
β_3	5	σ	0.5
β_4	2.5		

4.4 SIMULATION

Simulation is conducted to demonstrate the effectiveness of the proposed algorithm. The FWV model is formulated in as $m_i = 0.0056$kg,

$$\bar{I}_p = \begin{bmatrix} 5.75 \times 10^{-7} & 0 & 0 \\ 0 & 5.76 \times 10^{-7} & 0 \\ 0 & 0 & 9.97 \times 10^{-7} \end{bmatrix} \text{kg} \cdot \text{m}^2.$$

A group of 8 FWVs is considered for the formation problem, 4 of them are leaders labelled as 1, 2, 3, 4, and the other four are followers labelled as 5, 6, 7, 8. The trajectory of the formation is pre-set manually. The simulation parameters are presented in Table 4.1.

The initial positions of the FWVs are randomly set. The relative motion of the formation is desired to switch at 50s. In the first 50s, the leaders and the followers are required to form a square, respectively. In the last 50s, the shape of the leaders is required to change to a triangular pyramid, whereas the shape of followers is still required to be a square which rotates in the space.

Fig. 4.2 illustrates the 3-D trajectories of the FWVs. Figs. 4.3 and 4.4 show the translational tracking errors of the leaders and the followers, respectively. Figs. 4.5 and 4.6 plot the rotational tracking errors of the FWVs. Other simulation results are not presented here due to the page limitation. It can be observed that all the FWV states converge within $10s$.

As discussed in Remark 15, the desired distance $\boldsymbol{\delta}_{ij}$ can be designed fixed for a constant formation configuration, or smoothly varying for a time-varying configuration. Actually, it can also be designed for a configuration switch. In Figs. 4.3 and 4.4, $\boldsymbol{\delta}_{ij}$ for the rotational motion switches in the half time, therefore the errors of the formation have step changes accordingly. $\boldsymbol{\delta}_{ij}$ is designed smoothly for the translational motion to avoid sudden change, and the error signals are quite stable during the process, as shown in Figs. 4.7 and 4.8.

The estimation error of attitude angles is illustrated in Fig. 4.9. Given that the FWVs take off from the ground, their pitch angles and roll angles are near

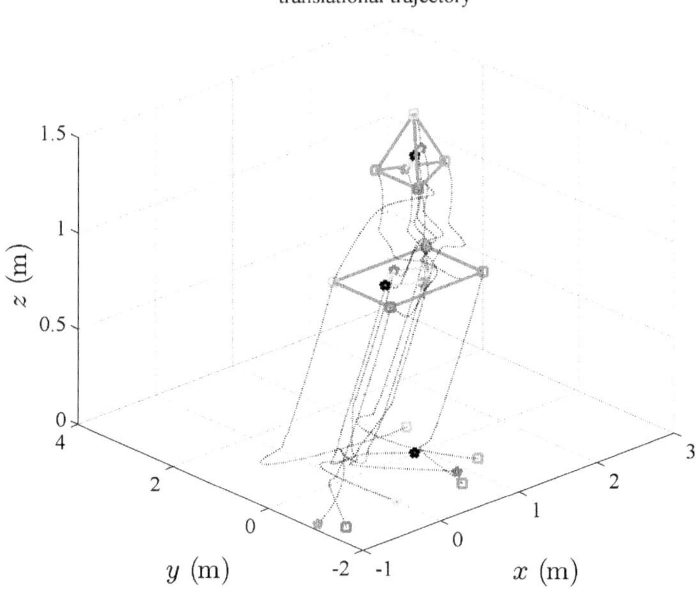

Figure 4.2 Translational trajectories of 8 FWVs with a time-varying formation.

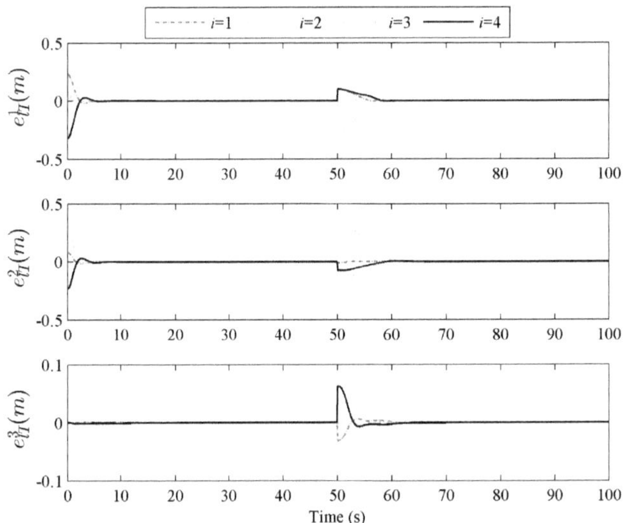

Figure 4.3 The translational errors of the leaders with switched δ_{iT}.

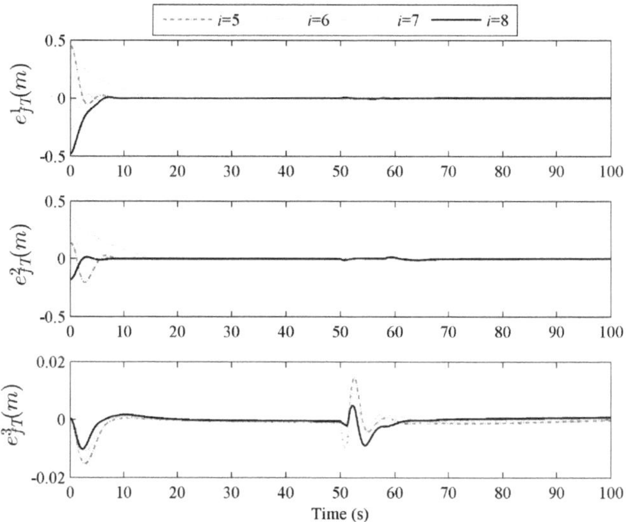

Figure 4.4 The translational errors of the followers with switched δ_{iT}.

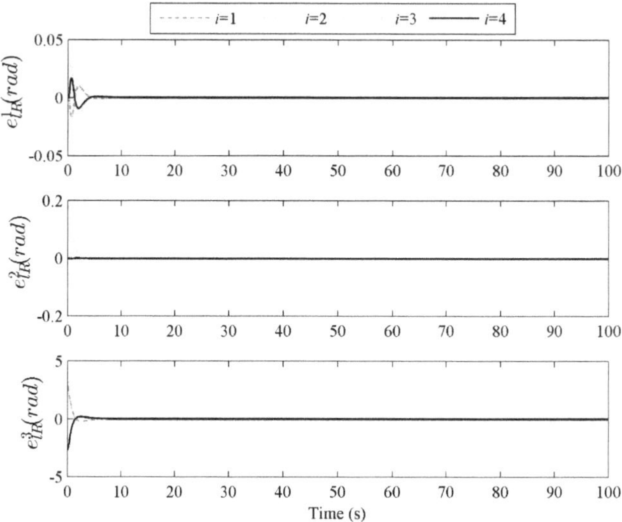

Figure 4.5 The rotational errors of the leaders.

zero. Their heading angles are set to be different initially. During flight, the FWVs estimate their attitude angles according to their translational tracking error. The estimated attitude angles converge to their true value in about $10s$. A small steady-state estimation error exists because the translational tracking

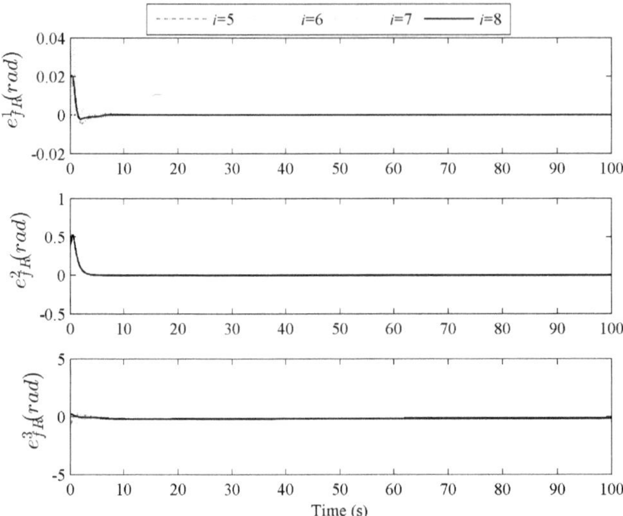

Figure 4.6 The rotational errors of the followers.

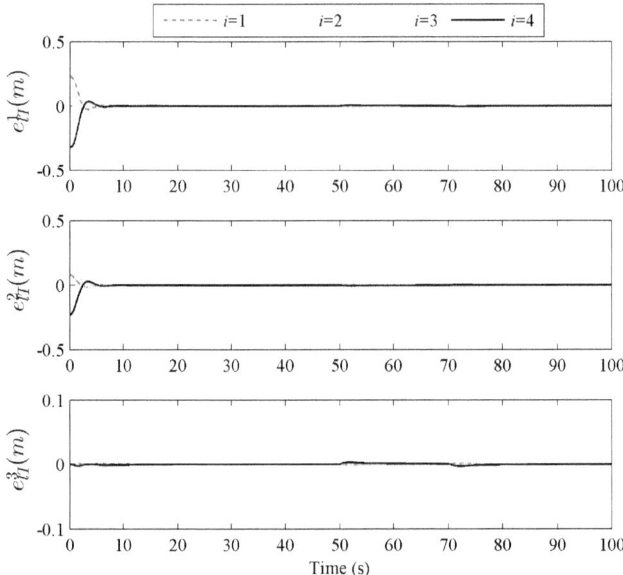

Figure 4.7 The rotational errors of the leaders with time-varying δ_{iT}.

error converges to zero, and no more correction can be made according to the translational tracking error. Similarly, the rotational tracking also has a small steady-state error because of the estimation error. Nevertheless, the estimation error and the rotational tracking error are small enough (about $0.01rad$).

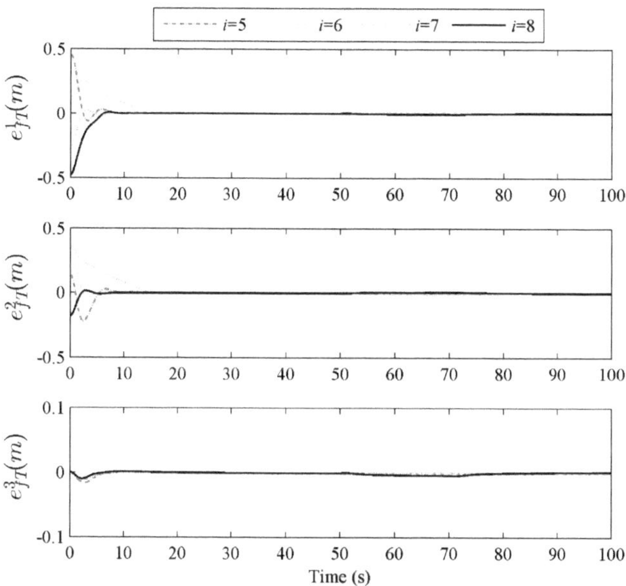

Figure 4.8 The rotational errors of the followers with time-varying δ_{iT}.

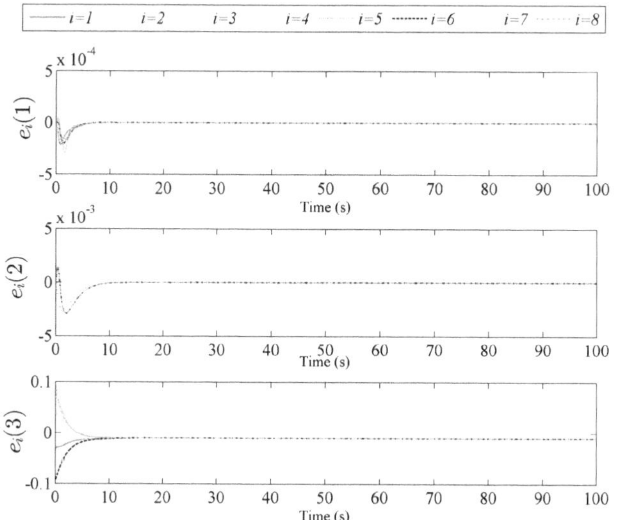

Figure 4.9 The estimation errors of the attitude angles.

4.5 CONCLUSION

In this chapter, an attitude estimation and adaptive neural network based formation control method have been proposed for the flapping-wing vehicle formation control problem. We first estimated the desired formation states according to communication typologies, then designed rotational and translational control algorithms for flapping-wing vehicles based on the neural network technique and estimated attitude angles, and the stability of the control algorithms. Based on the observation that the translational states are coupled with the rotational states, an attitude estimation algorithm is proposed based on the coupling information. The convergence of the estimation algorithm is also analyzed. Finally, we conducted a series of simulations, which demonstrated the effectiveness of the proposed methods.

4.6 APPENDIX

Matrix H_{IB} in (2) can be described by three matrices,

$$\begin{aligned} H_\psi &= \begin{bmatrix} \cos\psi & -\sin\psi & 0 \\ \sin\psi & \cos\psi & 0 \\ 0 & 0 & 1 \end{bmatrix}, \ H_\theta \begin{bmatrix} \cos\theta & 0 & \sin\theta \\ 0 & 1 & 0 \\ -\sin\theta & 0 & \cos\theta \end{bmatrix}, \\ H_\phi &= \begin{bmatrix} 1 & 0 & 0 \\ 0 & \cos\phi & -\sin\phi \\ 0 & \sin\phi & \cos\phi \end{bmatrix}. \end{aligned} \tag{4.81}$$

Similarly,

$$\begin{aligned} \bar{H}_\psi &= \begin{bmatrix} \cos\bar{\psi} & -\sin\bar{\psi} & 0 \\ \sin\bar{\psi} & \cos\bar{\psi} & 0 \\ 0 & 0 & 1 \end{bmatrix}, \ \bar{H}_\theta \begin{bmatrix} \cos\bar{\theta} & 0 & \sin\bar{\theta} \\ 0 & 1 & 0 \\ -\sin\bar{\theta} & 0 & \cos\bar{\theta} \end{bmatrix}, \\ \bar{H}_\phi &= \begin{bmatrix} 1 & 0 & 0 \\ 0 & \cos\bar{\phi} & -\sin\bar{\phi} \\ 0 & \sin\bar{\phi} & \cos\bar{\phi} \end{bmatrix} \end{aligned} \tag{4.82}$$

Denote the estimation error for each attitude angle as

$$\begin{aligned} \Delta\psi &= \bar{\psi} - \psi, \\ \Delta\theta &= \bar{\theta} - \theta, \\ \Delta\phi &= \bar{\phi} - \phi. \end{aligned} \tag{4.83}$$

The difference between (4.81) and (4.82) can be calculated,

$$\begin{aligned}
\Delta H_\phi &= \bar{H}_\phi - H_\phi \\
&= \begin{bmatrix} 0 & 0 & 0 \\ 0 & \cos\bar{\phi} - \cos\phi & -\sin\bar{\phi} + \sin\phi \\ 0 & \sin\bar{\phi} - \sin\phi & \cos\bar{\phi} - \cos\phi \end{bmatrix}, \\
\Delta H_\theta &= \bar{H}_\theta - H_\theta \\
&= \begin{bmatrix} \cos\bar{\theta} - \cos\theta & 0 & \sin\bar{\theta} - \sin\theta \\ 0 & 0 & 0 \\ -sin\bar{\theta} + \sin\theta & 0 & \cos\bar{\theta} - \cos\theta \end{bmatrix}, \\
\Delta H_\psi &= \bar{H}_\theta - H_\theta \\
&= \begin{bmatrix} \cos\bar{\psi} - \cos\psi & -sin\bar{\psi} + \sin\psi & 0 \\ \sin\bar{\psi} - \sin\psi & \cos\bar{\psi} - \cos\psi & 0 \\ 0 & 0 & 0 \end{bmatrix}.
\end{aligned} \tag{4.84}$$

Substitute (4.81) to H_{IB} and (4.82) to $\bar{H}_{IB}$, notice that $\bar{H}_{IB}^{-1} = \bar{H}_{IB}^T$, we have

$$\begin{aligned}
&H_{IB}\bar{H}_{IB}^{-1} - I \\
=&H_\psi H_\theta H_\phi \bar{H}_\phi^T \bar{H}_\theta^T \bar{H}_\phi^T - I \\
=&H_\psi H_\theta (I + H_\phi \Delta H_\phi^T) \bar{H}_\theta^T \bar{H}_\phi^T - I \\
=&H_\psi \Delta H_\psi^T + H_\psi H_\theta \Delta H_\theta^T H_\psi^T + H_\psi H_\theta H_\phi \Delta H_\phi^T H_\theta^T H_\psi^T.
\end{aligned} \tag{4.85}$$

$H_\psi \Delta H_\psi^T$, $H_\theta \Delta H_\theta^T$, and $H_\phi \Delta H_\phi^T$ are as follows,

$$\begin{aligned}
H_\phi \Delta H_\phi^T &= \begin{bmatrix} 0 & 0 & 0 \\ 0 & \cos\Delta\phi - 1 & \sin\Delta\phi \\ 0 & -\sin\Delta\phi & \cos\Delta\phi - 1 \end{bmatrix} \\
&= \Pi_\phi^{\text{diag}} + \Pi_\phi^{\text{ss}}, \\
H_\theta \Delta H_\theta^T &= \begin{bmatrix} \cos\Delta\theta - 1 & 0 & -\sin\Delta\theta \\ 0 & 0 & 0 \\ \sin\Delta\theta & 0 & \cos\Delta\theta - 1 \end{bmatrix} \\
&= \Pi_\theta^{\text{diag}} + \Pi_\theta^{\text{ss}}, \\
H_\psi \Delta H_\psi^T &= \begin{bmatrix} \cos\Delta\psi - 1 & \sin\Delta\psi & 0 \\ -\sin\Delta\psi & \cos\Delta\psi - 1 & 0 \\ 0 & 0 & 0 \end{bmatrix} \\
&= \Pi_\psi^{\text{diag}} + \Pi_\psi^{\text{ss}}.
\end{aligned} \tag{4.86}$$

where $\Pi_\psi^{\text{diag}}, \Pi_\theta^{\text{diag}}, \Pi_\phi^{\text{diag}}$ are the diagnal matirces for ψ, θ and ϕ, respectively, $\Pi_\psi^{\text{ss}}, \Pi_\theta^{\text{ss}}, \Pi_\phi^{\text{ss}}$ are the corresponding skew-symmetric matrices. Then we have

$$\begin{aligned} &\boldsymbol{x}^T(H_{IB}\bar{H}_{IB}^{-1} - I)\boldsymbol{x} \\ =&\boldsymbol{x}^T(\Pi_\psi^{\text{diag}} + H_\psi \Pi_\theta^{\text{diag}} H_\psi^T + H_\psi H_\theta \Pi_\phi^{\text{diag}} H_\theta^T H_\psi^T)\boldsymbol{x} \\ &+ \boldsymbol{x}^T(\Pi_\psi^{\text{ss}} + H_\psi \Pi_\theta^{\text{ss}} H_\psi^T + H_\psi H_\theta \Pi_\phi^{\text{ss}} H_\theta^T H_\psi^T)\boldsymbol{x}. \end{aligned} \tag{4.87}$$

Notice that $H_\psi \Pi_\theta^{\text{ss}} H_\psi^T$ and $H_\psi H_\theta \Pi_\phi^{\text{ss}} H_\theta^T H_\psi^T$ are also skew-symmetric,

$$\boldsymbol{x}^T(\Pi_\psi^{\text{ss}} + H_\psi \Pi_\theta^{\text{ss}} H_\psi^T + H_\psi H_\theta \Pi_\phi^{\text{ss}} H_\theta^T H_\psi^T)\boldsymbol{x} = 0. \tag{4.88}$$

Since $H_\psi H_\psi^T = I$ and $H_\theta H_\theta^T = I$,

$$\begin{aligned} &\boldsymbol{x}^T(\Pi_\psi^{\text{diag}} + H_\psi \Pi_\theta^{\text{diag}} H_\psi^T + H_\psi H_\theta \Pi_\phi^{\text{diag}} H_\theta^T H_\psi^T)\boldsymbol{x} \\ =&\boldsymbol{x}^T \Pi_\psi^{\text{diag}} \boldsymbol{x} + \boldsymbol{x}^T \Pi_\theta^{\text{diag}} \boldsymbol{x} + \boldsymbol{x}^T \Pi_\phi^{\text{diag}} \boldsymbol{x} \\ =&\boldsymbol{x}^T(\Pi_\psi^{\text{diag}} + \Pi_\theta^{\text{diag}} + \Pi_\phi^{\text{diag}})\boldsymbol{x}. \end{aligned} \tag{4.89}$$

The range of cosine function is $[-1, 1]$, and

$$\begin{aligned} \Pi_H^{rmdiag} =&\Pi_\psi^{\text{diag}} + \Pi_\theta^{\text{diag}} + \Pi_\phi^{\text{diag}} \\ =&\text{diag}(\cos\Delta\psi + \cos\Delta\theta - 2, \\ &\cos\Delta\phi + \cos\Delta\psi - 2, \\ &\cos\Delta\phi + \cos\Delta\theta - 2), \end{aligned} \tag{4.90}$$

which has no positive eigenvalue. Combine (4.87), (4.88), and (4.90), we have

$$\begin{aligned} &\boldsymbol{x}^T(H_{IB}\bar{H}_{IB}^{-1} - I)\boldsymbol{x} \\ =&\boldsymbol{x}^T(\Pi_\psi^{\text{diag}} + H_\psi \Pi_\theta^{\text{diag}} H_\psi^T + H_\psi H_\theta \Pi_\phi^{\text{diag}} H_\theta^T H_\psi^T)\boldsymbol{x} \leqslant 0. \end{aligned} \tag{4.91}$$

Therefore $\boldsymbol{x}^T(H_{IB}\bar{H}_{IB}^{-1} - I)\boldsymbol{x} \leqslant 0$ always holds, and $\boldsymbol{x}^T(H_{IB}\bar{H}_{IB}^{-1} - I)\boldsymbol{x} = 0$ exists only if $\boldsymbol{x} = 0$ or

$$\begin{aligned} \cos\Delta\psi + \cos\Delta\theta = 2, \\ \cos\Delta\phi + \cos\Delta\psi = 2, \\ \cos\Delta\phi + \cos\Delta\theta = 2, \end{aligned} \tag{4.92}$$

which means $\Delta\psi = 0$, $\Delta\theta = 0$, and $\Delta\phi = 0$, namely $\bar{\boldsymbol{q}}_R = \boldsymbol{q}_R$. Thus completes the proof of the first item.

The second item is quite straght forward. As analyzed above, the matrix $H_{IB}\bar{H}_{IB}^{-1} - I$ can be written as the sum of a diagnal matrix Π_H^{diag} and a skew-symmetric matrix Π_H^{ss},

$$H_{IB}\bar{H}_{IB}^{-1} - I = \Pi_H^{\text{diag}} + \Pi_H^{\text{ss}}, \tag{4.93}$$

where Π_H^{diag} is presented in (4.90) and $\lambda_{\max}(\Pi_H^{\text{diag}}) = 0$. Then,

$$\boldsymbol{x}^T \Pi_H^{\text{diag}} \boldsymbol{y} \leqslant \lambda_{\max}(\Pi_H^{\text{diag}}) \boldsymbol{x}^T \boldsymbol{y} = 0. \tag{4.94}$$

For any matrix skew-symmetric matrix $P \in \mathbb{R}^{N \times N}$, vector $\boldsymbol{x}$ and $\boldsymbol{y}$,

$$\begin{aligned} \boldsymbol{x}^T P \boldsymbol{y} &= \sum_{i=1}^{N} \sum_{j=1}^{N} P(i,j)(\boldsymbol{x}(i)\boldsymbol{y}(j) - \boldsymbol{x}(j)\boldsymbol{y}(i)) \\ &= \sum_{i=1}^{N} \sum_{j=1}^{N} P(i,j) \frac{\boldsymbol{x}^2(i) + \boldsymbol{x}^2(j) + \boldsymbol{y}^2(i) + \boldsymbol{y}^2(j)}{2} \\ &\leqslant \max(P(i,j)) \cdot (\boldsymbol{x}^T \boldsymbol{x} + \boldsymbol{y}^T \boldsymbol{y}). \end{aligned} \tag{4.95}$$

where $P(i,j)$ is the element in the i-th row and j-th column of matrix P, $\boldsymbol{x}(i)$ and $\boldsymbol{y}(i)$ are the i-th element in vector $\boldsymbol{x}$ and $\boldsymbol{y}$, respectively.

Based on the observation of H_{IB}, every element in the skew-symmetric matrix Π_H^{ss} is the sum of product of the sine and cosine functions. Every element in Π_H^{ss} is bounded, therefore $\max(\Pi_H^{\text{ss}}(i,j))$ has an upper bound. We define the upper bound of $\Pi_H^{\text{ss}}(i,j)$ as ϵ_h,

$$\begin{aligned} \boldsymbol{x}^T (H_{IB} \bar{H}_{IB}^{-1} - I) \boldsymbol{y} &= \boldsymbol{x}^T (\Pi_H^{\text{diag}} + \Pi_H^{\text{ss}}) \boldsymbol{y} \\ &= \boldsymbol{x}^T \Pi_H^{\text{ss}} \boldsymbol{y} \\ &\leqslant \max(\Pi_H^{\text{ss}}(i,j))(\boldsymbol{x}^T \boldsymbol{x} + \boldsymbol{y}^T \boldsymbol{y}) \\ &\leqslant \epsilon_h (\boldsymbol{x}^T \boldsymbol{x} + \boldsymbol{y}^T \boldsymbol{y}) \end{aligned} \tag{4.96}$$

thus complete the proof of item 2.

5 Layered Affine Formation Control of Networked Uncertain Systems: a Fully Distributed Approach over Directed Graphs

In this chapter, distributed formation control is presented for networked Euler-Lagrange systems (ELSs) over a directed interaction topology. This problem is defined by a layered framework in which information flow both among the leaders and among the followers are described by different layers. To empower the formation to make a variety of geometric transformations, we present the necessary and sufficient condition for affine maneuverability under a directed graph. Unlike most existing results using a diagonal stabilizing matrix to achieve the stabilizability of affine formation, this fully distributed approach is feasible without any global information. Next, we propose an adaptive control law for agents in each layer, where the closed-loop errors are driven to a neighborhood of the origin in finite time. Adaptive neural networks are integrated to tackle the model uncertainties in ELSs by updating the norm of the weight matrix, which can simplify the control design and alleviate the computational burden compared with traditional ones. Simulation results are given to show the effectiveness of the proposed approach.

5.1 INTRODUCTION

Multi-agent system control has been an active area with existing works focusing on interesting topics, such as the consensus problem [203, 362, 401, 515], the tracking problem [39, 189, 283, 377], the containment problem [210, 357], and the formation problem [50,110,168,260,275,488]. Generally, the formation control consists of formation shape control and formation maneuver control: (i) control a group of agents to achieve a desired geometric pattern; and (ii) maneuver the whole group to realize translation, rotation, scale, shear, and other geometric transformations continuously.

In typical multi-agent formation control, some of the agents are designated as leaders while the rest of the agents serve as followers. The leaders track their desired trajectories, and the followers track the leaders' positions with prescribed offsets to achieve a proper formation shape [360]. In fact, for

DOI: 10.1201/9781003298618-5

leader-follower formation control, the dynamics and the distributed communication of the leaders should also be considered to design an implementable control algorithm. Moreover, in advanced multi-agent missions, such as drone (or manned aircraft) cruise, aircraft carrier fleets, satellite network arrays, and advanced synthetic aperture radars, for better categorization and implementation, the agents are usually divided into different layers according to the agents' task areas, equipped payloads, and levels of intelligence. For example, in light of the cooperative exploration task of unmanned aerial vehicles (UAVs) and unmanned ground vehicles (UGVs), the UAVs with a broad vision, can independently navigate and form a desired formation configuration, while the UGVs can autonomously converge to the ground area covered by the UAVs' formation. In this case, the control algorithms of the two groups of agents cannot be designed by considering all the agents as a whole spare group or a single layer. Thus, a layered structure should be proposed to articulate the problem, where the agents of different layers can be described by heterogeneous models and various communication graphs. Most existing formation formulations, however, consider the whole agent group as a single layer, and assume that there is no communication among leaders or use desired trajectories to describe the leaders [50, 110, 168, 275, 488].

Using two-layer and multilayer frameworks, the formation problems are investigated in [255, 256], where the agents' group are divided into different layers to describe the behavior of agents. The communications among the layers are constructed, and the leaders in the first layer can form a certain geometric pattern coordinately. However, the second layer in [255] can only guarantee the containment of the followers, while the exact positions of followers cannot be easily predefined due to their dependence on the communication topology. On the other hand, a distributed finite-time estimator is essential to implement the control design in [255, 256], but the estimator is discontinuous and limited under undirected graphs. Another drawback is the lack of maneuverability. In this chapter, we aim to define a new framework named layered affine formation (LAF) and remove the deficiency of our previous approach.

According to a recent survey [360], the formation control can be categorized into three types: position-, displacement-, and distance-based. For the position-based approach, with respect to a global coordinate frame, each agent is able to acquire its position and achieve the target formation together with other agents. Different from the position-based method, the displacement-based one can constrain the displacements of inter-agents. In light of the distance-based method, a local coordinate associated with inter-agent distance is accessible to each agent. A translation or rotation can be obtained in [453], but it cannot fit for a scaling transformation. Using complex Laplacians, the result in [170] considers a group of agents moving in the plane, while achieving a translation, a rotation, or a scaling movement simultaneously. In [291], the affine formation framework in arbitrary dimensions is then addressed, holding a capability to flexibly transform the formation configurations. In [546], the affine formation problem is further investigated, maneuvering a multi-agent system to accomplish a series of formation shapes, while the formation maneuver control in [546] only considers undirected topologies, and the directed case has not been solved yet.

Inspired by the above ideas, we extend the two-layer and multilayer frameworks reported in our very recent work [255, 256] to a newly defined layered affine formation problem under directed graphs. The extension is rather challenging. First, the target containment positions in the second layer from [255] are related to the connectivity graph and cannot be flexibly designed. Thus a steerable affine transformation should be proposed to enhance the previous second layer as well as make the previous framework maneuverable. Second, the undirected problems [255,256] require a connected graph to drive agents in each layer to a desired formation shape, but it is not implementable for practical tasks over directed topologies. Thus, the affine maneuverability under a directed graph should be investigated and applied as a general case. Moreover, in [255,256], a finite-time estimator is proposed for each agent to acquire estimations of the target position and velocity information. To tackle the discontinuity induced by the estimator, it is important to design the controller without the aid of estimators. Third, unlike classical real Laplacian matrices that always have all eigenvalues in the right plane, an affine (or complex) Laplacian may have eigenvalues in the left plane [291, 293, 298]. To implement formation control via an affine (or complex) Laplacian of the directed graph, it is noteworthy that a stabilizing matrix is utilized in [291, 293, 298] to shift the eigenvalues of a Laplacian matrix to the open right plane, which requires global information (centralized computation) of the whole group of agents. To make the approach in this chapter fully distributed, the stabilizing matrix with global information cannot be applied. In addition, the symmetric property of the Laplacian matrix no longer exists. Without the aid of the symmetric property and stabilizing matrix, to achieve high-speed convergence, a new control law needs to be proposed, which is totally different from the case in [255].

The main contributions of this chapter are summarized below.

(i) A novel affine formation problem is defined by a layered framework. While many previous approaches to leader-follower formation control focus on the communication among the followers, this chapter considers information flow both among the leaders and among the followers by different layers which is more consistent with practical multi-agent missions. With the proposed framework, more flexible formation networks can be achieved, where the mechanism of the information flow among agents can be explored and designed naturally. Due to the inherent nature of the layered structure, this control design has potential advantages of characterizing networked heterogeneous agents associated with various mobile communication topologies to meet the requirements of the state-of-the-art multi-agent missions [172], e.g., the cooperative mission of UAVs and UGVs. In addition, there also exists numerical evidence that the layered structure can enhance the robustness and the synchronizability of the networked systems [31].

(ii) Different from stationary target formation in [291] and affine formation maneuver over undirected graphs in [546], based on the property of the affine transformation, the necessary and sufficient condition for affine maneuverability under a directed graph is proposed as a general extension. Agents in each layer can be steered to the desired affine formation shapes in arbitrary dimensions if the conditions are satisfied. Unlike most existing works using a diagonal stabilizing matrix with global information [291,293,298,444,528], the design in this chapter is still feasible without any stabilizing matrix, which is fully distributed.

(iii) An LAF control law is designed for agents in each layer, where the closed-loop errors converge to a small region of the origin in finite time. Moreover, the designed practical finite-time controller is singularity-free. Compared with the result in [431], the LAF control law is a generalization as not only the sliding surface but all the systems states are analyzed. Compared with widely studied first- or second-order linear systems, Euler-Lagrange systems (ELSs) can describe a wild class of practical systems, such as underwater robots, autonomous vehicles, and spacecraft. To deal with the inherent non-linearity and the model uncertainty in ELSs, neural networks (NNs) are incorporated by adaptively updating the norm of the weight matrix. Compared with the standard adaptive law which updates the whole weight matrix [179, 255, 262], this approach can lighten the computational burden.

5.2 LAYERED FORMATION PROBLEM FORMULATIONS

5.2.1 AGENT DYNAMICS

Consider a group of N agents in $\mathbb{R}^d$ and assume $d \geq 2$ and $N \geq d+2$. Denote the first N_l agents as the leaders and the rest $N_f = N - N_l$ as the followers. Thus the agents' set is $\mathcal{V} = \{1, 2, \ldots, N\}$, the leaders' subset $\mathcal{V}_l = \{1, 2, \ldots, N_l\}$, and the followers' subset $\mathcal{V}_f = \mathcal{V} \backslash \mathcal{V}_l$. The ith agent is modeled by the following ELS

$$M_i\left(q_i\right) \ddot{q}_i + C_i\left(q_i, \dot{q}_i\right) \dot{q}_i + g_i\left(q_i\right) = \tau_i(t), \tag{5.1}$$

where $q_i \in \mathbb{R}^d$, $i \in \mathcal{V}$ denotes the position vector of agent i, $M_i\left(q_i\right) \in \mathbb{R}^{d \times d}$ is the symmetric positive definite inertia matrix, $C_i\left(q_i, \dot{q}_i\right) \in \mathbb{R}^{d \times d}$ is the vector of Coriolis and centrifugal matrix, and $g_i\left(q_i\right) \in \mathbb{R}^d$ denotes the gravitational force. Define $\boldsymbol{q} = [q_1^T, q_2^T, \ldots, q_N^T]^T$, $\boldsymbol{q_l} = [q_1^T, q_2^T, \ldots, q_{N_l}^T]^T$, and $\boldsymbol{q_f} = [q_{N_l+1}^T, q_{N_l+2}^T, \ldots, q_N^T]^T$. We have $\boldsymbol{q} = [\boldsymbol{q_l}^T, \boldsymbol{q_f}^T]^T$.

The following properties of the ELS are given.

Property 3. *[150] Matrix $M_i(q_i)$, $i \in \mathcal{V}$, is bounded as*

$$0 < k_{\underline{m}} \|q_i\|^2 \le q_i^T M_i(q_i) q_i \le k_{\overline{m}} \|q_i\|^2, \ \forall q_i \in \mathbb{R}^d, \tag{5.2}$$

where $k_{\underline{m}}$ and $k_{\overline{m}}$ are positive constants.

Property 4. *[150] Matrix $\dot{M}_i(q_i) - 2C_i(q_i, \dot{q}_i)$ is skew-symmetric.*

We make an assumption on the measurability of the ELSs.

Assumption 7. *The position and the velocity of ELSs* (5.1) *are measurable.*

Then, the following necessary assumption are introduced since the NNs are valid over a compact set, which will be detailed in Section 5.4.

Assumption 8. *The initial position and velocity of each agent in the first layer originate from a compact set, i.e., $\boldsymbol{q_l}(0)$, $\dot{\boldsymbol{q}}_{\boldsymbol{l}}(0) \in \Omega_{l0} \subset \mathbb{R}^{dN_l}$.*

Assumption 9. *The initial position and velocity of each agent in the second layer originate from a compact set, i.e., $\boldsymbol{q_f}(0)$, $\dot{\boldsymbol{q}}_{\boldsymbol{f}}(0) \in \Omega_{f0} \subset \mathbb{R}^{dN_f}$.*

5.2.2 GRAPH THEORY

In this section, we utilize a communication graph $\mathcal{G} = (\mathcal{V}, \mathcal{E})$ to characterize the underlying information flow among the agents in each layer, where $\mathcal{E} \subseteq \mathcal{V} \times \mathcal{V}$ and $\mathcal{V} = \{1, 2, \ldots, k\}$, $k \in \mathbb{Z}^+$ denote the edge set and the vertex set, respectively. Denote $\mathcal{N}_i$ as the in-neighbor set of node v_i, where $\mathcal{N}_i = \{j : (j, i) \in \mathcal{E}\}$.

We introduce two kinds of Laplacian matrices. First, an ordinary Laplacian matrix is defined as

$$L(i,j) = \begin{cases} -a_{ij} & \text{if } i \neq j \text{ and } j \in \mathcal{N}_i, \\ 0 & \text{if } i \neq j \text{ and } j \notin \mathcal{N}_i, \\ \sum\limits_{k \in \mathcal{N}_i} a_{ik} & \text{if } i = j, \end{cases}$$

where $a_{ij} \in \mathbb{R}$ is a positive weight with respect to edge (i, j).

Another special Laplacian matrix named signed Laplacian matrix is introduced. A signed Laplacian matrix L^s of a directed graph is defined as

$$L^s(i,j) = \begin{cases} -\omega_{ij} & \text{if } i \neq j \text{ and } j \in \mathcal{N}_i, \\ 0 & \text{if } i \neq j \text{ and } j \notin \mathcal{N}_i, \\ \sum\limits_{k \in \mathcal{N}_i} \omega_{ik} & \text{if } i = j, \end{cases}$$

where $\omega_{ij} \in \mathbb{R}$ may be a positive or negative real weight attributed on edge (i, j).

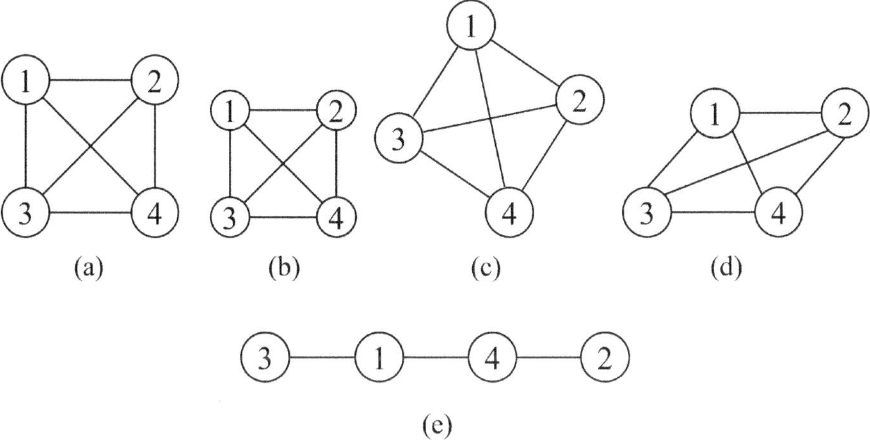

Figure 5.1 Affine transformation types of a nominal configuration (a) by scaling (b), rotating (c), shearing (d), or collinear (e).

We next present some useful properties of the affine transformation.

The affine transformation is a general linear transformation, which may be translation, rotation, scale, shear or a combination of them. A formation $(\mathcal{G}, \boldsymbol{q})$ can be defined as a directed graph $\mathcal{G}$ with its vertex i mapped to q_i. Then the nominal formation associated to $\mathcal{G}$ is represented as $(\mathcal{G}, \boldsymbol{r})$, and $\boldsymbol{r} = [r_1^T, r_2^T, \ldots, r_N^T]^T = [r_l^T, r_f^T]^T \in \mathbb{R}^{dN}$ is a constant vector and named as nominal configuration. The affine image of the nominal configuration can be defined as [291]

$$\mathcal{A}(\boldsymbol{r}) = \left\{\boldsymbol{q} \in \mathbb{R}^{dN} : \boldsymbol{q} = (I_N \otimes A)\,\boldsymbol{r} + \mathbf{1}_N \otimes b,\ \forall A \in \mathbb{R}^{d\times d}, \forall b \in \mathbb{R}^d\right\}, \quad (5.3)$$

where (A, b) is the affine transformation, and an illustration of several types are shown in Fig. 5.1. The formal definition of nominal formation is given as follows.

Definition 24. *The nominal formation $(\mathcal{G}, \boldsymbol{q})$ has the expression of*

$$\boldsymbol{q}(t) = [I_N \otimes A(t)]\,\boldsymbol{r} + \mathbf{1}_N \otimes b(t),$$

where $A(t) \in \mathbb{R}^{N\times N}$ and $b(t) \in \mathbb{R}^d$ are continuous.

In the sequel, the notion of affine maneuverability is formally defined that will be used throughout this chapter.

Definition 25. *The nominal formation $(\mathcal{G}, \boldsymbol{r})$ is said to be affinely maneuverable if for any $\boldsymbol{q} = [\boldsymbol{q_l}^T, \boldsymbol{q_f}^T]^T \in \mathcal{A}(\boldsymbol{r})$ in $\mathbb{R}^{Nd}$, $\boldsymbol{q_f}$ can be determined by $\boldsymbol{q_l}$ uniquely such that*

$$\boldsymbol{q_f} = -\left(\left(L_{f1}^s\right)^{-1} L_{f2}^s \otimes I_d\right)\boldsymbol{q_l}.$$

In Definition 25, affine maneuverability is introduced to represent that the target positions of the followers can be uniquely determined by the positions of the leaders.

5.2.3 PROBLEM FORMULATION

An LAF framework has two layers: the leading layer and the following layer. In the first layer, the leaders are supposed to move together with a desired reference simultaneously, while achieving a specified time-varying affine formation based on the relative positions of their neighbors. According to the formation in the first layer, the agents in the second layer are supposed to form a second specified formation and can be affinely maneuvered via information flow from the first layer. Finally, the layered formation with various configurations is constructed. The mathematical formulation of an LAF will be formally addressed in Problem 2.

Based on the LAF framework, we next introduce a new defined layered Laplacian matrix. We also reveal that the defined layered Laplacian matrix is a useful tool to describe the communication flow of a LAF framework.

We assume the information flow in both layers is directional, where the leaders only interact with the leaders and do not access the information from the followers. For convenience, define a virtual agent 0, which denotes the center of nominal affine formation. Then, the associated $L^{\dagger} \in \mathbb{R}^{(N_l+N_f+1)\times(N_l+N_f+1)}$ is defined as a layered Laplacian matrix that contains both the ordinary Laplacian blocks and the signed Laplacian ones, which can be rewritten as follows

$$L^{\dagger} = \left[\begin{array}{c|c|c} 0 & 0_{1\times N_l} & 0_{1\times N_f} \\ \cline{3-3} L_{l2} & \multicolumn{1}{|c|}{L_{l1}} & 0_{N_l\times N_f} \\ \cline{1-2} 0_{N_f\times 1} & L^s_{f2} & L^s_{f1} \end{array}\right], \tag{5.4}$$

where $L_{l1} \in \mathbb{R}^{N_l\times N_l}$ represents the communication topology for the leaders in the first layer, $L_{l2} \in \mathbb{R}^{N_l\times 1}$ the information flow from the virtual agent to the leaders, $L^s_{f1} \in \mathbb{R}^{N_f\times N_f}$ the interaction topology of the followers in the second layer, and $L^s_{f2} \in \mathbb{R}^{N_f\times N_l}$ the information flow between the leaders in the first layer and the followers in the second layer. More specifically, we can divide the layered Laplacian matrix $L^{\dagger}$ into an ordinary Laplacian matrix $L \in \mathbb{R}^{(N_l+1)\times(N_l+1)}$ for the first layer

$$L = \begin{bmatrix} 0 & 0^{1\times N_l} \\ L_{l2} & L_{l1} \end{bmatrix},$$

and a signed Laplacian matrix $L^s \in \mathbb{R}^{(N_l+N_f)\times(N_l+N_f)}$ for the second layer

$$L^s = \begin{bmatrix} 0^{N_l\times N_l} & 0^{N_l\times N_f} \\ L^s_{f2} & L^s_{f1} \end{bmatrix}.$$

Let $\mathcal{G}_l$ denote the directed graph associated with N_l+1 agents labeled from 0 to N_l. A relevant assumption and a useful lemma are given.

Assumption 10. *For the N_l $(N_l \geq d+1)$ agents in the first layer, there exists a directed spanning tree in $\mathcal{G}_l$, where the root of $\mathcal{G}_l$ is the agent 0.*

Lemma 5.1

For the first layer, if $\mathcal{G}_l$ contains a directed spanning tree, then the following statements hold:

(i) L_{l1} is non-singular;
(ii) All eigenvalues of L_{l1} have positive real parts;
(iii) Each entry of $-L_{l1}^{-1}L_{l2}$ is nonnegative and each row sum of $-L_{l1}^{-1}L_{l2}$ is equal to one.

Proof. The proof is similar to that in [338] and [340]. □

■

5.3 AFFINE MANEUVERABILITY OF DIRECTED GRAPH

In this section, we will present the directed graphical condition of the affine maneuverability, which is central to the LAF problem.

In order to obtain the affine maneuverability, we need to introduce a term affine independence of the given position set $\{q_i\}_{i=1}^{N}$ of N agents in $\mathbb{R}^d$.

Definition 26. *Define the configuration matrix as $Q(\boldsymbol{q}) = [q_1, q_2, \ldots, q_N]^T \in \mathbb{R}^{N\times d}$ and the corresponding augmented matrix as $\bar{Q}(\boldsymbol{q}) = [Q(\boldsymbol{q}), \mathbf{1}_N] \in \mathbb{R}^{N\times(d+1)}$. Then, $\{q_i\}_{i=1}^{N}$ are called affinely independent if and only if the rows of $\bar{Q}(\boldsymbol{q})$ are linearly independent.*

According to Definition 26, there exist at most $d+1$ positions that are affinely independent in $\mathbb{R}^d$ as $\bar{Q}(\boldsymbol{q})$ only has $d+1$ rows.

We next make an assumption of the nominal formation.

Assumption 11. *Assume that the configuration $\boldsymbol{r}$ of the given nominal formation $(\mathcal{G}, \boldsymbol{r})$ of N agents is generic.*

A configuration $\boldsymbol{r}$ in $\mathbb{R}^d$ is generic if its coordinates $r_1, \ldots, r_N$ are algebraically independent over the integers [159]. In other words, a generic configuration $\boldsymbol{r}$ cannot satisfy any non-trivial polynomial equation with integer coefficients. Assumption 11 means that there is no degeneracy in the nominal configuration $\boldsymbol{r}$, i.e., no three or more points on one line, no three or more lines crossing one point, etc. Next, in order to get the necessary and sufficient graphical condition for affine maneuverability, another three useful lemmas about the directed graph are introduced from results in [291].

Lemma 5.2

[291] A directed graph $\mathcal{G}$ has a spanning κ-tree if and only if $\mathcal{G}$ is κ-rooted. ■

Lemma 5.3

[291] For a generic signed Laplacian L^s associated to a directed graph $\mathcal{G}$, if $\mathcal{G}$ is κ-rooted, then after removing κ rows corresponding to κ roots and any κ columns, the remaining block is nonsingular. ■

Lemma 5.4

[291] Suppose a directed graph $\mathcal{G}$ has N nodes with $N \geq d+2$ and $\boldsymbol{r} = [r_1^T, r_2^T, \ldots, r_N^T]^T$ is generic. Then for graph $\mathcal{G}$ and $\boldsymbol{q} \in \mathcal{A}(\boldsymbol{r})$, we have $(L^s \otimes I_d)\boldsymbol{q} = 0$ if and only if $\mathcal{G}$ is $(d+1)$-rooted. ■

Based on the above lemmas and assumptions, we present the necessary and sufficient condition for the affine maneuverability.

Theorem 5.1

Under Assumption 11, the given nominal formation $(\mathcal{G}, \boldsymbol{r})$ of N agents is affinely maneuverable if and only if the leaders' subset $\mathcal{V}_l$ has $d+1$ leaders and every follower in $\mathcal{V}_f$ is $(d+1)$-reachable from $\mathcal{V}_l$. ■

Proof. (Sufficiency) If there exists a leaders' subset $\mathcal{V}_l$ having $d+1$ leaders and each follower in $\mathcal{V}_f$ is $(d+1)$-reachable from $\mathcal{V}_l$, $\mathcal{G}$ is $(d+1)$-rooted. It then follows from Lemma 5.2 that the directed graph $\mathcal{G}$ has a $(d+1)$-spanning tree. In view of the leader-follower structure under consideration, these roots belong to the leaders' subset $\mathcal{V}_l$ and satisfy the definition of κ-spanning tree. Then an associated signed Laplacian matrix L^s has $d+1$ rows with all elements being zeros corresponding to these roots, and L^s can be rewritten as the following blocks

$$L^s = \begin{bmatrix} 0_{(d+1)\times(d+1)} & 0_{(d+1)\times(N-d-1)} \\ L_{f2}^s & L_{f1}^s \end{bmatrix}.$$

Based on Lemma 5.3, it can be obtained that L_{f1}^s is nonsingular. Next, according to Lemma 5.4, we have $(L^s \otimes I_d)\boldsymbol{q} = 0$. Then, we can derive

$(L_{f2}^s \otimes I_d)\boldsymbol{q_l} + (L_{f1}^s \otimes I_d)\boldsymbol{q_f} = 0$, where $\boldsymbol{q_f}$ can be uniquely determined by $\boldsymbol{q_f} = -((L_{f1}^s)^{-1} L_{f2}^s \otimes I_d)\boldsymbol{q_l}$ if and only if L_{f1}^s is nonsingular. Based on Definition 25, the affine maneuverability of the given nominal formation $(\mathcal{G}, \boldsymbol{r})$ of N agents can be achieved. The sufficiency is concluded here.

(Necessity) If the affine formation of $\boldsymbol{q}$ with a directed interaction graph $\mathcal{G}$ is affinely maneuverable, then it is trivial that there exists a signed Laplacian L^s associated to $\mathcal{G}$ satisfying $(L^s \otimes I_d)\,\boldsymbol{q} = 0$ for $\boldsymbol{q} \in \mathcal{A}(\boldsymbol{r})$. Under Assumption 11, according to Lemma 5.4, $\mathcal{G}$ is $(d+1)$-rooted. Then, there must exist exactly $d+1$ leaders corresponding to $d+1$ roots as the leaders cannot receive information from the followers. Therefore each follower must be $(d+1)$-reachable from the leaders' subset $\mathcal{V}_l$, which completes the proof. □

The subsequent section will design proper control laws based on Theorem 5.1, and the following assumption is made.

Assumption 12. *Every follower in $\mathcal{V}_f$ is $(d+1)$-reachable from $\mathcal{V}_l$.*

5.4 LAYERED FORMATION CONTROL DESIGN

In this section, we propose adaptive NN based control laws for the first layer and the second layer, respectively, where the closed-loop errors converge to a small neighborhood of the origin in finite time. The LAF control problem to be solved in this section is formally stated below.

Problem 2. *Given the initial formation $\mathcal{G}(\boldsymbol{q}(0))$ and feasible constant affine constraints $Q(\boldsymbol{r})$, design $\tau_i(t)$ for agent $i \in \mathcal{V}$ based on the relative positions and velocities with its neighbors such that $\boldsymbol{q_l} \to -(L_{l1}^{-1} L_{l2} \otimes I_d)q_0 + \boldsymbol{p}$, and $\boldsymbol{q_f} \to -((L_{f1}^s)^{-1} L_{f2} \otimes I_d)\boldsymbol{q_l}$ with bounded errors, where $\boldsymbol{p} = [p_1^T, p_2^T, \ldots, p_{N_l}^T]^T$, and $p_i(t)$ denotes the displacement decomposed from the nominal affine formation.*

In this problem definition, the existences of L_{l1}^{-1} and $(L_{f1}^s)^{-1}$ are guaranteed by Assumptions 10 and 12, Lemma 5.1, and Theorem 5.1. For agent i, the displacement $p_i(t)$ represents the desired relative position between agent i and virtual agent 0. For the convenience of the later analysis, define $p_0 = 0$, and let $\boldsymbol{e_l}$ and $\boldsymbol{e_f}$ denote formation tracking errors for the first and second layers, respectively

$$
\begin{aligned}
\boldsymbol{e_l} &= \boldsymbol{q_l} + \left(L_{l1}^{-1} L_{l2} \otimes I_d\right) q_0 - \boldsymbol{p}, \\
\boldsymbol{e_f} &= \boldsymbol{q_f} + \left(\left(L_{f1}^s\right)^{-1} L_{f2} \otimes I_d\right) \boldsymbol{q_l}.
\end{aligned}
$$

5.4.1 CONTROL DESIGN FOR THE FIRST LAYER

Define the following auxiliary variables

$$z_{1i} = \sum_{j\in\mathcal{N}_i} a_{ij}\left(q_i - p_i - q_j + p_j\right),\ i \in \mathcal{V}_l, \tag{5.5}$$

$$z_{2i} = \dot{q}_i - \alpha_{1i},\ i \in \mathcal{V}_l. \tag{5.6}$$

Inspired by the approach in [480] to avoid the singularity problem, the virtual control α_{1i} is designed as

$$\alpha_{1i} = \frac{1}{\beta_i}\left(-k_{1i}\xi_i + \sum_{j\in\mathcal{N}_i} a_{ij}\left(\dot{q}_j - \dot{p}_j\right)\right) + \dot{p}_i, \tag{5.7}$$

$$\xi_{i,k} = \begin{cases} \mathrm{sig}_{\mathrm{c}}^{\ell}(z_{1i,k}), & \text{if } |z_{1i,k}| > \varepsilon, \\ \iota_1 z_{1i,k} + \iota_2 \mathrm{sig}_{\mathrm{c}}^{2}(z_{1i,k}), & \text{if } |z_{1i,k}| \le \varepsilon, \end{cases} \tag{5.8}$$

where $\xi_{1i,k}$ is the kth $(k = 1, ..., d)$ element of a new defined auxiliary variable ξ_{1i}, k_{1i} and ε are positive constants, ℓ is defined as $0.5 < \ell = \ell_1/\ell_2 < 1$ with positive odd integers ℓ_1 and ℓ_2, $\iota_1 = (2-\ell)\,\varepsilon^{\ell-1}$, $\iota_2 = (\ell-1)\,\varepsilon^{\ell-2}$, and $\beta_i = \sum_{j\in\mathcal{N}_i} a_{ij}$. The classical sign function $\mathrm{sig}_{\mathrm{c}}^{\ell}$ is in line with (2.28)–(2.30).

The choice of ι_1 and ι_2 is to make function ξ_i and its time derivative continuous, which further makes α_{1i} and $\dot{\alpha}_{1i}$ continuous. We can verify that there is no singularity problem in $\dot{\alpha}_{1i}$, since $\dot{\xi}_i$ has the following form

$$\dot{\xi}_{i,k} = \begin{cases} \ell\,\mathrm{sig}\left(z_{1i,k}\right)^{\ell-1} \dot{z}_{1i,k}, & \text{if } |z_{1i,k}| > \varepsilon, \\ \iota_1 \dot{z}_{1i,k} + 2\iota_2 \left|z_{1i,k}\right| \dot{z}_{1i,k}, & \text{if } |z_{1i,k}| \le \varepsilon. \end{cases}$$

Also, it is trivial to show that β_i, $i \in \mathcal{V}_l$ is a positive constant as long as Assumption 10 holds. Thus, $1/\beta_i$ in (5.7) is nonsingular.

Here, we the consider the case of $|z_{1i,k}| > \varepsilon$. Analysis of the case of $|z_{1i,k}| \le \varepsilon$ can be found in Remark 23.

According to (5.5), substituting (5.6) and (5.7) into $\dot{z}_{1i}$ yields

$$\dot{z}_{1i} = -k_{1i}\mathrm{sig}_{\mathrm{c}}^{\ell}(z_{1i}) + \beta_i z_{2i}. \tag{5.9}$$

Next, according to (5.1), the derivative of z_{2i} takes the form

$$\dot{z}_{2i} = M_i^{-1}\left(q_i\right)\left[\tau_i - C_i\left(q_i, \dot{q}_i\right)\dot{q}_i - g_i\left(q_i\right)\right] - \dot{\alpha}_{1i}. \tag{5.10}$$

Note that we can not directly use $M_i\left(q_i\right)$, $C_i\left(q_i,\dot{q}_i\right)$ and $g_i\left(q_i\right)$ to design a control law due to the model uncertainties. Here, we apply NNs to approximate the model uncertainties [246]. Define $Z_i = [q_i^T, \dot{q}_i^T, \alpha_{1i}^T, \dot{\alpha}_{1i}^T]^T$, and over a compact set $\Omega_{Z_i} \subset \mathbb{R}^{4d}$, NNs can approximate any continuous function $f_i(Z_i)$ (or any piece-wise continuous function as shown in [422,423]) to arbitrary any accuracy as [146, 539],

$$f_i\left(q_i, \dot{q}_i, \alpha_{1i}, \dot{\alpha}_{1i}\right) = W_i^{*T}\phi_i\left(q_i, \dot{q}_i, \alpha_{1i}, \dot{\alpha}_{1i}\right) + \epsilon_i, \tag{5.11}$$

where $f_i(q_i, \dot{q}_i, \alpha_{1i}, \dot{\alpha}_{1i}) = g_i(q_i) + C_i(q_i, \dot{q}_i)\alpha_{1i} + M_i(q_i)\dot{\alpha}_{1i}$ represents the model uncertainties, $W_i^* \in \mathbb{R}^{\mu_l \times d}$ the ideal constant weight matrix, μ_l the number of NN nodes, ϵ_i the approximation error, and $\phi_i = [\varphi_{i1}, \varphi_{i2}, \ldots, \varphi_{i\mu_l}]^T \in \mathbb{R}^{\mu_l}$ the activation function. The elements of the activation function are given as the Gaussian functions

$$\varphi_{ih}(Z_i) = \exp\left[\frac{-(Z_i - \mu_h)^T (Z_i - \mu_h)}{\eta_h{}^2}\right], \; h = 1, \ldots, \mu_l, \tag{5.12}$$

where η_h is the Gaussian function width and μ_h is the receptive field center [146].

In addition, ϵ_i is bounded over the compact set, i.e., $\|\epsilon_i(Z_i)\| \leq \bar{\epsilon}_i, \forall Z_i \in \Omega_{z_i} \subset \mathbb{R}^{4d}$ with $\bar{\epsilon}_i > 0$ as an unknown constant (for details, refer to [141]).

The proposed adaptive LAF control algorithm for the first layer is

$$\tau_i = -\beta_i z_{1i} - k_{2i}\mathrm{sig}_{\mathrm{c}}^{\ell}(z_{2i}) + \frac{\hat{\psi}_i}{2v_i^2}\|\phi_i(Z_i)\|^2 z_{2i}, \tag{5.13}$$

$$\dot{\hat{\psi}}_i = \gamma_i\left(\frac{\|z_{2i}\|^2}{2v_i^2}\|\phi_i(Z_i)\|^2 - \sigma_i\hat{\psi}_i^{\ell}\right), \tag{5.14}$$

where k_{2i}, γ_i, v_i, and σ_i are positive constants, and $\hat{\psi}_i$ is the estimation of ψ_i defined as $\psi_i = \|W_i^*\|^2$, $i \in \mathcal{V}_l$.

Remark 19. *Note that only the estimation of the norm of weight matrix W_i^* is utilized in adaptive law (5.14). Compared with the classical NN adaptive law, e.g., the updating scheme in [255], the algorithm (5.14) only needs to update a scalar $\hat{\psi}_i$ rather than a matrix $\hat{W}_i$, which can reduce the amount of calculation to some extent.*

Define the following variables as

$$\mathbf{z_{l1}} = [z_{11}^T, z_{12}^T, \ldots, z_{1N_l}^T]^T, \tag{5.15}$$

$$\mathbf{z_{l2}} = [z_{21}^T, z_{22}^T, \ldots, z_{2N_l}^T]^T, \tag{5.16}$$

$$\hat{\psi}_l = [\hat{\psi}_1, \hat{\psi}_2, \ldots, \hat{\psi}_{N_l}]^T, \tag{5.17}$$

$$\tilde{\psi}_l = [\tilde{\psi}_1, \tilde{\psi}_2, \ldots, \tilde{\psi}_{N_l}]^T, \tag{5.18}$$

where $\tilde{\psi}_i$ is the modified weight error defined as $\tilde{\psi}_i = \psi_i - \hat{\psi}_i$.

We propose the following lemma for the ease of later analysis.

Lemma 5.5

For variables $\hat{\psi}_i$ and $\tilde{\psi}_i$, the following inequality holds

$$\tilde{\psi}_i\hat{\psi}_i^{\ell} \leq -\kappa_1\tilde{\psi}_i^{\ell+1} + \kappa_2\psi_i^{\ell+1}, \tag{5.19}$$

where $\psi_i = \tilde{\psi}_i + \hat{\psi}_i$, and κ_1 and κ_2 are two positive constants defined as

$$\kappa_1 = \frac{2^{\ell-1} - 2^{(\ell-1)(\ell+1)}}{(1+\ell)^2},$$
$$\kappa_2 = \frac{1}{(1+\ell)^2}\left(1 - 2^{\ell-1} + \frac{2^{\ell-1}}{1+\ell} + \frac{\ell}{1+\ell}\right).$$

■

Proof. See Appendix 5.7. □

Then, the main result for the first layer is given.

Theorem 5.2

In the first layer, considering the network of ELSs (5.1) under the LAF control algorithm (5.13) and adaptive law (5.14), if Assumptions 7, 8, 10 and 11 hold, the closed-loop systems are semi-globally practical finite-time stable, i.e., the closed-loop errors $\boldsymbol{e}_l$, $\boldsymbol{z}_{l1}$, $\boldsymbol{z}_{l2}$, and $\tilde{\psi}_l$ are driven into the following compact sets Ω_{e_l}, $\Omega_{z_{l1}}$, $\Omega_{z_{l2}}$, and $\Omega_{\tilde{\psi}_l}$ in finite time:

$$\Omega_{e_l} := \left\{\boldsymbol{e}_l \in \mathbb{R}^{dN_l} \middle| \|\boldsymbol{e}_l\| \le D_l \left\|L_{l1}^{-1} \otimes I_d\right\|\right\},$$
$$\Omega_{z_{l1}} := \left\{\boldsymbol{z}_{l1} \in \mathbb{R}^{dN_l} \middle| \|\boldsymbol{z}_{l1}\| \le D_l\right\},$$
$$\Omega_{z_{l2}} := \left\{\boldsymbol{z}_{l2} \in \mathbb{R}^{dN_l} \middle| \|\boldsymbol{z}_{l2}\| \le \frac{D_l}{\min_{i\in\mathcal{V}_l}\left(\sqrt{\lambda_{\min}\left(M_i(q_i)\right)}\right)}\right\},$$
$$\Omega_{\tilde{\psi}_l} := \left\{\tilde{\psi}_l \in \mathbb{R}^{N_l} \middle| \left\|\tilde{\psi}_l\right\| \le \max_{i\in\mathcal{V}_l}\left(\sqrt{\gamma_i} D_l\right)\right\},$$

where D_l is a positive constant defined in the proof. ■

Proof. See Appendix 5.8. □

Remark 20. *The concept of the practical finite-time stability is first proposed in [431], and it addresses the property that the states of the closed-loop system reach a neighborhood of the origin in finite time. This result guarantees the bounded motion within the neighborhood of the sliding surface. Unfortunately, the stability of the dynamics of the sliding mode is not addressed in [431]. Thus, Theorem 5.2 is a generalization of [431] in the sense that the practical finite-time stability of the dynamics of all the system states is analyzed.*

Remark 21. *Note that the exponential stability and the uniformly ultimately bounded stability can also achieve the performance similar to the proposed* *practical finite-time stable result. It is noteworthy that the practical finite-time stability can enhance the control performance by flexibly adjusting the finite convergence time instant and the bounded set using parameters η, θ, and λ. In addition, when the accurate system model is available, we can rewrite the control law* (5.13) *as a model-based one taking the form*

$$\begin{aligned}\tau_i = &- \beta_i z_{1i} - k_{2i}\mathrm{sig}_{\mathrm{c}}^{\ell}\left(z_{2i}\right) + M_i\left(q_i\right)\dot{\alpha}_{1i} \\ &+ C_i\left(q_i, \dot{q}_i\right)\alpha_{1i} + g_i\left(q_i\right).\end{aligned} \tag{5.20}$$

Then, inequality (5.40) *becomes*

$$\dot{V}_l \leq -\xi_l\rho_l V_l^{\frac{1+\ell}{2}} - (1-\xi)\,\rho_l V_l^{\frac{1+\ell}{2}}, \tag{5.21}$$

which directly proves that the closed-loop errors will be driven to the origin in finite time.

Remark 22. *For Theorem 5.2, the control input τ_i and the feed-forward adaption $\hat{\psi}_i$ are bounded, which can be verified noting the fact that ψ_i is a positive constant and the closed-loop errors $\boldsymbol{z_{1i}}$, $\boldsymbol{z_{2i}}$, and $\hat{\psi}_i$ have been proved bounded.*

Remark 23. *When $|z_{1i,k}| \leq \varepsilon$, which means that the closed-loop errors are already bounded in a small set $z_{1i,k} \in \{z_{1i,k} \in \mathbb{R} |\, |z_{1i,k}| \leq \varepsilon\}$, the closed-loop system is still overall semi-globally practical finite-time stable by choosing a small enough ε. The reason is that the system errors can always be driven into the bounded set $\Omega_{z_{l1}}$ in finite time, if ε is chosen much smaller than the bound D_l of $\boldsymbol{z_{l1}}$ in Theorem 5.2. Actually, in the case of $|z_{1i,k}| \leq \varepsilon$, by replacing $\mathrm{sig}\left(z_{1i,k}\right)^{\ell}$ with $\iota_1 z_{1i,k} + \iota_2\,\mathrm{sig}\left(z_{1i,k}\right)^2$ into the Lyapunov functions, it is trivial to prove that the closed-loop system is also semi-globally practical finite-time stable with proper mathematical manipulations. For more details, please refer to [480], where this switching concept is first proposed. The discussion in this remark still holds for the control design in the second layer.*

5.4.2 CONTROL DESIGN FOR THE SECOND LAYER

For agent i in the second layer, define the following auxiliary variables

$$z_{1i} = \sum_{j\in\mathcal{N}_i} \omega_{ij}\left(q_i - q_j\right),\ i \in \mathcal{V}_f, \tag{5.22}$$

$$z_{2i} = \dot{q}_i - \alpha_{1i},\ i \in \mathcal{V}_f. \tag{5.23}$$

The virtual control α_{1i} is given as

$$\alpha_{1i} = \frac{1}{\beta_i}\left(-k_{1i}\xi_i + \sum_{j\in\mathcal{N}_i}\omega_{ij}\dot{q}_j\right), \tag{5.24}$$

$$\xi_{i,k} = \begin{cases} \mathrm{sig}_{\mathrm{c}}^{\ell}(z_{1i,k}), & \text{if } |z_{1i,k}| > \varepsilon, \\ \iota_1 z_{1i,k} + \iota_2\mathrm{sig}_{\mathrm{c}}^2(z_{1i,k}), & \text{if } |z_{1i,k}| \leq \varepsilon, \end{cases} \tag{5.25}$$

where $\beta_i = \sum_{j \in \mathcal{N}_i} \omega_{ij}$, and k_{1i} is a positive constant. Similar to the design in the first layer, $\xi_{i,k}$ $(k = 1, ..., d)$ is proposed to avoid the singularity, making α_{1i} and $\dot{\alpha}_{1i}$ continuous.

Note that $1/\beta_i$ in (5.24) may be singular when $\beta_i = 0$. We next present $\beta_i \neq 0$ using the following lemma.

Lemma 5.6

β_i, $i \in \mathcal{V}_f$ is a nonzero constant as long as Assumption 11 holds.

Proof. Under Assumption 11, based on Theorem 5.1, suppose that agent i, $i \in \mathcal{V}_f$ is $(d+1)$-reachable from $\mathcal{V}_l$. Then, we know that agent i is affinely maneuverable. Without loss of generality, assume that $d = 3$. Thus, the position of agent i can be uniquely determined by the positions of its neighbor agents j, k, l, and m. Then, we can treat this problem as a simple affine maneuverability with 4 leaders labeled as j, k, l, and m, and 1 follower labeled as i. From Theorem 5.1, agent i can be affinely maneuvered by $\boldsymbol{q_f} = ((L^s_{f1})^{-1} L^s_{f2} \otimes I_d)\boldsymbol{q_l}$, where $\boldsymbol{q_l} = \left[q_j^T, q_k^T, q_l^T, q_m^T\right]$. Since there exists only 1 follower, L^s_{f1} becomes a scalar and must be nonsingular (non-zero) according to Theorem 5.1. Therefore, we have $\beta_i = \sum_{j \in \mathcal{N}_i} \omega_{ij} = -L^s_{f1} \neq 0$. □

■

Similar to Section 5.4.1, we here consider the case of $|z_{1i,k}| > \varepsilon$. The analysis when $|z_{1i,k}| \leq \varepsilon$ can be given following Remark 23.

We proposed the adaptive LAF control algorithm for the second layer

$$\tau_i = -\beta_i z_{1i} - k_{2i}\mathrm{sig}_{\mathrm{c}}^{\ell}(z_{2i}) + \frac{\hat{\psi}_i}{2v_i^2} \|\phi_i(Z_i)\|^2 z_{2i}, \tag{5.26}$$

$$\dot{\hat{\psi}}_i = \gamma_i \left(\frac{\|z_{2i}\|^2}{2v_i^2} \|\phi_i(Z_i)\|^2 - \sigma_i \hat{\psi}_i^{\ell} \right), \tag{5.27}$$

where k_{2i}, γ_i, and σ_i are positive constants, $i \in \mathcal{V}_f$.

Before presenting the main result for the second layer, define the following variables as $\boldsymbol{z_{f1}} = [z^T_{1(N_l+1)}, \ldots, z^T_{1N}]^T$, $\boldsymbol{z_{f2}} = [z^T_{2(N_l+1)}, \ldots, z^T_{2N}]^T$, $\hat{\psi}_f = [\hat{\psi}_{N_l+1}, \ldots, \hat{\psi}_N]^T$, and $\tilde{\psi}_f = [\tilde{\psi}_{N_l+1}, \ldots, \tilde{\psi}_N]^T$.

Theorem 5.3

In the second layer, considering the network of ELSs (5.1) under the LAF control algorithm (5.26) and adaptive law (5.27), if Assumptions 7–12 hold, the closed-loop systems are semi-globally practical finite-time stable, i.e., the closed-loop errors $\boldsymbol{e_f}$, $\boldsymbol{z_{f1}}$, $\boldsymbol{z_{f2}}$, and $\tilde{\psi}_f$ are driven into the following compact sets Ω_{e_f}, $\Omega_{z_{f1}}$, $\Omega_{z_{f2}}$, and $\Omega_{\tilde{\psi}_f}$ in finite time:

$$\begin{aligned}
\Omega_{e_f} &:= \left\{\boldsymbol{e_f} \in \mathbb{R}^{dN_f} \mid \|\boldsymbol{e_f}\| \leq D_f \left\|(L_{f1}^s)^{-1} \otimes I_d\right\|\right\}, \\
\Omega_{z_{f1}} &:= \left\{\boldsymbol{z_{f1}} \in \mathbb{R}^{dN_f} \mid \|\boldsymbol{z_{f1}}\| \leq D_f\right\}, \\
\Omega_{z_{f2}} &:= \left\{\boldsymbol{z_{f2}} \in \mathbb{R}^{dN_f} \mid \|\boldsymbol{z_{f2}}\| \leq \frac{D_f}{\min\limits_{i \in \mathcal{V}_f}\left(\sqrt{\lambda_{\min}\left(M_i(q_i)\right)}\right)}\right\}, \\
\Omega_{\tilde{\psi}_f} &:= \left\{\tilde{\psi}_f \in \mathbb{R}^{N_l} \mid \left\|\tilde{\psi}_f\right\| \leq \max_{i \in \mathcal{V}_f}\left(\sqrt{\gamma_i} D_f\right)\right\},
\end{aligned}$$

where D_f is defined in the proof of Theorem 5.3. ■

Proof. See Appendix 5.9. □

According to Problem 2, combining Theorems 5.1, 5.2, and 5.3, the LAF is achieved in finite time with bounded formation tracking errors. Similar to the analysis in Remark 22, we know the control input τ_i and the feed-forward adaption $\hat{\psi}_i$ for the second layer are also bounded.

Remark 24. *From Theorems 5.2 and 5.3, it is clear that the LAF control laws are fully distributed and feasible without any global stabilizing matrix utilized as [291, 293, 298]. More specifically, the proposed controllers only require the non-singularity of L_{l1} and L_{f1}^s, and are independent of the signs of eigenvalues of the matrix L_{l1} and L_{f1}^s. It is worth mentioning that this approach can also be applied to the results in [291,293,298] to relieve their dependence on stabilizing matrices.*

Remark 25. *Note that in this layered approach, the ELS is able to describe a wide class of practical systems, which has advantages of characterizing heterogeneous agents of different layers, e.g., by using ELSs to model UAVs and UGVs in two layers, respectively, this approach can represent a cooperative mission of UAVs and UGVs as shown in the introduction.*

5.5 SIMULATION

To illustrate the performance of the LAF control law, consider a group of 7 networked satellites. The relative orbital motion of the ith satellite and the initial elements of the near-circular orbit are the same as those in [255]. We evaluate the adaptive LAF control laws (5.13) and (5.26), where the layered communication graph and nominal formation are given in Fig. 5.2.

The control parameters for simulation are given in Table 5.1. The continuous and smooth formation functions $\boldsymbol{p}(t)$ are constructed by proper designed

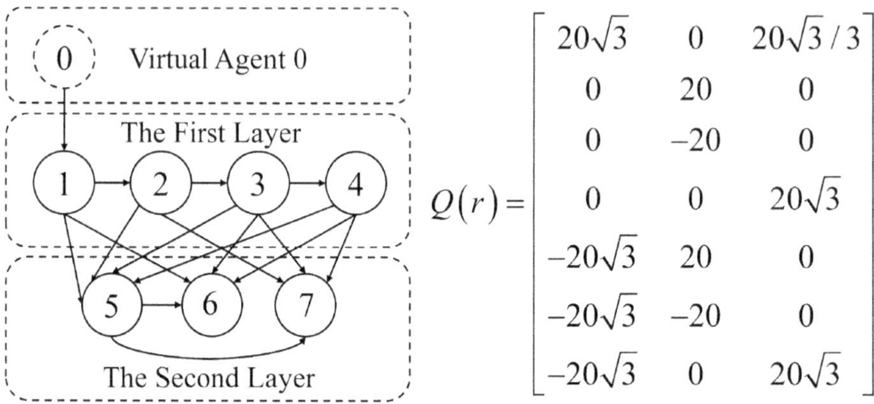

Figure 5.2 The layered communication graph and nominal formation.

Table 5.1
Parameters for simulation

Parameters	Values	Parameters	Values
k_{1i}	0.1	σ_i	5
k_{2i}	5	$\hat{\psi}_i(0)$	0
η_h	10	μ_h	random$[-1, 1]$
γ_i	4	μ_l	256
υ_i	10	ℓ	11/13
ε	0.001	d	3
$q_0(t)$	$[2t, 0, 0]^T$	$q_1(0)$	$[40, -5, -10]^T$
$q_2(0)$	$[20, 5, -5]^T$	$q_3(0)$	$[20, 5, -5]^T$
$q_4(0)$	$[30, -10, 0]^T$	$q_5(0)$	$[0, 40, 0]^T$
$q_6(0)$	$[0, -40, 10]^T$	$q_7(0)$	$[0, 0, 20]^T$

sigmoid functions, which is omitted here. The layered affine Laplacian matrix (5.4) is given in (5.28).

$$L_0^\dagger = \left[\begin{array}{ccccc|ccc} 0 & 0 & 0 & 0 & 0 & 0 & 0 & 0 \\ -1 & 1 & 0 & 0 & 0 & 0 & 0 & 0 \\ 0 & -1 & 1 & 0 & 0 & 0 & 0 & 0 \\ 0 & 0 & -1 & 1 & 0 & 0 & 0 & 0 \\ 0 & 0 & 0 & -1 & 1 & 0 & 0 & 0 \\ \hline 0 & 1 & -\frac{4}{3} & -\frac{1}{3} & -\frac{1}{3} & 1 & 0 & 0 \\ 0 & 1 & 0 & -\frac{5}{3} & -\frac{1}{3} & -\frac{1}{3} & \frac{4}{3} & 0 \\ 0 & 0 & 1 & 0 & -1 & -1 & 0 & 1 \end{array}\right]. \tag{5.28}$$

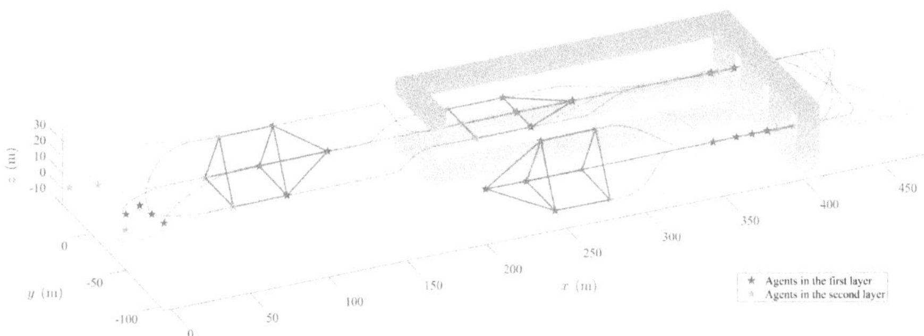

Figure 5.3 Trajectories of satellites. (The blue lines connecting the agents are given to sketch the formation shape.)

$$L_1^\dagger = \left[\begin{array}{ccccc|ccc} 0 & 0 & 0 & 0 & 0 & 0 & 0 & 0 \\ -1 & 1 & 0 & 0 & 0 & 0 & 0 & 0 \\ 0 & -1 & 1 & 0 & 0 & 0 & 0 & 0 \\ 0 & 0 & -1 & 1 & 0 & 0 & 0 & 0 \\ 0 & 0 & 0 & -1 & 1 & 0 & 0 & 0 \\ \hline 0 & -1 & 1 & 1 & 0 & -1 & 0 & 0 \\ 0 & -1 & 0 & 1 & 1 & 0 & -1 & 0 \\ 0 & 0 & -1 & 0 & 0 & 2 & 0 & -1 \end{array}\right]. \tag{5.29}$$

$$L_2^\dagger = \left[\begin{array}{ccccc|ccc} 0 & 0 & 0 & 0 & 0 & 0 & 0 & 0 \\ -1 & 1 & 0 & 0 & 0 & 0 & 0 & 0 \\ 0 & -1 & 1 & 0 & 0 & 0 & 0 & 0 \\ 0 & 0 & -1 & 1 & 0 & 0 & 0 & 0 \\ 0 & 0 & 0 & -1 & 1 & 0 & 0 & 0 \\ \hline 0 & -1 & 2 & 0 & 0 & -1 & 0 & 0 \\ 0 & -1 & 0 & 2 & 0 & 0 & -1 & 0 \\ 0 & 0 & 1 & 0 & 1 & \frac{1}{2} & 0 & -\frac{1}{2} \end{array}\right]. \tag{5.30}$$

The trajectories of satellites are given in Fig. 5.3, where the layered formation shape is maintained precisely. The simulation also indicates that the proposed control scheme can guarantee a safe journey for multi-agent systems in the constrained space via formation evolution, while the detail on obstacle avoidance is not within the scope of this chapter. Fig. 5.4 shows that the tracking errors of 7 satellites converge to a neighborhood of the origin within 80 seconds. The corresponding control inputs of the satellites are given in Fig. 5.5, which are acceptable for satellites in practice. In addition, we give another two simulation examples with the layered affine Laplacian matrices $L_1^\dagger$ and $L_2^\dagger$ defined in (5.29) and (5.30). Figs. 5.6 and 5.7 present the trajectories of 7 satellites starting from random initial states, where the nominal configurations are successfully achieved. It can be observed that $L_1^\dagger$ and $L_2^\dagger$ have eigenvalues located in the left-half complex plane. Thus, the feasibility of this

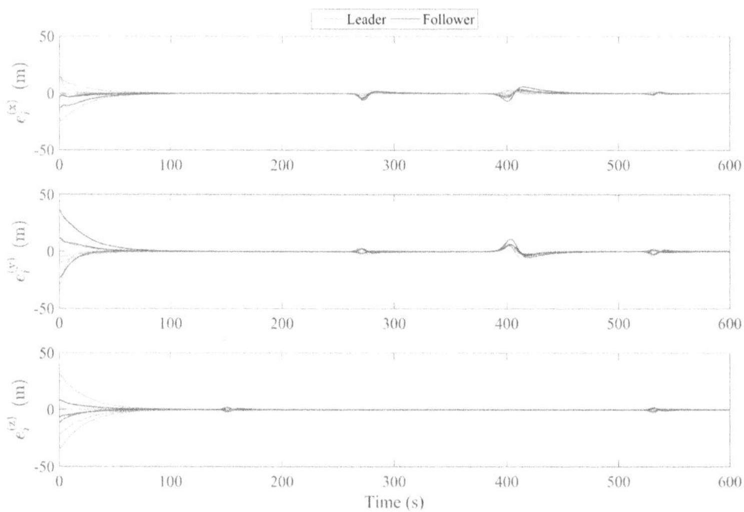

Figure 5.4 Formation errors of satellites.

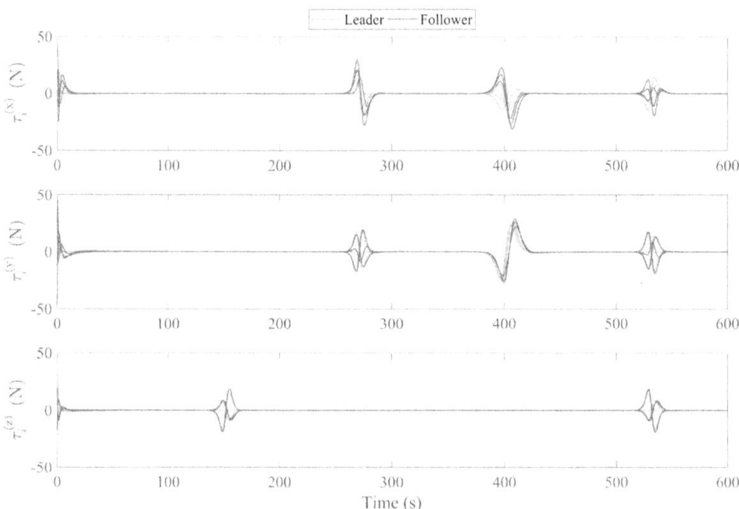

Figure 5.5 Control forces of satellites.

result is independent of the signs of the eigenvalues associated with Laplacian matrices. The above numerical results have shown that the proposed LAF control scheme is effective and feasible to achieved any formation as long as it is an affine transformation of the nominal formation. (These simulation examples are for demonstration purpose only, which aims to show the control performance and may not match practical satellite formation missions).

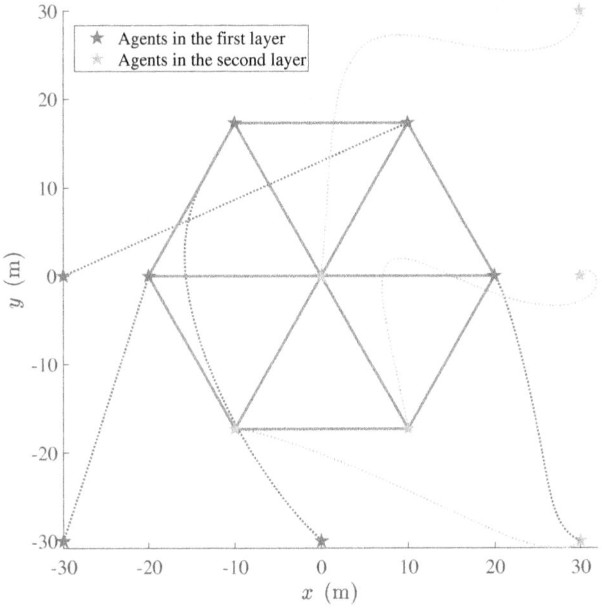

Figure 5.6 Affine formation example 1 with eigenvalues in the left plane. (The blue lines connecting the agents are given to sketch the formation shape.)

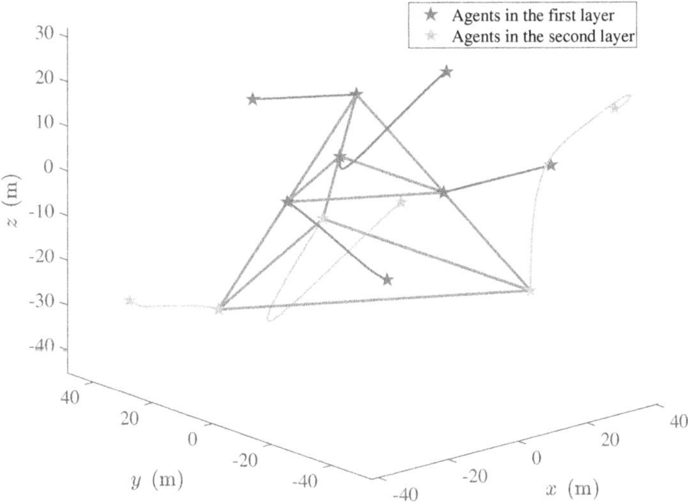

Figure 5.7 Affine formation example 2 with eigenvalues in the left plane. (The blue lines connecting the agents are given to sketch the formation shape.)

5.6 CONCLUSION

In this chapter, a layered affine formation problem has been defined and solved, where the formation shape and the information flow among the layers can be designed naturally. Under a directed graph, the necessary and sufficient condition for the affine maneuverability has been given. Adaptive neural network based layered affine formation control laws have been designed for networked Euler-Lagrange systems, where the closed-loop errors are driven into a bounded compact set in finite time. The adaptive law is designed by updating the norm of the weight matrix and requires less computation. In addition, the control laws are fully distributed, and no stabilizing matrix with global information is required. This approach provides a new insight into the leader-follower formation problems with a layered framework, where we can construct more flexible formation configurations and achieve translation, rotation, scale, or shear maneuver control.

5.7 APPENDIX A

Before moving on, a useful lemma is recalled.

Lemma 5.7

[388, 491] For $x \in \mathbb{R}$, $y \in \mathbb{R}$, and $r = \frac{r_1}{r_2} \in (0,1)$, where $r_1, r_2 > 0$ are positive odd integers, then $x^r(y-x) \leq \frac{1}{1+r}\left(y^{1+r} - x^{1+r}\right)$ and $|x^r - y^r| \leq 2^{1-r}|x-y|^r$. ■

Then, according to Lemma 5.7, we know

$$\begin{aligned}
\tilde{\psi}_i\hat{\psi}_i^r &= \left(\psi_i - \hat{\psi}_i\right)\hat{\psi}_i^r \\
&\leq \frac{1}{1+r}\left(\psi_i^{1+r} - \hat{\psi}_i^{1+r}\right) \\
&= \frac{1}{1+r}\left(\psi_i^{1+r} - \left|\psi_i - \tilde{\psi}_i\right|\left|\hat{\psi}_i - \tilde{\psi}_i\right|^r\right) \\
&\leq \frac{1}{1+r}\left(\psi_i^{1+r} - 2^{r-1}\left(\psi_i - \tilde{\psi}_i\right)\left(\psi_i^r - \tilde{\psi}_i^r\right)\right) \\
&\leq \frac{1}{1+r}\left(\psi_i^{1+r} - 2^{r-1}\psi_i^{1+r} - 2^{1+r}\tilde{\psi}_i^{1+r}\right. \\
&\quad \left. + 2^{r-1}\psi_i\tilde{\psi}_i^r + 2^{r-1}\tilde{\psi}_i\psi_i^r\right).
\end{aligned} \tag{5.31}$$

Based on Young's inequality, we have

$$2^{r-1}\psi_i\tilde{\psi}_i^r \leq 2^{r-1}\frac{r}{1+r}\tilde{\psi}_i^{1+r} + 2^{r-1}\frac{1}{1+r}\psi_i^{1+r} \tag{5.32}$$

$$2^{r-1}\tilde{\psi}_i\psi_i^r \leq \frac{r}{1+r}\left(2^{r-1}\tilde{\psi}_i\right)^{1+r} + \frac{1}{1+r}\psi_i^{1+r} \tag{5.33}$$

Substituting (5.32) and (5.33) into (5.31) yields

$$\begin{aligned}\tilde{\psi}_i\hat{\psi}_i^r \leq & \frac{1}{1+r}\left(1 - 2^{r-1} + \frac{r}{1+r} + \frac{2^{r-1}}{1+r}\right)\psi_i^{r+1} \\ & - \frac{1}{(1+r)^2}\left(2^{r-1} - 2^{(r-1)(r+1)}\right)\tilde{\psi}_i^{1+r}.\end{aligned} \tag{5.34}$$

Next, we have consider

$$\tilde{\psi}_i\hat{\psi}_i^r \leq -\kappa_1\tilde{\psi}_i^{r+1} + \kappa_2\psi_i^{r+1}, \tag{5.35}$$

with two positive constants κ_1 and κ_2 defined in Lemma 5.5. This completes the proof.

5.8 APPENDIX B

Consider the Lyapunov function

$$V_l = \sum_{i\in\mathcal{V}_l}\left(\frac{1}{2}z_{1i}^T z_{1i} + \frac{1}{2}z_{2i}^T M_i(q_i)z_{2i} + \frac{1}{2\gamma_i}\tilde{\psi}_i^2\right). \tag{5.36}$$

According to (5.9)–(5.11) and Property 4, the time derivative of V_l is given as

$$\begin{aligned}\dot{V}_l = & \sum_{i\in\mathcal{V}_l}\Big[-z_{1i}^T k_{1i}\mathrm{sig}_{\mathrm{c}}^{\ell}(z_{1i}) - \frac{1}{\gamma_i}\tilde{\psi}_i\dot{\tilde{\psi}}_i + \beta_i z_{1i}^T z_{2i} \\ & - z_{2i}^T\left(\tau_i - W_i^{*T}\phi_i(Z_i) - \epsilon_i(Z_i)\right)\Big].\end{aligned} \tag{5.37}$$

Based on Young's inequality, we have $z_{2i}^T W_i^{*T}\phi_i(Z_i) \leq \|z_{2i}\|^2\|\phi_i\|^2\psi_i/(2\upsilon_i^2) + \upsilon_i^2/2$, and $z_{2i}^T\epsilon_i(Z_i) \leq \nu_i\|z_{2i}\|^{1+\ell} + \ell/((1+\ell)^{\frac{\ell+1}{\ell}}\nu_i^{\frac{1}{\ell}})\bar{\epsilon}^{\frac{\ell+1}{\ell}}$, where υ_i and ν_i are two positive constants. Then, under control law (5.13), we can further write (5.37) as

$$\begin{aligned}\dot{V}_l \leq & \sum_{i\in\mathcal{V}_l}\Big[-z_{1i}^T k_{1i}\mathrm{sig}_{\mathrm{c}}^{\ell}(z_{1i}) - z_{2i}^T k_{2i}\mathrm{sig}_{\mathrm{c}}^{\ell}(z_{2i}) - \frac{1}{\gamma_i}\tilde{\psi}_i\dot{\tilde{\psi}}_i \\ & + \frac{\upsilon_i^2}{2} + \frac{\tilde{\psi}_i}{2\upsilon_i^2}\|\phi_i(Z_i)\|^2\|z_{2i}\|^2 + \nu_i\|z_{2i}\|^{1+\ell} \\ & + \frac{\ell}{(1+\ell)^{\frac{\ell+1}{\ell}}\nu_i^{\frac{1}{\ell}}}\bar{\epsilon}^{\frac{\ell+1}{\ell}}\Big].\end{aligned} \tag{5.38}$$

Next, substituting adaptive law (5.14) into (5.38) yields

$$\dot{V}_l \leq \sum_{i\in\mathcal{V}_l}\Bigg[-z_{1i}^T k_{1i}\mathrm{sig}_{\mathrm{c}}^{\ell}(z_{1i}) - z_{2i}^T k_{2i}\mathrm{sig}_{\mathrm{c}}^{\ell}(z_{2i}) + \nu_i\|z_{2i}\|^{1+\ell} \\ +\frac{v_i^2}{2} + \sigma_i\tilde{\psi}_i\hat{\psi}_i^{\ell} + \frac{r}{(1+\ell)^{\frac{\ell+1}{\ell}}\nu_i^{\frac{1}{\ell}}}\epsilon^{\frac{\ell+1}{\ell}}\Bigg]. \tag{5.39}$$

By utilizing Lemma 5.5 and Property 3, (5.39) becomes

$$\dot{V}_l \leq \sum_{i\in\mathcal{V}_l}\Bigg[-k_{1i}\|z_{1i}\|^{1+\ell} - (k_{2i}-\nu_i)\|z_{2i}\|^{1+\ell} - \kappa_1\sigma_i\tilde{\psi}_i^{\ell+1} \\ +\frac{v_i^2}{2} + \kappa_2\sigma_i\psi_i^{\ell+1} + \frac{\ell}{(1+\ell)^{\frac{\ell+1}{\ell}}\nu_i^{\frac{1}{\ell}}}\epsilon^{\frac{\ell+1}{\ell}}\Bigg] \\ \leq -\xi_l\rho_l V_l^{\frac{1+\ell}{2}} - (1-\xi_l)\rho_l V_l^{\frac{1+\ell}{2}} + C_l, \tag{5.40}$$

where $\xi_l \in (0,1]$, and

$$\rho_l = \min\left(2^{\frac{1+\ell}{2}}k_{1i}, \frac{2^{\frac{1+\ell}{2}}(k_{2i}-\nu_i)}{\lambda_{\max}(M_i(q_i))}, 2^{\frac{1+\ell}{2}}\kappa_1\sigma_i\gamma_i^{\frac{1+\ell}{2}}\right), \\ C_l = \frac{v_i^2}{2} + \kappa_2\sigma_i\psi_i^{\ell+1} + \frac{\ell}{(1+\ell)^{\frac{\ell+1}{\ell}}\nu_i^{\frac{1}{\ell}}}\epsilon^{\frac{\ell+1}{\ell}}.$$

Based on Lemma 2.7 and (5.36), we know that the closed-loop errors $\boldsymbol{z_{l1}}$, $\boldsymbol{z_{l2}}$, and $\tilde{\psi}_l$ will converge to the compact set $\Omega_{z_{l1}}$, $\Omega_{z_{l2}}$, and $\Omega_{\tilde{\psi}_l}$ in finite time $T_l \leq 2V_l^{\frac{1-\ell}{2}}(0)/((1-\ell)\rho_l\xi)$, where $\Omega_{z_{l1}}$, $\Omega_{z_{l2}}$, and $\Omega_{\tilde{\psi}_l}$ have the forms as given in Theorem 5.2 with $D_l = \sqrt{2}(C_l/((1-\xi_l)\rho_l))^{\frac{1}{1+\ell}}$.

We next show the formation tracking error $\boldsymbol{e_l}$ in the first layer converges to a bounded set in finite time T_l. Rewrite (5.5) as

$$\boldsymbol{z_{l1}} = (L_{l1}\otimes I_d)(\boldsymbol{q_l}-\boldsymbol{p}) + (L_{l2}\otimes I_d)q_0. \tag{5.41}$$

According to Assumption 10 and Lemma 5.1, L_{l1} is nonsingular, and (5.41) becomes

$$\boldsymbol{q_l} + \left(L_{l1}^{-1}L_{l2}\otimes I_d\right)q_0 - \boldsymbol{p} = \left(L_{l1}^{-1}\otimes I_d\right)\boldsymbol{z_{l1}}. \tag{5.42}$$

Since each row sum of $-L_{l1}^{-1}L_{l2}$ is equal to one, considering (5.42), we can obtain

$$\boldsymbol{e_l} = \left(L_{l1}^{-1}\otimes I_d\right)\boldsymbol{z_{l1}} \leq D_l\left\|L_{l1}^{-1}\otimes I_d\right\|,$$

which completes the proof for the first layer.

5.9 APPENDIX C

Consider the Lyapunov function

$$V_f = \sum_{i\in\mathcal{V}_f} \left(\frac{1}{2} z_{1i}^T z_{1i} + \frac{1}{2} z_{2i}^T M_i(q_i) z_{2i} + \frac{1}{2\gamma_i} \tilde{\psi}_i^2 \right), \tag{5.43}$$

whose time derivative is given as

$$\begin{aligned} \dot{V}_f =& \sum_{i\in\mathcal{V}_f} \Big[-z_{1i}^T k_{1i} \mathrm{sig}_{\mathrm{c}}^{\ell}(z_{1i}) - \frac{1}{\gamma_i} \tilde{\psi}_i \dot{\tilde{\psi}}_i + \beta_i z_{1i}^T z_{2i} \\ & - z_{2i}^T \left(\tau_i - W_i^{*T} \phi_i(Z_i) - \epsilon_i(Z_i) \right) \Big] . \end{aligned} \tag{5.44}$$

Based on Young's inequality, under control law (5.26), we further have

$$\begin{aligned} \dot{V}_f \leq & \sum_{i\in\mathcal{V}_f} \Big[-z_{1i}^T k_{1i} \mathrm{sig}_{\mathrm{c}}^{\ell}(z_{1i}) - z_{2i}^T k_{2i} \mathrm{sig}_{\mathrm{c}}^{\ell}(z_{2i}) - \frac{1}{\gamma_i} \tilde{\psi}_i \dot{\tilde{\psi}}_i \\ & + \frac{\upsilon_i^2}{2} + \frac{\tilde{\psi}_i}{2\upsilon_i^2} \|\phi_i(Z_i)\|^2 \|z_{2i}\|^2 + \nu_i \|z_{2i}\|^{1+\ell} \\ & + \frac{\ell}{(1+\ell)^{\frac{\ell+1}{\ell}} \nu_i^{\frac{1}{\ell}}} \bar{\epsilon}^{\frac{\ell+1}{\ell}} \Big], \end{aligned} \tag{5.45}$$

where υ_i and ν_i are two positive constants.

Next, substituting adaptive law (5.27) into (5.45) yields

$$\begin{aligned} \dot{V}_f \leq & \sum_{i\in\mathcal{V}_f} \Big[-z_{1i}^T k_{1i} \mathrm{sig}_{\mathrm{c}}^{\ell}(z_{1i}) - z_{2i}^T k_{2i} \mathrm{sig}_{\mathrm{c}}^{\ell}(z_{2i}) + \nu_i \|z_{2i}\|^{1+\ell} \\ & + \frac{\upsilon_i^2}{2} + \sigma_i \tilde{\psi}_i \hat{\psi}_i^{\ell} + \frac{\ell}{(1+\ell)^{\frac{\ell+1}{\ell}} \nu_i^{\frac{1}{\ell}}} \bar{\epsilon}^{\frac{\ell+1}{\ell}} \Big]. \end{aligned} \tag{5.46}$$

According to Lemma 5.5 and Property 3, we can rewrite (5.45) as

$$\begin{aligned} \dot{V}_f \leq & \sum_{i\in\mathcal{V}_f} \Big[-k_{1i} \|z_{1i}\|^{1+\ell} - (k_{2i} - \nu_i) \|z_{2i}\|^{1+\ell} - \kappa_1 \sigma_i \tilde{\psi}_i^{\ell+1} \\ & + \frac{\upsilon_i^2}{2} + \kappa_2 \sigma_i \psi_i^{\ell+1} + \frac{\ell}{(1+\ell)^{\frac{\ell+1}{\ell}} \nu_i^{\frac{1}{\ell}}} \bar{\epsilon}^{\frac{\ell+1}{\ell}} \Big] \\ \leq & - \xi_f \rho_f V_f^{\frac{1+\ell}{2}} - (1-\xi_f) \rho_f V_f^{\frac{1+\ell}{2}} + C_f, \end{aligned}$$

where $\xi_f \in (0, 1]$, and

$$\begin{aligned} \rho_f =& \min \left(2^{\frac{1+\ell}{2}} k_{1i}, \frac{2^{\frac{1+\ell}{2}} (k_{2i} - \nu_i)}{\lambda_{\max}(M_i(q_i))}, 2^{\frac{1+\ell}{2}} \kappa_1 \sigma_i \gamma_i^{\frac{1+\ell}{2}} \right), \\ C_f =& \frac{\upsilon_i^2}{2} + \kappa_2 \sigma_i \psi_i^{\ell+1} + \frac{\ell}{(1+\ell)^{\frac{\ell+1}{\ell}} \nu_i^{\frac{1}{\ell}}} \bar{\epsilon}^{\frac{\ell+1}{\ell}}. \end{aligned} \tag{5.47}$$

By using Lemma 2.7 and (5.43), the closed-loop errors $\boldsymbol{z_{f1}}$, $\boldsymbol{z_{f2}}$, and $\tilde{\psi}_f$ will converge to the compact set $\Omega_{z_{f1}}$, $\Omega_{z_{f2}}$, and $\Omega_{\tilde{\psi}_f}$ in finite time T_f, where $\Omega_{z_{f1}}$, $\Omega_{z_{f2}}$, and $\Omega_{\tilde{\psi}_f}$ take the forms as shown in Theorem 5.3 with $D_f = \sqrt{2}(C_f/((1-\xi_f)\,\rho_f))^{\frac{1}{1+\ell}}$. Further, according to Lemma 2.7, we have $T_f \leq 2V_f^{\frac{1-\ell}{2}}(0)/((1-\ell)\,\rho_f\xi_f)$.

Next, rewrite (5.22) as

$$\boldsymbol{z_{f1}} = (L^s_{f1} \otimes I_d)\boldsymbol{q_f} + (L^s_{f2} \otimes I_d)\boldsymbol{q_l}. \tag{5.48}$$

We know that the formation tracking error $\boldsymbol{e_f}$ in the second layer converges to Ω_{e_f} in finite time T_f, as $\boldsymbol{e_f} = \left((L^s_{f1})^{-1} \otimes I_d\right)\boldsymbol{z_{f1}}$, which completes the proof for the second layer.

6 Cooperative Circumnavigation Control of Networked Microsatellites

This chapter addresses the trajectory analysis, mission design, and control law for multiple microsatellites to cooperatively circumnavigate a host spacecraft. This cooperative circumnavigation (CCN) problem is defined to drive a group of networked microsatellites to a predefined planar ellipse concerning a host spacecraft while maintaining a geometric formation configuration. We first design several potential functions to guide the microsatellites to the given planar elliptical orbit with a proper radius. Next, the affine Laplacian matrix is introduced to characterize the desired formation shape of microsatellites. Based on the potential functions and the Laplacian matrix, a CCN control law is finally proposed. Then, simulation results of multiple microsatellites with Earth-orbiting mission scenarios are given, where the natural trajectory motion is incorporated which consumes nearly zero-fuel.

6.1 INTRODUCTION

Satellite formation flying has been an active area of research due to its importance in space engineering [21, 186, 224, 413, 421], which holds advantages over a single satellite as it offers stronger robustness, higher efficiency, simpler design procedure, and easier maintenance of space systems. In addition, There are academic results of satellite formation control using different techniques, such as the finite-time control theory [196, 255, 453], barrier Lyapunov functions [262], event-trigger based-control [59], potential functions [194], etc.

The concepts of microsatellites and networked systems, such as, optical stellar interferometers, synthetic aperture radar, and other space formation flight missions, e.g., GRACE mission, Prisma mission, Aero-4 mission, THEMIS mission, and mission of EO-1 in formation with LandSat-7, TechSat21 [7, 142, 154, 239], make satellite formation garner even more attention, which also bring satellite formation from academic research interest to practical missions.

As one of the most advanced enabling technologies for satellite missions, the cooperative circumnavigation (CCN) problem is defined to drive a group of networked microsatellites to a predefined planar ellipse with respect to a host

DOI: 10.1201/9781003298618-6

spacecraft while maintaining a geometric formation configuration [383]. There are several technical challenges that the CCN missions envisage: high-precision guidance, distributed communication, and geometric formation shape control. Another significant challenge is to develop an efficient formation maintenance control law which enables the microsatellites to maintain a desired relative orbit with respect to the host spacecraft with minimum fuel after the launching of multiple microsatellites.

Inspired by the vector-field technique in [342], we first design several potential functions to guide the microsatellites to the given planar elliptical orbit with a proper radius and a circulating rate. Next, to characterize the desired geometric formation shape of microsatellites in line with the orbit, the affine Laplacian matrix is introduced, which can preserve collinearity and ratios of distances corresponding to a nominal formation [291,505,506,511,546]. Based on the potential functions and the affine Laplacian matrix, a CCN control law is then proposed. Different from the classical formation tracking problems addressed in [21,59,186,194,196,224,255,262,413,421,453], using the proposed scheme, the networked mircosatellites can achieve elliptical motion around a host spacecraft with desired spaced formation in 3-D space. The significance of this problem also lies in the orbit design which consumes nearly zero-fuel without the need of designating specific waypoints if a proper phase is given. Compared with the vector-field control in [342, 383], we give the Lyapunov stability analysis of the positions and velocities of networked microsatellites simultaneously, while the method to describe the networked communication and the nominal formation shape is also fully addressed.

6.2 PRELIMINARIES AND PROBLEM FORMULATION

6.2.1 MOTION DYNAMICS OF MICROSATELLITES

Suppose that there are N microsatellites, where we denote $\mathcal{V}_c = \{1, 2, \ldots, N\}$, $N \in \mathbb{Z}^+$. We first introduce an inertial frame, which is fixed to the center of the Earth. The motion dynamics of each microsatellite is given as

$$\ddot{r}_i + \frac{\mu}{\|r_i\|^3} r_i = \frac{1}{m_i} F_i,\ i \in \mathcal{V}_c, \tag{6.1}$$

where $r_i \in \mathbb{R}^3$ is the position vector of the ith microsatellite in the inertial frame, μ is the gravitational parameter, m_i is the mass of the ith microsatellite, and $F_i \in \mathbb{R}^3$ is the external force contributed by control inputs or dynamic perturbations.

The dynamics in (6.1) gives an elliptical orbit with respect to the center of the Earth (which is approximately thc center of gravity). Next, we present the dynamics of relative motion between a microsatellite and a host spacecraft in a circumnavigation mission.

The microsatellite is modeled as a point mass, whose motion is described relative to the host spacecraft by the local-vertical-local-horizontal (LVLH)

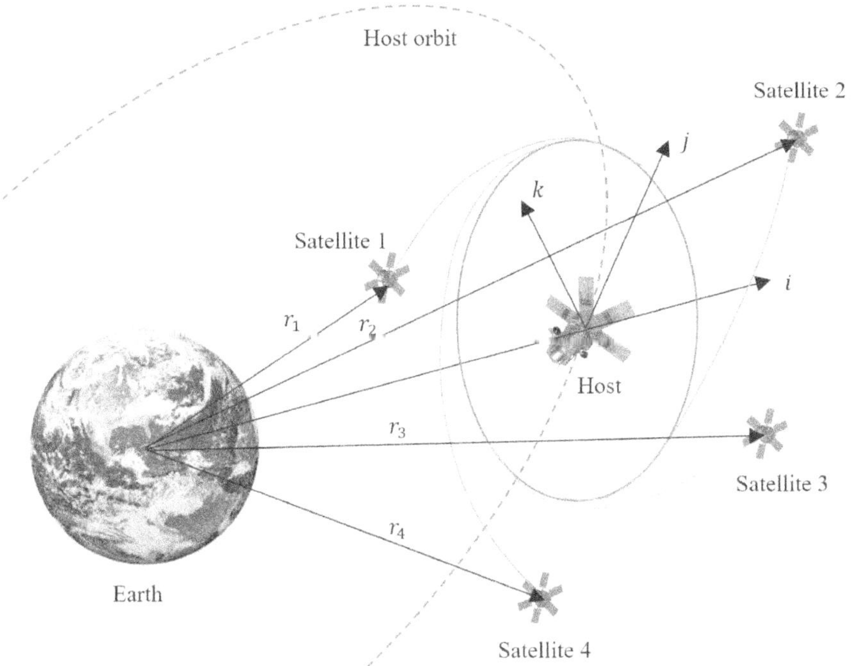

Figure 6.1 Circumnavigation of a host spacecraft with 4 microsatellites.

frame with unit axis vectors i, j, and k, aligned with the radial track, the in-track, and the out-of-plane orbital positions, respectively. Here, in a typical manner, Fig. 6.1 shows the coordinate system used. Again in a typical manner, the host spacecraft is assumed to be in a circular orbit, and to maintain that orbit in free flight, around a spherical Earth. The microsatellites are, on the other hand, in slightly elliptical orbits but can be considered to remain close to the host spacecraft; a situation which is appropriate when we compare their orbits about the host spacecraft to the significantly larger overall radii of their orbits around the Earth. We can use the classical Clohessy-Wiltshire (CW) to describe the relative motion as [18]

$$\begin{aligned} m_i\left(\ddot{x}_i - 2n\dot{y}_i - 3n^2 x_i\right) &= u_{i,x} + d_{i,x}, \\ m_i\left(\ddot{y}_i + 2n\dot{x}_i\right) &= u_{i,y} + d_{i,x}, \\ m_i\left(\ddot{z}_i + n^2 z_i\right) &= u_{i,z} + d_{i,x}, \end{aligned} \tag{6.2}$$

where $\rho_i = [x_i, y_i, z_i]^T$ is the relative vector for satellite i in the LVLH frame, r_0 is the radius of the host spacecraft's orbit, $n = \sqrt{\mu/r_0^3}$ is the mean motion with respect to the host spacecraft's orbit, $u_i = [u_{i,x}, u_{i,y}, u_{i,z}]^T \in \mathbb{R}^3$ and $d_i = [d_{i,x}, d_{i,y}, d_{i,z}]^T \in \mathbb{R}^3$ are the control input and the perturbation force expressed in the relative frame, respectively. We reformulate the relative motion (6.2) as

$$\begin{aligned} \dot{\rho}_i &= v_i, \\ \dot{v}_i &= A_\rho \rho_i + A_v v_i + B_i \left(u_i + d_i\right), \; i \in \mathcal{V}_c, \end{aligned} \tag{6.3}$$

where A_ρ, A_v, and B_i are given by

$$A_\rho = \begin{bmatrix} 3n^2 & 0 & 0 \\ 0 & 0 & 0 \\ 0 & 0 & -n^2 \end{bmatrix}, \; A_v = \begin{bmatrix} 0 & 2n & 0 \\ -2n & 0 & 0 \\ 0 & 0 & 0 \end{bmatrix}, \; B_i = \frac{I_3}{m_i}.$$

In this chapter, we assume that there are no external forces, i.g., $d_i = [0,0,0]^T$.

Remark 26. *We note that there exists an interesting stability property for (6.3). The dynamics in the $i-j$ plane have complex eigenvalues with zero real part, which results in stable elliptical orbits in the $i-j$ plane. Furthermore, under these circumstances, it is known that the resulting motion in the moving coordinate frame is an ellipse whose projection in the orbit plane is an ellipse having a semi-major axis twice its semi-minor axis [440].*

6.2.2 PROBLEM FORMULATION

Problem 3. *Given N microsatellites with full control authority, develop a control law that drives each microsatellite to a formation with the following requirements:*

i Each microsatellite must reach a formation plane defined by normal $\alpha \in \mathbb{R}^3$ with $\|\alpha\| = 1$, i.e.,

$$\alpha^T \left(\rho_i - \rho_h\right) = 0, \; i \in \mathcal{V}_c, \tag{6.4}$$

where ρ_h stands for the position of the host spacecraft in the LVLH frame.

ii The microsatellite formation circulate at a predefined rate around an ellipse with a semi-major axis of length l_M and a semi-minor axis of length l_m, with each axes direction defined by β and γ respectively.

iii The microsatellites achieve the formation with a desired phase

$$\theta_{ij} = \cos^{-1}\left(\frac{\left(\rho_i - \rho_h\right)^T \left(\rho_j - \rho_h\right)}{\|(\rho_i - \rho_h)\| \, \|(\rho_j - \rho_h)\|}\right), \tag{6.5}$$

where microsatellites $i, j \in \mathcal{V}_c$, and $i \neq j$.

6.2.3 COMMUNICATION TOPOLOGY

We utilize a communication graph $\mathcal{G} = (\mathcal{V}_c, \mathcal{E})$ to characterize the underlying information flow among N microsatellites, where $\mathcal{E} \subseteq \mathcal{V}_c \times \mathcal{V}_c$ and $\mathcal{V}_c =$

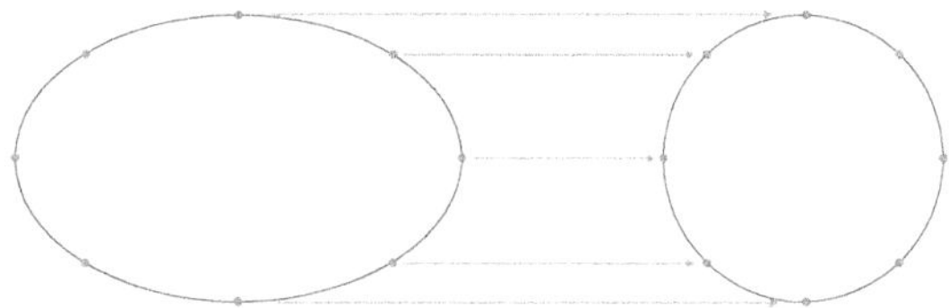

Figure 6.2 Scaling to make the desired elliptical formation a circle.

$\{1, 2, \ldots, N\}$ denote the arc set and the vertex set, respectively. Denote $\mathcal{N}_i$ as the in-neighbor set of node i, where $\mathcal{N}_i = \{j : (j, i) \in \mathcal{E}\}$.

Then, note the following affine Laplacian matrix introduced in [291, 506].

Definition 27. *An affine Laplacian matrix L^s associated to a graph has both positive and negative real off-diagonal entries. The matrix L^s of a directed graph is defined as follows*

$$L^s(i,j) = \begin{cases} -\omega_{ij} & \text{if } i \neq j \text{ and } j \in \mathcal{N}_i, \\ 0 & \text{if } i \neq j \text{ and } j \notin \mathcal{N}_i, \\ \sum\limits_{k \in \mathcal{N}_i} \omega_{ik} & \text{if } i = j, \end{cases} \tag{6.6}$$

where $\omega_{ij} \in \mathbb{R}$ may be a positive or negative real weight attributed on the edge (i, j), and L^s is normally a nonsymmetric matrix.

6.3 CIRCUMNAVIGATION AND CONTROL DESIGN

For ease of analyze, a scaling matrix S is used to scale the original elliptical formation to a circle as shown in Fig. 6.2. We rewrite the dynamics in (6.3) as

$$\begin{aligned} \dot{q}_i &= p_i, \\ \dot{p}_i &= SA_\rho \rho_i + SA_v v_i + SB_i u_i, \end{aligned} \tag{6.7}$$

where a new set of coordinates are given as

$$q_i = S\rho_i,\ p_i = Sv_i, \text{ and } S = r_h \left[\beta/l_M, \gamma/l_m, \alpha/r_h\right]^T,$$

and the prescribed radius of the circle corresponding to the elliptical formation is r_h.

Corresponding to first and second requirements introduced in Problem 3, we design potential functions $P_{i,\alpha}$, P_{i,r_h}, and $P_{i,v}$ respectively

$$P_{i,\alpha} = \frac{1}{2}\left(\alpha^T (q_i - q_h)\right)^2, \tag{6.8}$$

$$P_{i,r_h} = \frac{1}{2}\left(\|G_\alpha (q_i - q_h)\| - r_h\right)^2, \tag{6.9}$$

$$P_{i,v} = \frac{1}{2}(p_i - p_i^*)^T (p_i - p_i^*), \tag{6.10}$$

where $q_h = S\rho_h$, $p_i^* = n \left\| G_\alpha (q_i - q_h) \right\| (\alpha \times \phi_{ih}^\alpha)$, $G_\alpha = I_3 - \alpha\alpha^T / \|\alpha\|^2$. The partial derivatives of $P_{i,\alpha}$, P_{i,r_h}, and $P_{i,v}$ with respect to q_i are respectively given as

$$\frac{\partial P_{i,\alpha}}{\partial q_i} = \alpha^T (q_i - q_h) \alpha, \tag{6.11}$$

$$\frac{\partial P_{i,r_h}}{\partial q_i} = \left(\left\| G_\alpha (q_i - q_h) \right\| - r_h \right) \phi_{ih}^\alpha, \tag{6.12}$$

$$\frac{\partial P_{i,v}}{\partial q_i} = p_i - p_i^*, \tag{6.13}$$

where $\phi_{ih}^\alpha = G_\alpha (q_i - q_h) / \left\| G_\alpha (q_i - q_h) \right\|$.

To meet the third requirement introduced in Problem 3, we next present some useful properties of the affine transformation and nominal formation.

The affine transformation is a general linear transformation, which may be translation, rotation, scaling, shear or a combination of them. A formation $(\mathcal{G}, q)$ can be defined as a directed graph $\mathcal{G}$ with its vertex i mapped to q_i. Then the nominal formation associated to $\mathcal{G}$ is represented as $(\mathcal{G}, f)$, and $f = [f_1^T, f_2^T, \ldots, f_N^T]^T \in \mathbb{R}^{3N}$ is a constant vector and named as nominal configuration. The affine image of the nominal configuration can be defined as [291]

$$\begin{aligned} \mathcal{A}(f) = \{ & q \in \mathbb{R}^{3N} : q = (I_N \otimes A) f + \mathbf{1}_N \otimes b, \\ & A \in \mathbb{R}^{3\times 3},\ b \in \mathbb{R}^3 \}, \end{aligned} \tag{6.14}$$

where $q = \left[q_1^T, q_2^T, \ldots, q_N^T \right]^T$, and (A, b) is the affine transformation with $A(t) \in \mathbb{R}^{3\times 3}$ and $b(t) \in \mathbb{R}^3$.

Remark 27. *The affine formation was first introduced in [291], which preserves collinearity and ratios of distances with respect to a target configuration. Please refer to [291, 546] for more information on the stabilizability of nominal formation and the necessary and sufficient graphical condition of affine formation.*

Next, a lemma about the directed graph is introduced.

Lemma 6.1

[291] Suppose a directed graph $\mathcal{G}$ has N nodes with $N \geq d + 2$ and f is generic. Then for graph $\mathcal{G}$ and $q \in \mathcal{A}(r)$, the nominal formation is stabilizable if and only if $\mathcal{G}$ is $(d + 1)$-rooted. ■

Based on Lemma 6.1, we can use the nominal formation to characterize the circling formation of microsatellites with a desired phase (which is the third requirement in Problem 3), since the affine transformation preserves translation, rotation and scaling. Next, the following auxiliary variables are defined

$$z_{1i} = \sum_{j\in\mathcal{N}_i} \omega_{ij} (q_i - q_j), \tag{6.15}$$

$$z_{2i} = p_i - \sigma_i, \ i \in \mathcal{V}_c. \tag{6.16}$$

The virtual control σ_i takes the following form

$$\begin{aligned}\sigma_i = & - \Theta_i^\dagger \left[\left(\frac{\partial P_{i,\alpha}}{\partial q_i} + \frac{\partial P_{i,r_h}}{\partial q_i} \right)^T \left(-\dot{q}_h + \frac{1}{\bar{\omega}_i} \sum_{j\in N_i} \omega_{ij} p_j \right) \right] \\ & + \frac{1}{\bar{\omega}_i} \sum_{j\in N_i} \omega_{ij} p_j - \Theta_i^\dagger k_{1i} V_{\chi i},\end{aligned} \tag{6.17}$$

where

$$\bar{\omega}_i = \sum\nolimits_{j\in\mathcal{N}_i} \omega_{ij}, \tag{6.18}$$

$$\Theta_i = \left(\bar{\omega}_i z_{1i} + \frac{\partial P_{i,\alpha}}{\partial q_i} + \frac{\partial P_{i,r_h}}{\partial q_i} \right)^T, \tag{6.19}$$

$$V_{\chi i} = \frac{1}{2} z_{1i}^T z_{1i} + P_{i,\alpha} + P_{i,r_h} + P_{i,v}, \tag{6.20}$$

k_{1i} is a positive control gain, and $\Theta_i^\dagger$ is the Moore-Penrose inverse of Θ_i defined as

$$\Theta_i \Theta_i^\dagger = \begin{cases} 0, & \Theta_i = [0, 0, \ldots, 0]^T, \\ 1, & \text{otherwise}, \end{cases} \tag{6.21}$$

We next formally propose the CCN control law

$$u_i = B_i^{-1} S^{-1} \left(s_i + \dot{\sigma}_i - S A_\rho \rho_i - S A_v v_i \right), \tag{6.22}$$

$$\begin{aligned} s_i = & - \bar{\omega}_i z_{1i} - k_{2i} z_{2i} - \frac{\partial P_{i,\alpha}}{\partial q_i} - \frac{\partial P_{i,r_h}}{\partial q_i} \\ & - \left(z_{2i}^T \right)^\dagger \left(\left(\frac{\partial P_{i,v}}{\partial p_i} \right)^T (\dot{p}_i - \dot{p}_i^*) \right), \end{aligned} \tag{6.23}$$

where k_{2i} is a positive constant.

We then give the main result for the CCN.

Theorem 6.1

Consider the networked microsatellite systems (6.2) with control law (6.22), If the initial condition satisfies $\|G_\alpha (q_i(0) - q_h(0))\| > 0$, i.e., $(q_i(0) - q_h(0))$ is not parallel to α, then the closed-loop system are exponentially stable, and ultimately the objectives in Problem 3 would be achieved. ■

Table 6.1
Parameters for simulation

Parameters	Values	Parameters	Values
k_{1i}	0.001	α	$[0,0,1]^T$
k_{2i}	0.001	β	$[1,0,0]^T$
m_i	10 kg	γ	$[0,1,0]^T$
l_M	1000 m	n	0.001
l_m	500 m	ρ_h	$[0,0,0]^T$ m
r_h	500 m		

Proof. For details, see Appendix 6.5. □

Lemma 6.2

For Theorem 6.1, no singularity exists in ϕ_{ih}^{α}, i.e., $\|G_\alpha (q_i - q_h)\| > 0, \forall t \geq 0$.

Proof. According to the proof of Theorem 6.1, we can get $\dot{V} \leq -kV$ with $k = \min\{k_{1i}, \lambda_{\min}(K_{2i})\}$. Integrate the both sides yielding

$$\frac{1}{2}(\|G_\alpha (q_i - q_h)\| - r_h)^2 \leq V(t) \leq V(0)\, e^{-kt}, \tag{6.24}$$

which shows that $\|G_\alpha (q_i - q_h)\|$ converges to r_h exponentially. Since r_h is a positive constant, we can get that $\|G_\alpha (q_i - q_h)\| > 0$ as long as $\|G_\alpha (q_i(0) - q_h(0))\| > 0, \forall t \geq 0$. □

■

Remark 28. *From a practical perspective, the collision avoidance is important for the whole complex systems. One possible solution is to apply the following artificial potential function for agent i [416],*

$$P_{i,j}^{c}(\tilde{\rho}_{i,j}) = \begin{cases} 0, & \|\tilde{\rho}_{i,j}\| \geq r_d, \\ c_{i,j} \ln \frac{r_d}{\|\tilde{\rho}_{i,j}\|}, & \|\tilde{\rho}_{i,j}\| < r_d, \end{cases} \tag{6.25}$$

where $\tilde{\rho}_{i,j}$ is defined as $\tilde{\rho}_{i,j} = \rho_i - \rho_j$ $(\forall i, j \in \mathcal{V}_c,\ i \neq j)$, r_d is the radius of the detection range centered at the agent, $c_{i,j}$ is a positive constant gain, and $c_{i,j} = c_{j,i}$.

To illustrate the full power of the proposed CCN law, a group of 8 microsatellites are considered in the LVLH frame. The parameters for simulation are given in Table 6.1, which are similar to those in [383]. The initial positions

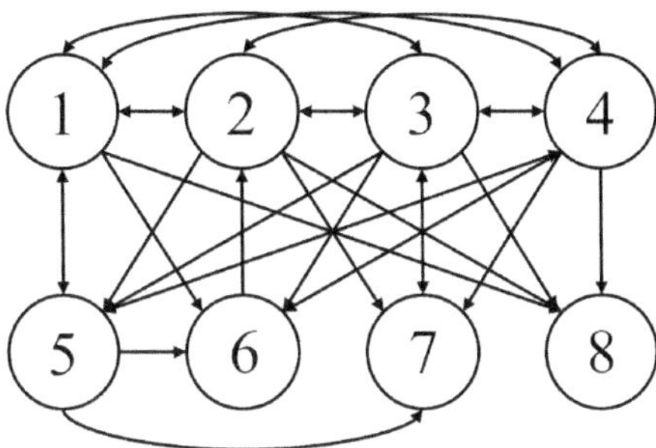

Figure 6.3 The communication topology.

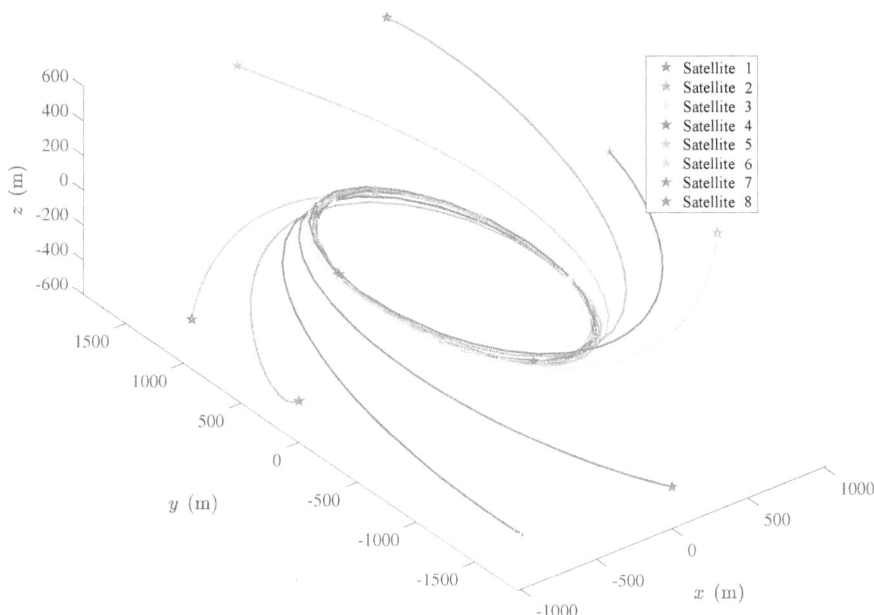

Figure 6.4 Trajectories of microsatellites (3D view)

of microsatellites are chosen around the host spacecraft, which satisfy the initial condition given in Theorem 6.1. The communication topology is given in Fig. 6.3. The detailed Laplacian matrix L^s is omitted here. The satellite trajectories are given in Figs. 6.4 and 6.5. Figs. 6.6 and 6.7 show that the formation errors of 8 microsatellites converge to the origin within 1.5×10^4

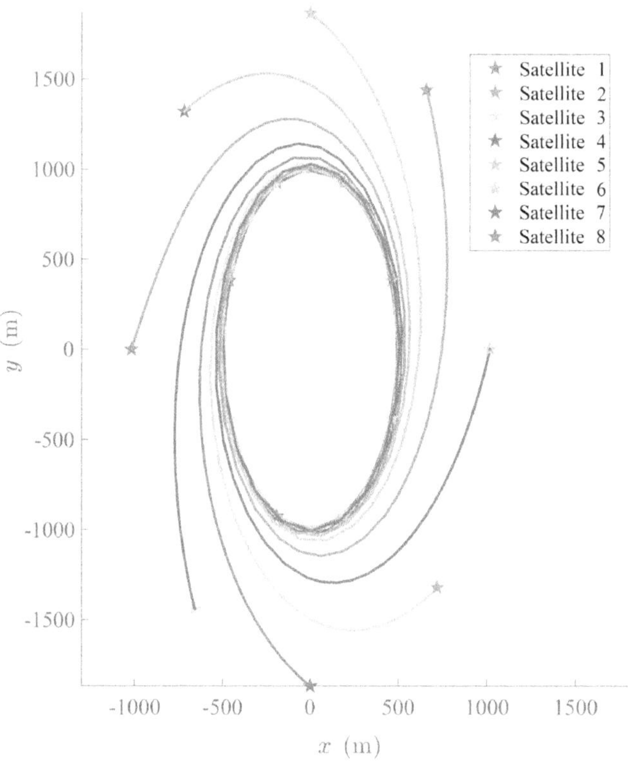

Figure 6.5 Trajectories of microsatellites (side view).

seconds. The corresponding control inputs of the first 50 seconds are given in Figs. 6.8 and 6.9, which are acceptable in practice. All control inputs converge to zero after microsatellites achieving CCN. Next, a three-dimensional CCN mission with two groups of networked microsatellites is given in Fig. 6.10, of which the simulation parameters are omitted. In addition, another two simulation results are presented to fully show the performance of the CCN control law, of which the simulation parameters are omitted. In Fig. 6.10, a three-dimensional CCN mission with two groups of networked microsatellites is given. Next, a moving host spacecraft is considered in Fig. 6.11, where the CCN is achieved by 8 microsatellites. The above numerical results have shown that the approach in this chapter is effective and feasible for a microsatellite mission of CCN of a host spacecraft.

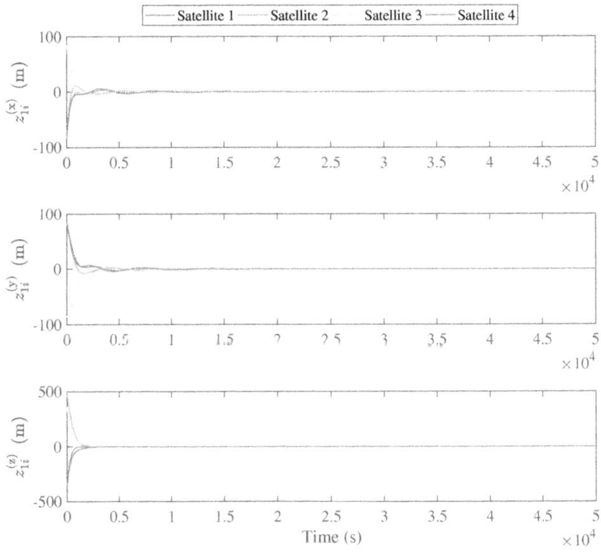

Figure 6.6 Formation errors of microsatellite 1–4.

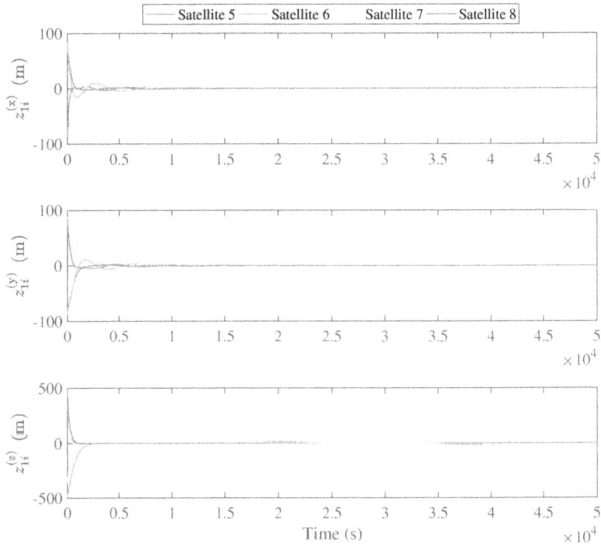

Figure 6.7 Formation errors of microsatellite 5–8.

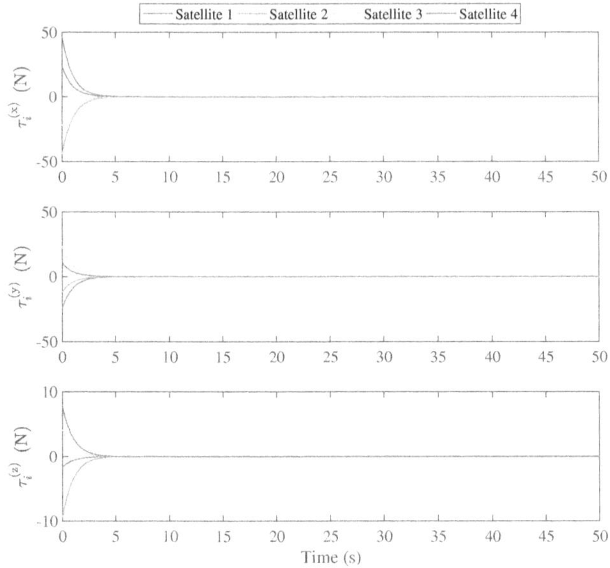

Figure 6.8 Control forces of microsatellite 1–4.

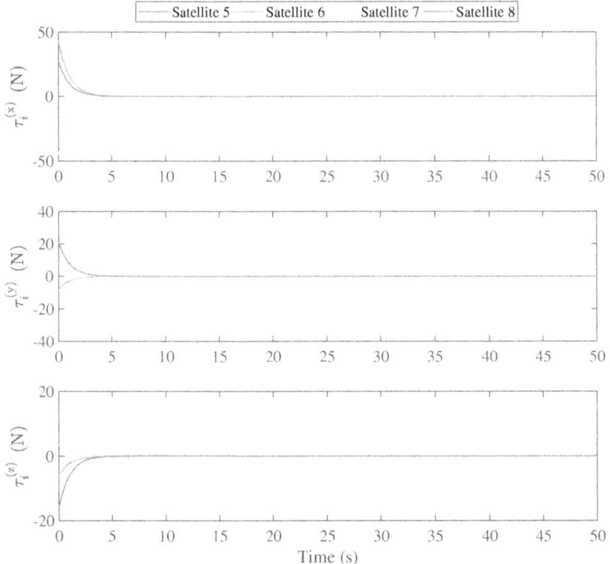

Figure 6.9 Control forces of microsatellite 5–8.

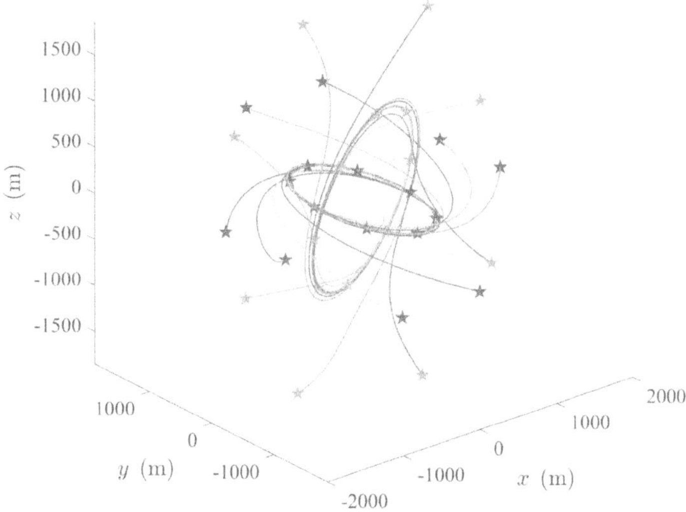

Figure 6.10 Cooperative circumnavigation with 2 groups of microsatellites. The blue stars and the red stars indicate two groups of macrosatellits, respectively.

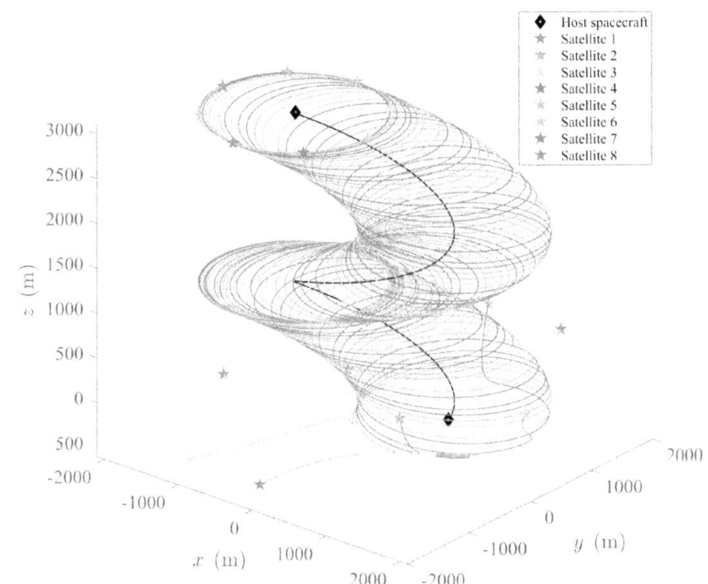

Figure 6.11 Cooperative circumnavigation of a moving host spacecraft by 8 microsatellites.

6.4 CONCLUSION

In this chapter, considering a host spacecraft, a cooperative circumnavigation problem has been defined and solved via a series of networked microsatellites. Three artificial potential functions and the affine transformations have been utilized to design the cooperative circumnavigation control law, where the closed-loop errors converge to the origin exponentially. This approach not only provides a new insight into the satellite formation problems with a cooperative circumnavigation framework but also holds advantages in practical missions as the fuel-free natural trajectories of networked microsatellites are integrated to the control design.

6.5 APPENDIX

Construct Lyapunov function

$$V=\sum_{i\in\mathcal{V}_c}\left(\frac{1}{2}z_{1i}^Tz_{1i}+\frac{1}{2}z_{2i}^Tz_{2i}+P_{i,\alpha}+P_{i,r_h}+P_{i,v}\right). \tag{6.26}$$

Note that $\frac{\partial P_{i,\alpha}}{\partial q_i}=-\frac{\partial P_{i,\alpha}}{\partial q_h}$, $\frac{\partial P_{i,r_h}}{\partial q_i}=-\frac{\partial P_{i,r_h}}{\partial q_h}$, and $\frac{\partial P_{i,v}}{\partial q_i}=-\frac{\partial P_{i,v}}{\partial q_i^*}$. Then, the derivative of (6.26) takes the form

$$\begin{aligned}\dot{V}=\sum_{i\in\mathcal{V}_c}&\left(\left(\frac{\partial P_{i,\alpha}}{\partial q_i}+\frac{\partial P_{i,r_h}}{\partial q_i}\right)^T\left(\frac{1}{\bar{\omega}_i}\left(\dot{z}_{1i}+\sum_{i\in N_i}\omega_{ij}p_j\right)-\dot{q}_h\right)\right.\\&\left.+z_{1i}^T\dot{z}_{1i}+z_{2i}^T\dot{z}_{2i}+\left(\frac{\partial P_{i,v}}{\partial q_i}\right)^T(\dot{p}_i-\dot{p}_i^*)\right).\end{aligned} \tag{6.27}$$

Taking the derivative of both sides of (6.15), according to (6.16), we can get

$$\dot{z}_{1i}=\bar{\omega}_i\left(z_{2i}+\sigma_i\right)-\sum_{i\in\mathcal{N}_i}\omega_{ij}p_j. \tag{6.28}$$

Substituting (6.17) and (6.28) into (6.27) yields

$$\begin{aligned}\dot{V}=\sum_{i\in\mathcal{V}_c}&\left(\left(\bar{\omega}_iz_{1i}+\frac{\partial P_{i,\alpha}}{\partial q_i}+\frac{\partial P_{i,r_h}}{\partial q_i}\right)^T\left(z_{2i}-\Theta_i^\dagger k_{1i}V_{\chi i}\right.\right.\\&\left.\left.-\Theta_i^\dagger\left(\frac{\partial P_{i,\alpha}}{\partial q_i}+\frac{\partial P_{i,r_h}}{\partial q_i}\right)^T\left(\frac{1}{\bar{\omega}_i}\sum_{j\in N_i}\omega_{ij}p_j-\dot{q}_h\right)\right)\right.\\&+\left(\frac{\partial P_{i,\alpha}}{\partial q_i}+\frac{\partial P_{i,r_h}}{\partial q_i}\right)^T\left(\frac{1}{\bar{\omega}_i}\sum_{i\in N_i}\omega_{ij}p_j-\dot{q}_h\right)\\&\left.+z_{2i}^T\dot{z}_{2i}+\left(\frac{\partial P_{i,v}}{\partial q_i}\right)^T(\dot{p}_i-\dot{p}_i^*)\right).\end{aligned} \tag{6.29}$$

Taking the derivative of both sides of (6.16), based on the dynamics (6.7) and the CCN control law (6.22), we know

$$\begin{aligned}\dot{z}_{2i} = s_i = & -\bar{\omega}_i z_{1i} - k_{2i} z_{2i} - \frac{\partial P_{i,\alpha}}{\partial q_i} - \frac{\partial P_{i,r_h}}{\partial q_i} \\ & - \left(z_{2i}^T\right)^{\dagger} \left(\frac{\partial P_{i,v}}{\partial p_i}\right)^T (\dot{p}_i - \dot{p}_i^*) .\end{aligned} \tag{6.30}$$

Next, substitute (6.30) into (6.29), and we further have

$$\begin{aligned}\dot{V} = & \sum_{i \in \mathcal{V}_c} \left(\left(\bar{\omega}_i z_{1i} + \frac{\partial P_{i,\alpha}}{\partial q_i} + \frac{\partial P_{i,r_h}}{\partial q_i} \right)^T z_{2i} + \left(\frac{\partial P_{i,v}}{\partial q_i} \right)^T (\dot{p}_i - \dot{p}_i^*) \right. \\ & + z_{2i}^T \left(-\bar{\omega}_i z_{1i} - k_{2i} z_{2i} - \frac{\partial P_{i,\alpha}}{\partial q_i} - \frac{\partial P_{i,r_h}}{\partial q_i} \right. \\ & \left. \left. - \left(z_{2i}^T\right)^{\dagger} \left(\frac{\partial P_{i,v}}{\partial p_i}\right)^T (\dot{p}_i - \dot{p}_i^*) \right) - k_{1i} V_{\chi i} \right) .\end{aligned} \tag{6.31}$$

According to (6.20) and the property of $\left(z_{2i}^T\right)^{\dagger}$, we next have $\dot{V} \leq -\min\{k_{1i}, k_{2i}\} V$. By using Lyapunov stability theory [230], it can be concluded that z_{1i}, z_{2i}, $\alpha^T (q_i - q_h)$, $\left\| \left(I_3 - \frac{\alpha \alpha^T}{\|a\|^2} \right) (q_i - q_h) \right\| - r_h$, and $p_i - p_i^*$ will converge to the origin exponentially, which concludes the proof of Theorem 6.1.

7 Fully Distributed Cooperative Circumnavigation of Networked Unmanned Aerial Vehicles

In this chapter, cooperative circumnavigation (CCN) is studied for groups of networked unmanned aerial vehicles (UAVs) under a directed interaction topology. The CCN drives UAVs to given planar ellipses with desired spatial formation. A first contribution is that based on affine transformations, CCN control design is structured, which can deploy UAVs on spatial orbits with different radii concerning a moving target in 3-D space. Second, no global information, such as the formation center, the desired radius, the angular velocity of circumnavigation, *etc.*, is needed for any follower UAVs since they are completely maneuvered by the leader UAVs.

7.1 INTRODUCTION

Networked unmanned systems cooperative control has become a stirring research trend due to its promising potential [37, 50, 110, 168, 242, 253, 378, 488, 518, 569]. Cooperative circumnavigation (CCN), an enabling technology for unmanned system tasks, is to steer multiple vehicles along a common circle concerning a stationary or dynamic center while maintaining a geometric formation configuration. Its applications include source localization, environmental exploration, and robotic security and surveillance. A similar but different collective circular motion control problem is introduced in [70, 105, 329, 348, 398], in which the formation center is unspecified and its value is determined by the initial positions of the vehicles.

Compared with the collective circular motion, more approaches focus on the CCN problem as its center can serve as a host or a target of interest in the practical networked unmanned system missions. The CCN envisages a series of technological challenges: distributed intercommunication, high-precision guidance, and formation control of the geometric configuration. Progress has been made on these challenges [185, 237, 260, 342, 383, 395, 412, 426, 532, 533, 557]. With fixed cruising speed, the circumnavigation problem of a group of mobile vehicles moving in a plane is investigated in [426], where two types of

DOI: 10.1201/9781003298618-7

motion are discussed. In [412], a vector field-based result is given to drive mobile vehicles on an arbitrarily shaped closed curve. Without direct distance measurements, a static circular formation control scheme for networked unmanned systems under velocity constraints is proposed in [533]. However, the results in [412, 426, 533] are confined in the plane. In [185], a self-propelled particle-based approach is studied in 3-D space, but it cannot tackle the circumnavigation problem with a moving target. Based on a cyclic pursuit strategy, a cooperative control law is addressed for a target-capturing task in [237], while however, requires the vehicles to be ordered. Considering a dynamic target, the circumnavigation problem of networked robots is investigated in [342, 383], but it is limited to an undirected graph. In [557], a target-enclosing control law is designed for multiple vehicles using bearing measurements. Based on the affine Laplacian, the circumnavigation of multiple satellites are explored in [260]. Unfortunately, the target information is required by all vehicles in [260, 557]. In addition, acceleration information is utilized in the circumnavigation control design [260], inevitably limiting possible practical implementations. Assuming that only one unmanned system can acquire the target information, a distributed control algorithm is designed in [532], where evenly-spaced formation can be achieved. Note that these works only consider a circular motion of the vehicles, and the circumnavigating radii of different vehicles are identical. Note that in the works above and most existing results, it is assumed that parts of the global information, e.g., the target position (formation center), the geometric configuration, the formation radius, among others., are known to all unmanned vehicles.

On the other hand, due to the inherent potential and the structural nature of the unmanned aerial vehicles (UAVs) [181, 182, 284, 500], they are well suited for the CCN task. Inspired by the above views, a CCN problem for networked UAVs, in which the information of the target and the orbits is only known to a subset of the UAVs (called the leader UAVs while the rest are the follower UAVs). To model the target geometric configuration of the elliptical circumnavigation in 3-D space, affine Laplacian is employed to characterize the communication topology and the nominal formation shape, where ratios of distances and collinearity with respect to nominal formation can be preserved [254, 291, 546]. CCN control design is then articulated under the condition that the follower UAVs can only receive their neighbors' displacements.

The contributions of this chapter lie in two aspects: more general formation configurations and less requirement on global information. First, in terms of the formation shape, the CCN control is capable of achieving an elliptical travel corresponding to a moving target in 3-D space, with arbitrary spaced formation. In contrast, most existing results focus on circular formation [185, 237, 342, 383, 412, 426, 532, 533, 557], a fixed formation center [185], formation in 2-D space [412, 426, 533], or an evenly-spaced formation shape [532]. To perform a more complex task over a larger area, it is noteworthy that the

CCN control can deploy the leader and follower UAVs on different orbits with different radii.

The second contribution is the alleviation of the follower UAVs' requirement on global information. Note that most related results [185, 237, 260, 342, 383, 412, 426, 532, 533, 557] assume that the formation center, desired spacing with neighbors, the desired radius, the angular velocity of the circumnavigation or any other global information are known to all vehicles. In this approach, the follower UAVs are completely maneuvered by the leader UAVs so that they do not need any global information. Moreover, although the affine formation is utilized, the extension is nontrivial as the stabilizing matrix derived by central computation [291, 546] is no longer needed for the proposed CCN control law. In addition, some works are addressed in the case of an undirected graph [342,383,546]. Our result is thus more general and nontrivial as only the displacements with neighbors are required for follower UAVs under a directed graph.

7.2 PROBLEM FORMULATION

7.2.1 DYNAMIC MODEL OF UNMANNED AERIAL VEHICLES

Consider a group of N UAVs in $\mathbb{R}^d$, and define $\mathcal{V} = \{1, 2, \ldots, N\}$. Denote the first N_l UAVs as the leader UAVs and the rest $N_f = N - N_l$ as the follower UAVS. Thus the leader UAVs' subset is $\mathcal{V}_l = \{1, 2, \ldots, N_l\}$ and the follower UAVS' subset $\mathcal{V}_f = \{N_l + 1, N_l + 2, \ldots, N\}$. Here, we assume that $d = 3$, i.e., the UAVS move in 3-D space. In light of the circumnavigation control level, we only consider the position control of the networked UAVs. The translational motion of the ith UAV is modeled as [47, 320],

$$\begin{aligned}
m_i\ddot{\rho}_{x,i} &= -U_i(\cos\phi_i\cos\psi_i\sin\theta_i + \cos\phi_i\sin\psi_i), \\
m_i\ddot{\rho}_{y,i} &= -U_i(\cos\phi_i\sin\psi_i\sin\theta_i - \sin\phi_i\cos\psi_i), \\
m_i\ddot{\rho}_{z,i} &= -U_i(\cos\phi_i\cos\theta_i) + m_i g,
\end{aligned} \tag{7.1}$$

where m_i is the mass, $\rho_i = [\rho_{x,i}, \rho_{y,i}, \rho_{z,i}]^T$ is the coordinate in the inertial frame, ψ_i, θ_i and ϕ_i denote respectively, the roll, pitch, and yaw associated with the inertial frame, U_i is the total trust generated by motors, and g represents acceleration due to gravity.

Let R_i be the rotation matrix given by

$$R_i = \begin{bmatrix} \cos\psi_i\cos\theta_i & \cos\psi_i\sin\theta_i\sin\phi_i - \sin\psi_i\cos\phi_i & \cos\psi_i\sin\theta_i\cos\phi_i + \sin\psi_i\sin\phi_i \\ \sin\psi_i\cos\theta_i & \sin\psi_i\sin\theta_i\sin\phi_i + \cos\psi_i\cos\phi_i & \sin\psi_i\sin\theta_i\cos\phi_i - \sin\phi_i\cos\psi_i \\ -\sin\theta_i & \cos\theta_i\sin\phi_i & \cos\theta_i\cos\phi_i \end{bmatrix}. \tag{7.2}$$

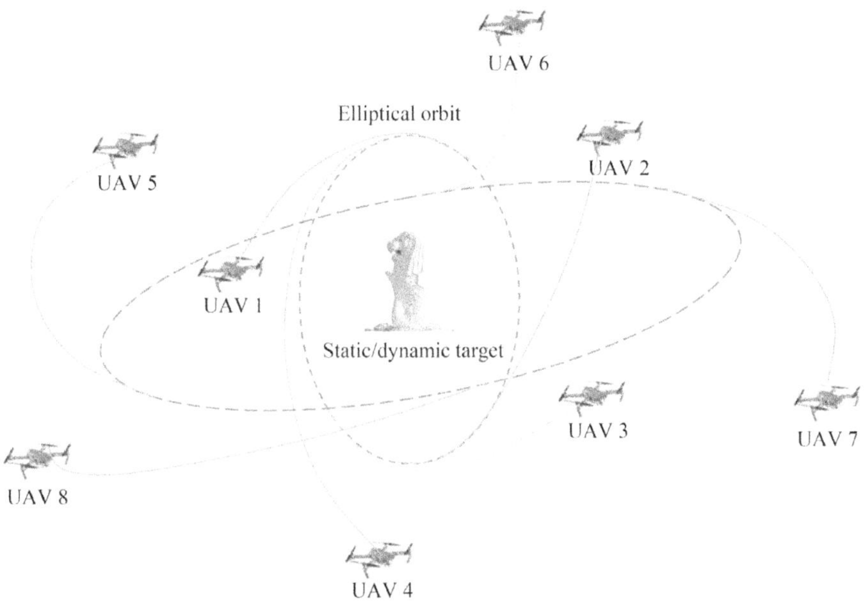

Figure 7.1 Circumnavigation of a static/dynamic target with 8 UAVs (how to assign the leader and follower UAVs will be addressed in the next subsection).

After some mathematical manipulations, the motion dynamics reads [47],

$$m_i \begin{bmatrix} \ddot{\rho}_{x,i} \\ \ddot{\rho}_{y,i} \\ \ddot{\rho}_{z,i} \end{bmatrix} = R_i F_i + \begin{bmatrix} 0 \\ 0 \\ -m_i g \end{bmatrix}, \tag{7.3}$$

where the external input F_i is chosen as

$$F_i = m_i R_i^T [u_{x,i}, u_{y,i}, u_{z,i} + g]^T. \tag{7.4}$$

Taking F_i into (7.1), the total thrust becomes

$$\begin{aligned} U_i =& m_i(\cos\psi \sin\theta \sin\phi + \sin\psi \sin\phi) u_{x,i} \\ &+ m_i(\sin\psi \sin\theta \cos\phi - \sin\phi \cos\psi) u_{y,i} \\ &+ m_i(\cos\theta \cos\phi) u_{z,i}. \end{aligned} \tag{7.5}$$

Define $u_i = [u_{x,i}, u_{y,i}, u_{z,i}]^T$, and we aim to explore the control design u_i to achieve cooperative circumnavigation.

7.2.2 PROBLEM REQUIREMENTS

Given N UAVs and a target, as shown in Fig. 7.1, the object of the CCN problem is to design a controller driving the UAVs to spatial formation with the following *requirements*:

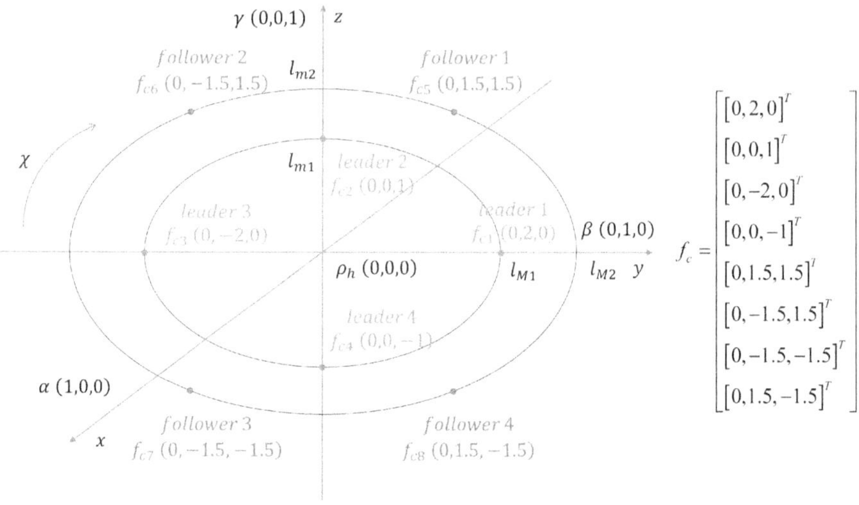

Figure 7.2 An example of the CCN problem with the nominal configuration f_c.

i. The formation is centered at a static or moving target. Let ρ_h stand for the target's position. Each UAV reaches the target plane subject to the normal $\alpha = [\alpha_x, \alpha_y, \alpha_z]^T \in \mathbb{R}^3$ with $\|\alpha\| = 1$, i.e.,

$$\alpha^T (\rho_i - \rho_h) = 0, \ i \in \mathcal{V}. \tag{7.6}$$

ii. With a predefined rate χ, the formation formed by the UAVs circulates around elliptical orbits. The leader UAVs are on an orbit where l_M is the length of the semi-major axis and l_m the length of the semi-minor axis. The axis directions are respectively defined by β and γ. The follower UAVs can be deployed on the same orbit with the leader UAVs' or other desired orbits according to the mission objective.

iii. The UAVs form specific formation corresponding to a constant vector $f_c = \left[f_{c1}^T, f_{c2}^T, \ldots, f_{cN}^T\right]^T$ named as the nominal configuration, where each $f_{ci} \in \mathbb{R}^3$, $i \in \mathcal{V}$. As the UAVs circulate on the orbits at the rate χ, the nominal configuration f_c can be generated by the desired formation at any time instant, e.g., by choosing different time instants t_1 and t_2, we may have different nominal configurations f_c' and f_c'' while they denote exactly the same nominal formation of the CCN problem.

To better explain the problem requirements, especially for the nominal configuration f_c in *Requirement iii*, an illustration on how to derive vector f_c from a given formation shape is shown in Fig. 7.2.

Remark 29. *Different from the circumnavigation problems in [342, 383, 533] and other existing results using the predefined separation angles of the networked unmanned systems regarding the center to formulate the formation* *shape, the nominal configuration f_c is applied to directly describe the formation shape which is easy-implemented, and it helps to design more general formation configurations. Note that only Requirement i has a formal mathematical formulation at this stage. We will focus on deriving mathematical formulations for Requirements ii and iii at the beginning of Section 7.3.*

7.2.3 PRELIMINARIES FROM GRAPH THEORY

A graph $\mathcal{G} = (\mathcal{V}, \mathcal{E})$ is applied to describe the underlying communication among the N UAVs. Let $\mathcal{E} \subseteq \mathcal{V} \times \mathcal{V}$ denote the arc set, $\mathcal{V} = \{1, 2, \ldots, N\}$ the vertex set, and $\mathcal{N}_i$ the in-neighbor set of node i, where $\mathcal{N}_i = \{j : (j, i) \in \mathcal{E}\}$.

We present an affine Laplacian matrix L^s with multiple blocks to associate the nominal configuration in *Requirement iii* with the graphical topology. We define

$$L^s(i,j) = -\omega_{ij}, \tag{7.7}$$

if $i \neq j$ and $j \in \mathcal{N}_i$;

$$L^s(i,j) = \sum\nolimits_{k \in \mathcal{N}_i} \omega_{ik}, \tag{7.8}$$

if $i = j$; and

$$L^s(i,j) = 0, \tag{7.9}$$

if $i \neq j$ and $j \notin \mathcal{N}_i$, where $\omega_{ij} \in \mathbb{R}$ is a positive or negative real weight. We can write $L^s \in \mathbb{R}^{N \times N}$ as

$$L^s = \begin{bmatrix} L_l^s & 0_{N_l \times N_f} \\ L_{f2}^s & L_{f1}^s \end{bmatrix}, \tag{7.10}$$

where L_l^s characterizes the information flow among leader UAVs, L_{f1}^s the information flow among the follower UAVs, and L_{f2}^s the information flow between the leader UAVs and the follower UAVs. In this chapter, the communication among the groups of UAVs is directed.

Properties of affine transformations and nominal formation are provided. An affine transformation is the motion of translation, rotation, scaling, shearing or their combinations. Let $(\mathcal{G}, f_c)$ denote the nominal formation associated to $\mathcal{G}$, where $f_c = [f_{c1}^T, f_{c2}^T, \ldots, f_{cN}^T]^T \in \mathbb{R}^{3N}$ is the nominal configuration. An affine image with respect to a nominal configuration is defined as

$$\begin{aligned} \mathcal{A}(f) = \{ & p \in \mathbb{R}^{3N} : p = (I_N \otimes A) f_c + \mathbf{1}_N \otimes b, \\ & A \in \mathbb{R}^{3 \times 3},\ b \in \mathbb{R}^3 \}, \end{aligned} \tag{7.11}$$

where $p = \left[p_1^T, p_2^T, \ldots, p_N^T\right]^T$. Define $p_l = [p_1^T, \ldots, p_{N_l}^T]^T$, and $p_f = [p_{N_l+1}^T, \ldots, p_N^T]^T$. Then, a definition on affine maneuverability and lemmas on affine formation under a directed graph are introduced.

Definition 28. *[254] The nominal formation $(\mathcal{G}, f_c)$ is said to be affinely maneuverable if for any $\rho = [\rho_l^T, \rho_f^T]^T \in \mathcal{A}(f_c)$ in $\mathbb{R}^{3N}$, ρ_f can be determined by ρ_l uniquely such that*

$$\rho_f = -\left(\left(L_{f1}^s\right)^{-1} L_{f2}^s \otimes I_d\right) \rho_l. \tag{7.12}$$

Lemma 7.1

[291] For graph $\mathcal{G}_l$ and $\rho \in \mathcal{A}(f_c)$ with a generic nominal configuration f_c, we have

$$(L_l^s \otimes I_d)\rho_l = 0, \tag{7.13}$$

iff $\mathcal{G}_l$ is $(d+1)$-rooted. ■

Lemma 7.2

[254] A generic nominal formation $(\mathcal{G}, f_c)$ of N UAVs is affinely maneuverable and L_{f1}^s is nonsingular iff the leader UAVs' set $\mathcal{V}_l$ contains a subset $\mathcal{V}_l^*$ with $d+1$ leader UAVs and every follower UAV in $\mathcal{V}_f$ is $(d+1)$-reachable from $\mathcal{V}_l^*$.
■

Let $\mathcal{G}_l$ denote the communication graph among N_l leader UAVs associated with L_l^s. Next, based on Lemmas 7.1 and 7.2, the following assumption is made.

Assumption 13. *$\mathcal{G}_l$ is $(d+1)$-rooted, and $\mathcal{V}_l$ contains a subset $\mathcal{V}_l^*$ with $d+1$ leader UAVs such that each follower UAV in $\mathcal{V}_f$ is $(d+1)$-reachable from $\mathcal{V}_l^*$.*

Remark 30. *Assumption 13 is the necessary and sufficient condition of affine maneuverability as in Lemma 7.2. In the rest of this chapter, we assume that this condition is naturally held. For more details on affine formation, please refer to [254, 291, 546], where the calculation of the affine Laplacian matrix from a given nominal configuration f_c is also provided.*

Remark 31. *The concept of affine formation is first introduced in [291]. It is capable of preserving ratios and collinearity of distances in accordance with a specific geometric configuration. More details of the stabilizability of nominal formation, the calculation of the Laplacian matrix from the nominal formation f_c, and the necessary and sufficient graphical condition of affine formation, are discussed in [254, 291, 546].*

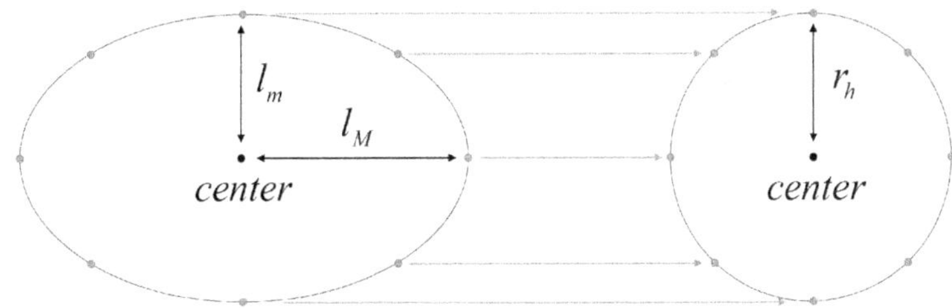

Figure 7.3 Scaling to make the desired elliptical formation a circle.

7.3 CIRCUMNAVIGATION AND CONTROL DESIGN

In this section, we first give mathematical formulations of *Requirements ii* and *iii* in Section 7.2.2, and the problem formulation of CCN is then addressed accordingly. Next, CCN control laws for the leader and follower UAVs are designed, respectively.

Inspired by the idea in Section 7.2.3 that shearing is a kind of the affine transformation, we introduce a scaling matrix S for the leader UAVs to shear the original elliptical formation to a circle as in Fig. 7.3, for ease of the analysis of the elliptical formation in *Requirement ii*.

Let r_h denote the radius of the circle after scaling. We then rewrite the motion dynamics (7.3) as

$$\dot{q}_i = p_i,\ \dot{p}_i = Su_i,\ i \in \mathcal{V}_l, \tag{7.14}$$

where new coordinates take the forms

$$q_i = S\rho_i,\ p_i = S\dot{\rho}_i, \tag{7.15}$$
$$q_h = S\rho_h,\ p_h = S\dot{\rho}_h, \tag{7.16}$$

and $S = [(r_h/l_M)\,\beta, (r_h/l_m)\,\gamma, \alpha]^T$.

For convenience, we define $q_l = [q_1^T, q_2^T, \ldots, q_{N_l}^T]^T$.

Requirement ii can be described as follows by using the scaled coordinates,

$$\|G_\alpha q_{ih}\| = r_h, \tag{7.17}$$
$$p_i = \chi \|G_\alpha q_{ih}\| \left(\alpha \times \phi_{ih}^\alpha\right), \tag{7.18}$$

where

$$q_{ih} = q_i - q_h, \tag{7.19}$$
$$G_\alpha = I_3 - \alpha\alpha^T, \tag{7.20}$$
$$\phi_{ih}^\alpha = G_\alpha q_{ih} / \|G_\alpha q_{ih}\|. \tag{7.21}$$

More specifically, (7.17) and (7.18) show that the UAVs are on orbits with desired radii r_h after scaling, and the UAVs circulate at a desired rate χ.

Remark 32. *It worth pointing out that S contains global information, such as the axes of the elliptical orbits and the normal vector of the formation plane. Scaling matrix S cannot be used in control design for the follower UAVs in the following subsections.*

Note that the affine formation introduced in Section 7.2.3 can preserve the geometric pattern of circumnavigation when the formation rotates, which motivates us to apply the affine nominal configuration to characterize *Requirement iii.* Next, based on Lemma 7.1, Lemma 7.2, and *Requirements i-iii*, the definition of CCN problem is formally given:

Problem 4. *Given an initial formation $\mathcal{G}(\rho(0))$ and a feasible nominal configuration f_c of circumnavigation, design $u_i(t)$ for leader UAV i, $i \in \mathcal{V}_l$ such that*

$$\left(L_l^s \otimes S\right) \rho_l \rightarrow 0, \tag{7.22}$$

and $u_i(t)$ for follower UAV i, $i \in \mathcal{V}_f$ such that

$$\rho_f \rightarrow -((L_{f1}^s)^{-1} L_{f2}^s \otimes I_3)\rho_l, \tag{7.23}$$

while for $\forall i \in \mathcal{V}$, the following holds simultaneously,

$$\|G_\alpha q_{ih}\| \rightarrow r_h, \tag{7.24}$$

$$\alpha^T\left(\rho_i - \rho_h\right) \rightarrow 0, \tag{7.25}$$

$$p_i \rightarrow \chi \|G_\alpha q_{ih}\| \left(\alpha \times \phi_{ih}^\alpha\right). \tag{7.26}$$

Remark 33. *Since the nominal configuration is not necessarily confined on a single orbit, we can deploy the UAVs on orbits with different radii by properly designing vector f_c.*

7.3.1 CIRCUMNAVIGATION FOR LEADER UAVS

In what follows, we address the CCN control design for the leader UAVs. In this subsection, we consider UAV i with $i \in \mathcal{V}_l$.

To fulfill *Requirements i* and *ii*, according to Problem 4, two potential functions $P_{i,\alpha}$ and P_{i,r_h} are respectively proposed,

$$P_{i,\alpha} = \frac{1}{2}\left(\alpha^T q_{ih}\right)^2, \tag{7.27}$$

$$P_{i,r_h} = \frac{1}{2}(\|G_\alpha q_{ih}\| - r_h)^2. \tag{7.28}$$

Remark 34. *Function $P_{i,\alpha}$ is introduced to stick the leader UAVs on the desired formation plane with regard to the target. Function P_{i,r_h} together with the scaling matrix S help to guide the leader UAVs to the predefined elliptical orbit around the target.*

The partial derivatives of $P_{i,\alpha}$ and P_{i,r_h} with respect to q_{ih} are given as

$$\frac{\partial P_{i,\alpha}}{\partial q_{ih}} = \alpha^T q_{ih} \alpha, \tag{7.29}$$

$$\frac{\partial P_{i,r_h}}{\partial q_{ih}} = \left(\|G_\alpha q_{ih}\| - r_h\right) \phi_{ih}^\alpha. \tag{7.30}$$

To fulfill *Requirement iii*, according to Problem 4, the following auxiliary variables are designed

$$z_i = \sum_{j \in \mathcal{N}_i} \omega_{ij} \left(q_i - q_j\right). \tag{7.31}$$

From (7.31), we know

$$\dot{z}_i = \varpi_i \dot{q}_{ih} - \sum_{j \in \mathcal{N}_i} \omega_{ij} \dot{q}_{jh}, \tag{7.32}$$

where $\varpi_i = \sum_{j \in \mathcal{N}_i} \omega_{ij}$.

Next, a desired velocity is designed as

$$p_{id} = -k_1 \frac{\partial P_{i,\alpha}}{\partial q_{ih}} - k_2 \frac{\partial P_{i,r_h}}{\partial q_{ih}} - k_3 \nu_i + p_h + \mu_i, \tag{7.33}$$

where k_1, k_2 and k_3 are positive control gains, the first and the second terms are the negative gradients of potential functions $P_{i,\alpha}$ and P_{i,r_h}, respectively, the third term

$$\nu_i = -\left(1/\varpi_i\right) z_i, \tag{7.34}$$

is the virtual control, the fourth term is the scaled velocity of the target, and the last term

$$\mu_i = \chi \left\|G_\alpha q_{ih}\right\| \left(\alpha \times \phi_{ih}^\alpha\right), \tag{7.35}$$

denotes the desired circulating velocity for leader UAV i corresponding to the predefined constant rate χ.

Let $\tilde{p}_i = p_i - p_{id}$ present the velocity tracking error for UAV i, and we formally propose the CCN control law for leader UAVs,

$$u_i = S^{-1}\left(-k_4 \tilde{p}_i + \dot{p}_{id} - \lambda_i (\tilde{p}_i^T)^\dagger\right), \tag{7.36}$$

where k_4 is a positive control gain, λ_i is an auxiliary variable taking the form,

$$\begin{aligned} \lambda_i = & \frac{\mu_i}{2} \|\varpi_i\|^2 - k_1 \varpi_i z_i^T \frac{\partial P_{i,\alpha}}{\partial q_{ih}} - k_2 \varpi_i z_i^T \frac{\partial P_{i,r_h}}{\partial q_{ih}} \\ & - z_i^T \sum_{j \in \mathcal{N}_i} \omega_{ij} \left(\dot{q}_j - \dot{q}_h\right), \end{aligned} \tag{7.37}$$

and $\left(\tilde{p}_i^T\right)^\dagger$ denotes the Moore-Penrose inverse of $\tilde{p}_i^T$ as follows,

$$\tilde{p}_i^T \left(\tilde{p}_i^T\right)^\dagger = \begin{cases} 0, & \tilde{p}_i^T = [0, 0, \ldots, 0]^T, \\ 1, & \text{otherwise.} \end{cases} \tag{7.38}$$

7.3.2 CIRCUMNAVIGATION FOR FOLLOWER UAVS

In this subsection, we study the CCN control for the follower UAVs. Consider UAV $i \in \mathcal{V}_f$. Similar to the design for the leader UAVs, define the following variables

$$z_i = \sum\nolimits_{j \in \mathcal{N}_i} \omega_{ij} \left(\rho_i - \rho_j\right), \tag{7.39}$$

$$\tilde{v}_i = \dot{\rho}_i - v_{zi}, \tag{7.40}$$

where $v_{zi} = -(k_5/\varpi_i)z_i$ and k_5 is a positive control gain.

According to Assumption 13 and Lemma 7.2, the follower UAVs can be completely maneuvered by the leader UAVs. We design the following CCN control law for the follower UAVs

$$u_i = -k_6\tilde{v}_i - \varpi_i z_i + \dot{v}_{zi} + \left(\tilde{v}_i^T\right)^{\dagger} z_i^T \sum_{j \in \mathcal{N}_i} \omega_{ij} v_j, \tag{7.41}$$

where k_6 is positive control gain.

It is clear that the global information is not used in the CCN control law (7.41), where the velocity of follower UAV i, the velocities and accelerations of follower UAV i's neighbors, and the displacements with follower UAV i's neighbors are used in the control law (7.41), and no global information is required.

7.3.3 MAIN RESULT OF COOPERATIVE CIRCUMNAVIGATION CONTROL

Theorem 7.1

Under Assumption 13, using CCN control laws (7.36) and (7.41), the closed-loop systems of N_l leader UAVs and N_f follower UAVs are asymptotically stable and exponentially stable, respectively, and the requirements in Problem 4 can be ultimately accomplished, if the initial conditions of the leader UAVs satisfy

$$\left\| G_\alpha \left(q_i(0) - q_h(0)\right) \right\| > 0, \tag{7.42}$$

i.e., $\left(q_i(0) - q_h(0)\right)$, $i \in \mathcal{V}_l$ is not parallel to α, and the control gains are chosen as

$$k_1 > \frac{k_3^2}{2} + \frac{1}{2},\ k_2 > \frac{k_3^2}{2} + \frac{1}{2}, \tag{7.43}$$

$$k_3 > \frac{1}{\varpi_M^2} + 1,\ k_4 > \frac{\varpi_M^2}{2} + 1, \tag{7.44}$$

where $\varpi_M = \max_{i \in \mathcal{V}_l} \{\varpi_i\}$ and $\varpi_m = \min_{i \in \mathcal{V}_l} \{\varpi_i\}$. ■

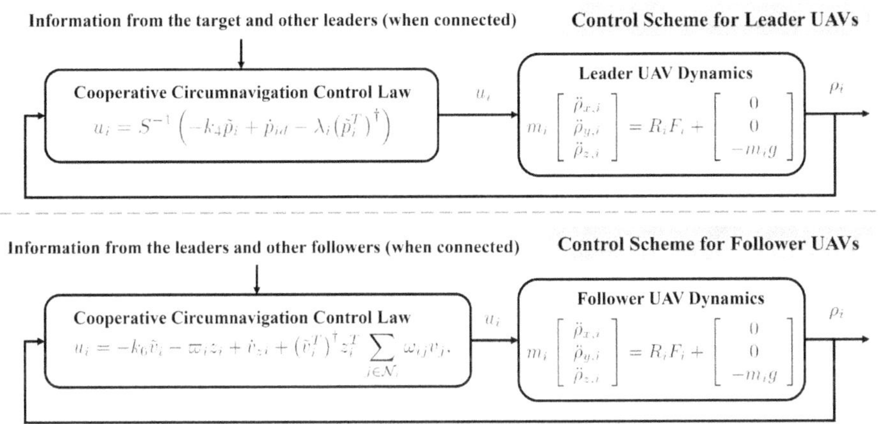

Figure 7.4 The CCN control scheme.

Proof. For details, see Appendix 7.6. □

For greater clarity, a CCN control scheme is provided in Fig. 7.4, where the control structures of both leader and follower UAVs are illustrated.

Remark 35. *Compared with the results in [185, 237, 260, 342, 383, 412, 426, 532, 533, 557]. the proposed approach define a more general CCN problem, where nearly any desired spaced formation can be achieved on different elliptical orbits around a moving target in 3-D space. Under a directed graph, the CCN problem is solved by CCN control laws* (7.36) *and* (7.41), *and no global information is required by follower UAVs.*

Remark 36. *Some practical issues such as time-synchronized convergence, communication delays, external disturbances, actuator faults, model uncertainties are not considered in this approach, as we mainly focus on the level of cooperative circumnavigation. Actually, elegant techniques on these topics are available in the literature, to name a few, the technique of convergence at the same time [253], bearing-based similar formation [36], the enhanced time-delay control design [216], the exogenous disturbance rejection observer [529], the sliding-mode-based fault-tolerant scheme [52, 540], the multi-layer formation structure [256], the admittance-based techniques [279], etc. And we would like to explore the CCN problem with the consideration of these practical matters in possible future works.*

Remark 37. *Onc possible drawback of this approach is the problem of the local minimum as the potential functions $P_{i,\alpha}$ and P_{i,r_h} are utilized. Although no local minimum occurs during a large number of simulations, the formal proof for the avoidance of the local minimum remains open, and we leave it in the future work.*

Table 7.1
Parameters for simulation

Parameters	Values	Parameters	Values
k_1	0.3	k_2	0.2
k_3	1	k_4	0.3
k_5	0.5	k_6	0.5
ρ_h	$[0, 0, 0]^T$	r_h	10
α	$[0, \sqrt{2}/2, \sqrt{2}/2]^T$	χ	0.1
β	$[0, \sqrt{2}/2, -\sqrt{2}/2]^T$	l_M	5
γ	$[1, 0, 0]^T$	l_m	5
m_i	1kg	g	9.8 m/s^2

7.4 SIMULATION

A group of 16 UAVs with 6 leader UAVs and 10 follower UAVs in a series of scenarios are considered in the numerical simulations. Table 7.1 covers the simulation parameters. Note that in Table 7.1, the control gain k_i, $i \in \{1, 2, 3, 4\}$ does not necessarily meet the inequalities in Theorem 7.1, since these are sufficient constraints which are likewise conservative. In practice, control engineers might first select proper control parameters based on the conditions in Theorem 7.1 to guarantee the achievement of CCN. Then they can suitably adjust the parameters to yield an acceptable CCN performance.

The detailed value of the Laplacian matrix L^s is given in the sequel:

$$L_l^s = \begin{bmatrix} 1.2071 & 0.7071 & -0.5 & 0 & 1 & 0 \\ -0.5 & 0.7071 & 0.2071 & 0 & 0 & 1 \\ -0.5 & -0.7071 & -0.2071 & 0 & 0 & 1 \\ -0.7071 & 0.2071 & 0 & 0.5 & 1 & 0 \\ 1 & -0.5858 & 0.4142 & 0 & 0.8284 & 0 \\ 1 & 0 & 0.4142 & 0.5858 & 0 & 0.8284 \end{bmatrix},$$

$$L_{f1}^s = \begin{bmatrix} 1 & 0 & 0 & 0 & 0 & 0 \\ 0 & 2 & 0 & 0 & 0 & 0 \\ 0 & 0 & 2 & 0 & 0 & 0 \\ 0 & 0 & 0 & -1 & 0 & 0 \\ 0 & 0 & 0 & 0 & -4.8284 & 0 \\ 0 & 0 & 0 & 0 & 0 & -4.8284 \end{bmatrix},$$

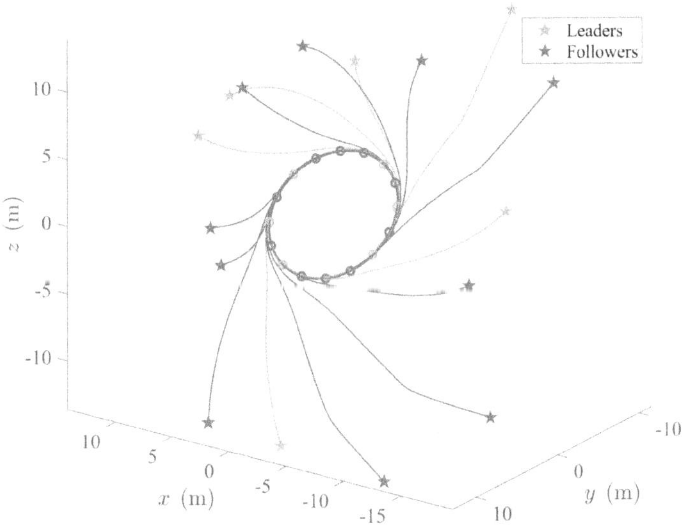

Figure 7.5 Circumnavigation of a static target by 16 networked UAVs.

$$L_{f2}^{s} = \begin{bmatrix} -1.7071 & 1.7071 & 0 & 0 & 1 & 0 \\ 1 & -2 & 3 & 0 & 0 & 0 \\ 1 & 0 & -1 & 2 & 0 & 0 \\ -1.7071 & -0.2929 & 0 & 0 & 1 & 0 \\ 1 & -3.4142 & -2.4142 & 0 & 0 & 0 \\ 1 & 0 & -2.4142 & -3.4142 & 0 & 0 \end{bmatrix}.$$

Fig. 7.5 shows the UAV trajectories, where a circular formation with a static center is achieved. The corresponding formation errors and control inputs are presented in Figs. 7.6 and 7.7. In addition, to illustrate the full power of this approach, another three simulation results are presented. In Fig. 7.8, 6 leader UAVs and 6 follower UAVs are assigned to circumnavigate a static target on different radii of ellipses. Based on the nominal configuration presented in Fig. 7.8, a moving target is then considered. The trajectories of the UAVs and the target are shown in Fig. 7.9. Next, with 24 UAVs on three different orbits, the circumnavigation of a stationary target is shown in Fig. 7.10. These simulation results show that the proposed controller is feasible for the CCN problem of a stationary or moving target with multiple elliptical orbits, where the follower UAVs do not need any global information.

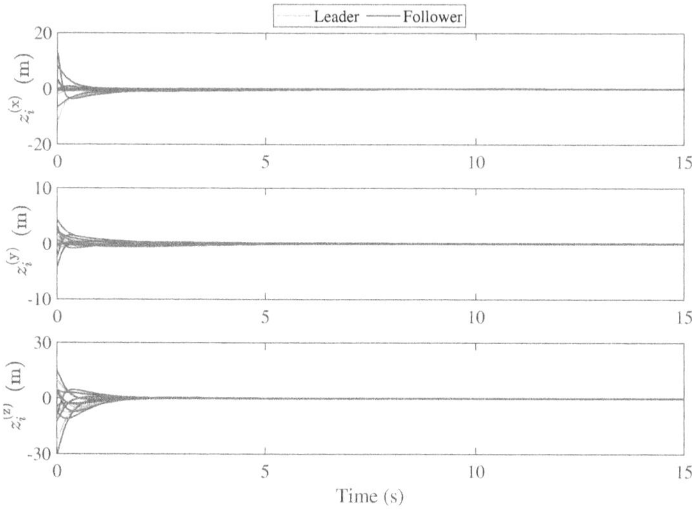

Figure 7.6 Formation errors of 16 networked UAVs.

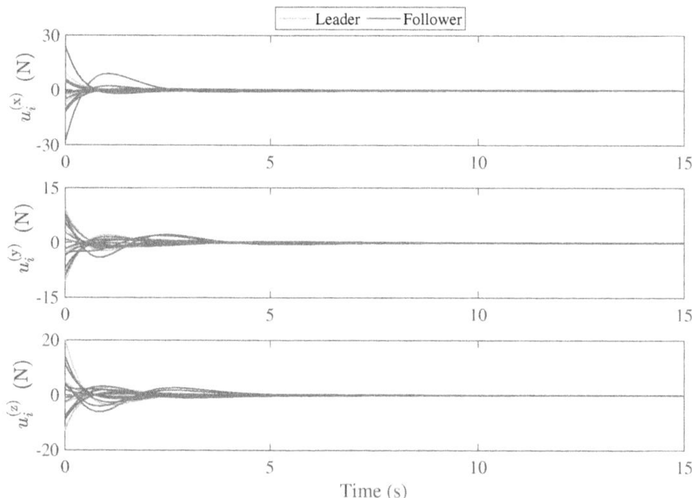

Figure 7.7 Control inputs of 16 networked UAVs.

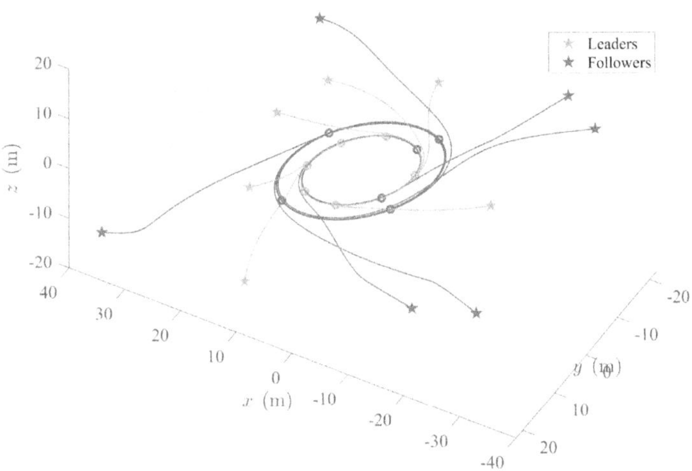

Figure 7.8 Circumnavigation of a static target by 12 networked UAVs on orbits with different radii.

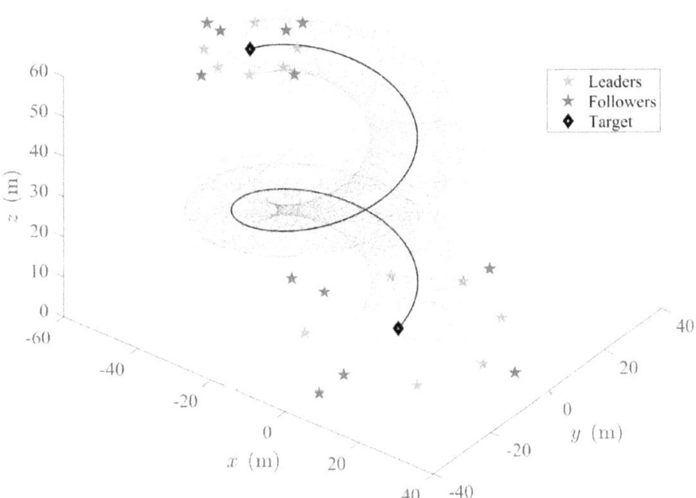

Figure 7.9 Circumnavigation of a moving target by 12 networked UAVs.

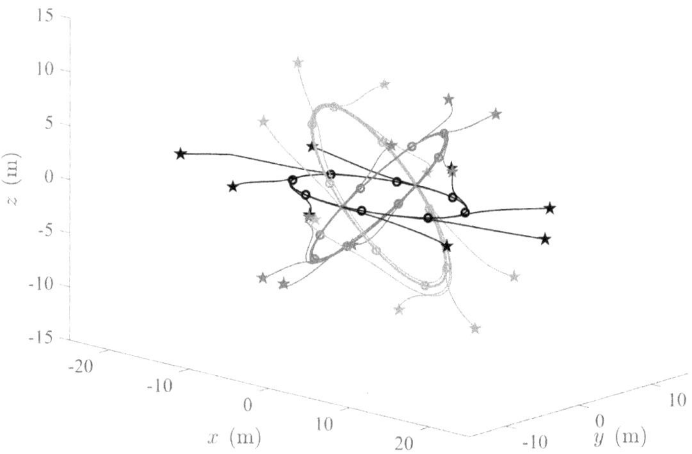

Figure 7.10 Circumnavigation of a static target by 24 networked UAVs on three different orbits.

7.5 CONCLUSION

For a moving or static target, a cooperative circumnavigation control problem for networked unmanned aerial vehicles has been articulated. The proposed control design enables simultaneous circumnavigation on multiple obits with arbitrary different radii. The followers need no global information to accomplish the circumnavigation. The leader and follower are asymptotically stable and exponentially stable, respectively.

This approach provides new insight into the circumnavigation problems, where more general geometric formation shapes can be achieved with less requirement on global information and central computation. Possible future directions include cooperative circumnavigation control problem for uncertain networked systems under disturbances and cyber attacks.

7.6 APPENDIX

In this proof, we first analyze the stability of the N_l leader UAVs and then the N_f follower UAVs.

Step 1: Analysis of Leader UAVs.

For the leader UAVs, take the following Lyapunov function

$$V_1 = \sum_{i \in \mathcal{V}_l} \left(P_{i,\alpha} + P_{i,r_h} + \frac{1}{2} z_i^T z_i + \frac{1}{2} \tilde{p}_i^T \tilde{p}_i \right), \tag{7.45}$$

where the first and the second terms on the right of the equation denote the potential functions introduced in Section 7.3 to fulfill *Requirements i* and *ii*, the third term and the last term represent the general position errors and velocity errors of the circumnavigation, respectively.

Then, the time derivative of (7.45) takes the form

$$\begin{aligned}\dot{V}_1 = \sum_{i\in\mathcal{V}_l} \Bigg(& \left(\frac{\partial P_{i,\alpha}}{\partial q_{ih}}\right)^T \dot{q}_{ih} + \left(\frac{\partial P_{i,r_h}}{\partial q_{ih}}\right)^T \dot{q}_{ih} + z_i^T \dot{z}_i \\ & + \tilde{p}_i^T \left(Su_i - \dot{p}_{id}\right)\Bigg). \end{aligned} \tag{7.46}$$

Following (7.31) and (7.32), we can rewrite (7.46) as the following form,

$$\begin{aligned}\dot{V}_1 = \sum_{i\in\mathcal{V}_l} \Bigg(& \left(\left(\frac{\partial P_{i,\alpha}}{\partial q_{ih}}\right)^T + \left(\frac{\partial P_{i,r_h}}{\partial q_{ih}}\right)^T\right) \left(p_{id} + \tilde{p}_i - p_h\right) \\ & + z_i^T \left(\varpi_i \left(p_{id} + \tilde{p}_i - p_h\right) - \sum_{j\in\mathcal{N}_i} \omega_{ij} \left(\dot{q}_j - \dot{q}_h\right)\right) \\ & + \left(p_i - p_{id}\right)^T \left(Su_i - \dot{p}_{id}\right)\Bigg). \end{aligned} \tag{7.47}$$

From the fact $\alpha \perp G_\alpha q_{ih}$, $\alpha \perp \alpha \times \phi_{ih}^\alpha$, and $G_\alpha q_{ih} \perp \alpha \times \phi_{ih}^\alpha$, it can be verified that

$$\frac{\partial P_{i,\alpha}}{\partial q_{ih}} \perp \frac{\partial P_{i,r_h}}{\partial q_{ih}}, \tag{7.48}$$

$$\frac{\partial P_{i,\alpha}}{\partial q_{ih}} \perp \mu_i, \tag{7.49}$$

$$\frac{\partial P_{i,r_h}}{\partial q_{ih}} \perp \mu_i. \tag{7.50}$$

The properties in (7.48)–(7.50) substantially reduce the complexity of the whole system design. Otherwise, the control algorithms proposed in this chapter would have contained more coupling terms, which evidently hinders the practicability.

Utilizing (7.48)–(7.50), we next have

$$\begin{aligned}\dot{V}_1 = \sum_{i\in\mathcal{V}_l} \Bigg(& \left(\frac{\partial P_{i,\alpha}}{\partial q_{ih}}\right)^T \left(-k_1 \frac{\partial P_{i,\alpha}}{\partial q_{ih}} - k_3 p_{zi} + \tilde{p}_i\right) \\ & + \left(\frac{\partial P_{i,r_h}}{\partial q_{ih}}\right)^T \left(-k_2 \frac{\partial P_{i,r_h}}{\partial q_{ih}} - k_3 p_{zi} + \tilde{p}_i\right) \\ & + z_i^T \left(\varpi_i \left(p_{id} + \tilde{p}_i - p_h\right) - \sum_{j\in\mathcal{N}_i} \omega_{ij} \left(\dot{q}_j - \dot{q}_h\right)\right) \\ & + \left(p_i - p_{id}\right)^T \left(Su_i - \dot{p}_{id}\right)\Bigg). \end{aligned} \tag{7.51}$$

Using Young's inequality, (7.51) becomes

$$\begin{aligned}\dot{V}_1 =& \sum_{i\in\mathcal{V}_l}\Bigg(-\left(k_1-\frac{k_3^2}{2}-\frac{1}{2}\right)\left\|\frac{\partial P_{i,\alpha}}{\partial q_{ih}}\right\|^2+\left(1+\frac{\|\varpi_i\|^2}{2}\right)\|\tilde{p}_i\|^2\\ &-\left(k_3-\frac{1}{\|\varpi_i\|^2}-1\right)\|z_i\|^2-\left(k_2-\frac{k_3^2}{2}-\frac{1}{2}\right)\left\|\frac{\partial P_{i,r_h}}{\partial q_{ih}}\right\|^2\\ &+\frac{\mu_i}{2}\|\varpi_i\|^2-k_1\varpi_i z_i^T\frac{\partial P_{i,\alpha}}{\partial q_{ih}}-k_2\varpi_i z_i^T\frac{\partial P_{i,r_h}}{\partial q_{ih}}\\ &-\varpi_i z_i^T\sum_{j\in\mathcal{N}_i}\omega_{ij}\left(\dot{q}_j-\dot{q}_h\right)+\tilde{p}_i^T\left(Su_i-\dot{p}_{id}\right)\Bigg),\end{aligned} \tag{7.52}$$

Substituting the CCN control law (7.36) into (7.52) yields

$$\begin{aligned}\dot{V}_1 \leq& -k_{\min}\sum_{i\in\mathcal{V}_l}\Bigg(\left\|\frac{\partial P_{i,\alpha}}{\partial q_{ih}}\right\|^2+\left\|\frac{\partial P_{i,r_h}}{\partial q_{ih}}\right\|^2\\ &+\|z_i\|^2+\|\tilde{p}_i\|^2\Bigg),\end{aligned} \tag{7.53}$$

$$\begin{aligned}k_{\min} =& \min_{i\in\mathcal{V}_l}\Bigg\{k_1-\frac{k_3^2}{2}-\frac{1}{2},k_2-\frac{k_3^2}{2}-\frac{1}{2},\\ &k_3-\frac{1}{\varpi_m^2}-1,k_4-\frac{\varpi_M^2}{2}-1\Bigg\},\end{aligned} \tag{7.54}$$

which implies the closed-loop follower UAV systems are asymptotically stable [230].

To ensure $k_{\min}>0$, we can choose the control gains

$$k_1 > k_3^2/2+1/2, \tag{7.55}$$

$$k_2 > k_3^2/2+1/2, \tag{7.56}$$

$$k_3 > 1/\varpi_M^2+1, \tag{7.57}$$

$$k_4 > \varpi_M^2/2+1, \tag{7.58}$$

We next prove that $(L_l^s\otimes I_3)\,q_l=0$ in Problem 4 is guaranteed. Rewrite (7.31) in a compact form

$$z_l=(L_l^s\otimes I_3)\,q_l, \tag{7.59}$$

which straightforwardly shows that

$$(L_l^s\otimes I_3)\,q_l\to 0,\ t\to\infty. \tag{7.60}$$

According to (7.45) and the definition of $P_{i,\alpha}$, P_{i,r_h} and $\tilde{p}_i$, we can also get that, $\forall i\in\mathcal{V}_l$,

$$\|G_\alpha q_{ih}\|\to r_h, \tag{7.61}$$

$$\alpha^T(\rho_i-\rho_h)\to 0, \tag{7.62}$$

$$p_i\to\chi\,\|G_\alpha q_{ih}\|\,(\alpha\times\phi_{ih}^\alpha). \tag{7.63}$$

This concludes the proof for the leader UAVs.

Step 2: Analysis of Follower UAVs.

For the N_f follower UAVs, we take the following Lyapunov function

$$V_2 = \sum_{i \in \mathcal{V}_f} \left(\frac{1}{2} z_i^T z_i + \frac{1}{2} \tilde{v}_i^T \tilde{v}_i \right). \tag{7.64}$$

Under CCN control law (7.41), it is trivial to verify that the time derivative of V_2 satisfies

$$\dot{V}_2 = \sum_{i \in V_f} \left(-k_5 \|z_i\|^2 - k_6 \|\tilde{v}_i\|^2 \right) \leq -k_f V_2, \tag{7.65}$$

with $k_f = \min\{k_5, k_6\}$.

By using Lyapunov stability theory [230], it can be concluded that the closed-loop system is exponentially stable. We next show

$$\rho_f = -((L_{f1}^s)^{-1} L_{f2}^s \otimes I_3)\rho_l, \tag{7.66}$$

in Problem 4 is guaranteed. Rewrite z_i in a compact form

$$z_f = \left(L_{f1}^s \otimes I_3\right) \rho_f + \left(L_{f2}^s \otimes I_3\right) \rho_l. \tag{7.67}$$

According to Lemma 7.2, L_{l1} is nonsingular, and (7.67) becomes

$$\left(L_{f1}^s \otimes I_3\right)^{-1} z_f = \rho_f + \left(\left(L_{f1}^s\right)^{-1} L_{f2}^s \otimes I_3\right) \rho_l, \tag{7.68}$$

which shows that

$$\rho_f \to -((L_{f1}^s)^{-1} L_{f2}^s \otimes I_3)\rho_l, \tag{7.69}$$

as $t \to \infty$. Thus, the follower UAVs have been completely maneuvered by the leader UAVs. Based on Lemma 7.2 and the property of affine maneuverability, we know $\forall i \in \mathcal{V}_f$,

$$\|G_\alpha q_{ih}\| \to r_h, \tag{7.70}$$

$$\alpha^T \left(\rho_i - \rho_h\right) \to 0, \tag{7.71}$$

$$p_i \to \chi \|G_\alpha q_{ih}\| \left(\alpha \times \phi_{ih}^u\right), \tag{7.72}$$

which completes the proof for the follower UAVs.

Part II

Formation Control in Constrained Space

8 Styled-Velocity Flocking of Autonomous Vehicles: A Systematic Design

This chapter develops a measure-theoretic approach for generating the collective behavior of flocking of planar autonomous vehicles under all-to-all communication scheme. To obtain flocking protocol, we formulate the notion of *state of styled-velocity flocking* for the target state of a design problem instead of considering solution-*ansatses*. This notion captures a broader class of flocking behaviors of non-parallel motions, as well as embodies the two important performances of cohesion maintenance and collision avoidance. To pose a control design problem, a measure-valued transition equation is obtained to induce an auxiliary one-body continuum-model from the exact multi-body particle-model of the collective system. The design problem is then formulated on the auxiliary continuum-model, and a systematic design procedure is presented. The state of styled-velocity flocking of the particle-model under the obtained flocking protocol is then verified with an important result on the continuous dependence on initial data of solutions of the measured-valued transition equation. Application to elliptical flocking of mobile robots is presented to illustrate the novelty of the presented theory.

8.1 INTRODUCTION

Complex systems of identical units evolving together are of common interest among disciplines. From gases of interacting molecules [235], biological colonies [27], [125] to distributed computing networks [244], it is an intriguing fact that meaningful group behaviors can be tailored from certain primitive behaviors of the group members. This presents the control area the challenging problem of obtaining a common strategy for group members, under which a desired group behavior emerges. For mobile agents, a special group behavior termed flocking has been actively studied in the control literature [44,46,87,164,204,348,363,425,460]. The corresponding common strategy in such a case is a control algorithm termed *flocking protocol.*

A common approach to obtain flocking protocol is to consider a solution-*ansatz*, i.e., a heuristic flocking protocol, for verification study, e.g., [87, 204, 363]. As the solution-*ansatzes* invoke preparation studies such as modeling and simulation [405, 468] which are not always available, we are motivated to develop in this chapter a control framework obtaining the desired flocking protocol by systematic design. For this principal study, we assume all-to-all

DOI: 10.1201/9781003298618-8

communication scheme, i.e., each agent exchanges its state with all other agents.

The two typical difficulties for such development are due to the multiple goals of the flocking motion, and the high-dimensionality of the collective system. Indeed, flocking is understood as the collective coherent motion of large numbers of self-propelled organisms [463]. For a well-behaved flocking, collision avoidance is naturally invoked [363]. As such, the multiple goals of flocking control are apparently the desired group motion, cohesion maintenance, and collision avoidance.

Previous works achieving the above multiple goals includes [363] and [86]. We observe that the solution-*ansatzes* proposed in these works are much dedicated to agents of point-mass model and *parallel flocking*, i.e., all agents move at the same velocity. Attempts for flocking of mobile robots have been made for either parallel flocking [460] or circular formation [220, 348, 425]. However, these results are partial in the sense that collision avoidance are not theoretically verified, i.e., a part of the multiple goals of flocking is not achieved.

The control approach developed in this chapter overcomes the above limitations. In particular, the desired flocking protocol guaranteeing the multiple goals of flocking is to be obtained by systematic design, the agents are not limited to point-masses, and the allowed flocking regimes are broaden to curved flocking, for which the agents may differ in velocity. To this end, we specify a target state of the collective system embodying the above multiple goals, and aim at driving the collective system to this state by means of control design. We formulate such desired target state results in the notion of *state of styled-velocity flocking.*

To formulate such notion of *state of styled-velocity flocking*, we first interpret the desired spatial pattern as the consensus on the *styled-velocity* defined as a transformation of the physical velocity. We then interpret the violation of cohesion maintenance and collision avoidance (CMCA) by the unboundedness of a coordination function. Combining these together, we obtain a mathematical formulation of the *state of styled-velocity flocking* of the collective system to pose a standard control problem. This constitutes a conceptual contribution of the current chapter.

The flocking control problem then becomes driving the collective system of mobile robots to the target state of the state of styled-velocity flocking. The high-dimensionality of the collective system appeals for a measure-theoretic approach, which not only is effective but also shed further light on the problem. Thus, we present in this chapter a measure-theoretic approach based on a fact from [354] that: there is a measure-valued transition equation that captures the trajectories of the exact particle-model, described by the collection of all equations of motion of the individual agents [87, 204, 363], and the trajectories of an approximate continuum-model or mean-field model [247, 441] of the collective system, described by an evolutionary equation of a one-particle distribution function [44, 46, 164], for its Dirac measure-valued solutions and its continuous measure-valued solutions, respectively. This suggests our underlying principle that the flocking state of the particle-model can be verified by relevant properties of the continuum-model through the measure-valued transition equation.

Accordingly, we formulate a measure-valued transition equation describing the collective dynamics of mobile robots, from which an auxiliary continuum-model for protocol design is induced. Our control framework then consists of a design problem, rendering a desired dissipation property for the auxiliary continuum-model, and a verification problem, showing that the Dirac measure-valued solutions of the transition equation preserve the dissipative behavior of the continuous measured-valued solutions, and the desired flocking is achieved.

In view of the existing mean-field approaches, e.g., [44, 46, 164], the prominent feature making our framework original is that it does not only verify the flocking behavior on the exact particle-model but also relax the requirement that the number of agents is sufficiently large, and hence allows a broader class of collective systems. This is possible since our verification problem concludes the desired dissipation properties for the particle-model. Indeed, we shall not consider the continuum-model as the limit of the particle-models as usual [44–46, 164]. Instead, we consider the Dirac measure-valued solution from the particle-model as the limit of continuous measure-valued solutions from the auxiliary continuum-model. Accordingly, we shall verify the desired dissipation properties for the particle-model without requiring a sufficiently large number of agents. Clearly, these together constitute the methodological contribution of the current chapter.

Moreover, our design procedure is Lyapunov-based. Thus, our qualitative analysis of the closed-loop system obtains the regularity of the transition mappings as a result. For a comparison, existing works on the measure-theoretic approach to dynamical systems include [44, 164, 247, 354, 466]. While [466] uses the regularity of the transition operators as a property to limit the class of admissible feedback control strategies for a computational approach, the solution-ansatzes in [44, 164, 247, 354] already embody the desired regularity. Accordingly, from the general perspective of the measure-theoretic approach to dynamical systems [44, 164, 247, 354, 466], the regularity study in this chapter constitutes an important technical enhancement for the area.

Finally, to illustrate the novelty of the proposed theory, we present simulation results on elliptical flocking of mobile robots.

8.2 PROBLEM FORMULATION

Given a collective system of N identical autonomous planar vehicles labeled by the numbers $1, \ldots, N$ whose respective equations of motion are [460]

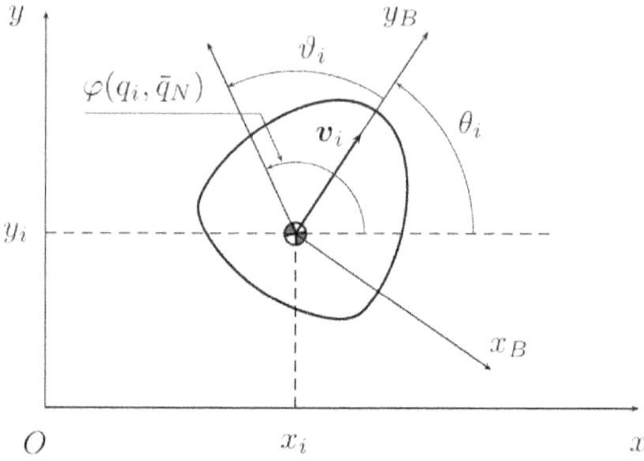

Figure 8.1 Configuration of the i-th wheeled mobile robot.

$$\begin{aligned} \dot{q}_i &= v_i e(\theta_i) \\ \dot{\theta}_i &= w_i \qquad i = 1, \ldots, N \\ \dot{v}_i &= a_i \end{aligned} \tag{8.1}$$

where, as shown in Fig. 8.1, $q_i = [x_i, y_i]^T \in \mathbb{R}^2$ and $\theta_i \in \mathbb{R}$ are respectively the position and the heading angle of the i-th vehicle in the inertial frame Oxy; $\boldsymbol{v}_i = v_i \boldsymbol{e}(\theta_i)$ with $v_i \in \mathbb{R}$ the linear speed, and $\boldsymbol{e}(\theta_i)$ the unit vector $[cos\theta_i, sin\theta_i]^T$; and w_i, $a_i \in \mathbb{R}$ are control inputs. Hereafter, $\bar{q}_N = \text{col}(q_1, \ldots, q_N)$, and $q_C = \sum_{i=1}^{N} q_i/N$ is the center of mass.

Our general goal for the collective system (8.1) is to design *all-to-all communication* controls $p_i = [w_i, a_i]^T$, $i = 1, \ldots, N$, i.e., the collective state (q_i, θ_i, v_i), $i = 1, \ldots, N$ is available for the design of all p_i's, of the same structure such that the emergent collective behavior of the N vehicles forms a desired flocking motion. For our control approach, we have the following subsection defining styled-velocity flocking as a state of the system (8.1) for control goal.

Remark 38. *It is of practical importance to study flocking under limited communication. The complete solution to this problem invokes the necessary and challenging problem of connectivity preserving [534]. Equipping the agents the ability of managing communication links, one has an additional control freedom to study limited communication flocking as the coupled problems of all-to-all communication flocking and connectivity preserving [534]. Accordingly, all-to-all communication flocking is of principal importance, and hence, for the current development of a design methodology, we limit ourself to all-to-all communication scheme without reducing the contribution of the chapter.*

8.2.1 THE STATE OF STYLED-VELOCITY FLOCKING

The general objective of the group motion is to form a desired group formation. In parallel flocking, e.g., [363], the desired group formation generates parallel trajectories, and hence the agents can be considered to move in a linear velocity flow. Generalizing to agents moving in nonlinear velocity flows, it is relevant to call it curved flocking. In this chapter, we are interested in the class of curved flocking in which an agent i occupying the location q_i is considered in its right motion if its velocity orientation matches a given reference $\varphi(q_i, \bar{q}_N) + c$ with c a common bias. As the velocity orientation of the nonholonomic vehicle i is also its heading angle θ_i, the desired group formation is achieved if all the orientation mismatches $\theta_i - (\varphi(q_i, \bar{q}_N) + c)$ vanish. Thus, allowing arbitrary bias c, our goal for this class of curved flocking is to have consensus on the orientation mismatches $\vartheta_i \stackrel{\text{def}}{=} \theta_i - \varphi(q_i, \bar{q}_N)$.

In parallel flocking, all agents move in the same direction, and hence, for a cohesive motion, it is relevant to have consensus on linear speeds v_i's, see, e.g., [363]. In curved flocking, the vehicles may vary their directions and hence consensus on v_i's may results in loss of cohesion. To resolve this issue, we shall take the trajectory curvature into account and aim at consensus on $\nu_i \stackrel{\text{def}}{=} s(q_i, q_C)v_i$ with $s(q_i, q_C)$ a design function. Throughout this chapter, s is selected to satisfy $s(q, q') \geq c_s > 0, \forall q, q' \in \mathbb{R}^2$ with c_s a constant.

From the above consideration, we shall encode the desired group formation by the consensus on ϑ_i's and the consensus on ν_i's. In addition, CMCA is understood that all agents remain in a moving ball and all inter-agent distances remain non-zero [363]. In this chapter, we shall capture such CMCA performance by the boundedness of an auxiliary function U of all vehicle positions called *coordination function*. For the nonholonomic structure of (8.1), we further lend the desired boundedness of U to $\boldsymbol{e}(\theta_i)\nabla_{q_i}U(\cdot) = 0, \forall i$.

Definition 29. *A flocking theme is a triple* (U, φ, s), *where* $U : \mathbb{R}^{2\times N} \to \mathbb{R}^+, \varphi : \mathbb{R}^2 \times \mathbb{R}^{2\times N} \to \mathbb{R}$, *and* $s : \mathbb{R}^2 \times \mathbb{R}^2 \to \mathbb{R}$ *are real-valued functions.*

Definition 30 (State of Styled-Velocity Flocking)**.** *Given a flocking theme* (U, φ, s). *Let* $\boldsymbol{v}_{s,i} = \nu_i \boldsymbol{e}(\vartheta_i), \nu_i = s(q_i, q_C)v_i, \vartheta_i = \theta_i - \varphi(q_i, \bar{q}_N)$ *defined as styled-velocity. The collective system* (8.1) *is said to be in the state of styled-velocity flocking of the theme* (U, φ, s) *if i) there is consensus on* $\boldsymbol{v}_{s,i}$, *i.e.,* $\boldsymbol{v}_{s,i} = \boldsymbol{v}_{s,j}, \forall i, j$ *and ii) the coordinate* $\bar{q}_N$ *satisfies* $U(\bar{q}_N) < \infty$ *and* $\boldsymbol{e}(\theta_i) \cdot \nabla_{q_i}U(\bar{q}_N) = 0, \forall i = 1, \ldots, N$.

Definition 31. *A collective motion* $(q_i(t), \theta_i(t), v_i(t)), i = 1, \ldots, N, t \geq t_0$ *of the system* (8.1) *is said to have the CMCA property if for each initial condition* $(q_i(t_0), \theta_i(t_0), v_i(t_0)), i = 1, \ldots, N$, *there are* $D_0 > 0$ *and* $d_0 > 0$ *such that* $D_0 \geq \|q_i(t) - q_j(t)\| > d_0, \forall t > t_0, \forall i, j = 1, \ldots, N, i \neq j$

Remark 39. *It can be seen that the CMCA property defined in Definition 31 is equivalent to the CMCA in [363]. In this chapter, we are interested in using inter-agent distances* $\|q_i(t) - q_j(t)\|$ *instead of the distances* $\|q_i(t) - q_c(t)\|$ *to a moving origin* $q_c(t)$ *as in [363]. This is because we shall achieve CMCA by using the coordination function* U *whose boundedness implies that all inter-agent distances are bounded and non-zero. Furthermore, to achieve the boundedness of* U*, we aim at* $\mathbf{e}(\theta_i)\cdot\nabla_{q_i}U(\bar{q}_N) = 0, \forall i$ *as formulated in Definition 30 instead of* $\nabla_{q_i}U(\bar{q}_N) = 0$ *as in, e.g., [86, 363]. This is because our agents are nonholonomic vehicles and* $\nabla_{q_i}U(\bar{q}_N)$ *may play the role of centrifugal force in collective motions of nonlinear trajectories.*

Remark 40. *To capture flocking with parallel trajectories, we can select* $\varphi(\cdot) \equiv 0, s(\cdot) \equiv 1$*, and design a function* U *such that* $\mathbf{e}(\theta_i)\cdot\nabla_{q_i}U(\bar{q}_N) = 0, \forall i$ *guarantees the desired coordination. To capture flocking with circular trajectories, we can select* $\varphi(q_i, \bar{q}_N) = \varphi_r(q_i) - \pi/2$ *and* $s(q_i, q_o) = 1/r_i$ *and design* U *such that* $\mathbf{e}(\theta_i)\cdot\nabla_{q_i}U(\bar{q}_N) = 0, \forall i$ *implies that all agents have the same possibly varying angular velocity, where* $\Phi_r(q_i)$ *is the argument of the vector* $q_i - q_o, q_o$ *is the interested center of the flocking, and* $r_i = \|q_i - q_o\|$.

Remark 41. *the property* $\mathbf{e}(\theta_i)\cdot\nabla_{q_i}U(\bar{q}_N) = 0$ *in Definition 30 can be used in the collective motion requiring* $\nabla_{q_i}U(\bar{q}_N) = 0$*. Indeed this can be achieved by rendering* $\mathbf{e}(\theta_i)\cdot\nabla_{q_i}U(\bar{q}_N) = 0$ *and adjusting* θ_i *to force* $\mathbf{e}(\theta_i)$ *to point away from the normal vector of* $\nabla_{q_i}U(\bar{q}_N)$.

8.2.2 MEASURE-THEORETIC FORMULATION

The underlying principle of our framework is to transfer the design task on the multi-body model (8.1) to an auxiliary continuum-model whose one-body nature facilitates systematic design. To this end, we shall obtain in this subsection a measure-theoretic description of the dynamics of (8.1) from which the desired auxiliary continuum-model is induced.

Let us begin with a Lyapunov formulation desired for the flocking state in Definition 30. As $\boldsymbol{v}_{s,i}(\bar{q}_N) = \nu_i \mathbf{e}(\vartheta_i)$ by Definition 30, the consensus on $\boldsymbol{v}_{s,i}(\bar{q}_N)$ is equivalent to the virtual control of the q_i-dynamics, we define

$$\zeta_i = \nu_i - s\left(q_i, q_C\right)\frac{1}{N}\sum_{j=1}^{N}\alpha\left(q_i, q_j\right) \tag{8.2}$$

with $\alpha(q_i, q_j)$ a design function to be specified. Clearly, if the sum in (8.2) approaches zero, then consensus on ζ_i implies consensus on ν_i. Thus, we shall achieve the consensus condition in Definition 30 by means of consensus on $\eta_i \stackrel{\text{def}}{=} [\vartheta_i, \zeta_i]^T$.

Let $X_i = \left[q_i^T, \eta_i^T\right]^T$, and $\bar{X} = \operatorname{col}(X_1, \ldots, X_N)$. To achieve the specifications in Definition 30, it is relevant to have the following function converge along the trajectories of (8.1):

$$V(\bar{X}) = \frac{1}{4}\frac{1}{N}\sum_{i=1}^{N}\frac{1}{N}\sum_{j=1}^{N}\|\eta_i - \eta_j\|^2 + U(\bar{q}_N) \tag{8.3}$$

where the first term is to achieve consensus on η_i and the last term is to achieve the coordination goal. We now further assume the following structure for $U(\bar{q}_N)$:

$$U(\bar{q}_N) = \frac{1}{N}\sum_{i=1}^{N}U_0(q_i, q_C) + \frac{1}{2}\frac{1}{N}\sum_{i=1}^{N}\frac{1}{N}\sum_{j=1}^{N}U_1(q_i, q_j) \tag{8.4}$$

where $U_0(q_i, q_C)$ is to coordinate the i-th vehicle with respectto (w.r.t) the center of mass q_C , and $U_1(q_i, q_j)$ is to coordinate the i-th vehicle w.r.t the j-th vehicle.

Let $B_i(R, \epsilon) = \{(q_1, q_2) : R \geq \|q_1 - q_2\| \geq d_i + \epsilon\}$.

Assumption 14. *There are $d_i \geq 0$, $i = 0, 1$ such that i) $U_i(q_1, q_2) = 0$ if $\|q_1 - q_2\| \leq d_i$; ii) for any $R > \epsilon > 0$, $U_i(q_1, q_2)$ and its derivatives up to the second order are bounded and continuous on $B_i(R, \epsilon)$; and iii) $U_i(q_1, q_2) \to \infty$ as either $\|q1 - q2\| \to d_i$ or $\|q_1 - q_2\| \to \infty$. In addition, $U_1(q_1, q_2)$ is symmetric.*

Assumption 15. *There is $\epsilon > 0$ such that the initial configuration $\bar{q}_N(t_0)$ of the system (8.1) satisfies $(q_i(t_0), q_j(t_0)) \in B_0(1/\epsilon, \epsilon) \cup B_1(1/\epsilon, \epsilon), \forall i, j = 1, \ldots, N, i \neq j$. The functions φ and s are differentiable, and $1/s(q_1, q_2)$ is Lipschitz continuous with constant L_s.*

Remark 42. *As we are expressing (8.4) in terms of Dirac measures, we do not exclude the case $j = i$ as usual, e.g., [363, 460]. Accordingly, we have Assumption 14 ensuring $U_i(q_1, q_2), i = 1, 2$ well-defined for all $q_1, q_2 \in \mathbb{R}^2$ and $U_i(q, q) = 0$, $\forall q \in \mathbb{R}^2, i = 1, 2$. This is of no limitation as CMCA is to maintain $q_i(t) = q_j(t), \forall i, j = 1, \ldots, N, i \neq j$.*

We have a measure-theoretic representation of V given by (8.3) and (8.4) as follows. Let $\mathfrak{B}$ be the set of Borel sets in $\mathbb{R}^4$ [525]. For $\bar{X} = \mathrm{col}(X_1, \ldots, X_N) \in \mathbb{R}^{4N}$, define the following discrete measures $\delta_{X_i}, \delta_{\bar{X}} : \mathfrak{B} \to [0, 1]$:

$$\delta_{X_i}(M) = \begin{cases} 1 & \text{if } X_i \in M \\ 0 & \text{if } X_i \notin M, \end{cases} \qquad M \in \mathfrak{B}$$

$$\delta_{\bar{X}} M = \frac{1}{N}\sum_{i=1}^{N}\delta_{\bar{X}_1} M \tag{8.5}$$

Let $\mathfrak{M}$ be the set all measures $\mu : \mathfrak{B} \to [0,1]$.Throughout this chapter, we make the conventions that $X = \mathrm{col}(q,\eta) = [q^T, \vartheta, \zeta]^T \in \mathbb{R}^4$ and $Y = \mathrm{col}(q^*,\eta^*) = [q^{*T}, \vartheta^*, \zeta^*]^T \in \mathbb{R}^4$ with $q,\eta,q^*,\eta^* \in \mathbb{R}^2$ and $\vartheta,\zeta,\vartheta^*,\zeta^* \in \mathbb{R}$. For a measure $\mu \in \mathfrak{M}$, we define

$$q_\mu = \int_{\mathbb{R}^4} q\mu(dX) \in \mathbb{R}^2 \tag{8.6}$$

where the integral is of Lebesgue type [525].

Clearly, using the measure $\delta_{\bar{X}}$, we have $q_C = q_{\delta_{\bar{X}}}$, and the function $V(\bar{X})$ given by (8.3), (8.4) has the following equivalent expression:

$$\begin{aligned} V(\bar{X}) =& \frac{1}{2}\int_{\mathbb{R}^4}\left(\int_{\mathbb{R}^4}\left(\frac{1}{2}\left\|\eta-\eta^*\right\|^2 + U_1\left(q,q^*\right)\right)\delta_{\bar{X}}(dY)\right)\times\delta_{\bar{X}}(dX) \\ &+\int_{\mathbb{R}^4} U_0\left(q, q_{\delta_{\bar{X}}}\right)\delta_{\bar{X}}(dX). \end{aligned} \tag{8.7}$$

The above formulation (8.7) suggests that $V(\bar{X})$ can be studied in terms of the measure $\delta_{\bar{X}}$. To this end, we shall derive a measure-theoretic description of the dynamics (8.1) of $\bar{X} = \mathrm{col}(X_1, \ldots, X_N)$. In particular, we are interested in the dynamics of $\delta_{\bar{X}(t)}$, i.e., a transition equation whose solution is the Dirac measure-valued mapping $\mu_\delta^\dagger : \mathbb{R} \to \mathfrak{M}, t \mapsto \delta_{\bar{X}(t)}$.

Note that $X = [q^T, \vartheta, \zeta]^T \in \mathbb{R}^4$ and $Y = [q^{*T}, \vartheta^*, \zeta^*]^T \in \mathbb{R}^4$. For a measure-valued function $\mu^\dagger : \mathbb{R} \to \mathfrak{M}$ and a design function p_β, let $\sum(X,\mu^\dagger)$ denote the system given by

$$\dot{X} = \Gamma\left(X,\mu^\dagger\right) \tag{8.8}$$

$$\begin{aligned} &\stackrel{\text{def}}{=}\mathrm{col}\left(\left(\frac{\zeta}{s\left(q,q_{\mu^\dagger(t)}\right)} + \int_{\mathbb{R}^4}\alpha\left(q,q^*\right)\mu^\dagger(t)(dY)\right)\times e(\vartheta+\varphi(q)),\right. \\ &\left.\int_{\mathbb{R}^4} p_\beta(X,Y)\mu^\dagger(t)(dY)\right) \end{aligned} \tag{8.9}$$

Our next objective is to obtain an initial design for $p_i = [w_i, a_i]^T$ under which the collective dynamics (8.1) has the measure-theoretic description as $\sum_N = \left\{\sum(X_i,\mu_\delta^\dagger) : i = 1,\ldots,N\right\}$. As $s(q_i,q_C) = s(q,q_{\delta_{\bar{X}}}) \geq c_s > 0$, we shall obtain a_i through the design of p_i° defined as

$$p_i^\circ \stackrel{\text{def}}{=} \begin{bmatrix} w_i \\ a_i^\circ \end{bmatrix}, \quad a_i^\circ = s\left(q_i, q_{\delta_{\bar{X}}}\right)a_i. \tag{8.10}$$

Let $s_i \stackrel{\text{def}}{=} s(q_i, q_C)$. From (8.1) and (8.2), the dynamics of $\vartheta_i = \theta_i - \varphi(q_i, \bar{q}_N)$ and ζ_i is

$$\begin{aligned}
\dot{\vartheta}_i &= w_i - \sum_{j=1}^{N} v_j e(\theta_j) \cdot \nabla_{q_j} \varphi(q_i, \bar{q}_N) \\
\dot{\zeta}_i &= s_i a_i + \dot{s}_i v_i - \frac{1}{N} \frac{d}{dt} \left(s_i \sum_{j=1}^{N} \alpha(q_i, q_j) \right) \\
&= a_i^\circ + v_i \sum_{j=1}^{N} v_j e(\theta_j) \cdot \nabla_{q_j} s(q_i, q_C) \\
&= -\frac{1}{N} \sum_{k=1}^{N} v_k e(\theta_k) \cdot \nabla_{q_k} \left(s(q_i, q_C) \sum_{j=1}^{N} \alpha(q_i, q_j) \right). \qquad (8.11)
\end{aligned}$$

To make (8.11) become $\sum(X_i, \mu_\delta^\dagger)$, let us design p_i° as

$$p_i^\circ = \hat{p}_i(\bar{X}) + \int_{\mathbb{R}^4} p_\beta(X_i, Y) \delta_{\bar{X}}(dY) \qquad (8.12)$$

where p_β is to be designed later, and $\hat{p}_i$ is given by

$$\hat{p}_i(\bar{X}) = \begin{bmatrix} \sum_{j=1}^{N} v_j \boldsymbol{e}(\theta_j) \cdot \nabla_{q_j} \varphi(q_i, \bar{q}_N) \\ -v_i \sum_{j=1}^{N} v_j \boldsymbol{e}(\theta_j) \cdot \nabla_{q_j} s_i + \frac{1}{N} \sum_{k=1}^{N} v_k \boldsymbol{e}(\theta_k) \cdot \nabla_{q_k} \left(s_i \sum_{j=1}^{N} \alpha(q_i, q_j) \right) \end{bmatrix} \qquad (8.13)$$

Substituting (8.12) and (8.13) into (8.11) for dynamics of $\eta_i = [\vartheta_i, \zeta_i]^T$, and (8.2) into (8.1) for dynamics of q_i, we obtain

$$\begin{aligned}
\dot{q}_i &= \left(\frac{\zeta_i}{s(q_i, q_{\delta_{\bar{X}}})} + \int_{\mathbb{R}^4} \alpha(q_i, q^*) \delta_{\bar{X}}(dY) \right) e(\vartheta_i + \varphi(q_i)) \\
\dot{\eta}_i &= \int_{\mathbb{R}^4} p_\beta(X_i, Y) \delta_{\bar{X}}(dY) \qquad (8.14)
\end{aligned}$$

which, in view of (8.8), (8.9), is $\sum(X_i, \mu_\delta^\dagger)$. We note that though the design (8.13) of $\hat{p}_i(\bar{X})$ invokes $\alpha(\cdot, \cdot)$ which has not been defined, we shall design α and p_β as inputs of the system (8.14) without invoking $\hat{p}_i(\cdot)$, and hence there will arise no circular argument.

For a measure-valued mapping $\mu^\dagger$, let $X(t; t_0, X_0, \mu^\dagger)$ denote the state, if exists, at a time t of $\sum(X_i, \mu_\delta^\dagger)$ whose state at t_0 is X_0. Let X_{null} be a fictitious state. Then, the *two-parameter transition operator* $\mathcal{T}_{\mu^\dagger}(t, t_0) : \mathbb{R}^4 ß \mathbb{R}^4 \cup \{X_{\text{null}}\}$

of the system $\sum(X_i, \mu_\delta^\dagger)$ is defined as

$$\mathcal{T}_{\mu^\dagger}(t, t_0) X_0 = \begin{cases} X(t; t_0, X_0, \mu^\dagger) & \text{if } X(t; t_0, X_0, \mu^\dagger) \text{ exists} \\ X_{\text{null}} & \text{otherwise.} \end{cases} \tag{8.15}$$

We shall call $\mathcal{T}_{\mu^\dagger}$ the *transition operator* of $\sum(X_i, \mu_\delta^\dagger)$ as well.

For a measure v, let $v \circ \mathcal{T}_{\mu^\dagger}(t_0, t)$ be the measure composition defined as $v \circ \mathcal{T}_{\mu^\dagger}(t_0, t)(M) = v(\mathcal{T}_{\mu^\dagger}(t_0, t)M), M \in \mathfrak{B}$. Consider the following measure-valued transition equation:

$$\mu^\dagger(t) = \mu^\dagger(t_0) \circ \mathcal{T}_{\mu^\dagger}(t_0, t). \tag{8.16}$$

We have the following result arguing that $\mu_\delta^\dagger$ can be a Dirac measure-valued solution of (8.16). We are not assuming the regularity of $\mathcal{T}_{\mu_\delta^\dagger}(t_0, t)$ as in, e.g., [[354], Lemma 1] and [[441], Theorem 5.1]. Instead, we shall use the desired regularity as a hypothesis to be verified in Theorem 8.2.

Definition 32 (Regularity). *A two-parameter operator $\mathcal{T}(\cdot, \cdot) : \mathbb{R}^4 \to \mathbb{R}^4 \cup \{X_{null}\}$ is said to be regular on $[t_0, T)$ w.r.t a set Ω if, for all $t \in [t_0, T)$ and $X_0 \in \Omega, \mathcal{T}(t, t_0)X_0 \neq X_{null}, \mathcal{T}(t_0, t_0)X_0$, and $\mathcal{T}$ is bijective on $[t_0, T)$, w.r.t Ω, i.e., $\mathcal{T}(t, t_0)X_0 = (\mathcal{T}(t_0, t))^{-1}X_0, \forall t \in [t_0, T), \forall X_0 \in \Omega$. If $T = \infty$, then $\mathcal{T}$ is said to be regular w.r.t Ω. If $T = \infty$ and $\Omega = \mathbb{R}^4$, then $\mathcal{T}$ is said to be regular.*

Proposition 1. *Suppose that the transition operator $\mathcal{T}_{\mu_\delta^\dagger}$ of $\sum(X_i, \mu_\delta^\dagger)$ is regular on $[t_0, T)$ w.r.t $\{X_1(t_0), \ldots, X_N(t_0)\}$. Then, $\mu_\delta^\dagger : [t_0, T) \to \mathfrak{M}$ is a solution of* (8.16).

Proof. From definition (8.15), the regularity of $\mathcal{T}_{\mu_\delta^\dagger}$ implies that, for $t \in [t_0, T), i = 1, \ldots, N, \mathcal{T}_{\mu_\delta^\dagger}(t, t_0)X_i(t_0)$ is the state $X_i(t)$ of (8.14). Hence, $\delta_{X_i(t)}(M) = 1$ if and only if $\mathcal{T}_{\mu_\delta^\dagger}(t, t_0)X_i(t_0) \in M$, or equivalently, if and only if $X_i(t_0) \in \mathcal{T}_{\mu_\delta^\dagger}(t_0, t)M$, i.e., $\delta_{X_i(t0)}(\mathcal{T}_{\mu_\delta^\dagger}(t_0, t)M) = 1$. Thus, $\delta_{X_i(t)} = \delta_{X_i(t_0)} \circ \mathcal{T}_{\mu_\delta^\dagger}(t_0, t), \forall t \in [t_0, T), \forall i = 1, \ldots, N$. This and the expression (8.5) indicate that $\mu_\delta^\dagger$ satisfies (8.16) for $t \in [t_0, T)$. □

By Proposition 1, the solutions of (8.1) subject to (8.12), (8.13) generate Dirac measure-valued solutions of (8.16). Adopting [[354], Lemma 1], we have the following result inducing a continuum equation whose solutions generate continuous measure-valued solutions of (8.16). Recall that a real-valued function $f : \mathbb{R}^4 \to \mathbb{R}$ is said to be the density of a measure μ if

$$\mu M = \int_M f(X) dX, \forall M \in \mathfrak{B}. \tag{8.17}$$

Hereafter, we shall write $\boldsymbol{e}(q, \vartheta)$ in place of $\boldsymbol{e}(\vartheta + \varphi(q))$ for brevity.

Proposition 2. *Given an absolutely continuous measure μ_0 whose density function is f_0. Let $f : [t_0, T) \times \mathbb{R}^4 \to \mathbb{R}, (t, X), \mapsto f(t, X)$ be a weak solution [354] of the equation*

$$\frac{\partial f(t,X)}{\partial t} + \nabla_q \cdot \left(f(t,X) \left(\frac{\zeta}{s\left(q, q_{\mu^\dagger(t)}\right)} + \int_{\mathbb{R}^4} f(t,Y)\alpha\left(q,q^*\right) dY \right) e(q,\vartheta) \right)$$
$$+ \nabla_\eta \cdot \left(f(t,X) \int_{\mathbb{R}^4} f(t,Y) p_\beta(X,Y) dY \right) = 0 \qquad (8.18)$$

8.2.3 THE MEASURE-THEORETIC CONTROL PROBLEM

Propositions 1–2 indicate that the particle-model (8.1) under (8.12), (8.13) and the evolutionary (8.18) respectively generate Dirac measure-valued solutions $\mu_\delta^\dagger$ and continuous measure-valued solutions $\mu^\dagger$ for the measure-valued transition equation (8.16). Accordingly, based on the continuous dependence on initial data of solutions of (8.16), to be proved in Section IV, the design of flocking protocol can be addressed on (8.18).

In our framework, we consider (8.18) as an auxiliary model for design rather than an approximate continuum-model of the particle-model (8.1), (8.12), (8.13) as in, e.g., [44–46, 164].Accordingly, we have a freedom to limit the class of solutions of (8.18) by the following condition.

Assumption 16. *The function $f(t,\cdot)$ is a continuous solution of (8.18) with the boundary condition that $f(t,\cdot)$ vanishes at infinity and satisfies the following identity at $t = t_0$:*

$$\int_{\mathbb{R}^4} f(t,X) dX = 1. \qquad (8.19)$$

Hereafter, $\mathfrak{M}_\delta^\dagger$ is the class of all Dirac measure-valued mappings $\mu_\delta^\dagger : \mathbb{R} \to \mathfrak{M}, t \mapsto \delta_{\bar{X}(t)}$ with $\bar{X}(t)$ the trajectory of the collective system $\{\sum(X_i, \mu_\delta^\dagger), i = 1, \ldots, N\}$ given by (8.8),(8.9) with initial configuration satisfying Assumption 15, and $\mathfrak{M}_f^\dagger$ is the class of measure-valued mappings $\mu^\dagger$ induced by the solutions $f(t,\cdot)$ of (8.18) satisfying Assumption 16, i.e.,

$$\mu^\dagger(t)(M) = \int_M f(t,X) dX, \quad \forall M \in \mathfrak{B}. \qquad (8.20)$$

Let $\mathfrak{M}^\dagger \overset{\text{def}}{=} \mathfrak{M}_\delta^\dagger \cup \mathfrak{M}_f^\dagger$.Generalizing (8.7) for $\mu^\dagger \in \mathfrak{M}^\dagger$, we have the following function:

$$\mathcal{V}(t,\mu^\dagger) = \frac{1}{2} \int_{\mathbb{R}^4 \times \mathbb{R}^4} (\frac{1}{2} \|\eta - \eta^*\|^2 + U_1(q,q^*)) \mu^\dagger(t)(dY) \times \mu^\dagger(t)(dX)$$
$$+ \int_{\mathbb{R}^4} U_0(q, q_{\mu^\dagger(t)}) \mu^\dagger(t)(dX). \qquad (8.21)$$

where U_0 and U_1 are those given in (8.4). Clearly, if $\mu^\dagger \in \mathfrak{M}^\dagger_\delta$, then $\mathcal{V}(t,\mu^\dagger) = V(\bar{X}(t))$, with $V(\bar{X}(t))$ given by (8.7). For $\mu^\dagger \in \mathfrak{M}^\dagger_f$, $\mathcal{V}(,t,\mu^\dagger)$ can be computed using definition (8.17) as

$$\begin{aligned}\mathcal{V}(t,\mu^\dagger) =& \frac{1}{2}\int_{\mathbb{R}^4\times\mathbb{R}^4}(\frac{1}{2}\|\eta-\eta^*\|^2 + U_1(q,q^*))f(t,X)\times f(t,Y)dXdY\\ &+\int_{\mathbb{R}^4} f(t,X)U_0(q,q_{\mu^\dagger(t)})dX.\end{aligned} \tag{8.22}$$

To formulate our design goal, let us further express $\mathcal{V}(t,\mu^\dagger) = \mathcal{V}_q(t,\mu^\dagger) + \mathcal{V}_\eta(t,\mu^\dagger)$, where

$$\begin{aligned}\mathcal{V}_q(t,\mu^\dagger) =& \frac{1}{2}\int_{\mathbb{R}^4\times\mathbb{R}^4} U_1(q,q^*)\mu^\dagger(t)(dY)\mu^\dagger(t)(dX)\\ &+\int_{\mathbb{R}^4} U_0(q,q_{\mu^\dagger(t)})\mu^\dagger(t)(dX)\\ \mathcal{V}_\eta(t,\mu^\dagger) =& \frac{1}{4}\int_{\mathbb{R}^4\times\mathbb{R}^4}\|\eta-\eta^*\|^2\mu^\dagger(t)(dY)\mu^\dagger(t)(dX)\end{aligned} \tag{8.23}$$

Let $U^t_{\mu^\dagger}(q,q^*) = U_0(q,q_{\mu^\dagger(t)}) + U_1(q,q^*)$. Consider the following auxiliary functions:

$$\begin{aligned}\mathcal{W}_q(t,\mu^\dagger) =& \int_{\mathbb{R}^4}\Big(\int_{\mathbb{R}^4} \boldsymbol{e}(q,\vartheta)\cdot\Big(\nabla_q U^t_{\mu^\dagger(t)}(q,q^*)\\ &+\nabla_{q_{\mu^\dagger(t)}}U_0(q^*,q_{\mu^\dagger(t)})\Big)\mu^\dagger(t)(dY)\Big)^2\mu^\dagger(t)(dX)\\ \mathcal{W}_\eta(t,\mu^\dagger) =& \frac{1}{2}\left\|\int_{\mathbb{R}^4}\eta\mu^\dagger(t)(dX)\right\|^2\end{aligned} \tag{8.24}$$

We observe that the properties i) and ii) in Definition 30 can be achieved asymptotically if $\mathcal{V}_q(t,\mu^\dagger_d elta)$ remains bounded and $\mathcal{V}_\eta(t,\mu^\dagger_d elta)$ and $\mathcal{W}_\eta(t,\mu^\dagger_\delta)$ converge. Furthermore, such desired convergence can be achieved if there is a nonnegative function whose dissipation rate is a relevant combination of $\mathcal{V}_\eta(t,\mu^\dagger_d elta)$ and $\mathcal{W}_\eta(t,\mu^\dagger_\delta)$. Accordingly, let us consider the functional

$$\begin{aligned}\bar{\mathcal{V}}(t,\mu^\dagger) &\stackrel{\text{def}}{=} \mathcal{V}(t,\mu^\dagger) + \mathcal{W}_\eta(t,\mu^\dagger)\\ &= \mathcal{V}_q(t,\mu^\dagger) + \mathcal{V}_\eta(t,\mu^\dagger) + \mathcal{W}_\eta(t,\mu^\dagger)\end{aligned} \tag{8.25}$$

and aim at the following dissipation inequality for $\mu^\dagger \in \mathfrak{M}^\dagger$:

$$\dot{\bar{\mathcal{V}}}(t,\mu^\dagger) \le -k_1\mathcal{V}_\eta(t,\mu^\dagger) - k_2\mathcal{W}_q(t,\mu^\dagger) \tag{8.26}$$

with k_1 and k_2 the design parameters. We have the following problems.

$\mathcal{P}1)$ **Design Problem***Mean-filed flocking control*: Given design parameters $k_1 > 0$ and $k_2 > 0$, find, if possible, the interaction rules $\alpha(q, q^*)$ and $p_\beta(X, Y)$ such that the dissipation inequality (8.26) holds true for all $\mu^\dagger \in \mathfrak{M}_f^\dagger$

$\mathcal{P}2)$ **Verification Problem**:verify that under the interaction rules $\alpha(q, q^*)$ and $p_\beta(X, Y)$ obtained as solution of the Problem $\mathcal{P}1$, the collective motion of the N mobile robots (8.1) asymptotically reaches the state of styled-velocity flocking with CMCA in the sense of Definitions 30–31.

Remark 43. *From the perspectives of the mean-field approach to multi-body problems and measure-theoretic approach to dynamical systems, e.g., [44, 46, 164], [35, 45, 247, 354, 441, 466], the above design and verification problems make the current chapter original. Indeed, in Problem P1, we do not assume a sufficiently large number of mobile robots, and hence the continuum model* (8.18) *cannot be considered as an effective model of the original system* (8.1) *as usual, e.g., [44, 46, 164], [35, 45, 354, 441]. Instead, the model* (8.18) *is an auxiliary model to obtain α and p_β whose effectiveness with the original system* (8.1) *is addressed by Problem $\mathcal{P}2$. Without sufficiently large number of mobile robots, our study on the continuous dependence on initial data of solutions of* (8.16) *in Section IV is not to verify the convergence of solutions of the exact particle-models to the solution of the mean-field model as $N \to \infty$ as in, e.g., [44, 46, 164], but is to verify the convergence of solutions of the mean-field model to the solution of the exact particle-model. This does not only make Assumption 16 no limitation to our theory but also relax the condition that N is sufficiently large, e.g., [44–46, 164].*

8.3 FLOCKING CONTROL DESIGN

The main objective of this section is to present a systematic design procedure to obtain $\alpha(q, q^*)$ and $p_\beta(X, Y)$ solving Problem $\mathcal{P}1$. In particular, the dissipation inequality (8.26) is to be satisfied. Our design strategy is to design $\alpha(q, q^*)$ to render the dissipation rate $-k_2\mathcal{W}_2(t, \mu^\dagger)$, and design $p_\beta(X, Y)$ to complete (8.26).

Proposition 3. *Let $f(t, .)$ be a solution of* (8.18) *satisfying Assumption 16. Then $f(t, .)$ satisfies* (8.19) *for all $t > t_0$, and the following identity holds true:*

$$\int_{\mathbb{R}^k} \nabla_\chi \cdot (f(t, X)Q(X))\, dv = 0 \tag{8.27}$$

for any K-dimensional subvector χ of $X = [q^T, \eta^T]^T$ and any continuous function $Q(X)$.

Proof. With the boundary condition that $f(t, X)$ vanishes at infinity, the well-known divergence theorem [409] applies directly to yield the identity (8.27).

Using (8.27) to integrate (8.18) against X, it follows that $f(t,.)$ satisfies (8.19) for all $t \geq t_0$. □

To satisfy (8.26), we have the following subsections computing the time derivative of the functional $\bar{\mathcal{V}}(t,\mu^\dagger),\mu^\dagger \in \mathfrak{M}_f^\dagger$ given by (8.25),(8.23), and (8.24). The Computation is made component-wise with the functionals $\mathcal{V}_q(t,\mu^\dagger)$, $\mathcal{V}_\eta(t,\mu^\dagger)$, and $\mathcal{W}_\eta(t,\mu^\dagger)$.

In the following, for measure integrations invoking the density function $f(t,X)$ and without specifying the domain of integration, we understand that the domain of integration is the whole space $\mathbb{R}^4$ or $\mathbb{R}^4 \times \mathbb{R}^4$. We denote $\boldsymbol{f}(X,Y) \stackrel{\text{def}}{=} f(t,X)f(t,Y)$ for brevity.

We recall the convention that $X = \text{col}(q,\vartheta,\zeta), Y = \text{col}(q^*,\vartheta^*,\zeta^*)$, and in view of (8.2), denote

$$v(q) = \frac{\zeta}{s(q,q_{\mu^\dagger(t)})} + \int f(t,Y)\alpha(q,q^*)dY \tag{8.28}$$

8.3.1 COMPUTATION OF $\dot{\mathcal{V}}_Q\left(T,\mu^\dagger\right)$,$\mu^\dagger \in \mathfrak{M}_F^\dagger$

By Assumption 14, $U_1(q,q^*)$ is symmetric. Thus, expressing (8.23) via the density function $f(t,\cdot)$, we have

$$\begin{aligned}\dot{\mathcal{V}}_q\left(t,\mu^\dagger\right) = &\int f(t,X)\left(\nabla_{q_{\mu^\dagger(t)}}U_0\left(q,q_{\mu^\dagger(t)}\right)\cdot\frac{\partial q_{\mu^\dagger(t)}}{\partial t}\right)dX\\ &+\int\frac{\partial f(t,X)}{\partial t}U_0\left(q,q_{\mu^\dagger(t)}\right)dX\\ &+\int\frac{\partial f(t,X)}{\partial t}f(t,Y)U_1\left(q,q^*\right)dXdY.\end{aligned} \tag{8.29}$$

In addition, by definition (8.6), $q_{\mu^\dagger(t)}$ is time-dependent and is independent of X and using (8.6), we have

$$\begin{aligned}&\int f(t,X)\left(\nabla_{q_{\mu^\dagger(t)}}U_0\left(q,q_{\mu^\dagger(t)}\right)\cdot\frac{\partial q_{\mu^\dagger(t)}}{\partial t}\right)dX\\ &=\left(\int f(t,X)\nabla_{q_{\mu^\dagger(t)}}U_0\left(q,q_{\mu^\dagger(t)}\right)dX\right)\cdot\frac{\partial q_{\mu^\dagger(t)}}{\partial t}\\ &=\left(\int f(t,Y)\nabla_{q_\mu(t)}U_0\left(q^*,q_{\mu^\dagger(t)}\right)dY\right)\\ &\quad\cdot\int\frac{\partial f(t,X)}{\partial t}qdX \stackrel{\text{def}}{=} J_{\mu^\dagger}(t)\cdot\int\frac{\partial f(t,X)}{\partial t}qdX\end{aligned} \tag{8.30}$$

where, as $q_{\mu^\dagger(t)}$ is a time function, we have changed the integration variable $X \to Y$ in defining $J_{\mu^\dagger}(t)$ for later use. Now, for any differentiable and scalar function $\psi(q)$, multiplying both sides of (8.18) by $\psi(q)$ and using (8.28), we have

$$\begin{aligned}\frac{\partial f(t,X)}{\partial t}\psi(q) =& f(t,X)v(q)\boldsymbol{e}(q,\vartheta)\cdot\nabla_q\psi(q)\\ &-\nabla_q\cdot\left(f(t,X)\psi(q)v(q)\boldsymbol{e}(q,\vartheta)\right)\\ &-\nabla_\eta\cdot\left(f(t,X)\psi(q)\int p_\beta(X,Y)f(t,Y)dY\right)\end{aligned} \tag{8.31}$$

Taking integrals of both sides of (8.31) against X, and then noting that, due to (8.27), the integrals against q and η of the last two terms in the last equation of (8.31) vanish, we obtain

$$\int\frac{\partial f(t,X)}{\partial t}\psi(q)dX = \int f(t,X)v(q)\boldsymbol{e}(q,\vartheta)\cdot\nabla_q\psi(q)dX \tag{8.32}$$

Now, writing $q=[x,y]^T$, and then applying (8.32) for $\psi(q)=x$ and $\psi(q)=y$, we obtain

$$\int\frac{\partial f(t,X)}{\partial t}qdX = \int f(t,X)v(q)\boldsymbol{e}(q,\vartheta)dX \tag{8.33}$$

As the identity (8.19) holds true by Proposition 3, we have

$$\int\frac{\partial f(t,X)}{\partial t}U_0(q,q_{\mu^\dagger(t)})dX = \int\frac{\partial f(t,X)}{\partial t}f(t,Y)U_0(q,q_{\mu^\dagger(t)})dXdY \tag{8.34}$$

Thus, using notation $U^t_{\mu^\dagger}(q,q^*)=U_0(q,q_{\mu^\dagger(t)})+U_1(q,q^*)$, and substituting (8.33) into (8.30), and then substituting the result together with (8.34) into (8.29), we obtain

$$\begin{aligned}\dot{\mathcal{V}}_q\left(t,\mu^\dagger\right) =& \int f(t,X)v(q)\boldsymbol{e}(q,\vartheta)\cdot J_{\mu^\dagger}(t)dX\\ &+\int\frac{\partial f(t,X)}{\partial t}f(t,Y)U^t_{\mu^\dagger}\left(q,q^*\right)dXdY.\end{aligned} \tag{8.35}$$

Applying (8.32) for $\psi(q)=U^t_{\mu^\dagger}(q,q^*)$, we have

$$\int\frac{\partial f(t,X)}{\partial t}U^t_{\mu^\dagger}(q,q^*)dX = \int f(t,X)v(q)\boldsymbol{e}(q,\vartheta)\cdot\nabla_q U^t_{\mu^\dagger}(q,q^*)dX \tag{8.36}$$

Substituting (8.36) into (8.35), and then rearranging the order of integrations, we arrive at

$$\dot{\mathcal{V}}_q\left(t,\mu^\dagger\right) = \int f(t,X)v(q)\boldsymbol{e}(q,\vartheta)\cdot\left(J_{\mu^\dagger}(t)+\int f(t,Y)\nabla_q U^t_{\mu^\dagger}\left(q,q^*\right)dY\right)dX. \tag{8.37}$$

8.3.2 COMPUTATION OF $\dot{\mathcal{V}}_\eta\left(T, \mu^\dagger\right), \mu^\dagger \in \mathfrak{M}_F^\dagger$

Expressing (8.23) via the density function $f(t, \cdot)$, we have the following decomposition for $\mathcal{V}_\eta(t, \mu^\dagger)$ using (8.24) and (8.19):

$$\begin{aligned}\mathcal{V}_\eta\left(t, \mu^\dagger\right) &= \frac{1}{4}\int \boldsymbol{f}(X,Y)\left(\|\eta\|^2 - 2\eta\cdot\eta^* + \|\eta^*\|^2\right)dXdY \\ &= \frac{1}{2}\int f(t,X)\|\eta\|^2 dX - \mathcal{W}_\eta\left(t,\mu^\dagger\right) \end{aligned} \tag{8.38}$$

The time derivative of the first term in the last equation of (8.38) is computed as follows. Multiplying both sides of (8.18) by$\|\eta\|^2/2$, and then rearranging the result, we have

$$\begin{aligned}\frac{\partial(f(t,X))}{\partial t}\frac{\|\eta\|^2}{2} =& -\nabla_q\cdot\left(f(t,X)v(q)\boldsymbol{e}(q,\vartheta)\frac{\|\eta\|^2}{2}\right) - \nabla_\eta \\ &\cdot\left(f(t,X)\frac{\|\eta\|^2}{2}\int p_\beta(X,Y)f(t,Y)dY\right) \\ &+\eta\cdot\left(f(t,X)\int p_\beta(X,Y)f(t,Y)dY\right)\end{aligned} \tag{8.39}$$

Taking the integrals of both side of (8.39) against X, and then noting that, due to (8.27), the integral against X of the first two terms in the last equation of (8.39) vanishes, we obtain

$$\int\frac{\partial(f(t,X))}{\partial t}\frac{\|\eta\|^2}{2}dX = \int \boldsymbol{f}(X,Y)\eta\cdot p_\beta(X,Y)dXdY \tag{8.40}$$

Thus, taking the time derivatives of both sides of (8.38), and then using (8.40), we arrive at

$$\mathcal{V}_\eta\left(t,\mu^\dagger\right) = \int \boldsymbol{f}(X,Y)\eta\cdot p_\beta(X,Y)dXdY - \dot{\mathcal{W}}_\eta(t,\mu^\dagger). \tag{8.41}$$

8.3.3 CONTROL SPECIFICATION

Taking the time derivative of $\bar{\mathcal{V}}\left(t,\mu^\dagger\right)$ given by (8.25), and then using (8.37) and (8.41), we obtain

$$\begin{aligned}\dot{\bar{\mathcal{V}}}\left(t,\mu^\dagger\right) =& \int f(t,X)v(q)e(q,\vartheta) \\ &\cdot\left(J_{\mu^\dagger}(t) + \int f(t,Y)\nabla_q U_{\mu^\dagger}^t\left(q,q^*\right)dY\right)dX \\ &+\int \boldsymbol{f}(X,Y)\eta\cdot p_\beta(X,Y)dXdY\end{aligned} \tag{8.42}$$

Substituting $J_{\mu^\dagger}(t)$ defined in (8.30) and the expression (8.28) of $v(q)$ into (8.42), we obtain

$$\begin{aligned}\dot{\mathcal{V}}\left(t,\mu^\dagger\right) = & \int f(t,X)\frac{\zeta}{s(\cdot)}e(q,\vartheta)\cdot\int f(t,Y) \\ & \times\left(\nabla_{q_{\mu^\dagger(t)}}U_0\left(q^*,q_{\mu^\dagger(t)}\right)+\nabla_q U^t_{\mu^\dagger}\left(q,q^*\right)\right)dYdX \\ & +\int f(t,X)\left(\int_{\mathbb{R}^4} f(t,Y)\alpha\left(q,q^*\right)dY\right) \\ & \times\int f(t,Y)e(q,\vartheta) \\ & \cdot\left(\nabla_{q_{\mu^\dagger(t)}}U_0(q^*,q_{\mu^\dagger(t)})+\nabla_q U^t_{\mu^\dagger}\left(q,q^*\right)\right)dYdX \\ & +\int \boldsymbol{f}(X,Y)\eta\cdot p_\beta(X,Y)dXdY \end{aligned} \tag{8.43}$$

where $s(\cdot)$ is $s(q,q_{\mu^\dagger(t)})$. Accordingly, the design

$$\alpha(q,q^*) = -k_2\boldsymbol{e}(q,\vartheta)\cdot\left(\nabla_{q_{\mu^\dagger(t)}}U_0\left(q^*,q_{\mu^\dagger(t)}\right)+\nabla_q U^t_{\mu^\dagger}\left(q,q^*\right)\right) \tag{8.44}$$

leads to

$$\begin{aligned}\dot{\mathcal{V}}\left(t,\mu^\dagger\right) = & \int f(t,X)\frac{\zeta}{s(\cdot)}\int f(t,Y)e(q,\vartheta) \\ & \cdot\left(\nabla_{q_{\mu^\dagger(t)}}U_0\left(q^*,q_{\mu^\dagger(t)}\right)+\nabla_q U^t_{\mu^\dagger}\left(q,q^*\right)\right)dYdX \\ & +\int \boldsymbol{f}(X,Y)\eta\cdot p_\beta(X,Y)dXdY - k_2\mathcal{W}_q(t,\mu^\dagger)\end{aligned} \tag{8.45}$$

where $\beth_q(t,\mu^\dagger)$ is defined by (8.24). To design p_β, let us consider the following structure.

$$p_\beta(X,Y) = p^\circ_\beta(X,Y) + \begin{bmatrix} 0 \\ a_\zeta(X,Y)\end{bmatrix}. \tag{8.46}$$

As $\eta = [\vartheta,\zeta]^T$, we eliminate the first term in the right-hand-side of (8.45) by the design

$$a_\zeta(X,Y) = -\frac{1}{s(q,q_{\mu^\dagger(t)})}\boldsymbol{e}(q,\vartheta)\cdot\left(\nabla_{q_{\mu^\dagger(t)}}U_0\left(q^*,q_{\mu^\dagger(t)}\right)+\nabla_q U^t_{\mu^\dagger}\left(q,q^*\right)\right) \tag{8.47}$$

Substituting (8.47) into (8.46), and then Substituting the resulting $p_beta(X,Y)$ into (8.45),we arrive at

$$\dot{\mathcal{V}}\left(t,\mu^\dagger\right) = \int \boldsymbol{f}(X,Y)\eta\cdot p^\circ_\beta(X,Y)dXdY - k_2\mathcal{W}_q(t,\mu^\dagger) \tag{8.48}$$

Finally, let us specify $p^{\circ}_{\beta}(X,Y)$ such that the right-hand-side of (8.48) completes the dissipation inequality (8.26). To this end, let us exploit the face that

$$\int \boldsymbol{f}(X,Y)\frac{\|\eta\|^2}{2}dXdY = \int \boldsymbol{f}(X,Y)\frac{\|\eta^*\|^2}{2}dXdY \tag{8.49}$$

to express the function $\mathcal{V}_\eta(t,\mu^\dagger)$ defined in (8.23) as

$$\begin{aligned}\mathcal{V}_\eta(t,\mu^\dagger) &= \frac{1}{4}\int \boldsymbol{f}(X,Y)\left(\|\eta\|^2 - 2\eta\cdot\eta^* + \|\eta^*\|^2\right)dXdY \\ &= \frac{1}{2}\int \boldsymbol{f}(X,Y)\left(\|\eta\|^2 - \eta\cdot\eta^*\right)dXdY \\ &= \frac{1}{2}\int \boldsymbol{f}(X,Y)\eta\cdot(\eta-\eta^*)\,dXdY \end{aligned} \tag{8.50}$$

Comparing (8.50) with the first term in the right-hand-side of (8.48), we have the design

$$p^{\circ}_{\beta}(X,Y) = -\frac{k_1}{2}(\eta-\eta^*). \tag{8.51}$$

Substituting (8.51) into (8.48), and then using equality (8.50), we have (8.26) satisfied.

In summary, by $\alpha(q,q^*)$ given by (8.44) and $p_\beta(X,Y)$ given by (8.46),(8.47), and (8.51), we solved the design problem $\mathcal{P}1$ in Section II-C, Let$U'(q_i,q_j) \overset{\text{def}}{=} \nabla_{qc}U_0(q_j,q_C) + \nabla_{q_i}(U_0(q_i,q_C)+U_1(q_i,q_j))$. Applying such design of α and p_β to $\mu^\dagger_\delta$, we obtain the protocol

$$\begin{aligned}\alpha(q_i,q_j) &= -k_2\boldsymbol{e}(q_i,\vartheta_i)\cdot U'(q_i,q_j) \\ p_\beta(X_i,X_j) &= -\frac{k_1}{2}(\eta_i-\eta_j) + \begin{bmatrix} 0 \\ -\frac{1}{s(q_i,q_C)}\boldsymbol{e}(q_i,\vartheta_i)\cdot U'(q_i,q_j)\end{bmatrix}.\end{aligned} \tag{8.52}$$

Substituting (8.52) and (8.13) into (8.12), and then using (8.10), we obtain the final flocking protocol $p_i = [w_i,a_i]^T$ for the particle-model (8.1). This completes our design procedure.

8.4 CONVERGENCE ANALYSIS

In this section, we address Problem $\mathcal{P}2$ completing our measure theoretic framework for styled-velocity flocking control. Specifically, we shall show that the Dirac measure-valued mappings $\mu^\dagger \in \mathfrak{M}^\dagger_\delta$ induced by the trajectories $\bar{X}(t)$ of the collective system (8.1) subject to the control $p_i = [w_i,a_i]^T$ given by (8.10), (8.12), (8.13), and (8.52) preserve the dissipative behavior (8.26) of the continuous measure-valued mappings $\mu^\dagger \in \mathfrak{M}^\dagger_f$ subject to the design (8.44), (8.46), (8.47), and (8.51). To this end, we establish the two important yet original results on the regularity of the transition operators $\mathcal{T}_{\mu^\dagger}$, $\mu^\dagger \in \mathfrak{M}_\delta^\dagger \cup \mathfrak{M}_f^\dagger$, and on the continuous dependence on the initial data of the solutions of (8.16).

8.4.1 THE REGULARITY OF TRANSITION OPERATORS

Let us recall that $U_{\mu^\dagger}^t(q, q^*) = U_0(q, q_{\mu^\dagger(t)}) + U_1(q, q^*)$. Define

$$W_\mu(q, \vartheta) = \int_{\mathbb{R}^4} e(q, \vartheta) \cdot \left(\nabla_{q_\mu} U_0\left(q, q_\mu\right) + \nabla_q \left(U_0\left(q, q_\mu\right) + U_1\left(q, q^*\right)\right)\right) \mu(dY) \tag{8.53}$$

and

$$\mathcal{V}_{\mu^\dagger}(t) = \frac{1}{2} \int_{\mathbb{R}^4} f(t, X) \|X\|^2 dX + \mathcal{V}_q\left(t, \mu^\dagger\right) \tag{8.54}$$

where $\mathcal{V}_q(t, \mu^\dagger)$ is given by (8.23).

Assumption 17. *the control $\alpha(q, q^*)$ is given by (8.44),the control $p_\beta(X, Y)$ has the structure (8.46) with α_ζ given by (8.47), and $p_\beta^\circ(X, Y)$ is Lipschitz continuous in the sense that*

$$\left\|p_\beta^\circ(X, Y)\right\| \le L_p \|X\| + L_p \|Y\| + L_p, \quad \forall X, Y \in \mathbb{R}^4 \tag{8.55}$$

with L_p a nonnegative constant. In addition, $\mathcal{W}_q(t, \mu^\dagger)$ defined by (8.24) is uniformly continuous with respect to t.

Assumption 18. *There is a constant $K > 0$ such that along the trajectory $[q^T(t), \vartheta(t), \zeta(t)]^T$ of the system (8.8), (8.9) with $\mu^\dagger \in \mathfrak{M}_f^\dagger$, we have*

$$\left|W_{\mu^\dagger(t)}(q(t), \vartheta(t))\right|^2 \le K\mathcal{W}_q(t, \mu^\dagger) \tag{8.56}$$

Remark 44. *Clearly, the flocking control $p_\beta^\circ(X, Y)$ designed in (8.51) fulfills the condition (8.55). By Assumption 17, we consider the general case of flocking control. The condition (8.55) allows $p_\beta(\cdot)$ to be designed for other objective instead of consensus on η_i. The following Theorem 8.3 indicates that the basic goal of CMCA is still achievable in such general case.*

Remark 45. *Property (8.56) is obviously satisfied for $K = N$, $\mu^\dagger \in \mathfrak{M}_\delta^\dagger, (q(t), \vartheta(t)) \in \{(q_i(t), \vartheta_i(t)) : i = 1, \dots, N\}$. Physically, this property expresses that the energy of a single vehicle is no greater than the total energy of the N vehicles. By Assumption 18, we are using a subclass of $\mathfrak{M}_f^\dagger$ for study. Again, as $\mathfrak{M}_f^\dagger$ is instrumental in verifying flocking behavior of (8.1), this assumption is of no limitation to our theory.*

Theorem 8.1

Let $\mu^\dagger \in \mathfrak{M}_f^\dagger$, i.e., $\mu^\dagger$ is the measure-valued mapping defined by (8.20) with $f(t,\cdot)$ be a continuous solution of (8.18) satisfying Assumption 17. Suppose that the system (8.8), (8.9) satisfies Assumptions 14, 17, and 18, and $\mathcal{V}_{\mu^\dagger}(t_0) < \infty$. Then, the two-parameter transition operator $\mathcal{T}_{\mu^\dagger}$ of (8.8), (8.9) is regular in the sense of Definition 32. ■

Proof. See Appendix A. □

By Theorem 8.1, the regularity of the transition operator $\mathcal{T}_{\mu^\dagger}$ is verified. As we are studying the convergence of $\mu^\dagger$ to $\mu_\delta^\dagger$, our next objective is to verify the regularity of the transition operator $\mathcal{T}_{\mu_\delta^\dagger}$ of the system $\sum(X, \mu_\delta^\dagger)$ subject to the same controls α and p_β satisfying Assumption 17. From (8.8), (8.9), the trajectory $\bar{X}(t)$ defining $\mu_\delta^\dagger$ is generated by $\sum(X_i, \mu_\delta^\dagger)$

$$\begin{aligned}
\dot{q}_i &= \left(\frac{\zeta_i}{s\left(q_i, q_C\right)} - \frac{k_2}{N} \sum_{j=1}^{N} e\left(q_i, \vartheta_i\right) \cdot U'\left(q_i, q_j\right) \right) e\left(q_i, \vartheta_i\right) \\
\dot{\eta}_i &= \frac{1}{N} \sum_{j=1}^{N} p_\beta^\circ \left(X_i, X_j\right) \\
&\quad + \frac{1}{N} \sum_{j=1}^{N} \begin{bmatrix} 0 \\ -\frac{1}{s(q_i, q_C)} e\left(q_i, \vartheta_i\right) \cdot U'\left(q_i, q_j\right) \end{bmatrix}
\end{aligned} \tag{8.57}$$

where $q_C = (\sum_{i=1}^N q_i)/N$ and $U'(q_i, q_j) \stackrel{\text{def}}{=} \nabla_{qc} U_0(q_j, q_C) + \nabla_{q_i}(U_0(q_i, q_C) + U_1(q_i, q_j))$

Theorem 8.2

Under Assumptions 14, 15, and 17, the transition operator $\mathcal{T}_{\mu_\delta^\dagger}$ of $\sum(X, \mu_\delta^\dagger)$ is regular w.r.t $\{X_1(t_0), \ldots, X_N(t_0)\}$. ■

Proof. See Appendix B. □

As a consequence of Theorem 8.2, the following theorem argues that the design of p_β° for coordination purpose can be separated from achieving CMCA.

Theorem 8.3

Suppose that Assumptions 14 and 15 holds true. Then, under the controls $\alpha(q, q^*)$ and $p_\beta(X, Y)$ satisfying Assumption 17, the collective system (8.1) is free of collision, and $\max_{i,j} \|q_i(t) - q_j(t)\| < \infty, \forall t \geq t_0$. ■

Proof. By proof of Theorem 8.2, $V(\bar{X}(t))$ defined by (8.87) remains finite. Hence, so does $U_1(q_i(t), q_j(t)), i \neq j$. By Assumptions 14 and 15, the conclusion of the theorem follows. □

By Theorems 8.1 and 8.2, we have the regularity of the transition operators. Hence, by Propositions 1 and 2, the set $\mathfrak{M}$ under Assumption 18 is contained in the solution space of (8.16). We have the following result on the continuous dependence on initial data of solutions of (8.16) which enables the desired performance $\mu_n^\dagger(t) \to \delta_{\bar{X}(t)}, n \to \infty$.

8.4.2 CONTINUOUS DEPENDENCE ON INITIAL DATA

For convergence analysis, let us recall relevant concepts from [30, 354].Let $\mathcal{C}_B$ be the set of all functions $\psi : \mathbb{R}^4 \to [0, 1]$ satisfying $|\psi(X) - \psi(Y)| \leq \|X - Y\|, \forall X, Y \in \mathbb{R}^4$.

Definition 33. *[354] A sequence of measures $\{\mu_n\}_{n=1}^\infty$ is said to converge weakly to a measure μ, written as $\mu_n \to \mu, n \to \infty$, if and only if*

$$\lim_{n\to\infty} \int_{\mathbb{R}^4} \varphi(X)\mu_n(dX) = \int_{\mathbb{R}^4} \varphi(X)\mu(dX), \forall \varphi \in \mathcal{C}_B. \tag{8.58}$$

The bounded Lipschitz metric p_ℓ of a space of measures $\mathfrak{M}$ is defined as

$$\rho_\ell(\mu, \nu) = \sup_{\psi\in\mathcal{D}} \left| \int_{\mathbb{R}^4} \psi(X)\mu(dX) - \int_{\mathbb{R}^4} \psi(X)\nu(dX) \right|, \mu, \nu \in \mathfrak{M}. \tag{8.59}$$

Let $\mathfrak{N}$ be a set of solutions of (8.16). A solution $\mu_*^\dagger$ of (8.16) is said to depend continuously on initial data with respect to $\mathfrak{N}$ if for any sequence of solutions $\{\mu_n\}_{n=1}^\infty \subset \mathfrak{N}$, we have the following implication:

$$\mu_n^\dagger(t_0) \to \mu_*^\dagger(t_0), n \to \infty \Rightarrow \mu_n^\dagger(t) \to \mu_*^\dagger(t), n \to, \forall t \geq t_0 \tag{8.60}$$

Theorem 8.4

Let $\mathfrak{M}_\delta^\dagger$ be the set of $\mu_\delta^\dagger(t)$ given by $\mu_\delta^\dagger(t) = \delta_{\bar{X}(t)}$, $\bar{X}(t)$ the trajectory of the system (8.57) subject to Assumptions 14, 15, and 17, and let $\mathfrak{M}_f^\dagger(H)$ be the set of measure-valued mappings $\mu^\dagger$ defined by (8.20) with $f(t,\cdot)$ solution of (8.18) subject to Assumptions 14, 16, 17, and 18, and $\mathcal{V}_{\mu^\dagger}(t_0) \leq H$. Then, the solutions $\mu_\delta^\dagger \in \mathfrak{M}_\delta^\dagger$ of (8.16) depend continuously on initial data with respect to $\mathfrak{M}_f^\dagger(H)$. ■

Proof. See Appendix C. □

Remark 46. *Continuous dependence on initial data of solutions of Vlasov-like equations in kinetic models of collective motion was studied in [35, 354]. Our proof of Theorem 8.4 in Appendix C adopts the common framework of using bounded Lipschitz distance and Gronwall-type lemmas. The main technical feature making our proof original is that these works do not address CMCA and hence bounded potential functions are used to grant global Lipschitz continuity for vector fields. Whereas, we are using unbounded coordination function U to penalize the loss of CMCA. Accordingly, by Theorem 8.4, we introduce the initially uniform boundedness condition $\mathcal{V}_{\mu^\dagger}(t_0) \leq H$ for our technique dealing with the lack of the global Lipschitz continuity.*

8.4.3 VALIDATION OF FLOCKING ALGORITHM

In this subsection, from the dissipation inequality (8.26) and the result on the continuous dependence on initial data in Theorem 8.4, we show that, under the designed flocking protocol given by (8.10), (8.12), (8.13), and (8.52), the collective motion of the N vehicles of model (8.1) asymptotically approaches the flocking state in Definition 30. We have the following theorem.

Theorem 8.5

Given N autonomous vehicles of model (8.1) and a flocking theme (U, φ, s). Suppose that $U(\bar{q}_N)$ is given by (8.4) and Assumptions 14 and 15 hold true. Then, under the controls $p_i = [w_i, a_i]^T, i = 1, \ldots, N$ given by (8.10)–(8.12), (8.13), and (8.52), the collective motion of the N vehicles approaches the flocking state in Definition 30 asymptotically with CMCA. ■

Proof. Recall that $\bar{X} = \text{col}(X_1, \ldots, X_N), X_i = [q_i^T, \eta_i^T]^T, \eta_i = [\vartheta_i, \zeta_i]^T$, and $\vartheta_i = \theta_i - \varphi(q_i)$. Let $\bar{X}(t0) = \bar{X}_0$ be an initial state of the system (8.1) satisfying Assumption 15, and let $\{f_n^{t_0}\}_{n=0}^{\infty}$ be the sequence of density functions such that the measures $\mu_n^{t_0}$ defined as

$$\mu_n^{t_0}(M) = \int_M f_n^{t_0}(X)dX, \quad M \in \mathfrak{B} \tag{8.61}$$

converge weakly to the Dirac measure $\mu_n^{t_0}(M) = \delta_{\bar{X}_0}$, and satisfy $\mathcal{V}_{\mu_n^\dagger}(t_0) \leq H, \forall n$, for some $H > 0$. For each $f_n^{t_0}$, let $f_n(t,\cdot)$ be the solution of (8.18) with initial condition $f_n(t_0,\cdot)=f_n^0(\cdot)$ subjects to Assumptions 14, 16, 17, and 18. By Theorems 8.1–8.2 and Propositions 1–2, the mappings $\mu_n^\dagger:[t_0,\infty)\to\mathfrak{M}, t\mapsto\mu_n^\dagger(t)$ defined as

$$\mu_n^\dagger(t)(M)=\int_M f_n(t,X)dX \tag{8.62}$$

as well as $\mu_\delta^\dagger:[t_0,\infty)\to\mathfrak{M}, t\mapsto\delta_{\bar{X}(t)}$ are solutions of (8.16), where $\bar{X}(t)$ is the trajectory through X_0, defined for all $t\geq t_0$ by Theorem 8.2, of the collective system (8.57) which is a transformation of (8.1) closed by the designed p_i. As $\mu_n^{t_0}\to\delta_{\bar{X}_0}, n\to\infty$, applying Theorem 8.4, we have $\mu_n^\dagger(t)\to\mu_\delta^\dagger(t)=\delta_{\bar{X}(t)}, n\to\infty, \forall t\geq t_0$.

Moreover, from Proof of Theorem 8.2, $V(\bar{X}(t))$ given by (8.87) remains finite for $t<\infty$. Accordingly, there are compact sets $\Omega_n(t)$ and $\Omega_n^*(t)$ containing $X_i(t), i=1,\ldots,N$ and bounded continuous functions $a_n^t(X,Y)$ and $b_n^t(X,Y)$ such that $a_n^t(X,Y)=\|\eta-\eta^*\|^2$ and $b_n^t(X,Y)=e(q,\vartheta)U_n'(q,q^*)$ for all $X=[q^T,\eta^T]^T\in\Omega_n(t), Y=[q^{*T},\eta^{*T}]^T\in\Omega_n^*(t)$ and

$$\begin{aligned}\mathcal{V}_\eta\left(t,\mu_n^\dagger\right)&\geq\frac{1}{4}\int_{\mathbb{R}^4\times\mathbb{R}^4}a_n^t(X,Y)\mu_n^t(dY)\mu_n^t(dX)\overset{\text{def}}{=}\mathcal{V}_\eta^\circ\left(t,\mu_n^\dagger\right)\\ \mathcal{W}_q\left(t,\mu_n^\dagger\right)&\geq\int_{\mathbb{R}^4}\left(\int_{\mathbb{R}^4}b_n^t(X,Y)\mu_n^t(dY)\right)^2\mu_n^t(dX)\\ &\overset{\text{def}}{=}\mathcal{W}_q^\circ\left(t,\mu_n^\dagger\right)\end{aligned} \tag{8.63}$$

where $\mu_n^t=\mu_n^\dagger(t)$ and $U'(q_i,q_j)\overset{\text{def}}{=}\nabla_{qC}U_0(q_j,q_C)+\nabla_{q_i}(U_0(q_i,q_C)+U_1(q_i,q_j))$.From the above construction of a_n^t and b_n^t and the convergence $\mu_n^\dagger\to\delta_{\bar{X}(t)}$, we have

$$\begin{aligned}\lim_{n\to\infty}\mathcal{V}_\eta^\circ\left(t,\mu_n^\dagger\right)&=\mathcal{V}_\eta^\circ\left(t,\mu_\delta^\dagger\right)=\frac{1}{4}\frac{1}{N}\sum_{i=1}^N\frac{1}{N}\sum_{j=1}^N\|\eta_i-\eta_j\|^2\\ \lim_{n\to\infty}\mathcal{W}_q^\circ\left(t,\mu_n^\dagger\right)&=\mathcal{W}_q^\circ\left(t,\mu_\delta^\dagger\right)\\ &=\frac{1}{N}\sum_{i=1}^N\left(\frac{1}{N}\sum_{j\neq i}e\left(q_i,\vartheta_i\right)\cdot U'\left(q_i,q_j\right)\right)^2.\end{aligned} \tag{8.64}$$

By (8.63), we shall prove that $\mathcal{V}_\eta^\circ(t,\mu_n^\dagger)$ and $\mathcal{W}_q^\circ(t,\mu_n^\dagger)$ both converge to zero as $t\to\infty$ by showing that so do both $\mathcal{V}_\eta^\circ(t,\mu_n^\dagger)$ and $\mathcal{W}_q^\circ(t,\mu_n^\dagger)$.Indeed, as designed, for every n, (8.26) holds true for $\mu^\dagger=\mu_n^\dagger$, i.e.,

$$\dot{\bar{\mathcal{V}}}\left(t,\mu_n^\dagger\right)=-k_1\mathcal{V}_\eta\left(t,\mu_n^\dagger\right)-k_2\mathcal{W}_q\left(t,\mu_n^\dagger\right). \tag{8.65}$$

As both $\mathcal{V}_\eta(t,\mu_n^\dagger)$ and $\mathcal{W}_q^\circ(t,\mu_n^\dagger)$ are nonnegative, (8.65) implies that $\bar{\mathcal{V}}(t,\mu_n^\dagger)$ is nonincreasing, and hence, by Barbalat's lemma [229], both $\mathcal{V}_\eta(t,\mu_n^\dagger)$

and $\mathcal{W}_q^\circ(t,\mu_n^\dagger)$ approach zero as $t\to\infty$. Accordingly

$$\lim_{t\to\infty}\mathcal{V}_\eta^\circ\left(t,\mu_\delta^\dagger\right)=\lim_{t\to\infty}\lim_{n\to\infty}\mathcal{V}_\eta^\circ\left(t,\mu_n^\dagger\right)\le\lim_{t\to\infty}\lim_{n\to\infty}\mathcal{V}_\eta\left(t,\mu_n^\dagger\right)=0$$
$$\lim_{t\to\infty}\mathcal{W}_q^\circ\left(t,\mu_\delta^\dagger\right)=\lim_{t\to\infty}\lim_{n\to\infty}\mathcal{W}_q^\circ\left(t,\mu_n^\dagger\right)\le\lim_{t\to\infty}\lim_{n\to\infty}\mathcal{W}_q\left(t,\mu_n^\dagger\right)=0 \quad (8.66)$$

Thus, in view of (8.64), we have

$$\lim_{t\to\infty}\frac{1}{2}\frac{1}{N}\sum_{i=1}^{N}\frac{1}{N}\sum_{j\neq i}\|\eta_i-\eta_j\|^2=0 \quad (8.67)$$

which indicates the asymptotic consensus on $\eta_i=[\vartheta_i,\zeta_i]^T$. Furthermore, in view of (8.52) and (8.64), $\mathcal{W}_q(t,\mu_\delta^\dagger)\to 0$ implies that $\sum_j\alpha(q_i,q_j)\to 0$. Accordingly, by definition (8.2), we have asymptotic consensus on ν_i, and hence, asymptotic consensus on the styled-velocity $\boldsymbol{v}_{s,i}$. This fulfills condition i) of Definition 30. The satisfaction of condition ii) of Definition 30 follows directly from (8.66) and the finiteness of $V(\bar{X}(t))$ given by (8.87) in Proof of Theorem 8.2. Thus, the state of the system (8.1) approaches the state of styled-velocity asymptotically.

Finally, in view of (8.25) and the expression (8.78) in Appendix A of $\mathcal{V}_{\mu^\dagger}(t)$ defined by (8.54), we have $\mathcal{V}_{\mu_n^\dagger}(t_0)\ge\bar{\mathcal{V}}(t_0,\mu_n^\dagger),\forall n$. As $\mathcal{V}_{\mu_0^\dagger}(t_0)\le H,\forall n$, this yields $H\ge\bar{\mathcal{V}}(t_0,\mu_n^\dagger),\forall n$. As argued above, (8.65) holds true for every n, and hence, by (8.25), $H\ge\bar{\mathcal{V}}(t,\mu_n^\dagger)\ge\bar{\mathcal{V}}_q(t,\mu_n^\dagger),\forall t\ge t_0,\forall n$. As verified above, $\mu_n^\dagger(t)\to\delta_{\bar{X}_n(t)},n\to\infty,\forall t\ge t_0$, and hence, in view of (8.4) and (8.23), $\mathcal{V}_q(t,\mu_n^\dagger)\to U(\bar{q}_N(t)),n\to\infty,\forall t\ge t_0$. These together imply that $U(\bar{q}_N(t)),t\ge t_0$ is bounded. By Assumption 14, such boundedness of $U(\bar{q}_N(t)),t\ge t_0$ implies CMCA. □

8.5 AN ELLIPTICAL FLOCKING OF MOBILE ROBOTS

In this section, we apply the presented theory to achieve elliptical flocking of mobile robots in the sense that the collective motion of the mobile robots stays cohesive without collision and the trajectories of the robots have elliptical shapes. Clearly, this problem remains open. To apply the presented theory to solve this problem, a flocking theme (U,φ,s) dedicated to elliptical flocking is to be specified. Let d_0,d_1,ρ_0, and ρ_1 be positive design parameters.

We first specify a coordination function $U(\bar{q}_N)$ of the structure (8.4) with $U_0(q_i,q_C)$ and $U_1(q_i,q_j)$ satisfying Assumption 14. Our desired behavior for the i-th vehicle is: moving away q_C (resp., q_j) if its distance to q_C (resp., q_j), $ri=\|q_i-q_C\|$ (resp., $d_{ij}=\|q_i-q_j\|$), is approaching d_0 (resp., d_1); moving toward q_C (resp., q_j) if $r_i>\rho_0$ (resp., $d_{ij}>\rho_1$); once $d_0<r_i\le\rho_0$ and $d_1<d_{ij}<\rho_1,\forall j\neq i$, the vehicle rests on its flocking motion. Thus, we specify

$$U_0(q_i, q_C) = \begin{cases} 0 & \text{if } r_i \leq d_0 \\ \left(\frac{h(r_i; d_0+\varepsilon, d_0)}{r_i - d_0} + h(r_i; \rho_0, \rho_0 + \varepsilon)(r_i - \rho_0)\right) & \text{if } r_i > d_0 \end{cases} \tag{8.68}$$

$$U_1(q_i, q_j) = \begin{cases} 0 & \text{if } d_{ij} \leq d_1 \\ \left(\frac{h(d_{ij}, d_1+\varepsilon, d_1)}{(d_{ij} - d_1)^2} + h(d_{ij}, \rho_1, \rho_1 + \varepsilon)(d_{ij} - \rho_1)\right) & \text{if } d_{ij} > d_1 \end{cases} \tag{8.69}$$

where $\varepsilon > 0$ is a small constant and $h(x; a, b)$ is the bump function defined by [464]

$$h(x; a, b) = g\left(\frac{x-a}{b-a}\right), \quad g(s) = \frac{f(s)}{f(s) + f(1-s)}$$
$$f(s) = \begin{cases} e^{-1/s} & \text{if } s > 0 \\ 0 & \text{if } s \leq 0 \end{cases} \tag{8.70}$$

To generate elliptical flocking, we choose the scheme that the vehicles move on elliptical orbits congruent to a basic ellipse, namely $(\mathcal{E}_B)$, of semi-major axis a and semi-minor axis b, centered at the center of mass $q_C = [x_C, y_C]^T$. According, we specify the desired orientation $\varphi(q_i, q_C)$ for the i-th vehicle occupying the location $q_i = [x_i, y_i]^T$ as follows. Consider the ellipse passing qi that is congruent to the $(\mathcal{E}_B)$. The equation of such ellipse is

$$\frac{(x - x_C)^2}{a^2} + \frac{(y - y_C)^2}{b^2} = \frac{(x_i - x_C)^2}{a^2} + \frac{(y_i - y_C)^2}{b^2}. \tag{8.71}$$

From the following expression of the unit vector tangent to the ellipse (8.71) at q_i:

$$\boldsymbol{e}_T(q_i, q_C) = \frac{1}{\frac{(x_i - x_C)^2}{a^4} + \frac{(y_i - y_C)^2}{b^4}} \begin{bmatrix} -\frac{y_i - y_C}{b^2} \\ \frac{x_i - x_C}{a^2} \end{bmatrix} \tag{8.72}$$

we may assign $\varphi(q_i, q_C) = \arg \boldsymbol{e}_T(q_i, q_C)$, the argument of $\boldsymbol{e}_T(q_i, q_C)$. Nevertheless, as we wish the vehicles rest on the region $d_0 < r_i \leq \rho_0$, we have the following design:

$$\varphi(q_i, q_C) = \arg(\boldsymbol{e}_T(q_i, q_C)) - k_\varphi \arctan\left((U_a'(q_i))^\perp \cdot \boldsymbol{e}_T(q_i, q_C) \|\nabla_{q_i} U_a(q_i)\|\right) \tag{8.73}$$

where $k_\varphi > 0$ is a design parameter, $(U_a'())^\perp$ is the counter-clockwise rotation by an angle of $\pi/2$ of $U_a'(q_i) = \sum_j (\nabla_{q_C} U_0(q_j, q_C) + \nabla_{q_i}(U_0(q_i, q_C) +$

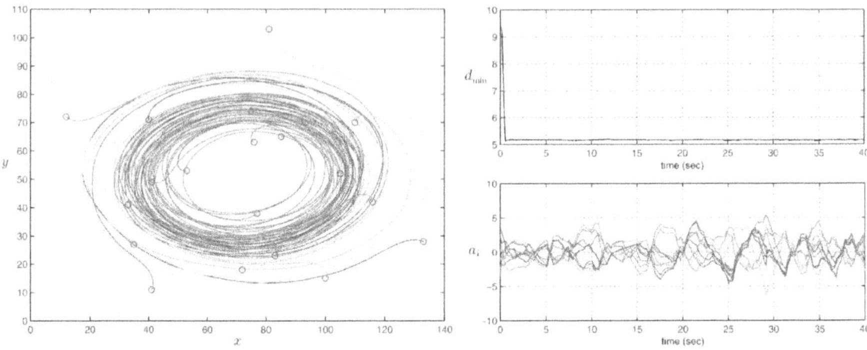

Figure 8.2 LHS: trajectories of 20 mobile robots; RHS: the time diagrams of the minimal distance among 20 mobile robots and their controls a_i's.

$U_1(q_i, q_j))$), and $U_a(q_i) = U_0(q_i, q_C) + \sum_j U_1(q_i, q_j)$. The second term in (8.73) is due to the performance that the i-th vehicle turns away q_C if r_i is approaching d_0, i.e., $(U_a'(q_i))^{\perp} \cdot \mathbf{e}^T(q_i, q_C) < 0$, turns toward q_C if it is going away the region $d_0 < r_i \geq \rho_0$, i.e., $(U_a'(q_i))^{\perp} \cdot \mathbf{e}^T(q_i, q_C) > 0$, and rests on its elliptical motion if $(U_a'(q_i))^{\perp} \cdot \mathbf{e}^T(q_i, q_C) = 0$.

Finally, we specify

$$
\begin{aligned}
s\left(q_i, q_C\right) &= c_s + k_s \kappa\left(q_i, q_C\right) \\
\kappa\left(q_i, q_C\right) &= \frac{ab\left(\left(x_i - x_C\right)^2 / a^2 + \left(y_i - y_C\right)^2 / b^2\right)}{\left(\left(x_i - x_C\right)^2 a^2 / b^2 + \left(y_i - y_C\right)^2 b^2 / a^2\right)^{\frac{3}{2}}}
\end{aligned}
\tag{8.74}
$$

where $c_s > 0, k_s > 0$ are design parameter and $\kappa(q_i, q_C)$ is the curvature of (8.71) at $q_i = [x_i, y_i]^T$. The physical insight for such design is that the trajectories closer to the center of mass have greater curvatures and hence the linear speeds should be smaller, and converse.

The result of our simulation with 20 robots is shown in Fig. 8.2. We selected: $k_1 = 2, k_2 = 1, a = 3, b = 2, \varepsilon = 0.2, d_0 = 10, \rho_0 = 30, d_1 = 5, \rho_1 = 10, k_\varphi = 0.2, c_s = 0.01$, and $k_s = 0.5$. The left-hand-side figure in Fig. 8.2 indicates that an elliptical ring congruent to the basic ellipse $(\mathcal{E}B)$ had been established, and the right-hand-side figure indicates that the minimal interagent distance $d_{\min}(t) = \min_{i,j} d_{ij}(t)$ is always greater that $d_1 = 5$, i.e., collision avoidance was guaranteed.

8.6 CONCLUSION

This chapter developed a measure-theoretic approach to obtain flocking protocol of mobile robots by means of control design. Addressing the general class of curved flocking motions, we formulated the notion of state of styled-velocity flocking as the desired state to formulate a control problem. The proposed framework considers the evolutionary equation of one-particle distribution function as an auxiliary model for design instead of an effective model of the original collective system. The desired properties of the original system were then derived using the continuous dependence on initial data of solutions of a measure-valued transition equation. Accordingly, the flocking problem was solved by systematic design without requiring a sufficiently large number of mobile robots. In view of [534], equipping the agents the ability of managing the communication links, practical conditions such as limited-communication can be addressed for applications.

8.7 APPENDIX A

The regularity of $\mathcal{T}_{\mu^\dagger}$ clearly follows from the existence and uniqueness of the solution of the differential (8.8), (8.9) whose expression using the density function $f(t,\cdot)$ is

$$\dot{X}(t) = \operatorname{col}\left(\left(\frac{\zeta(t)}{s\left(q(t), q_{\mu^\dagger(t)}\right)} + \int_{\mathbb{R}^4} f(t,Y) \times \alpha\left(q(t), q^*\right) dY\right) e(q(t), \vartheta(t)),\right.$$
$$\left.\int_{\mathbb{R}^4} f(t,Y) p_\beta(X(t),Y) dY\right) \tag{8.75}$$

Let $X(t;t_0,X_0)$ denote the solution of (8.75) with initial condition $X(t_0) = X_0$. For convenience, we use $X_t = [q_t^T, \eta_t^T]^T = [q_t^T, \vartheta_t, \zeta_t]^T$ to label $X(t;t_0,X_0)$ and $s_t \stackrel{\text{def}}{=} s(q(t), q_{\mu^\dagger}(t))$.

We shall prove the existence and uniqueness by showing that the following function remains finite for $t < \infty$:

$$V(t) = \frac{1}{2}\left\|X(t;t_0,X_0)\right\|^2 = \frac{1}{2}\left\|\left[q_t^T, \eta_t^T\right]^T\right\|^2. \tag{8.76}$$

Again, for measure integrations without specifying the domain of integration, we understand that the domain of integration is the whole space $\mathbb{R}^d$. By Assumption 17, $p_\beta(X,Y)$ has the structure (8.46) with α_ζ given by (8.47). Accordingly, with $\boldsymbol{e}()$ unit vector, $\alpha()$ given in (8.44), and $p_\beta^\circ()$ satisfying (8.55), the time derivative of $V(t)$ along trajectory of (8.75) is

$$\begin{aligned}
\dot{V}(t) =& q_t \cdot \left(\frac{\zeta_t}{s_t} - k_2 \int f(t,Y) e\left(q_t, \vartheta_t\right) \cdot \left(\nabla_{q_{\mu^\dagger(t)}} U_0(\cdot) + \nabla_q U_{\mu^\dagger}^t(\cdot)\right) dY\right) e\left(q_t, \vartheta_t\right) \\
&+ \eta_t \cdot \int f(t,Y) p_\beta^\circ\left(X_t, Y\right) dY \\
&- \frac{\zeta_t}{s_t} \int f(t,Y) e\left(q_t, \vartheta_t\right) \cdot \left(\nabla_{q_{\mu^\dagger(t)}} U_0(\cdot) + \nabla_q U_{\mu^\dagger}^t(\cdot)\right) dY \\
\leq& K_s \left\|X_t\right\|^2 + \frac{k_2^2}{2} \left\|q_t\right\|^2
\end{aligned}$$

$$
\begin{aligned}
&+\frac{1}{2}\left(\int f(t,Y)e\left(q_t,\vartheta_t\right)\cdot\left(\nabla_{q_{\mu^\dagger(t)}}U_0(\cdot)+\nabla_q U^t_{\mu^\dagger}(\cdot)\right)dY\right)^2\\
&+L_p\left\|\eta_t\right\|\int f(t,Y)\left(\|X_t\|+\|Y\|+1\right)dY+\frac{K_s^2}{2}\zeta_t^2\\
&+\frac{1}{2}\left(\int f(t,Y)e\left(q_t,\vartheta_t\right)\cdot\left(\nabla_{q_{\mu^\dagger(t)}}U_0(\cdot)+\nabla_q U^t_{\mu^\dagger}(\cdot)\right)dY\right)^2\\
\le&\left(K_s+2L_p+\frac{k_2^2+K_s^2}{2}\right)\|X_t\|^2+\frac{L_p}{2}+\frac{L_p}{2}\int f(t,Y)\|Y\|^2dY\\
&+W^2_{\mu^\dagger(t)}\left(q_t,\vartheta_t\right)
\end{aligned}
\tag{8.77}
$$

where $K_s=1/c_s$, c_s is the lower bound of $s(\cdot), U_0(\cdot)=U_0(q_t,q_{\mu^\dagger}(t)), U^t_{\mu^\dagger}(\cdot)=U^t_{\mu^\dagger}(q_t,q^*), W_{\mu^\dagger(t)}(q,\vartheta)$ is (8.53) with $\mu=\mu^\dagger(t)$, and we have applied the identity (8.19), the facts that $|\zeta_t|,\|\eta_t\|\le\|X_t\|$, and the Cauchy-Schwarz inequality.

We now show that the last two terms in (8.77) are finite through the finiteness of the functional $\mathcal{V}_{\mu^\dagger}(t)$ given by (8.54). Using the expression (8.38), we have

$$
V_{\mu^\dagger}(t)=\frac{1}{2}\int f(t,X)\left\|q\right\|^2dX+\mathcal{V}_\eta(t,\mu^\dagger)+\mathcal{W}_\eta(t,\mu^\dagger)+\mathcal{V}_q(t,\mu^\dagger). \tag{8.78}
$$

Applying (8.32) for $\psi(q)=\|q\|^2$, we have

$$
\frac{1}{2}\int\frac{\partial f(t,X)}{\partial t}\left\|q\right\|^2dX=\int f(t,X)v(q)e(q,\vartheta)\cdot qdX. \tag{8.79}
$$

Recall that $\boldsymbol{f}(X,Y)=f(t,X)f(t,Y)$ with $f(t,\cdot)$ solution of (8.18) satisfying Assumption 16. Taking the time derivative of $\mathcal{V}_{\mu^\dagger}(t)$ given by (8.78) and applying (8.79), (8.37), and (8.41) with $J_{\mu^\dagger}(t)$ replaced by its expression in (8.30), and then applying the designs (8.28), (8.44), (8.46), and (8.47), we obtain

$$
\begin{aligned}
\dot{\mathcal{V}}_{\mu^\dagger}(t)=&\int f(t,X)v(q)\boldsymbol{e}(q,\vartheta)\cdot qdX+\int f(t,X)f(t,Y)\times\eta\cdot p_\beta(X,Y)dXdY\\
&+\int f(t,X)v(q)\boldsymbol{e}(q,\vartheta)\\
&\times\int f(t,Y)\boldsymbol{e}(q,\vartheta)\cdot\left(\nabla_{q_{\mu^\dagger(t)}}U_0(\cdot)+\nabla_q U^t_{\mu^\dagger}(\cdot)\right)dYdX\\
\le&\int f(t,X)\left\|q\right\|\\
&\cdot\left|\frac{\zeta}{s(\cdot)}-k_2\int f(t,Y)\boldsymbol{e}(q,\vartheta)\cdot\left(\nabla_{q_{\mu^\dagger(t)}}U_0(\cdot)+\nabla_q U^t_{\mu^\dagger}(\cdot)\right)dY\right|dX\\
&+\int\boldsymbol{f}(X,Y)\times\eta\cdot p^\circ_\beta(X,Y)dXdY-k_2\int f(t,X)W^2_{\mu^\dagger(t)}(q,\vartheta)dX
\end{aligned}
\tag{8.80}
$$

where $s(\cdot) = s(q, q_{\mu^\dagger(t)}), U_0(\cdot) = U_0(q, q_{\mu^\dagger(t)}), U^t_{\mu^\dagger}(\cdot) = U^t_{\mu^\dagger}(q, q^*)$. Thus, applying (8.55), the property that $1/|s(\cdot)| \le K_s$, and Cauchy-Schwarz inequality, we further have

$$\begin{aligned}
\dot{V}_{\mu^\dagger}(t) \le & \int f(t,X) \|q\| (k_s |\zeta| + k_2 |W_{\mu^\dagger(t)}(q,\vartheta)|)dX \\
& - k_2 \int f(t,X) W^2_{\mu^\dagger(t)}(q,\vartheta)dX \\
& + \int \boldsymbol{f}(X,Y) \|\eta\| L_p(\|X\| + \|Y\| + 1)dXdY \\
\le & \int f(t,X) \left(K_s \|X\|^2 + \frac{k_2}{2} \|q\|^2\right) dX \\
& - \frac{k_2}{2} \int f(t,X) W^2_{\mu^\dagger(t)}(q,\vartheta)dX \\
& + \int \boldsymbol{f}(X,Y) \left(2L_p \|X\|^2 + \frac{L_p}{2} \|Y\|^2 + \frac{L_p}{2}\right) dXdY \\
\le & K_1 \mathcal{V}_{\mu^\dagger}(t) + K_2 - \frac{k_2}{2} \mathcal{W}_q(t,\mu^\dagger)
\end{aligned} \tag{8.81}$$

where K_1 and K_2 are appropriate constants, and$\mathcal{W}_q(t,\mu^\dagger)$ is given by (8.24).

Dropping the negative term $-k_2\mathcal{W}_q(t,\mu^\dagger)$ in the last inequality of (8.81), and then applying Gronwall's lemma [[14],Lemma 1.1], we obtain

$$\mathcal{V}_{\mu^\dagger}(t) \le \mathcal{V}_{\mu^\dagger}(t_0) e^{K_1(t-t_0)} + K_2 \int_{t_0}^t e^{K_1(t-s)} ds \tag{8.82}$$

Since $\mathcal{V}_{\mu^\dagger}(t_0) < \infty$, (8.82) indicates that $\mathcal{V}_{\mu^\dagger}(t)$ is finite. As shown by (8.54), the integral term in the last inequality of (8.77) is included in $\mathcal{V}_{\mu^\dagger}(t)$, and hence it is finite as well.

To prove the finiteness of the last term of (8.77), let us add $k_2\mathcal{W}_q(s,\mu^\dagger)/2$ to both sides of (8.81), and then take the integrals against t of the resulting inequality to obtain

$$\frac{k_2}{2} \int_{t_0}^t \mathcal{W}_q(s,\mu^\dagger)ds + \mathcal{V}_{\mu^\dagger}(t) - \mathcal{V}_{\mu^\dagger}(t_0) \le \int_{t_0}^t (K_1\mathcal{V}(s) + K_2)ds. \tag{8.83}$$

In addition, $\mathcal{W}_q(s,\mu^\dagger)$ is uniformly continuous by Assumption 17. Thus, (8.83) and (8.82) imply that$\mathcal{W}_q(s,\mu^\dagger) < \infty, \forall s \in [t_0, t]$, and hence, there is $b_{\mathcal{W}}(t)$ such that $\mathcal{W}_q(s,\mu^\dagger) < b_{\mathcal{W}}(t)$.

Consequently, using (8.56), we have

$$|W_{\mu^\dagger(t)}(q_t,\vartheta_t)|^2 \le K\mathcal{W}_q(t,\mu^\dagger) \le Kb_{\mathcal{W}}(t). \tag{8.84}$$

In summary, the last two terms of (8.77) are bounded by a time-varying parameter, namely $b_V(t)$. Thus, noting from (8.76) that $\|X_t\|^2 = 2V(t)$, we conclude from (8.77) that

$$\dot{V}(t) \leq k_V V(t) + b_V(t) \tag{8.85}$$

with k_V an appropriate constant. In view of (8.85), $b_V(t)$ can be assigned to be continuous. Applying Gronwall's lemma [[14], Lemma 1.1] to (8.85), we obtain

$$V(t) \leq V(t_0)\exp(k_V(t-t_0)) + \int_{t_0}^{t} b_V(s)\exp(k_V(t-s))ds \tag{8.86}$$

which shows that $V(t)$ remains finite, and hence $X(t)$ is well-defined for $t < \infty$. The uniqueness then follows from the theory of differential equations [171], and we have the regularity of $\mathcal{T}_{\mu^\dagger}$ accordingly.

8.8 APPENDIX B

Let $\bar{X}(t) = [X_1^T(t), \ldots, X_N^T(t)]^T$ be the trajectory of the collective system $\{\sum(X_i, \mu_\delta^\dagger) : i = 1, \ldots, N\}$ given by (8.57). We shall prove the theorem by showing the existence and uniqueness of $\bar{X}(t)$. Consider the following function:

$$V(\bar{X}) = \frac{1}{2}\frac{1}{N}\sum_{i=1}^{N}\|X_i\|^2 + \frac{1}{N}\sum_{i=1}^{N}U_0(q_i, q_C) + \frac{1}{2}\frac{1}{N}\sum_{i=1}^{N}\frac{1}{N}\sum_{j=1}^{N}U_1(q_i, q_j). \tag{8.87}$$

In the following computation, we denote $s_i = s(q_i, q_C), \boldsymbol{e}_i = \boldsymbol{e}(q_i, \vartheta_i), U_0(\cdot) = U_0(q_i, q_C), U_1(\cdot) = U_1(q_i, q_j)$, and $U'_{ij} = U'(q_i, q_j)$. The time derivative of $V(\bar{X}(t))$ along the trajectories $X_i(t)$ of (8.57) is

$$\begin{aligned}
\dot{V}(\bar{X}(t)) = &\sum_{i=1}^{N}\left(q_i \cdot \left(\frac{\zeta_i}{s_i} - \frac{k_2}{N}\sum_{j=1}^{N}\boldsymbol{e}_i \cdot U'_{ij}\right)\boldsymbol{e}_i + \frac{1}{N}\sum_{j=1}^{N}\eta_i \cdot p_\beta^\circ(X_i, X_j)\right. \\
&\left.-\frac{1}{N}\sum_{j=1}^{N}\frac{\zeta_i}{s_i}\boldsymbol{e}_i \cdot U'_{ij}\right) \\
&+\frac{1}{N}\sum_{i=1}^{N}\nabla_{q_i}U_0(\cdot)\cdot\left(\frac{\zeta_i}{s_i} - \frac{k_2}{N}\sum_{k=1}^{N}\boldsymbol{e}_i \cdot U'_{ik}\right)\boldsymbol{e}_i \\
&+\frac{1}{N}\sum_{i=1}^{N}\frac{1}{N}\sum_{j=1}^{N}\nabla_{q_C}U_0(\cdot)\cdot\left(\frac{\zeta_j}{s_j} - \frac{k_2}{N}\sum_{k=1}^{N}\boldsymbol{e}_j \cdot U'_{jk}\right)\boldsymbol{e}_j \\
&+\frac{1}{2}\frac{1}{N}\sum_{i=1}^{N}\frac{1}{N}\sum_{j=1}^{N}\nabla_{q_i}U_1(\cdot)\cdot\left(\frac{\zeta_i}{s_i} - \frac{k_2}{N}\sum_{k=1}^{N}\boldsymbol{e}_i \cdot U'_{ik}\right)\boldsymbol{e}_i \\
&+\frac{1}{2}\frac{1}{N}\sum_{i=1}^{N}\frac{1}{N}\sum_{j=1}^{N}\nabla_{q_j}U_1(\cdot)\cdot\left(\frac{\zeta_j}{s_j} - \frac{k_2}{N}\sum_{k=1}^{N}\boldsymbol{e}_j \cdot U'_{jk}\right)\boldsymbol{e}_j.
\end{aligned} \tag{8.88}$$

Relabeling the running indices $i \to j$ and $j \to i$, the sum in the fourth line of (8.88) is

$$\frac{1}{N}\sum_{i=1}^{N}\frac{1}{N}\sum_{j=1}^{N}\nabla_{q_C}U_0(q_j,q_C)\cdot\left(\frac{\zeta}{s_i}-\frac{k_2}{N}\sum_{k=1}^{N}\boldsymbol{e}_i\cdot U'_{ik}\right)\boldsymbol{e}_i \tag{8.89}$$

and, as $U_1(q_i,q_j)$ is symmetric so that $\nabla_{qi}U_1(q_i,q_j)=\nabla_{q_i}U_1(q_j,q_i)$, the last sum in (8.88) is equal to the sum in the fifth line of (8.88). Thus, substituting (8.89) into (8.88), we have

$$\begin{aligned}
\dot{V}(\bar{X}(t)) =& \sum_{i=1}^{N}\left(q_i\cdot\left(\frac{\zeta_i}{s_i}-\frac{k_2}{N}\sum_{j=1}^{N}\boldsymbol{e}_i\cdot U'_{ij}\right)\boldsymbol{e}_i\right.\\
&\left.+\frac{1}{N}\sum_{j=1}^{N}\eta_i\cdot p^{\circ}_{\beta}(X_i,X_j)-\frac{1}{N}\sum_{j=1}^{N}\frac{\zeta_i}{s_i}\boldsymbol{e}_i\cdot U'_{ij}\right)\\
&+\frac{1}{N}\sum_{i=1}^{N}\frac{1}{N}\sum_{j=1}^{N}\frac{\zeta_i}{s_i}U'_{ij}\cdot\boldsymbol{e}_i-\frac{1}{N}\sum_{i=1}^{N}\frac{1}{N}\sum_{j=1}^{N}U'_{ij}\cdot\boldsymbol{e}_i\left(\frac{k_2}{N}\sum_{k=1}^{N}\boldsymbol{e}_i\cdot U'_{ik}\right)\\
\leq& \sum_{i=1}^{N}\left(\|q_i\|\left(\left|\frac{\zeta_i}{s_i}\right|+\left|\frac{k_2}{N}\sum_{j=1}^{N}\boldsymbol{e}_i\cdot U'_{ij}\right|\right)\right.\\
&\left.+\frac{1}{N}\sum_{j=1}^{N}\|\eta_i\|\,L_p\left(\|X_i\|+\|X_j\|+1\right)\right)-k_2\frac{1}{N}\sum_{i=1}^{N}\left(\frac{1}{N}\sum_{k=1}^{N}\boldsymbol{e}_i\cdot U'_{ik}\right)^2
\end{aligned} \tag{8.90}$$

where we have canceled the last term in the second line of (8.90) by the sum in the third line, and applied (8.55).

Note that $1/s_i$ is bounded by K_s. From (8.90), by a simple computation with Cauchy-Schwarz inequality, we further have

$$\dot{V}\left(\bar{X}(t)\right)\leq k_V\left(\bar{X}(t)\right)+k_0 \tag{8.91}$$

with k_V and k_0 appropriate constants. Thus, Gronwall's lemma applies, and we have $V(\bar{X}(t))<\infty, \forall t\geq t_0$. This further implies that $X_i(t)=\mathcal{T}_{\mu_\delta^\dagger}(t,t_0)X_i(t_0)$ is well-defined for all $t\geq t_0$, and hence the desired regularity of $T_{\mu_\delta^\dagger}$ follows.

8.9 APPENDIX C

Let $\{\mu_n\}_{n=1}^{\infty}$ be a sequence in $\mathfrak{m}_F^\dagger$, and $\mu_\delta^\dagger\in\mathfrak{M}_\delta^\dagger$ such that, at $t=t_0$, $\{\mu_n^\dagger(t_0)\}_{n=1}^{\infty}$ converges weakly to $\mu_\delta^\dagger(t_0)$. We briefly denote $\mu_n^t=\mu_n^\dagger(t)$ and $\mu_\delta^t=\mu_\delta^\dagger(t)$. Our objective is to show that, for each $t\geq t_0, \mu_n^t\to\mu_\delta^t, n\to\infty$.

We shall use the bounded Lipschitz metric ρ_ℓ of $\mathfrak{M}$ to prove this convergence, i.e., we shall show that $\rho_\ell(\mu_n^t, \mu_\delta^t) \to 0$ as $n \to \infty$.

As $T_{\mu_\delta^\dagger}(t, t_0)X$ may be undefined for arbitrary $X \in \mathbb{R}^4$, our proof invokes the mollified transition mappings $T^\varepsilon_{\mu_\delta^\dagger}(t, t_0)$ specified as follows.

Let $\bar{X}(t) = [X_1^T(t), \ldots, X_N^T(t)]^T$ be the trajectory defining $\mu_\delta^\dagger$. In view of (8.91) and (8.87), $V(\bar{X}(\tau))$ and hence $U_0(q_i(\tau), q_C(\tau))$ and $U_1(q_i(\tau), q_j(\tau)), i \neq j = 1, \ldots, N$ are uniformly bounded with respect to $\tau \in [t_0, t]$. Thus, by Assumptions 14 and 15, there is a number $\varepsilon > 0$ depending on $\mu_\delta^\dagger(\tau), \tau \in [t_0, t]$ such that

$$\begin{aligned} d_0 + \varepsilon \leq \|q_i(\tau) - q_C(\tau)\| \leq \tfrac{1}{\varepsilon} \\ d_1 + \varepsilon \leq \|q_i(\tau) - q_j(\tau)\| \leq \tfrac{1}{\varepsilon} \end{aligned} \tag{8.92}$$

for all $i \neq j = 1, \ldots, N, \tau \in [t_0, t]$.

Then, $\tilde{\mathcal{T}}_{\mu_\delta^\dagger}(t, t_0)$ is defined as the transition mapping of the system

$$\dot{X} = \tilde{\Gamma}\left(X, \mu_\delta^\dagger\right) \tag{8.93}$$

where $\tilde{\Gamma}\left(X, \mu_\delta^\dagger\right)$ is the vector field $\Gamma\left(X, \mu_\delta^\dagger\right)$ given by (8.8), (8.9) with α and p_β respectively replaced by $\tilde{\alpha}$ and $\tilde{p}_\beta$ defined by

$$\begin{aligned} \tilde{\alpha}(q, q^*) &= -k_2 \boldsymbol{e}(q, \vartheta) \cdot \left(\nabla_{q_C(t)} \tilde{U}_0^\varepsilon(q^*, q_C(t)) + \nabla_q \tilde{U}_0^\varepsilon(q, q_C(t)) + \nabla_q \tilde{U}_1^\varepsilon(q, q^*)\right) \\ \tilde{p}_\beta(X, Y) &= p_\beta^\circ(X, Y) + \begin{bmatrix} 0 \\ \frac{\tilde{\alpha}(q, q^*)}{k_2 s(q, q_{\mu^t})} \end{bmatrix} \end{aligned} \tag{8.94}$$

where $\tilde{U}_i^\varepsilon, i = 0, 1$ are *mollifications* of $U_i, i = 0, 1$ satisfying the following conditions:

C1) $\tilde{U}_i^\varepsilon(q_1, q_2)$ and $\nabla_{qj} \tilde{U}_i^\varepsilon(q_1, q_2), j = 1, 2$ are globally bounded; and
C2) $\tilde{U}_i^\varepsilon(q_1, q_2) = U_i(q_1, q_2)$ if either $d_i + \varepsilon \leq \|q_1 - q_2\| \leq d_i + 1/\varepsilon$ or $\|q_1 - q_2\| \leq d_i$;

Clearly, from the above conditions and Assumption 14, there is a constant $\tilde{L}_\varepsilon$ such that $\tilde{\Gamma}(X, \mu_\delta^\dagger)$ is Lipschitz continuous in X with constant $\tilde{L}_\varepsilon$. Thus, $\tilde{\mathcal{T}}_{\mu_\delta^\dagger}(t, t_0)$ is regular. In addition, by C2) and (8.92), $\tilde{\mathcal{T}}_{\mu_\delta^\dagger}(\tau, s)X = \mathcal{T}_{\mu_\delta^\dagger}(\tau, s)X, \forall X \in \{X_i(s) : i = 1, \ldots, N\}, \forall [s, \tau] \subset [t_0, t]$.

Accordingly, as both $\mu_n^\dagger$ and $\mu_\delta^\dagger$ satisfy (8.16), we have

$$\begin{aligned} \rho(\mu_n^t, \mu_\delta^t) =& \rho_\ell\left(\mu_n^{t_0} \circ \mathcal{T}_{\mu_n^\dagger}(t_0, t), \mu_\delta^{t_0} \circ \mathcal{T}_{\mu_n^\dagger}(t_0, t)\right) \\ \leq& \rho_\ell\left(\mu_n^{t_0} \circ \mathcal{T}_{\mu_n^\dagger}(t_0, t), \mu_\delta^{t_0} \circ \tilde{\mathcal{T}}_{\mu_n^\dagger}(t_0, t)\right) \\ &+ \rho_\ell\left(\mu_n^{t_0} \circ \tilde{\mathcal{T}}_{\mu_n^\dagger}(t_0, t), \mu_\delta^{t_0} \circ \tilde{\mathcal{T}}_{\mu_n^\dagger}(t_0, t)\right) \end{aligned} \tag{8.95}$$

Again, in the following, for measure integrations without specifying the domain of integration, we understand that the domain of integration is the whole space $\mathbb{R}^d$. Let us evaluate the right-hand-side of (8.95) component-wise as follows.

Recall that $\nu \circ \mathcal{T}_{\mu^\dagger}(t_0,t)(M) = \nu(\mathcal{T}_{\mu^\dagger}(t_0,t)M)$ for Borel sets $M \in \mathfrak{B}$. As $\mathcal{T}_{\mu_n^\dagger}(t,t_0)$ is regular by Theorem 8.1, and so is $\tilde{\mathcal{T}}_{\mu_\delta^\dagger}(t,t_0)$, the first term in the last inequality of (8.95) can be computed via (8.59) as

$$\begin{aligned}
&\rho_\ell\left(\mu_n^{t_0} \circ \mathcal{T}_{\mu_n^\dagger}(t_0,t), \mu_n^{t_0} \circ \tilde{\mathcal{T}}_{\mu_\delta^\dagger}(t_0,t)\right)\\
&\quad = \sup_{\psi \in \mathcal{D}} \left| \int \psi(X)\left(\mu_n^{t_0} \circ \mathcal{T}_{\mu_n^\dagger}(t_0,t)\right)(dX) - \psi(X)\left(\mu_n^{t_0} \circ \tilde{\mathcal{T}}_{\mu_\delta^\dagger}(t_0,t)\right)(dX)\right|\\
&\quad = \sup_{\psi \in \mathcal{D}} \left| \int \left(\psi \circ \mathcal{T}_{\mu_n^\dagger}(t,t_0)(X) - \psi \circ \tilde{\mathcal{T}}_{\mu_\delta^\dagger}(t,t_0)(X)\right)\mu_n^{t_0}(dX)\right|
\end{aligned} \tag{8.96}$$

As $|\psi(X) - \psi(Y)| \leq \|X - Y\|$ for all $\psi \in \mathcal{D}$, (8.96) yields

$$\begin{aligned}
&\rho_\ell\left(\mu_n^{t_0} \circ \mathcal{T}_{\mu_n^\dagger}(t_0,t), \mu_n^{t_0} \circ \tilde{\mathcal{T}}_{\mu_\delta^\dagger}(t_0,t)\right) \leq \int \left\|\mathcal{T}_{\mu_n^\dagger}(t,t_0)X - \tilde{\mathcal{T}}_{\mu_\delta^\dagger}(t,t_0)X\right\| \mu_n^{t_0}(dX)\\
&= \int \left\| \int_{t_0}^t \left(\Gamma\left(X_n^\tau, \mu_n^\dagger\right) - \tilde{\Gamma}\left(\tilde{X}_\delta^\tau, \mu_\delta^\dagger\right)\right) d\tau \right\| \mu_n^{t_0}(dX)\\
&\leq \int \left\| \int_{t_0}^t \left(\Gamma\left(X_n^\tau, \mu_n^\dagger\right) - \tilde{\Gamma}\left(X_\delta^\tau, \mu_\delta^\dagger\right)\right) d\tau \right\| \mu_n^{t_0}(dX)\\
&\quad + \int \left\| \int_{t_0}^t \left(\tilde{\Gamma}\left(X_n^\tau, \mu_n^\dagger\right) - \tilde{\Gamma}\left(\tilde{X}_\delta^\tau, \mu_\delta^\dagger\right)\right) d\tau \right\| \mu_n^{t_0}(dX)\\
&\stackrel{\text{def}}{=} \mathcal{I}_1 + \mathcal{I}_2
\end{aligned} \tag{8.97}$$

where $X_n^\tau = [q_n^\tau, \vartheta_n^\tau]^T = \mathcal{T}_{\mu_n^\dagger}(\tau,t_0)X$, $\tilde{X}_\delta^\tau = [\tilde{q}_\delta^\tau, \tilde{\vartheta}_\delta^\tau]^T = \tilde{\mathcal{T}}_{\mu_\delta^\dagger}(\tau,t_0)X$, and we have integrated (8.8) with $\mu^\dagger = \mu_n^\dagger$ and (8.93) with $\mu^\dagger = \mu_\delta^\dagger$ to obtain expressions of $\mathcal{T}_{\mu_n^\dagger}(t,t_0)X$ and $\tilde{\mathcal{T}}_{\mu_\delta^\dagger}(t,t_0)X$, respectively.

Using Fubini's theorem and definitions (8.9), (8.93), and (8.94), we have

$$\begin{aligned}
\mathcal{I}_1 &\leq \int_{t_0}^t \left(\int \left\|\Gamma\left(X_n^\tau, \mu_n^\dagger\right) - \tilde{\Gamma}\left(X_n^\tau, \mu_\delta^\dagger\right)\right\| \mu_n^{t_0}(dX)\right) d\tau\\
&\leq \int_{t_0}^t \left(\int \|\zeta_n^\tau\| \left|s^{-1}\left(q_n^\tau, q_{\mu_n^\tau}\right) - s^{-1}\left(q_n^\tau, q_{\mu_\delta^\tau}\right)\right| \mu_n^{t_0}(dX)\right) d\tau\\
&\quad + \int_{t_0}^t \left(\int \left\| \left(\int \alpha\left(q_n^\tau, q^*\right) \mu_n^\tau(dY)\right)\right.\right.
\end{aligned}$$

$$
\begin{aligned}
&- \int \tilde{\alpha}\left(q_n^\tau, q^*\right) \mu_\delta^\tau(dY)\Big) \boldsymbol{e}\left(q_n^\tau, \vartheta_n^\tau\right)\Big\| \mu_n^{t_0}(dX)\Big) d\tau \\
&+ \int_{t_0}^{t} \left(\int \left| \int \frac{\alpha\left(q_n^\tau, q^*\right)}{k_2 s\left(q_n^\tau, q_{\mu_n^\tau}\right)} \mu_n^\tau(dY) - \int \frac{\tilde{\alpha}\left(q_n^\tau, q^*\right)}{k_2 s\left(q_n^\tau, q_{\mu_\delta^\tau}\right)} \mu_\delta^\tau(dY) \right| \mu_n^{t_0}(dX) \right) d\tau \\
&+ \int_{t_0}^{t} \left(\int \left\| \int p_\beta^\circ\left(X_n^\tau, Y\right) \mu_n^\tau(dY) - \int p_\beta^\circ\left(X_n^\tau, Y\right) \mu_\delta^\tau(dY) \right\| \mu_n^{t_0}(dX) \right) d\tau
\end{aligned} \tag{8.98}
$$

As $s_n^\tau \overset{\text{def}}{=} s(q_n^\tau, q_{\mu_n^\tau})$ and $s_\delta^\tau \overset{\text{def}}{=} s(q_n^\tau, q_{\mu_\delta^\tau})$ are independent of the variable of integration q^*, we have

$$
\begin{aligned}
&\left| \int \frac{\alpha\left(q_n^\tau, q^*\right)}{k_2 s\left(q_n^\tau, q_{\mu_n^\tau}\right)} \mu_n^\tau(dY) - \int \frac{\tilde{\alpha}\left(q_n^\tau, q^*\right)}{k_2 s\left(q_n^\tau, q_{\mu_\delta^\tau}\right)} \mu_\delta^\tau(dY) \right| \\
\leq &\frac{1}{k_2 s_n^\tau} \left| \int \alpha\left(q_n^\tau, q^*\right) \mu_n^\tau(dY) - \int \tilde{\alpha}\left(q_n^\tau, q^*\right) \mu_\delta^\tau(dY) \right| \\
&+ \frac{1}{k_2} \left| \frac{1}{s_n^\tau} - \frac{1}{s_\delta^\tau} \right| \left| \int \tilde{\alpha}\left(q_n^\tau, q^*\right) \mu_\delta^\tau(dY) \right|
\end{aligned} \tag{8.99}
$$

Since $\boldsymbol{e}(q_n^\tau, \vartheta_n^\tau)$ is a unit vector and $s(\cdot) \geq 1/c_s$, substituting (8.99) into (8.98), we obtain

$$
\begin{aligned}
\mathcal{I}_1 \leq &\int_{t_0}^{t} \left(\int \tilde{L}_s\left(X_n^\tau\right) \left| \frac{1}{s_n^\tau} - \frac{1}{s_\delta^\tau} \right| \mu_n^{t_0}(dX) \right) d\tau \\
&+ K_\alpha \int_{t_0}^{t} \left(\int \left| \int \alpha\left(q_n^\tau, q^*\right) \mu_n^\tau(dY) - \int \tilde{\alpha}\left(q_n^\tau, q^*\right) \mu_\delta^\tau(dY) \right| \mu_n^{t_0}(dX) \right) d\tau \\
&+ \int_{t_0}^{t} \left(\int \left\| \int p_\beta^\circ\left(X_n^\tau, Y\right) \mu_n^\tau(dY) - \int p_\beta^\circ\left(X_n^\tau, Y\right) \mu_\delta^\tau(dY) \right\| \mu_n^{t_0}(dX) \right) d\tau
\end{aligned} \tag{8.100}
$$

where

$$
\begin{aligned}
\tilde{L}_s\left(X_n^\tau\right) &= \|\zeta_n^\tau\| + \frac{1}{k_2} \left| \int \tilde{\alpha}\left(q_n^\tau, q^*\right) \mu_\delta^\tau(dY) \right| \\
K_\alpha &= 1 + \frac{1}{k_2 c_s}.
\end{aligned} \tag{8.101}
$$

Using (8.100), we shall evaluate $\mathcal{I}_1$ in terms of the distance $\rho_\ell(\mu_n^\tau, \mu_\delta^\tau)$. To this end, we have the following claim.

Claim C.1: For continuous functions $\varphi_n(X, Y)$ and $\varphi_\delta(X, Y)$ satisfying

$$
\left\| \int \varphi_\gamma(X, Y) \mu_\gamma^\tau(dY) \right\| < \infty, \quad \gamma \in \{n, \delta\} \tag{8.102}
$$

there is $b_X \geq 0$such that

$$\left\| \int \varphi_n(X,Y)\mu_n^\tau(dY) - \int \varphi_\delta(X,Y)\mu_\delta^\tau(dY) \right\| \leq b_X \rho_\ell(\mu_n^\tau, \mu_\delta^\tau). \tag{8.103}$$

Proof. Without loss of generality, we assume that φ_n and φ_δ are scalar functions. Due to the discrete nature of $\mu^\tau\delta$ and the finiteness given in (8.102), there is a bounded function $\varphi(X,Y)$ which is Lipschitz continuous with respect to Y such that

$$\begin{aligned} &\int \varphi(X,Y)\mu_n^\tau(dY) = \int \varphi_n(X,Y)\mu_n^\tau(dY) \\ &\varphi(X, X_i(\tau)) = \varphi_\delta(X, X_i(\tau)), \quad \forall i = 1,\ldots,N. \end{aligned} \tag{8.104}$$

A simple selection is $\varphi(X,Y) = \int \varphi_n(X,Y)\mu_n^\tau(dY), \forall Y \notin \{X_i(\tau) : i = 1,\ldots,N\}$
Thus, there is $b_X \geq 0$ such that $b_X^{-1}\varphi(X,Y) \in \mathcal{D}$ with respect to Y. Substituting (8.104) into (8.103), we obtain

$$\begin{aligned} &\left| \int \varphi_n(X,Y)\mu_n^\tau(dY) - \int \varphi_\delta(X,Y)\mu_\delta^\tau(dY) \right| \\ &= b_X \left| \int b_X^{-1}\varphi(X,Y)\mu_n^\tau(dY) - \int b_X^{-1}\varphi(X,Y)\mu_\delta^\tau(dY) \right| \\ &\leq b_X \rho_\ell(\mu_n^\tau, \mu_\delta^\tau) \end{aligned} \tag{8.105}$$

which implies that (8.103) holds true. □

From (8.82), we have

$$\mathcal{V}_{\mu_n^\dagger}(\tau) \leq \mathcal{V}_{\mu_n^\dagger}(t_0)e^{K_1(\tau - t_0)} + K_2 \int_{t_0}^{\tau} e^{K_1(\tau - s)}ds \tag{8.106}$$

for all $\tau \in [t_0, t], \mu_n^\dagger \in \mathfrak{M}_f^\dagger(H)$. Since $\mathcal{V}_{\mu_n^\dagger}(t_0) \leq H, \forall n$ and, in view of (8.81), K_1 and K_2 are independent of n, (8.106) implies that there is a constant $H_t \geq 0$ such that

$$\mathcal{V}_{\mu_n^\dagger}(\tau) \leq H_t, \quad \forall \tau \in [t_0, t], \forall n. \tag{8.107}$$

We continue to evaluate $\mathcal{I}_1$ as follows. First, by specification (8.94), the *mollification* $\tilde{\alpha}(\cdot,\cdot)$ is globally bounded. Thus, in view of (8.101), there is a constant $b_\alpha \geq 0$ such that

$$\tilde{L}_s(X_n^\tau) \leq \|\zeta_n^\tau\| + b_\alpha \leq \|X_n^\tau\| + b_\alpha. \tag{8.108}$$

As $1/s(\cdot,\cdot)$ is Lipschitz continous with constant Ls by Assumption 15, using (8.108), we obtain

$$\tilde{L}_s(X_n^\tau) \left| \frac{1}{s_n^\tau} - \frac{1}{s_\delta^\tau} \right| \leq L_s(\|X_n^\tau\| + b_\alpha) \left\| q_{\mu_n^\tau} - q_{\mu_\delta^\tau} \right\|. \tag{8.109}$$

On the other hand, by (8.107) and definition (8.54), we have

$$\begin{aligned}\left\|q_{\mu_n^\tau}\right\| &= \left\|\int q\mu_n^\tau(dX)\right\| \leq \int \|X\|^2\mu_n^\tau(dX) + \frac{1}{4} \\ &\leq \mathcal{V}_{\mu_n^\dagger}(\tau) + 1 < H_t + 1, \quad \forall\tau \in [t_0, t], \forall n \end{aligned} \tag{8.110}$$

By Claim C.1, there is a constant $b_q \geq 0$ such that

$$L_s \left\|q_{\mu_n^\tau} - q_{\mu_\delta^\tau}\right\| \leq b_q \rho_\ell \left(\mu_n^\tau, \mu_\delta^\tau\right) \tag{8.111}$$

As $\|q_{\mu_n^\tau}\| < H_t + 1$ for all n, b_q can be chosen such that (8.111) holds for all n. Thus, (8.109) and (8.111) yield

$$\int \tilde{L}_s\left(X_n^\tau\right)\left|\frac{1}{s_n^\tau} - \frac{1}{s_\delta^\tau}\right| \mu_n^{t_0}(dX) \leq b_q \rho_\ell\left(\mu_n^\tau, \mu_\delta^\tau\right) \int \left(\|X_n^\tau\| + b_\alpha\right)\mu_n^{t_0}(dX) \tag{8.112}$$

Second, using (8.44), (8.53), and (8.56), we have

$$\left|\int \alpha\left(q_n^\tau, q^*\right)\mu_n^\tau(dY)\right| = \left|W_{\mu_n^\tau}\left(q_n^\tau, \vartheta_n^\tau\right)\right| \leq \sqrt{K\mathcal{W}_q\left(\tau, \mu_n^\dagger\right)}. \tag{8.113}$$

By (8.82), (8.83), (8.107) and the uniform continuity of $\mathcal{W}_q(t, \mu_n^\dagger)$in Assumption 17, there is a $H_\alpha \geq 0$ such that $\mathcal{W}_q(\tau, \mu_n^\dagger) \leq H_\alpha^2/K$. Thus, (8.113) yields

$$\left|\int \alpha\left(q_n^\tau, q^*\right)\mu_n^\tau(dY)\right| \leq H_\alpha, \quad \forall\tau \in [t_0, t], \forall n \tag{8.114}$$

By Claim C.1, there is a $b_\alpha \geq 0$ such that

$$K_\alpha \int \left|\int \alpha\left(q_n^\tau, q^*\right)\mu_n^\tau(dY) - \int \tilde{\alpha}\left(q_n^\tau, q^*\right)\mu_\delta^\tau(dY)\right| \mu_n^{t_0}(dX) \leq b_\alpha \rho_\ell\left(\mu_n^\tau, \mu_\delta^\tau\right). \tag{8.115}$$

Since the boundedness in (8.114) is uniform with respect to n, we can choose b_α such that (8.115) is satisfied for all n.

And third, by condition (8.55) and exploiting (8.110), we have

$$\left\|\int p_\beta^\circ\left(X_n^\tau, Y\right)\mu_n^\tau(dY)\right\| \leq L_p\left\|X_n^\tau\right\| + \frac{L_p}{2}\bar{H}_t + \frac{3L_p}{2}, \quad \forall\tau \in [t_0, t], \forall n. \tag{8.116}$$

As the bound in (8.116) is linear in $\|X_n^\tau\|$, from the proof of Claim C.1, we can choose $b_p = a_p\|X_n^\tau\| + c_p$ such that

$$\begin{aligned}&\int \left\|\int p_\beta^\circ\left(X_n^\tau, Y\right)\mu_n^\tau(dY) - \int p_\beta^\circ\left(X_n^\tau, Y\right)\mu_\delta^\tau(dY)\right\| \mu_n^{t_0}(dX) \\ &\leq \rho_\ell\left(\mu_n^\tau, \mu_\delta^\tau\right)\int \left(a_p\left\|X_n^\tau\right\| + c_p\right)\mu_n^{t_0}(dX)\end{aligned} \tag{8.117}$$

Noting that $X_n^\tau = \mathcal{T}_{\mu_n^\dagger}(\tau, t_0)X$, we have

$$\int \|X_n^\tau\| \, \mu_n^{t_0}(dX) = \int \|X\| \, \mu_n^\tau(dX) \tag{8.118}$$

and hence, from (8.107) and definition (8.54), there is a number $\bar{H}_t \geq 0$ such that

$$\bar{H}_t \geq \max\left\{ \int (a_p \|X_n^\tau\| + c_p)\mu_n^{t_0}(dX), \int (\|X_n^\tau\| + b_\alpha)\mu_n^{t_0}(dX) \right\}, \forall n. \tag{8.119}$$

Substituting (8.112), (8.115), and (8.117) into (8.100), and using (8.119), we arrive at

$$\mathcal{I}_1 \leq C_t \int_{t_0}^{t} \rho_\ell\left(\mu_n^\tau, \mu_\delta^\tau\right) d\tau, \quad \forall n \tag{8.120}$$

where $C_t = b_q \bar{H}_t + b_\alpha + H_t$.

Now, using Fubini's theorem, we evaluate $\mathcal{I}_2$ defined in (8.97)

$$\begin{aligned} \mathcal{I}_2 &= \int \left\| \int_{t_0}^{t} \left(\tilde{\Gamma}\left(X_n^\tau, \mu_\delta^\dagger\right) - \tilde{\Gamma}\left(\tilde{X}_\delta^\tau, \mu_\delta^\dagger\right) \right) d\tau \right\| \mu_n^{t_0}(dX) \\ &\leq \int_{t_0}^{t} \left(\int \left\| \tilde{\Gamma}\left(X_n^\tau, \mu_\delta^\dagger\right) - \tilde{\Gamma}\left(\tilde{X}_\delta^\tau, \mu_\delta^\dagger\right) \right\| \mu_n^{t_0}(dX) \right) d\tau \end{aligned} \tag{8.121}$$

By specification (8.94) and Assumption 14, $\tilde{\Gamma}\left(X, \mu_\delta^\dagger\right)$ is Lipschitz continuous, i.e., there is a constant $L_\delta \geq 0$ such that

$$\left\| \tilde{\Gamma}\left(X, \mu_\delta^\dagger\right) - \tilde{\Gamma}\left(Y, \mu_\delta^\dagger\right) \right\| \leq L_\delta \|X - Y\|, \quad \forall X, Y. \tag{8.122}$$

Substituting (8.122) into (8.121) yields

$$\mathcal{I}_2 \leq \int_{t_0}^{t} \left(\int L_\delta \left\| \mathcal{T}_{\mu_n^\dagger}(\tau, t_0)\, X - \tilde{\mathcal{T}}_{\mu_\delta^\dagger}(\tau, t_0)\, X \right\| \mu_n^{t_0}(dX) \right) d\tau \tag{8.123}$$

Define

$$Q_n(\sigma) = \int \left\| \mathcal{T}_{\mu_n^\dagger}(\sigma, t_0)\, X - \tilde{\mathcal{T}}_{\mu_\delta^\dagger}(\sigma, t_0)\, X \right\| \mu_n^{t_0}(dX). \tag{8.124}$$

Recall that the right hand side of (8.97) is $\mathcal{I}_1 + \mathcal{I}_2$. Substituting (8.120) and (8.123) into (8.97) and using notation (8.124), we obtain

$$\begin{aligned} \rho_\ell\left(\mu_n^{t_0} \circ \mathcal{T}_{\mu_n^\dagger}(t_0, t), \mu_n^{t_0} \circ \tilde{\mathcal{T}}_{\mu_\delta^\dagger}(t_0, t)\right) &\leq Q_n(t) \\ &\leq C_t \int_{t_0}^{t} \rho_\ell\left(\mu_n^\tau, \mu_\delta^\tau\right) d\tau + L_\delta \int_{t_0}^{t} Q_n(\tau) d\tau \end{aligned} \tag{8.125}$$

As $Q_n(t_0) = 0$, appying Gronwall's lemma [[14], Corollary 1.4] for $Q_n(\cdot)$ in (8.125) yields

$$\rho_\ell\left(\mu_n^{t_0} \circ \mathcal{T}_{\mu_n^\dagger}(t_0, t), \mu_n^{t_0} \circ \tilde{\mathcal{T}}_{\mu_\delta^\dagger}(t_0, t)\right) \leq Q_n(t)$$
$$\leq C_t \int_{t_0}^{t} \rho_\ell\left(\mu_n^\tau, \mu_\delta^\tau\right) e^{L_\delta(t-\tau)} d\tau \quad (8.126)$$

By (8.126), we completed our estimate for the first term in the last inequality of (8.95).

As argued above, $\tilde{\Gamma}\left(X, \mu_\delta^\dagger\right)$ is Lipschitz continuous with constant L_δ. Thus, $\tilde{\mathcal{T}}_{\mu_\delta^\dagger}(t_0, t)\, X$ is Lipschitz continuous with constant $L_\delta^t = e^{L_\delta(t-t_0)}$. Hence, for $\psi \in \mathcal{D}$, we have $(L_\delta^t)^{-1}(\psi \circ \tilde{\mathcal{T}}_{\mu_\delta^\dagger}(t_0, t))(X)\psi\mathcal{D}$. Using (8.59), we have

$$\begin{aligned}
&\rho_\ell\left(\mu_n^{t_0} \circ \tilde{\mathcal{T}}_{\mu_\delta^\dagger}(t_0, t), \mu_\delta^{t_0} \circ \tilde{\mathcal{T}}_{\mu_\delta^\dagger}(t_0, t)\right) \\
=&L_\delta^t \sup_{\psi \in \mathcal{D}} \left| \int \left(L_\delta^t\right)^{-1} \left(\psi \circ \tilde{\mathcal{T}}_{\mu_\delta^\dagger}(t_0, t)\right)(X) \mu_n^{t_0}(dX) \right. \\
&\left. - \int \left(L_\delta^t\right)^{-1} \left(\psi \circ \tilde{\mathcal{T}}_{\mu_\delta^\dagger}(t_0, t)\right)(X) \mu_\delta^{t_0}(dX) \right| \\
\leq& L_\delta^t \rho_\ell\left(\mu_n^{t_0}, \mu_\delta^{t_0}\right).
\end{aligned} \quad (8.127)$$

Substituting (8.126) and (8.127) into (8.95), we arrive at

$$\rho_\ell\left(\mu_n^t, \mu_\delta^t\right) \leq L_\delta^t \rho_\ell\left(\mu_n^{t_0}, \mu_\delta^{t_0}\right) + C_t \int_{t_0}^{t} \rho_\ell\left(\mu_n^\tau, \mu_\delta^\tau\right) d\tau. \quad (8.128)$$

By Gronwall's lemma [[14], Corollary 1.4], we obtain

$$\rho_\ell\left(\mu_n^t, \mu_\delta^t\right) \leq L_\delta^t \rho_\ell\left(\mu_n^{t_0}, \mu_\delta^{t_0}\right) e^{C_t(t-t_0)} \quad (8.129)$$

As $\rho_\ell\left(\mu_n^{t_0}, \mu_\delta^{t_0}\right) \to 0, n \to \infty$, taking the limits as $n \to \infty$ of both sides of (8.129), we obtain $\rho_\ell\left(\mu_n^t, \mu_\delta^t\right) \to 0, n \to \infty$. Since the metric ρ_ℓ generates weak convergence, we conclude that $\mu_n^t ß \mu_\delta^t$ as $n \to \infty$.

9 Bioinspired Neurodynamics Based Formation Control for Unmanned Surface Vehicles with Line-of-Sight Range and Angle Constraints

In this chapter, a leader-follower formation control for a group of waterjet unmanned marine surface vehicles (USV) is proposed under the consideration of formation tracking errors constraints. To guarantee line-of-sight (LOS) range and angle tracking errors constraints, a time-varying tan-type barrier Lyapunov function (BLF) is used. In addition, the bioinspired neurodynamics is introduced to solve the traditional differential explosion problem which can not only avoid the differential of the virtual control but can limit the output in a certain range. Further, an observer is involved to overcome the leader's velocity unavailable problem and decrease the communication burden. Simulation results verify that under the proposed method, the LOS range and angle errors can converge into an arbitrary small neighborhood around the origin, while the requirements of the constraints are never violated during the maneuver.

9.1 INTRODUCTION

In recent years, formation control of underactuated marine surface vehicles has attracted more and more attention in the control science and engineering area, where multiple unmanned marine surface vehicles (USV) can complete tasks that a single vehicle can not do even with the advanced equipment, such as minesweeping, transport strategic materials and rescue, since the single vehicle has limited fuel and will increase the time to complete the task. The objectives of formation control are to achieve desired distance and relative angle among USV. By reviewing the available literatures, great progress has been made in the related research [55, 370, 380, 381, 496, 501]. A distributed

DOI: 10.1201/9781003298618-9

formation control was proposed for multiple underwater vehicles in three dimension space in [387]. Several potential functions are first designed in [260] to realize cooperative formation control. In [56], the prescribed performance function was employed to ensure the transient errors not too large for USV. A formation control scheme for the USV based on the minimal learning parameter method was developed, while the disturbance observer was used to deal with environment forces in [318]. An interesting work of multilayer formation control of multi-agent system was first proposed in [256]. In these works, three types of formation structures are commonly used, the leader-follower, the virtual structure and the behavior-based method, respectively. The leader-follower strategy is used for its convenience in this chapter. This technique has also been applied in the formation tracking results [89, 163, 451].

Currently, there are several challenges for the underactuated unmanned marine surface vehicles formation control, three of which will be considered. The first fundamental issue is the constraint problem. For the formation control, except for achieving the desired distance and angle, communication connectivity and collision avoidance should be considered during the whole operation process. This is especially true for underactuated vessels. Collision avoidance algorithms are hard to impose for these vehicles due to the stability problem of zero dynamics of the unactuated degree of freedom [109]. It is worth noticing that though the USV moves in a desired pattern, once the sea states changes suddenly and the disturbances become very large, the stability of the formation structure among USVs may be affected, which may lead to the communication interrupt or collision if the range and angle constraints are not considered. A controller for Euler-Lagrange systems was developed in [57] and a tan-type BLF was used to guarantee the error variables would not exceed the predefined bounds. [90, 176] developed a formation control law for a string of fully actuated USVs under collision and connectivity constraints. A symmetric barrier Lyapunov function was introduced in [177, 555] to prevent states from violating the constraints for a fully actuated marine surface vessels with multiple output constraints. An asymmetric barrier Lyapunov function was employed to deal with the output constraints problem for a marine vessel in [184, 241]. In [65], command filters were used to deal with the actuator constraints and avoid the computations of time derivatives in the backstepping procedure. Although the constraint problems were considered in these works, the systems being considered were all a full actuated system and the method can not be used directly in the underactuated systems. To guarantee the attitude constraints of spacecraft systems, the barrier Lyapunov function technology was used in [113, 261]. [217] addressed the line of sight range and angle constraints used a BLF for autonomous surface vessels formation control. Though the constraint problems has been considered in some literatures, the related researches are still very limited.

The second challenge is that the virtual control law differentiation terms are included in the controller, which lead to a very complex controller. [379]

designed a formation control law incorporating the dynamic surface control technique which simplified the control than the traditional backstepping method. [109] developed a formation control law for underactuated ships with limited communication ranges and guaranteed no collisions between any ships in the group with a novel method. A bounded neural network control law was constructed at the kinetic level with the aid of dynamic surface design method in [302]. The dynamic surface control technique integrated with a novel predictor based iterative neural network was used in [301] for the path following of marine surface vehicles. A path following control incorporated with Nussbaum function was introduced in [349] and a neural shunting model was used to solve the problem of "explosion of complexity". A Nussbaum type function was employed in the dynamic surface control design in [285]. However, these works did not consider the constraints problem in the formation control. In addition, the control schemes designed are complicated. So how to develop an effective control strategy for the underactuated marine surface vehicles to guarantee the constraints requirements and achieve desired formation structures while the control law can be easily used in a practical engineering is still a challenge.

In addition, the information exchange between the vehicles is a challenge problem. Most of current research assume that the full states of the leader can be accessed by the follower. Actually, due to the economic reason or efficiency, only the position and heading can be obtained by the follower, the velocity information is unavailable. This implies that a practical control method demanding less for communications should be developed. A fuzzy state observer was introduced to estimate the unmeasurable state variables in [267]. This chapter will use a high-gain observer to estimate the leader velocity.

Motivated by the above research, in this chapter, a formation control law for USV based on bioinspired neurodynamics method is developed. The main contributions of this chapter are summarized as follows: First, inspired by [445], three bioinspired neurodynamic shunting models derived from plasma membrane are introduced to avoid the complex differential calculations and simplify the control law, while the output of the bioinsipred neurodynamics model can be limited in a certain range. Second, with the communication connectivity and collision avoidance under the consideration, a tan-type BLF is introduced to guarantee the distance among USVs which is neither overstep the maximum communication distance nor less than the minimum collision distance. Third, to alleviate the computational burden in the control design process, the command filter is incorporated with the controller and an auxiliary system is introduced to eliminate the filter error. A high-gain observer is proposed to overcome the leader's velocity unavailable problem.

9.2 PRELIMINARIES AND PROBLEM FORMULATION

9.2.1 PRELIMINARIES

The bioinspired neurodynamics model was first put forward by Grossberg [160]. It can describe the online adaptive behavior of individuals. It was originally derived based on the membrane model proposed by Hodgkin and Huxley [188] for a patch of membrane using electrical elements. The dynamics of voltage across the membrane can be described in the membrane model using state equation technique as:

$$C_m \frac{dV_m}{dt} = -\left(E_p + V_m\right) g_p + \left(E_{Na} - V_m\right) g_{Na} - \left(E_k + V_m\right) g_k, \tag{9.1}$$

where V_m represents the membrane voltage, C_m is the membrane capacitance, and E_k, E_{Na} and E_p are the potassium potential of the membrane, the sodium ion and the passive leakage current of the Nernst potential (saturation potential). g_K, g_{Na} and g_P represent the potassium ion, sodium ion and passive channel conductance, respectively. The bioinspired neurodynamics model can be defined as following form:

$$\dot{V}_i = -AV_i + \left(B - V_i\right) f\left(\alpha_i\right) - \left(D + V_i\right) g\left(\alpha_i\right), \tag{9.2}$$

where

$$f\left(\alpha_i\right) = \begin{cases} \alpha_i, \alpha_i \geq 0, \\ 0, \alpha_i < 0, \end{cases} \quad g\left(\alpha_i\right) = \begin{cases} -\alpha_i, \alpha_i \leq 0, \\ 0, \alpha_i > 0. \end{cases} \tag{9.3}$$

The virtual control $\alpha_u, \alpha_r, \alpha_\psi$ are chosen as the input of the neural dynamic model, and the outputs V_i will substitute α_u, α_r and α_ψ.

Lemma 9.1

[184] Suppose the output $y(t)$ of a system and its first n derivatives are bounded such that $|y^{(k)}| < Y_K$ with the positive constants Y_K, we can consider the following linear system:

$$\begin{aligned} \varepsilon\dot{\pi}_i &= \pi_{i+1}, i = 1, \cdots, n-1, \\ \varepsilon\dot{\pi}_n &= -\bar{\lambda}_1\pi_n - \bar{\lambda}_2\pi_{n-1} - \cdots - \bar{\lambda}_{n-1}\pi_2 - \pi_1 + x_1(t), \end{aligned} \tag{9.4}$$

where ε is an arbitrary small positive constant and the parameters $\bar{\lambda}_1$ to $\bar{\lambda}_{n-1}$ are chosen so that the polynomial $s^n + \bar{\lambda}_1 s^{n-1} + ... + \bar{\lambda}_{n-1}s + 1$ is Hurwitz. Then, there exists positive constants $\Gamma_i, i = 1, ...n$ and t^*, for all $t \geqslant t^*$, we have

$$\mid \frac{1}{\varepsilon}\pi_{i+1} - y^{(k)} \mid \leq \Gamma_{i+1}, i = 0, \cdots, n-1, \tag{9.5}$$

where Γ_i are the decreasing function of the Y_K. ∎

Remark 47. *The high gain observer can guarantee that* $\frac{1}{\varepsilon^i}\pi_{i+1}$ *converges to* $y^{(k)}$ *with the bounded error* $\varepsilon\Gamma_{i+1}$ *if the function* $y(t)$ *and its ith derivative is bounded.*

9.2.2 USV KINEMATICS AND DYNAMICS

Consider a multi-USV system, neglecting the heave, roll and pitch motion. The kinematics model can be described as:

$$\begin{cases} \dot{x}_i = u_i \cos(\psi_i) - v_i \sin(\psi_i), \\ \dot{y}_i = u_i \sin(\psi_i) + v_i \cos(\psi_i), \\ \dot{\psi}_i - r_i, \end{cases} \tag{9.6}$$

the dynamics model subjected to external disturbances is presented as:

$$\begin{cases} \dot{u}_i = \frac{m_{22i}}{m_{11i}} v_i r_i - \frac{d_{11i}}{m_{11i}} u_i + \frac{1}{m_{11i}} \tau_{ui} + \frac{1}{m_{11i}} \tau_{wui}, \\ \dot{v}_i = \frac{m_{11i}}{m_{22i}} u_i r_i - \frac{d_{22i}}{m_{22i}} \nu_i + \frac{1}{m_{22i}} \tau_{wvi}, \\ \dot{r}_i = \frac{m_{11i} - m_{22i}}{m_{33i}} u_i v_i - \frac{d_{33i}}{m_{33i}} r_i + \frac{1}{m_{33i}} \tau_{ri} + \frac{1}{m_{33i}} \tau_{wri}, \end{cases} \tag{9.7}$$

where $\eta_i = [x_i, y_i, \psi_i]^T$ represents the position and yaw angle in the earth-fixed frame, $\nu_i = [u_i, v_i, r_i]^T$ denotes the surge, sway and yaw velocity in the body-fixed frame. m_{11i}, m_{22i} and m_{33i} are the combined inertia and added mass of a USV. Parameters d_{11i}, d_{22i} and d_{33i} are hydrodynamic damping coefficients. The control inputs of the ith USV are τ_{ui} and τ_{ri}. τ_{wui}, τ_{wvi} and τ_{wri} represent the external disturbances including wind, wave and ocean currents. We assume the disturbances is zero. The transformation matrix $J(\eta_i)$ is defined as

$$J(\eta_i) = \begin{bmatrix} \cos\psi_i & -\sin\psi_i & 0 \\ \sin\psi_i & \cos\psi_i & 0 \\ 0 & 0 & 1 \end{bmatrix}. \tag{9.8}$$

Remark 48. *For the waterjet USV, the sway direction has no direct input, so the control design method for the full-actuated system can not be directly applied to the underactuated system.*

9.2.3 CONTROL OBJECTIVES

In this chapter, we consider a control strategy to achieve the satisfactory desired distance and angle between USVs with the intervehicle communications. Under the designed control for each vehicle of a string of n USVs, the distance and angle errors can be constrained in a certain range. The line-of-sight (LOS) strategy is used to define the distance and angle between two USVs. The leader is defined as 0th USV, the ith follower USV is denoted by i.

Consider the direct network topology in this chapter, the whole network consisting of one leader and 4 followers. We assume that the position information of the leader is available to all followers. Communication graph among USV is shown in Fig. 9.1.

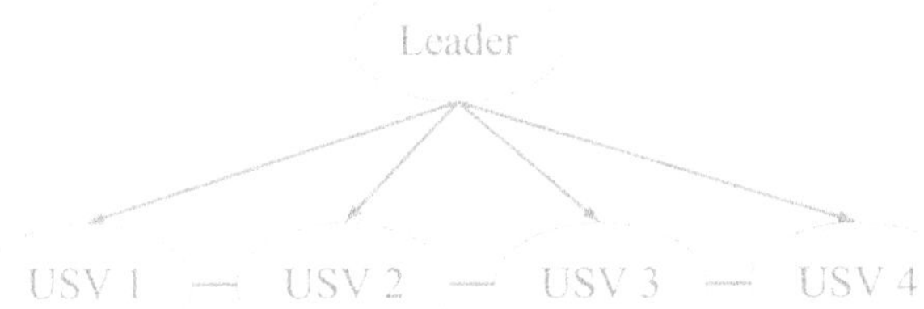

Figure 9.1 Communication graph among five USVs.

The formation tracking errors of LOS range and angle are defined as follows:

$$d_i(t) = \sqrt{(x_0 - x_i)^2 + (y_0 - y_i)^2}, \tag{9.9}$$

$$\varphi_i = \arctan(\frac{y_0 - y_i}{x_0 - x_i}). \tag{9.10}$$

The formation tracking errors are

$$\begin{cases} e_{di} = d_i - d_{i,des}, \\ e_{\varphi i} = \varphi_i - \varphi_{i,des}. \end{cases} \tag{9.11}$$

The control objectives are that the LOS range and angle should be satisfied with the following constraints:

$$|e_{di}| < k_{di}, \, |e_{\varphi i}| < k_{\varphi i}. \tag{9.12}$$

9.3 TAN-TYPE BARRIER LYAPUNOV FUNCTION

9.3.1 HIGH-GAIN OBSERVER

For convenience, the equation (9.6) can be rewritten as

$$\dot{\eta} = J(\eta)\nu \tag{9.13}$$

Note that the velocity of the leader may be not always accessible in practice, a high gain observer is used to estimate the unavailable states. According to the Lemma 9.1, a high gain observer is constructed as follows:

$$\begin{aligned} \varepsilon\dot{\pi}_1 &= \pi_2, \\ \varepsilon\dot{\pi}_2 &= -\bar{\lambda}_1\pi_2 - \pi_1 + \eta(t), \end{aligned} \tag{9.14}$$

where π_1 and π_2 denote the states of the high gain observer and ε can be any small positive constant.

Next, the estimation dynamic $\dot{\hat{\eta}}$ of the η can be written as

$$\begin{aligned} \hat{\eta} &= \pi_1, \\ \varepsilon\dot{\hat{\eta}} &= \pi_2. \end{aligned} \tag{9.15}$$

According to (9.6) and (9.15), we can obtain the estimate value $\hat{\nu}$ of ν

$$\varepsilon\hat{\nu} = J^T \pi_2. \tag{9.16}$$

Theorem 9.1

As the position and attitude information can be measured in the practice, by choosing a small parameter ε and a positive constant $\bar{\lambda}_1$ so that the polynomial $s^2 + \bar{\lambda}_1 s$ is Hurwitz, the velocity estimation error of the observer constructed by (9.14), (9.15) and (9.16) for the USV is bounded.

Proof. The estimation errors can be denoted by

$$\begin{aligned} \tilde{\eta} &= \hat{\eta} - \eta, \\ \tilde{\nu} &= \hat{\nu} - \nu. \end{aligned} \tag{9.17}$$

Then, it follows the Lemma 9.1 and equation (9.15), we have

$$\begin{aligned} &\|\tilde{\eta}\| = \|\hat{\eta} - \eta\| \leqslant \varepsilon\Gamma_1, \\ &\|\tilde{\nu}\| = \|\hat{\nu} - \nu\| = \|J^T(\frac{\pi_2}{\varepsilon} - \dot{\eta})\| \leqslant \varepsilon\|[\lambda_1\ \lambda_2\ \lambda_3]^T\| := \varepsilon\Gamma_2, \end{aligned} \tag{9.18}$$

where λ_1, λ_2 and λ_3 are some constants and Γ_1 and Γ_2 are positive constants. Thus, the estimate errors are bounded. □

■

9.3.2 KINEMATICS CONTROL

To guarantee the formation errors constraint on e_{di} and $e_{\varphi i}$, we choose a tan-type barrier Lyapunov function as follows:

$$V_{di} = \frac{k_{di}^2}{\pi}\tan(\frac{\pi e_{di}^2}{2k_{di}^2}), \tag{9.19}$$

$$V_{\varphi i} = \frac{k_{\varphi i}^2}{\pi}\tan(\frac{\pi e_{\varphi i}^2}{2k_{\varphi i}^2}). \tag{9.20}$$

Remark 49. *By the L'Hospital's rule, we can obtain that if $k_{di} \to \infty$, the Lyapunov function (9.19) will become $V_{di} = \frac{1}{2}e_{di}^2$. It means that if there is no constraint, the Lyapunov function can be simplified to the normal quadratic form. So our tan-type Lyapunov function can be used in the system without constraint requirements. It's different from $V = \frac{1}{2}\log\frac{k_b^2}{k_b^2 - e^2}$ and $V = \frac{k_b}{\pi}\tan^2(\frac{\pi e}{2k_b})$, which were used in many works.*

At this step, we aim to make d_i and φ_i approach to the desired value of $d_{i,des}$ and $\varphi_{i,des}$ so that the desired formation pattern can be achieved. Note that the LOS range and angle has the following dynamics:

$$\begin{aligned}\dot{d}_i =& u_0 \cos(\psi_0 - \varphi_i) - v_0 \sin(\psi_0 - \varphi_i)\\ &- u_i \cos(\psi_i - \varphi_i) + v_i \sin(\psi_i - \varphi_i),\\ \dot{\varphi}_i =& \frac{1}{d_i}[u_0 \sin(\psi_0 - \varphi_i) + v_{i-1} \cos(\psi_0 - \varphi_i)\\ &- u_i \sin(\psi_i - \varphi_i) - v_i \cos(\psi_i - \varphi_i)].\end{aligned} \tag{9.21}$$

The differential of LOS range error can be written as follows:

$$\begin{aligned}\dot{e}_{di} =& u_0 \cos(\psi_0 - \varphi_i) - v_0 \sin(\psi_0 - \varphi_i)\\ &- u_i \cos(\psi_i - \varphi_i) + v_i \sin(\psi_i - \varphi_i) - \dot{d}_{i,des}.\end{aligned} \tag{9.22}$$

We choose a virtual control signal α_{ui} and $\alpha_{\psi i}$. To avoid differential exploration and simplify control design, the bioinspired neurodynamics method is introduced. Let the virtual control go through a bioinspired neurodynamics model, respectively. We have

$$\begin{aligned}\dot{u}_{fi} &= -A_{ui} u_{fi} + (B_{ui} - u_{fi}) f(\alpha_{ui}) - (D_{ui} + u_{fi}) g(\alpha_{ui}),\\ \dot{\psi}_{fi} &= -A_{\psi i} \psi_{fi} + (B_{\psi i} - \psi_{fi}) f(\alpha_{\psi i}) - (D_{\psi i} + \psi_{fi}) g(\alpha_{\psi i}),\end{aligned} \tag{9.23}$$

where u_{fi} and ψ_{fi} are the output of the bioinspired neurodynamics, respectively.

The velocity tracking errors are defined as:

$$\begin{aligned} z_{ui} &= u_{fi} - \alpha_{ui},\\ e_{ui} &= u_i - u_{fi},\end{aligned} \tag{9.24}$$

and yaw angle errors as:

$$\begin{aligned} z_{\psi i} &= \psi_{fi} - \alpha_{\psi i},\\ e_{\psi i} &= \psi_i - \psi_{fi}.\end{aligned} \tag{9.25}$$

Substituting (9.24) and (9.25) into (9.22) yields

$$\begin{aligned}\dot{e}_{di} =& u_0 \cos(\psi_0 - \varphi_i) - v_0 \sin(\psi_0 - \varphi_i)\\ &- (e_{ui} + z_{ui} + \alpha_{ui}) \cos(e_{\psi i} + z_{\psi i} + \alpha_{\psi i}\\ &- \varphi_i) + v_i \sin(\psi_i - \varphi_i) - \dot{d}_{i,des}\\ =& - \alpha_{ui} \cos(\alpha_{\psi i} - \varphi_i) + \alpha_{ui} \sin(e_{\psi i}\\ &+ z_{\psi i}) \sin(\alpha_{\psi i} - \varphi_i) - \alpha_{ui} \cos(\alpha_{\psi i} - \varphi_i)(\cos(e_{\psi i}\\ &+ z_{\psi i}) - 1) - (e_{ui} + z_{ui}) \cos(\psi_i - \varphi_i)\\ &+ u_0 \cos(\psi_0 - \varphi_i) - v_0 \sin(\psi_0 - \varphi_i)\\ &+ v_i \sin(\psi_i - \varphi_i) - \dot{d}_{i,des}.\end{aligned} \tag{9.26}$$

For simplicity, we define

$$\begin{aligned} w_1 =& -\alpha_{ui}\cos(\alpha_{\psi i}-\varphi_i),\\ \Delta_{11} =&\alpha_{ui}\sin(e_{\psi i}+z_{\psi i})\sin(\alpha_{\psi i}-\varphi_i)\\ &-\alpha_{ui}\cos(\alpha_{\psi i}-\varphi_i)(\cos(e_{\psi i}+z_{\psi i})-1),\\ \Delta_{21} =&(e_{ui}+z_{ui})\cos(\psi_i-\varphi_i),\\ \Delta_{31} =&u_0\cos(\psi_0-\varphi_i)-v_0\sin(\psi_0-\varphi_i),\\ \Delta_{41} =&v_i\sin(\psi_i-\varphi_i). \end{aligned}$$

We can rewrite (9.26) as

$$\dot{e}_{di} = w_1+\Delta_{11}+\Delta_{21}+\Delta_{31}+\Delta_{41}-\dot{d}_{i,des}. \tag{9.27}$$

By differential (9.19), we can obtain

$$\begin{aligned} \dot{V}_{di} =&\frac{2k_{di}\dot{k}_{di}}{\pi}\tan\frac{\pi e_{di}^2}{2k_{di}^2}+\frac{e_{di}\dot{e}_{di}}{\cos^2(\frac{\pi e_{di}^2}{2k_{di}^2})}-\frac{e_{di}^2\dot{k}_{di}}{\cos^2(\frac{\pi e_{di}^2}{2k_{di}^2})k_{di}}\\ =&\frac{e_{di}}{\cos^2(\frac{\pi e_{di}^2}{2k_{di}^2})}(w_1+\Delta_{11}+\Delta_{21}+\Delta_{31}+\Delta_{41}-\dot{d}_{i,des})\\ &+\frac{2k_{di}\dot{k}_{di}}{\pi}\tan\frac{\pi e_{di}^2}{2k_{di}^2}-\frac{e_{di}^2\dot{k}_{di}}{\cos^2(\frac{\pi e_{di}^2}{2k_{di}^2})k_{di}}. \end{aligned} \tag{9.28}$$

The nominal stabilizing function w_{10} is chosen as:

$$\begin{aligned} w_{10} =&\dot{d}_{i,des}-k_d\frac{1}{e_{di}}\frac{k_{di}^2}{\pi}\sin(\frac{\pi e_{di}^2}{2k_{di}^2})\cos(\frac{\pi e_{di}^2}{2k_{di}^2})-\Delta_{31}\\ &-\Delta_{41}-k_{d1}e_{di}+k_\rho e_1, \end{aligned} \tag{9.29}$$

where e_1 is the state auxiliary system will be designed in the next step. Let w_{10} go through a command filter, and the filterting error is defined as: $\Delta w_1 = w_1 - w_{10}$. The auxiliary system e_1 is designed:

$$\dot{e}_1 = -k_1e_1+\Delta w_1. \tag{9.30}$$

In order to analysis the stability of the LOS range, a Lyapunov candidate function is designed as:

$$V_1 = V_{di}+\frac{1}{2}e_1^2. \tag{9.31}$$

The derivation of the tan-type BLF with respect to time leads to

$$\begin{aligned} \dot{V}_1 \le&\frac{e_{di}(\Delta_{11}+\Delta_{21}+k_\rho e_1)}{\cos^2(\frac{\pi e_{di}^2}{2k_{di}^2})}-(k_d-2k_{d1})\frac{k_{di}^2}{\pi}\tan(\frac{\pi e_{di}^2}{2k_{di}^2})\\ &-(k_1-\frac{1}{2})e_1^2+\frac{1}{2}\Delta w_1^2. \end{aligned} \tag{9.32}$$

The derivative of the LOS angle yields

$$\begin{aligned}\dot{e}_{\varphi i} =& \frac{1}{d_i}[-\alpha_{ui}\sin(\alpha_{\psi i}-\varphi_i) - \alpha_{ui}\cos(\alpha_{\psi i}-\varphi_i)\sin(e_{\psi i} \\ &+ z_{\psi i}) - \alpha_{ui}\sin(\alpha_{\psi i}-\varphi_i)(\cos(e_{\psi i}+z_{\psi i})-1) \\ &- (e_{ui}+z_{ui})\sin(\psi_i-\varphi_i) + u_0\sin(\psi_0-\varphi_i) \\ &+ v_0\cos(\psi_0-\varphi_i) - v_i\cos(\psi_i-\varphi_i)] - \dot{\varphi}_{i,des}.\end{aligned} \tag{9.33}$$

Similar to (19), let

$$\begin{aligned}w_2 =& -\alpha_{ui}\sin(\alpha_{\psi i}-\varphi_i), \\ \Delta_{12} =& -\alpha_{ui}\cos(\alpha_{\psi i}-\varphi_i)\sin(e_{\psi i}+z_{\psi i}) \\ &- \alpha_{ui}\sin(\alpha_{\psi i}-\varphi_i)(\cos(e_{\psi i}+z_{\psi i})-1), \\ \Delta_{22} =& -(e_{ui}+z_{ui})\sin(\psi_i-\varphi_i), \\ \Delta_{32} =& u_0\sin(\psi_0-\varphi_i) + v_0\cos(\psi_0-\varphi_i), \\ \Delta_{42} =& -v_i\cos(\psi_i-\varphi_i).\end{aligned}$$

The time derivative of $e_{\varphi i}$ can be rewritten as

$$\dot{e}_{\varphi i} = \frac{1}{d_i}(w_2+\Delta_{12}+\Delta_{22}+\Delta_{32}+\Delta_{42}) - \dot{\varphi}_{i,des}. \tag{9.34}$$

Derived from (9.20) yields

$$\begin{aligned}\dot{V}_{\varphi i} =& \frac{e_{\varphi i}}{\cos^2(\frac{\pi e_{\varphi i}^2}{2k_{\varphi i}^2})}(\frac{1}{d_i}(w_2+\Delta_{12}+\Delta_{22}+\Delta_{32}+\Delta_{42}) \\ &- \dot{\varphi}_{i,des}) + \frac{2k_{\varphi i}\dot{k}_{\varphi i}}{\pi}\tan(\frac{\pi e_{\varphi i}^2}{2k_{\varphi i}^2}) - \frac{\dot{k}_{\varphi i}}{k_{\varphi i}}\frac{e_{\varphi i}^2}{\cos^2(\frac{\pi e_{\varphi i}^2}{2k_{\varphi i}^2})},\end{aligned} \tag{9.35}$$

The nominal stabilizing function w_{20} is defined as:

$$\begin{aligned}w_{20} =& d_i(\dot{\varphi}_{i,des} - k_\varphi\frac{k_{\varphi i}^2}{\pi e_{\varphi i}}\sin(\frac{\pi e_{\varphi i}^2}{2k_{\varphi i}^2})\cos(\frac{\pi e_{\varphi i}^2}{2k_{\varphi i}^2}) - \Delta_{32} \\ &- \Delta_{42}) + k_{\varphi 1}e_{\varphi i} + k_\lambda e_2,\end{aligned} \tag{9.36}$$

where $k_{\varphi 1} = \sqrt{\left(\frac{\dot{k}_{\varphi i}}{k_{\varphi i}}\right)^2 + \varepsilon}$, $\epsilon > 0$ is a small constant, e_2 is the state auxiliary system similar to e_1 which will be designed in the next step.

The filtering error $\Delta w_2 = w_2 - w_{20}$, and the auxiliary system e_2 is designed as the following:

$$\dot{e}_2 = -k_2 e_2 + \Delta w_2. \tag{9.37}$$

The Lyapunov candidate function for LOS angle analysis is defined as follows:

$$V_2 = V_{\varphi i} + \frac{1}{2}e_2^2. \tag{9.38}$$

The derivation of V_2 is

$$\begin{aligned}\dot{V}_2 \leq & \frac{e_{\varphi i}(\Delta_{12}+\Delta_{22}+k_\lambda e_2)}{d_i \cos^2(\frac{\pi e_{\varphi i}^2}{2k_{\varphi i}^2})} - (k_\varphi - 2k_{\varphi 1})\frac{k_{\psi i}^2}{\pi}\tan(\frac{\pi e_{\varphi i}^2}{2k_{\varphi i}^2}) \\ & - (k_2 - \frac{1}{2})e_2^2 + \frac{1}{2}\Delta w_2^2.\end{aligned} \tag{9.39}$$

Virtual control can be designed as follows:

$$\begin{aligned}\alpha_{ui} &= -w_1 \cos(\alpha_{\psi i} - \varphi_i) - w_2 \sin(\alpha_{\psi i} - \varphi_i), \\ \alpha_{\psi i} &= \arctan \tfrac{w_2}{w_1} + \varphi_i.\end{aligned} \tag{9.40}$$

Next, the stabilizing function α_{ri} will be designed to stabilize $e_{\psi i}$, where $e_{\psi i} = \psi_i - \psi_{fi}$. The derivation of $e_{\psi i}$ can be calculated as

$$\dot{e}_{\psi i} = r_i - \dot{\psi}_{fi}. \tag{9.41}$$

The virtual yaw velocity is designed as:

$$\alpha_{ri} = -k_\psi e_{\psi i}^2 + \dot{\psi}_{fi}. \tag{9.42}$$

Let α_{ri} go through a bioinspired neurodynamics model, we can obtain

$$\dot{r}_{fi} = -A_{ri} r_{fi} + (B_{ri} - r_{fi}) f(\alpha_{ri}) - (D_{ri} + r_{fi}) g(\alpha_{ri}). \tag{9.43}$$

Define the yaw velocity error as:

$$\begin{aligned}z_{ri} &= r_{fi} - \alpha_{ri}, \\ e_{ri} &= r_i - r_{fi}.\end{aligned} \tag{9.44}$$

A Lyapunov function to analyze the yaw angle error is defined as:

$$V_\psi = \frac{1}{2} e_{\psi i}^2. \tag{9.45}$$

Derivation of (9.45) then gives

$$\begin{aligned}\dot{V}_\psi &= e_{\psi i}(r_i - \dot{\psi}_{fi}) \\ &= e_{\psi i}(e_{ri} + z_{ri} + \alpha_{ri} - \dot{\psi}_{fi}) \\ &= -k_\psi e_{\psi i}^2 + e_{\psi i}(e_{ri} + z_{ri}).\end{aligned} \tag{9.46}$$

9.3.3 DYNAMICS CONTROL

Define the velocity errors as follows:

$$\begin{aligned}\dot{e}_{ui} &= \dot{u}_i - \dot{u}_{fi}, \\ \dot{e}_{ri} &= \dot{r}_i - \dot{r}_{fi}.\end{aligned} \tag{9.47}$$

So the control input is designed as follow:

$$\begin{aligned}\tau_{ui} &= m_{11i}(-k_u e_{ui} + \dot{u}_{fi} - \tfrac{m_{22i}}{m_{11i}} v_i r_i), \\ \tau_{ri} &= m_{33i}(-k_r e_{ri} + \dot{r}_{fi} - \tfrac{m_{11i} - m_{22i}}{m_{33i}} u_i v_i).\end{aligned} \tag{9.48}$$

Remark 50. *Compared with the traditional method, such as [163], the control law is as follows:*

$$\begin{aligned}\tau_u =& m_1(-k_4 u_e + \dot{\alpha}_u - f_u(v) - \tau_{wu}),\\ \tau_r =& m_3(-k_5 r_e + \dot{\alpha}_r - f_r(v) - \tau_{wr}),\end{aligned} \tag{9.49}$$

and [217]:

$$\begin{aligned}\tau_u =& \frac{1}{b_{u,\min}}(-K_u u_e - K_{u0} sig^{\frac{1}{2}}(u_e) - \hat{\theta}_u^T F_u \tanh(\frac{\hat{\theta}_u^T F_u u_e}{\delta})\\ &- \hat{M}_u \tanh(\frac{\hat{M}_u u_e}{\delta}) - \frac{\rho_{Le,\cos}\Delta_{21}}{u_e}\tanh(\frac{\rho_{Le,\cos}\Delta_{21}}{u_e})\\ &- \dot{\sigma}_u \tanh(\frac{\dot{\sigma}_u u_e}{\delta}) - \frac{\rho_{Le,\cos}\Delta_{22}}{\rho_L u_e}\tanh(\frac{\rho_{Le,\cos}\Delta_{22}}{\rho_L \delta})),\\ \tau_r =& \frac{1}{b_{r,\min}}(-K_r u_r - \hat{\theta}_r^T F_r \tanh(\frac{\hat{\theta}_r^T F_r r_e}{\delta})\\ &- \hat{M}_r \tanh(\frac{\hat{M}_r r_e}{\delta}) - \dot{\sigma}_r \tanh(\frac{\dot{\sigma}_r r_e}{\delta})\\ &- \psi_e \tanh(\frac{\psi_e r_e}{\delta}) - K_{r0} sig^{\frac{1}{2}}(r_e)).\end{aligned} \tag{9.50}$$

As can be seen from (9.49) and (9.50), the derivation of virtual control α_u and α_r are needed in the control law, while we can know from (9.41) and (9.43), this kind of derivation is very complicated and it is difficult to give a specific explicit expression. The control law (9.48) proposed can avoid this complicated derivative operation.

9.3.4 STABILITY ANALYSIS

A complete Lyapunov function is constructed as

$$V = V_1 + V_2 + V_{\psi i} + \frac{1}{2}e_{ui}^2 + \frac{1}{2}e_{ri}^2 + \frac{1}{2}z_{ui}^2 + \frac{1}{2}z_{ri}^2 + \frac{1}{2}z_{\psi i}^2. \tag{9.51}$$

According to the above analysis, we have

$$\begin{aligned}V =& \frac{k_{di}^2}{\pi}\tan(\frac{\pi e_{di}^2}{2k_{di}^2}) + \frac{1}{2}e_1^2 + \frac{k_{\varphi i}^2}{\pi}\tan(\frac{\pi e_{\varphi i}^2}{2k_{\varphi i}^2}) + \frac{1}{2}e_2^2 + \frac{1}{2}e_{\psi i}^2\\ &+ \frac{1}{2}z_{\psi i}^2 + \frac{1}{2}(e_{ui}^2 + e_{ri}^2 + z_{ui}^2 + z_{ri}^2).\end{aligned} \tag{9.52}$$

Derivative of V we can obtain that

$$\begin{aligned}\dot{V} \leq& \frac{e_{di}(\Delta_{11} + \Delta_{21} + k_\rho e_1)}{\cos^2(\frac{\pi e_{di}^2}{2k_{di}^2})} - (k_d - 2k_{d1})\frac{k_{di}^2}{\pi}\tan(\frac{\pi e_{di}^2}{2k_{di}^2})\\ &- (k_1 - \frac{1}{2})e_1^2 + \frac{1}{2}\Delta w_1^2 + \frac{e_{\varphi i}(\Delta_{12} + \Delta_{22} + k_\lambda e_2)}{d_i \cos^2(\frac{\pi e_{\varphi i}^2}{2k_{\varphi i}^2})}\end{aligned}$$

$$
\begin{aligned}
&-(k_\varphi-2k_{\varphi 1})\frac{k_{\varphi i}^2}{\pi}\tan(\frac{\pi e_{\varphi i}^2}{2k_{\varphi i}^2})-(k_2-\frac{1}{2})e_2^2+\frac{1}{2}\Delta w_2^2 \\
&-k_\psi e_{\psi i}^2+e_{\psi i}(e_{ri}+z_{ri})-k_u e_{ui}^2-k_r e_{ri}^2+z_{ui}\dot{z}_{ui}+z_{ri}\dot{z}_{ri}+z_{\psi i}\dot{z}_{\psi i}.
\end{aligned} \tag{9.53}
$$

From the perspective of practical application, the control input and the velocities u_i, v_i and r_i are all bounded, respectively. $\dot{\alpha}_{ui}$, $\dot{\alpha}_{vi}$ and $\dot{\alpha}_{\psi i}$ are continuous and bounded functions, the maximum values are assumed to be Ω_1, Ω_2 and Ω_3, respectively. The bioinspired neurodynamics model can be written in another form as

$$
\dot{u}_{fi}=-(A_{ui}+f(\alpha_{ui})+g(\alpha_{ui}))u_{fi}+B_{ui}f(\alpha_{ui})-D_{ui}g(\alpha_{ui}). \tag{9.54}
$$

Let $A_{ui}+f(\alpha_{ui})+g(\alpha_{ui})=P_{ui}$, $D_{ui}=B_{ui}$, as $f(x)\geq 0$ and $g(x)\geq 0$. Then,

$$
\dot{z}_{ui}=\dot{u}_{fi}-\dot{\alpha}_{ui}=-P_{ui}u_{fi}+B_{ui}\alpha_{ui}-\dot{\alpha}_{ui}. \tag{9.55}
$$

Similar to the above, let $D_{ri}=B_{ri}$, $A_{ri}+f(\alpha_{ri})+g(\alpha_{ri})=P_{ri}$, $D_{\psi i}=B_{\psi i}$ and $A_{\psi i}+f(\alpha_{\psi i})+g(\alpha_{\psi i})=P_{\psi i}$. Then,

$$
\begin{aligned}
\dot{z}_{ri}&=\dot{r}_{fi}-\dot{\alpha}_{ri}=-P_{ri}r_{fi}+B_{ri}\alpha_{ri}-\dot{\alpha}_{ri}, \\
\dot{z}_{\psi i}&=\dot{r}_{\psi i}-\dot{\alpha}_{\psi i}=-P_{\psi i}\psi_{fi}+B_{\psi i}\alpha_{\psi i}-\dot{\alpha}_{\psi i}.
\end{aligned} \tag{9.56}
$$

So we can obtain that

$$
\begin{aligned}
\dot{V}\leq&\frac{e_{di}(\Delta_{11}+\Delta_{21}+k_\rho e_1)}{\cos^2(\frac{\pi e_{di}^2}{2k_{di}^2})}-(k_d-2k_{d1})\frac{k_{di}^2}{\pi}\tan(\frac{\pi e_{di}^2}{2k_{di}^2}) \\
&-(k_1-\frac{1}{2})e_1^2+\frac{1}{2}\Delta w_1^2+\frac{e_{\varphi i}(\Delta_{12}+\Delta_{22}+k_\lambda e_2)}{d_i\cos^2(\frac{\pi e_{\varphi i}^2}{2k_{\varphi i}^2})} \\
&-(k_\varphi-2k_{\varphi 1})\frac{k_{\varphi i}^2}{\pi}\tan(\frac{\pi e_{\varphi i}^2}{2k_{\varphi i}^2})-(k_2-\frac{1}{2})e_2^2+\frac{1}{2}\Delta w_2^2 \\
&-k_\psi e_{\psi i}^2+e_{\psi i}(e_{ri}+z_{ri})-k_u e_{ui}^2-k_r e_{ri}^2+z_{ui}(-P_{ui}u_{fi} \\
&+B_{ui}\alpha_{ui}-\Omega_1)+z_r(-P_{ri}r_{fi}+B_{ri}\alpha_{ri}-\Omega_2) \\
&+z_{\psi i}(-P_{\psi i}\psi_{fi}+B_{\psi i}\alpha_{\psi i}-\Omega_3).
\end{aligned} \tag{9.57}
$$

Let $B_{ui}=P_{ui}, B_{ri}=P_{ri}, B_{\psi i}=P_{\psi i}$. Then,

$$
\begin{aligned}
\dot{V}\leq&-(k_d-2k_{d1})\frac{k_{di}^2}{\pi}\tan(\frac{\pi e_{di}^2}{2k_{di}^2})-(k_1-\frac{1}{2})e_1^2 \\
&-(k_\varphi-2k_{\varphi 1})\frac{k_{\varphi i}^2}{\pi}\tan(\frac{\pi e_{\varphi i}^2}{2k_{\varphi i}^2})-(k_2-\frac{1}{2})e_2^2 \\
&-k_\psi e_{\psi i}^2-k_u e_{ui}^2-k_r e_{ri}^2-P_{ui}z_{ui}^2-P_{ri}z_{ri}^2-P_{\psi i}z_{\psi i}^2 \\
&+|z_{ui}|\,|\Omega_1|+|z_{ri}|\,|\Omega_2|+|z_{\psi i}|\,|\Omega_3| \\
&+\bar{\Delta}_1+\frac{1}{2}\Delta w_1^2+\bar{\Delta}_2+\frac{1}{2}\Delta w_2^2+|e_{\psi i}(e_{ri}+z_{ri})|.
\end{aligned} \tag{9.58}
$$

According to Young's inequality,

$$
\begin{aligned}
|z_{ui}|\,|\Omega_1| &\leq \frac{\varpi_1}{2} z_{ui}^2 + \frac{1}{2\eta_1}\Omega_1^2, \\
|z_{ri}|\,|\Omega_2| &\leq \frac{\varpi_2}{2} z_{ri}^2 + \frac{1}{2\eta_2}\Omega_2^2, \\
|z_{\psi i}|\,|\Omega_3| &\leq \frac{\varpi_3}{2} z_{\psi i}^2 + \frac{1}{2\eta_3}\Omega_3^2,
\end{aligned} \tag{9.59}
$$

where ϖ_1, ϖ_2 and ϖ_3 are positive constants. Similarly,

$$
|e_{\psi i}(e_{ri} + z_{ri})| \leq \frac{1}{2\xi_1} e_{\psi i}^2 + \frac{\xi_1}{2} e_{ri}^2 + \frac{1}{2\xi_2} e_{\psi i}^2 + \frac{\xi_2}{2} z_{ri}^2, \tag{9.60}
$$

where ξ_1 and ξ_2 are positive constants. Substituting (9.59) and (9.60) into (9.58), we can obtain

$$
\begin{aligned}
\dot{V} \leq & - l_1 \frac{k_{di}^2}{\pi} \tan(\frac{\pi e_{di}^2}{2k_{di}^2}) - l_2 e_1^2 - l_3 \frac{k_{\varphi i}^2}{\pi} \tan(\frac{\pi e_{\varphi i}^2}{2k_{\varphi i}^2}) \\
& - l_4 e_2^2 - m_1 e_{\psi i}^2 - k_u e_{ui}^2 - m_2 e_{ri}^2 - n_1 z_{ui}^2 \\
& - n_2 z_{ri}^2 - n_3 z_{\psi i}^2 + C,
\end{aligned} \tag{9.61}
$$

where

$$
\begin{aligned}
l_1 =& k_d - 2k_{d1}, l_2 = k_1 - \frac{1}{2}, l_3 = k_\varphi - 2k_{\varphi 1}, l_4 = k_2 - \frac{1}{2}, \\
m_1 =& k_\psi - \frac{1}{2\xi_1} - \frac{1}{2\xi_2}, m_2 = k_r - \frac{\xi_1}{2}, \\
n_1 =& P_{ui} - \frac{\varpi_1}{2}, n_2 = P_{ri} - \frac{\varpi_2}{2} - \frac{\xi_2}{2}, n_3 = P_{\psi i} - \frac{\varpi_3}{2}, \\
C =& \frac{1}{2\eta_1}\Omega_1^2 + \frac{1}{2\eta_2}\Omega_2^2 + \frac{1}{2\eta_3}\Omega_3^2 \\
& + \bar{\Delta}_1 + \bar{\Delta}_2 + \frac{1}{2}\Delta w_1^2 + \frac{1}{2}\Delta w_2^2.
\end{aligned}
$$

To guarantee the system stability, let

$$
\begin{aligned}
& k_d > 2k_{d1}, k_1 > \frac{1}{2}, k_2 > \frac{1}{2}, k_\varphi > 2k_{\varphi 1}, k_\psi > \frac{1}{2\xi_1} + \frac{1}{2\xi_2}, \\
& k_u > 0, k_r > \frac{\xi_1}{2}, P_{ui} > \frac{\varpi_1}{2}, P_{ri} > \frac{\varpi_2}{2} + \frac{\xi_2}{2} \text{ and } P_{\psi i} > \frac{\varpi_3}{2}.
\end{aligned}
$$

We define

$$
l = \min\{l_1, l_2, l_3, l_4, m_1, m_2, n_1, n_2, n_3, k_u\},
$$

we can rewrite (9.61) as

$$
\dot{V} \leq -lV + C. \tag{9.62}
$$

Hence, $0 \leq V(t) \leq (V(0) - C/l)e^{-lt} + C/l$.

Table 9.1
Initial states and control parameters

Parameter	Value	Parameter	Value
$\eta_0(0)$	$[40;0;0]$	$\upsilon_0(0)$	$[0.4;0;0]$
$\eta_1(0)$	$[0;30;0]$	$\upsilon_1(0)$	$[0.4;0;0]$
$\eta_2(0)$	$[0;-30;0]$	$\upsilon_2(0)$	$[0.4;0;0]$
$\eta_3(0)$	$[0;30;0]$	$\upsilon_3(0)$	$[0.4;0;0]$
$\eta_4(0)$	$[0;-30;0]$	$\upsilon_4(0)$	$[0.4;0;0]$
$d_{i,des1}$	30	$d_{i,des3}$	30
$\varphi_{i,des1}$	$\pi/6$	$\varphi_{i,des3}$	$\pi/6$
$d_{i,des2}$	30	$d_{i,des4}$	30
$\varphi_{i,des2}$	$-\pi/6$	$\varphi_{i,des4}$	$-\pi/6$
k_{d1}	30	k_{d2}	30
$k_{\varphi 1}$	$\pi/20$	$k_{\varphi 2}$	$\pi/20$
k_u	1	k_r	1
A_{ui}	8	A_{ri}	2
$A_{\psi i}$	6		

9.4 SIMULATION

In this section, a string of 5 USVs are used for the numerical simulations to demonstrate the effectiveness of the proposed method. The ship model parameters used to simulate are Cybership-II, a 1:70 scale supply vessel replica built in a marine control laboratory in the Norwegian University of Science and Technology [461]. The leader USV is denoted as 0 and the follower USVs are defined as follower 1–4. For comparison with quadratic Lyapunov function (QLF), the follower 1 and 2 are controlled by the BLF, and the follower 3 and 4 are controlled by the QLF. The initial states and formation requirements are listed in Table 9.1, and the control parameters are given. The simulation results are shown in Figs. 9.2–9.3. It can be seen that the formation tracking errors of the line of sight range and the angle for follower 1 and 2 converge to a neighborhood around zero while the errors are not violating the prescribed constraints. We can see from Figs. 9.2(a) and 9.3(a) that though the range errors of follower 3 and follower 4 converge to a neighborhood of 0, the converging speed of the LOS range error is lower than the follower 1 and follower 2, which are controlled by the BLF. Besides, the Figs. 9.2(b) and 9.3(b) show that the angle tracking errors of the follower 3 and follower 4 violated the constraint boundary while follower 1 and follower 2 achieved the satisfying performance.

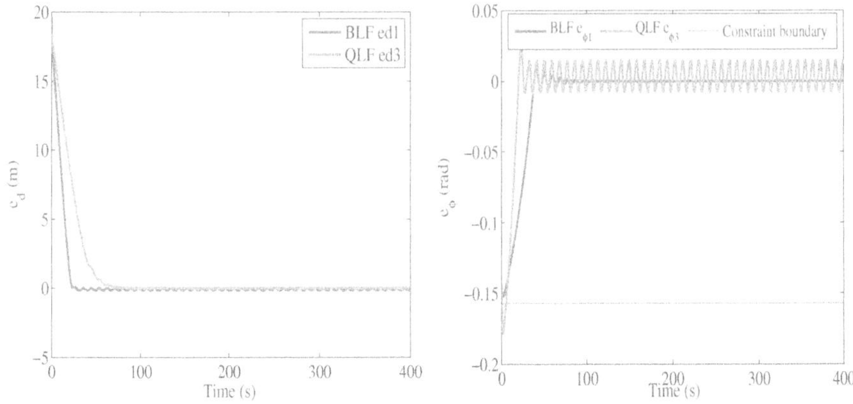

(a) LOS range tracking error of the follower 1 and the follower 3 (b) LOS angle tracking error of the follower 1 and the follower 3

Figure 9.2 LOS range and angle tracking errors of the follower 1 and the follower 3.

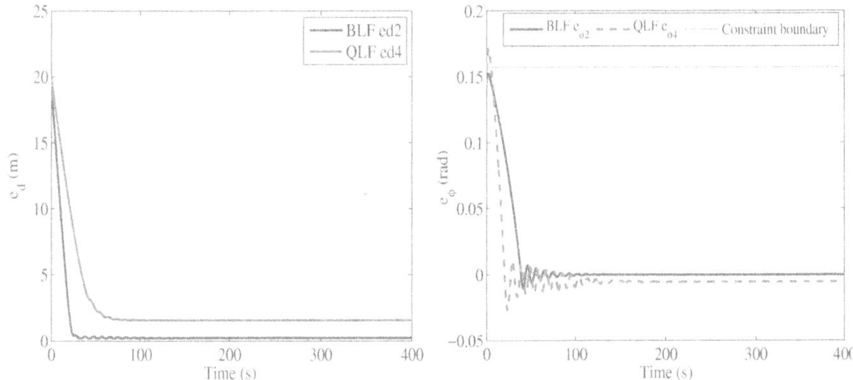

(a) LOS range tracking error of the follower 2 and the follower 4 (b) LOS angle tracking error of the follower 2 and the follower 4

Figure 9.3 LOS range and angle tracking error of the follower 2 and the follower 4.

9.5 CONCLUSIONS

In this chapter, a novel formation control strategy for unmanned marine surface vehicles with the output constraints is presented. We have proven that under the proposed control, the line-of-sight range and angle errors constraints are never violated. In addition, the process of the control design is simplified by means of a bioinspired neurodynamics method and desired formation pattern achieved without violation of the constraints including collision avoidance and connectivity distance. Simulation results confirm the performance of the proposed methods.

10 Velocity Free Platoon Formation Control for Unmanned Surface Vehicles with Output Constraints and Model Uncertainties

This chapter studies the velocity free platoon formation control for unmanned surface vehicles (USV) with the model uncertainties and output constraints. First, a reconstruction module is designed to estimate the velocity of the leader, which will accomplish in finite time and reduce the communication burden. Along with this, the model-based control combined with the symmetric barrier Lyapunov functions (BLF) method is designed to guarantee the output constraints. Then, the model uncertainties of USV are approximated by the neural networks (NN) and NN BLF control is developed. To achieve the desired formation pattern, the constraints including collision avoidance and communication distance are under consideration. Finally, we proved that our system is semiglobally uniformly ultimately bounded (SGUUB) and verified the effectiveness of this approach by simulations.

10.1 INTRODUCTION

In recent years, unmanned surface vehicles (USV) formation control has attracted large amounts of attention in control science and engineering area for different purposes such as mine clearance, patrol, investigation, transportation of strategic materials, coordinated operations, target rounding and marine science applications [129, 309, 352, 483]. The objectives of formation control are to achieve the desired distance and orientation among multiple agents. Meanwhile, to avoid collision and keep effective communication during the formation control, the relative distance among USV should be guaranteed to be within a reasonable interval. Therefore, it is necessary to deal with model uncertainties and constraints to design the control system. A motivating example is controlling a group of moving cars in the form of closely spaced vehicular platoons while ensuring safety [191]. Vehicular platoons could deal

DOI: 10.1201/9781003298618-10

well with traffic flow capacity and reduce the consume of the fuel in the automated highway systems. Therefore, many researchers extend the platoon formation control problem to the robot or autonomous vehicles [90,191,310].

Great progress has been made in related research [177,264,318,327,565]. A control scheme based on backstepping method technique and Lyapunov direct method for formation control of ships used leader-follower strategy was proposed in [104]. Based on [104], a control strategy based on the bioinspired method for the underactuated surface vehicles solved the problem of computation exploration and realized the constraints on virtual velocity was developed in [139]. However, the ship models they used were all simplified and the model uncertainties were not considered. The unmodeled dynamics and parameter uncertainties induced by hydrodynamic damping terms may lead to problems that make it difficult or impossible to build an accurate ship model and this affect the transient and steady-state performance which leads to collisions and gives rise to invalid connectivity in formation control. In [379], neural network (NN)-based dynamic surface control (DSC) technique was employed to solve uncertain local dynamics of leader-follower formation control. The adaptive NN were used in [261] to deal with model uncertainties. The extended dissipativity criterion for discrete-time NN with time-varying delay was established in [132]. A NN formation control for multiple USV cooperated with adaptive filtering methods were employed to extract the low-frequency content of the model uncertainties and ocean disturbances in [382]. The input constraint, model uncertainties, and external disturbances were considered for spacecraft formation control in [559]. Though the uncertain parameters and unmodeled dynamics or input constraints have been considered in above research, they have not taken the output constraints into consideration. The output and error constraints of the formation are very critical factors affecting formation stability, for the reason that if the distance between the two vehicles is too large that exceed the effective range of communication in the process of formation pattern establishing, the connectivity between each vehicle will invalid which will affect the formation pattern. On the other hand, if the distance between each other is too short, it will cause intervehicle collisions and damage the system. Therefore, it is necessary to consider the output constraint of formation control. To deal with the problem of constraint, some methods have also been proposed to achieve the desired performance [90,304,462,538]. For example, the exponentially decaying functions of time and formation error transformation were used in [90]. An output feedback control was proposed in [485] for a certain class of systems that guaranteed the convergence of generic bounded solutions to the desired set-point. Under these methods, the formation errors are always within the predefined regions. To deal with the issue of connectivity preserving and collision avoidance problems between each other, a linear transformed error surface was introduced in [371]. To prevent output constraints violation, a symmetric barrier Lyapunov function was employed in [555] and realized trajectory tracking with output constraints. A two layered optimal approach toward cooperative motion planning of USV in a constrained maritime environment was proposed in [29]. [433] integrated the modules of efficient optimal path planning, robust path following guidance and cooperative swarm aggregation approach toward development of a new hybrid framework for cooperative navigation of swarm of USV to enable optimal and autonomous operation in a maritime environment.

Also, the information exchange between vehicles is a challenging problem. Most of the current research assumes that the full states of the leader can be accessed by the follower. The adaptive finite-time decentralized control problem was addressed in [108] with the time-varying output-constraint. Under the consideration of input quantization and dynamic uncertainties, a decentralized adaptive NN tracking control was developed for the interconnected nonlinear system in [447]. A new estimate-based dynamic event-triggered output feedback control for networked control systems subjected to nonlinear uncertainties was proposed in [448]. The BLF was used to solve the problem of state constraints for USV and the input saturation was considered in [389]. All the references listed need the full-state information of the leader or the desired trajectory. Actually, due to the reason of economy or efficiency for the formation control problem, only the position and heading can be obtained by the follower and the velocity information is unavailable. This implies that a practical control method which is less demanding for communications should be developed. A fuzzy state observer was introduced to estimate the unmeasurable state variables in [267]. An event-triggered $H\infty$ tracking control was first established for nonlinear NCSs in [20], which leads to a considerable reduction of sending data from the controller to the remote operated vehicle (ROV). Unlike the existing studies in the previous literature, where the full states need to be known, this chapter address the problem of the unavailability of the leader's velocity to the follower, which is more practical in modern applications.

The value of BLF will increase to infinity when its arguments close to some specified values. So, keeping the BLF bounded could ensure that the constraints are never violated in the closed-loop system. Motivated by the above research and the property of BLF, a model-based formation control under the consideration of collision avoidance and effective communication distance constraints is proposed for USV different from [90, 371]. In this chapter, the errors combined with a BLF are directly used to design the formation control instead of complex error conversion through the guaranteed performance function, which reduces the amount of calculation and simplifies the design of the control.

Compared to existing work in this field, the main contributions of this chapter are threefold.

A BLF is used to tackle the problem of multiple USV constraints ensuring that the satisfactory formation performance with collision avoidance distance and connectivity constraints are never violated during the whole operation.

An adaptive NN control combined with a symmetric BLF is developed for autonomous USV to deal with the unmodeled dynamics and parameter uncertainties caused by hydrodynamic damping terms.

Unlike the existing studies in the previous literature, where the full states need to be known, this chapter address the problem of the unavailability of the leader's velocity to the follower using a reconstruction module, which is more practical in modern applications. One of the contributions thus lies in that this method can accurately reconstruct the target velocity in finite time.

10.2 PRELIMINARIES AND PROBLEM DESCRIPTION

10.2.1 PRELIMINARIES

Lemma 10.1

[555] For any constant $x \in \mathbb{R}^n$, if $|x| < k$, k is a constant, then the following inequality holds

$$\ln \frac{k^2}{k^2 - x^2} \leq \frac{x^2}{k^2 - x^2}. \tag{10.1}$$

■

Lemma 10.2

[57, 390] For an unknown continuous nonlinear function $f(x) : \mathbb{R}^m \to \mathbb{R}$, the radial basis function NN can be used to approximate it over a compact set $\Omega \subseteq \mathbb{R}^m$ as follows:

$$f(x) = W^{*T} H(x) + \varepsilon(x),$$

where $x \in \mathbb{R}^n$ is the input vector, $H(x) = [h_1(x), ...h_m(x)]^T$ is the radial basis function vector and $h_j = \exp(-\frac{\|\mathbf{x}-c_j\|^2}{2b_j^2})$. $c = [c_{ij}]_{n\times m}$ is the center and b_j is the width of the neural cell of the hidden layer. $\varepsilon(x)$ denotes the approximate error satisfying $\varepsilon(x) \leq \bar{\varepsilon}, \bar{\varepsilon}$ is an arbitrary small positive constant. The structure of the NN is presented in Fig. 10.1. RBF optimal weight vector is $W^* = [w_1^*, \cdots, w_m^*]^T$. The weight vector of W^* is calculated by

$$W^* = \arg \min_{\hat{W}} \{\sup_{x\in\Omega} |f(x) - \hat{W}^T H(x)|\},$$

where $\hat{W}$ is the estimate value of W^*, which is produced by an adaptive update law. ■

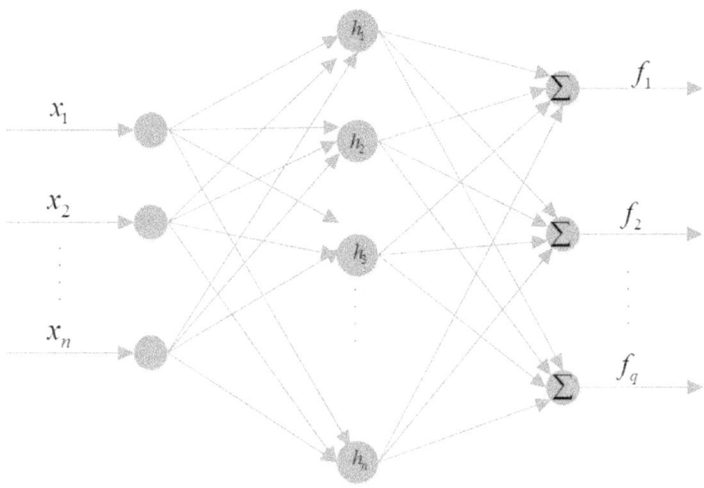

Figure 10.1 The structure of NN.

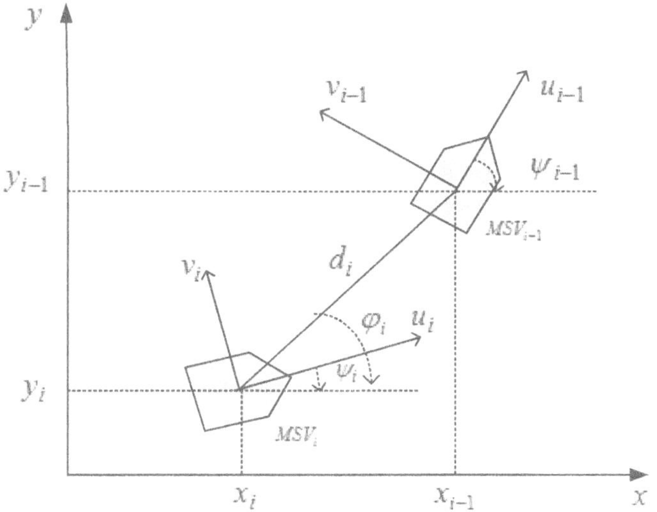

Figure 10.2 USV formation configuration.

10.2.2 USV MODEL

Consider a class of networked multiagent system consisting of $n + 1$ USV (labeled as 0 to n) moving in a special formation pattern as shown in Fig. 10.2. The motion of the $i - th$ USV can be described with kinematics

$$\dot{\eta}_i = J_i(\eta_i)\nu_i, \tag{10.2}$$

and dynamics

$$M_i\dot{\nu}_i = -C_i(\nu_i)\nu_i - D_i(\nu_i)\nu_i + w_i + \tau_i + \Delta_i, \tag{10.3}$$

where $\eta_i = [x_i, y_i, \psi_i]^T$ represents the position and yaw angle of the USV in the earth-fixed frame. $J_i(\eta_i)$ is a nonsingular transformation matrix defined as

$$J_i(\eta_i) = \begin{bmatrix} \cos(\psi_i) & -\sin(\psi_i) & 0 \\ \sin(\psi_i) & \cos(\psi_i) & 0 \\ 0 & 0 & 1 \end{bmatrix}. \tag{10.4}$$

$\nu_i = [u_i, v_i, r_i]^T$ denotes the surge, sway and yaw velocities with respect to body fixed frame. $M_i \in \mathbb{R}^{3\times 3}$ is a symmetric positive definite inertia matrix, $C_i(\nu_i)$ is the matrix of Coriolis and centripetal terms, $D_i(\nu_i)$ is the damping matrix, w_i is the external disturbances induced by wind, wave, and ocean currents, etc and Δ_i represents the unmodeled dynamics. The control inputs of the i-th USV is denoted by τ_i.

The matrices M_i, $C_i(\nu_i)$ and $D_i(\nu_i)$ are given below:

$$M_i = \begin{bmatrix} m_{11i} & 0 & 0 \\ 0 & m_{22i} & m_{23i} \\ 0 & m_{32i} & m_{33i} \end{bmatrix}, \tag{10.5}$$

$$D_i(\nu_i) = \begin{bmatrix} d_{11i} & 0 & 0 \\ 0 & d_{22i} & d_{23i} \\ 0 & d_{32i} & d_{33i} \end{bmatrix}, \tag{10.6}$$

$$C_i(\nu_i) = \begin{bmatrix} 0 & 0 & -m_{22i}v_i - m_{23i}r_i \\ 0 & 0 & m_{11i}u_i \\ m_{22i}v_i + m_{23i}r_i & -m_{11i}u_i & 0 \end{bmatrix}, \tag{10.7}$$

where

$$m_{11i} = m_i - X_{\dot{u}i}, m_{22i} = m_i - Y_{\dot{v}i}, \tag{10.8}$$
$$m_{23i} = m_{32i} = m_i x_{gi} - Y_{\dot{r}i}, m_{33i} = I_{zi} - N_{\dot{r}i}, \tag{10.9}$$
$$d_{11i}(u_i) = -(X_{ui} + X_{|ui|u_i}|u_i| + X_{uiuiui}u_i^2), \tag{10.10}$$
$$d_{22i}(v_i, r_i) = -(Y_{vi} + Y_{|vi|v_i}|v_i| + Y_{|ri|v_i}|r_i|), \tag{10.11}$$
$$d_{23i}(v_i, r_i) = -(Y_{ri} + Y_{|vi|r_i}|v_i| + Y_{|ri|r_i}|r_i|), \tag{10.12}$$
$$d_{32i}(v_i, r_i) = -(N_{vi} + N_{|vi|v_i}|v_i| + N_{|ri|v_i}|r_i|), \tag{10.13}$$
$$d_{33i}(v_i, r_i) = -(N_{ri} + N_{|vi|r_i}|v_i| + N_{|ri|r_i}|r_i|), \tag{10.14}$$

the mass of the ship i is denoted by m_i and the i-th ship's inertia matrix in the body-fixed frame is denoted by I_{zi}.

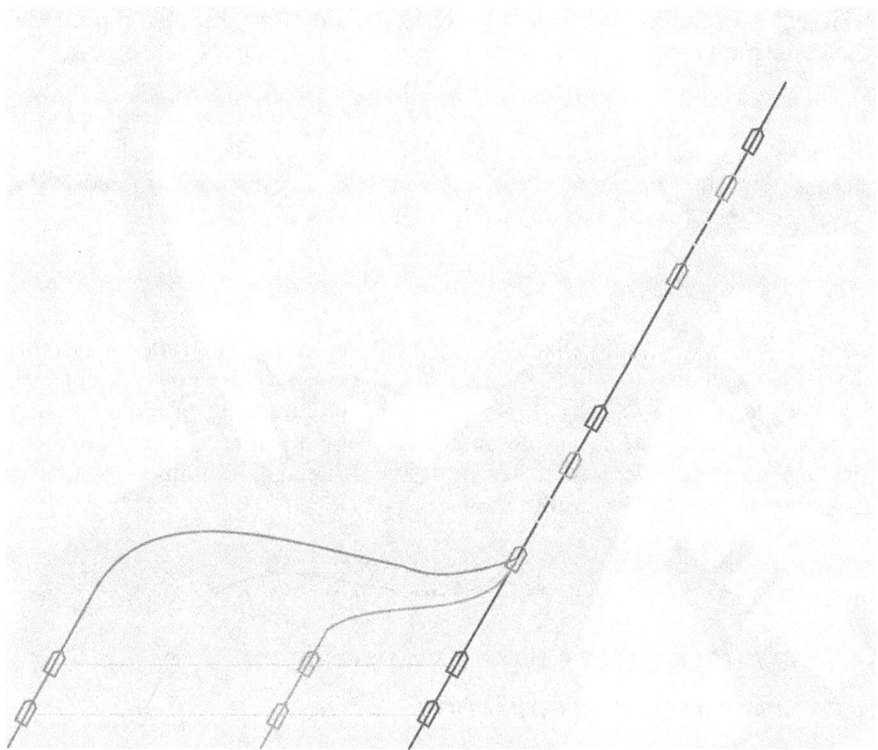

Figure 10.3 Platoon formation control graphic description in a real-time strait.

10.2.3 PLATOON FORMATION CONTROL WITH OUTPUT CONSTRAINTS

We can see a graphic description about the platoon formation control perform in real-time strait in Fig. 10.3. When USVs moves in a formation from a wide passage to a constrained channel, they have to change the pattern while keep connectivity and avoid collision. With our proposed method, the desired performance can be achieved. In this chapter, we consider a control strategy to achieve the satisfactory desired distance and angle among USVs with the intervehicle communications. Under the designed control for each vehicle of a string of $n+1$ USVs, the distance and angle errors can be constrained in a certain range. Line-of-sight (LOS) strategy [350] is used to define the distance and angle between the two USVs,

$$d_i = \sqrt{(x_{i-1} - x_i)^2 + (y_{i-1} - y_i)^2}, \tag{10.15}$$

$$\varphi_i = \arctan 2(y_{i-1} - y_i, x_{i-1} - x_i). \tag{10.16}$$

In order to avoid collision among vehicles, the desired distance during the whole moving process must satisfy the following equations:

$$d_{i,\min col} < d_i \leq d_{i,\max com}, \tag{10.17}$$

where $0 < d_{i,\min col} < d_{i,\max com}$. The minimum safety distance is denoted by $d_{i,\min col}$ and the maximum effective communication distance is denoted by $d_{i,\max com}$. Next, the formation errors of the USVs are designed [21]:

$$\begin{aligned} e_{di} &= d_i - d_{i,des}, \\ e_{\psi i} &= \psi_{i-1} - \psi_i, \end{aligned} \tag{10.18}$$

where $d_{i,des}$ is a desired distance between the two USVs, and $d_{i,des}$ satisfies the inequality $0 < d_{i,\min col} < d_{i,des} < d_{i,\max com}$. For convenience, we define $\hat{e}_{di}(t) = d_{i,col\min} - d_{i,des}$ as the minimum error distance and $\bar{e}_{di}(t) = d_{i,con\max} - d_{i,des}$ as the maximum error distance. To guarantee the output constraints, we need to ensure that all signals are bounded and that the errors defined in (8) are not violated, i.e.,

$$\begin{aligned} \hat{e}_{di} &< e_{di} < \bar{e}_{di}, \\ \hat{e}_{\psi i} &< e_{\psi i} < \bar{e}_{\psi i}. \end{aligned} \tag{10.19}$$

10.3 CONTROL DESIGN WITH A SYMMETRIC BLF

10.3.1 LEADER VELOCITY ESTIMATION

Note that the velocity of the leader may be not always accessible in practical applications, in which case a fast convergent observer is designed to estimate the unavailable states. According to (10.2), let the unknown leader's velocity $\dot{\eta}_l$ pass through a first order filter:

$$\dot{\phi} = -k\phi + k\dot{\eta}_l, \tag{10.20}$$

where $k \in \mathbb{R}$ is a positive scalar. Suppose that $\dot{\eta}_l$ is a differentiable function with time-derivative χ, the dynamics in (10.20) can be rewritten in the form of a linear system as follows:

$$\begin{aligned} \dot{x}_1 &= -kx_1 + kx_2, \\ \dot{x}_2 &= \chi, \\ y &= x_1, \end{aligned} \tag{10.21}$$

where x_1 and x_2 are equal to ϕ and $\dot{\eta}_l$, respectively, χ is the unknown input while y is the measurable output of the linear system. The problem of estimating $\dot{\eta}_l$ can be formulated as observing the states of the linear system (10.21). The following observer can therefore be designed to exactly reconstruct the linear system (10.21) with a finite-time:

$$\begin{aligned} \dot{\hat{x}}_1 &= -k\hat{x}_1 + k\hat{x}_2 - k_1\mathrm{sign}_{\mathrm{c}}(\tilde{x}_1) - k_2\tilde{x}_1, \\ \dot{\hat{x}}_2 &= -k_3\tilde{x}_1 - k4\lfloor\vartheta\rceil^{\frac{p}{q}} - k_5\mathrm{sign}_{\mathrm{c}}(\vartheta), \end{aligned} \tag{10.22}$$

where $\tilde{x}_1 = \hat{x}_1 - x_1$ and $\tilde{x}_2 = \hat{x}_2 - x_2$ are the estimation errors and are denoted by $\tilde{x} = [\tilde{x}_1^T, \tilde{x}_2^T]^T, k_i \in \mathbb{R}, i = 1, ..., 5$ are positive observer gains to be defined later. The variables ϑ is defined as $\vartheta = k_1 \text{sign}_c(\tilde{x}_1)$, p and q are positive odd integers such that $p < q$. The classical sign function sign_c has been introduced in (2.28)–(2.30). The observer designed above ensures that the estimate value can converge to its real value in a finite time, provided that the observer gains are selected as: $k_1 > \max\{\underline{k}_1, \overline{k}_1\}$ and $k_5 k - \gamma > 0$, where $\underline{k}_1 = (k\frac{k_4\sqrt{3}+\sqrt{3}k_5+\gamma}{\lambda_{\min}(P)} + \delta_0)^{\frac{q}{q+p}}$ and $\overline{k}_1 = (\frac{k(\sqrt{3}k_5+\gamma)+\delta_0\lambda_{\min}(P)}{\sqrt{3}k_5+\gamma+\delta_0\lambda_{\min}(P)})^{\frac{q}{p}}$ with $\delta_0 > 0$ is a positive scalar, and P is a positive definite matrix given as

$$P = \begin{bmatrix} (k+k_2)I_3 & -k_3 I_3 \\ k_3 I_3 & 0_3 \end{bmatrix}. \tag{10.23}$$

Proof. Detailed proof can be found in [195]. □

10.3.2 CONTROL DESIGN

In case that matrices M_i, $C_i(\nu_i)$ and $D_i(\nu_i)$ are all known. Let $z_{1i} = [z_{11i}, z_{12i}]^T = [e_{di}, e_{\psi i}]^T$, and $z_{2i} = \nu_i - \alpha_i = [z_{21i}, z_{22i}, z_{23i}]^T$, where $\alpha_i = [\alpha_{1i}, \alpha_{2i}, \alpha_{3i}]^T$ is a stabilizing function which will be designed later. Consider the symmetric BLF candidate as

$$V_{1i} = \frac{1}{2}\log\frac{k_{ai}^2}{k_{ai}^2 - e_{di}^2} + \frac{1}{2}\log\frac{k_{bi}^2}{k_{bi}^2 - e_{\psi i}^2}, \tag{10.24}$$

where k_{ai} and k_{bi} are positive constants used to constraint e_{di} and $e_{\psi i}$, i.e., $|e_{di}| < k_{ai}, |e_{\psi i}| < k_{bi}$, respectively. Differentiating of V_{1i} with respect to time we have

$$\dot{V}_{1i} = \frac{e_{di}\dot{e}_{di}}{k_{ai}^2 - e_{di}^2} + \frac{e_{\psi i}\dot{e}_{\psi i}}{k_{bi}^2 - e_{\psi i}^2}. \tag{10.25}$$

According to (8), differentiating e_{di} and $e_{\psi i}$ with respect to time, we can obtain

$$\begin{aligned} \dot{e}_{di} =& -(z_{21i} + \alpha_{1i})\cos(\psi_i - \varphi_i) \\ &+ (z_{22i} + \alpha_{2i})\sin(\psi_i - \varphi_i) + \dot{x}_{i-1}\cos\varphi_i + \dot{y}_{i-1}\sin\varphi_i, \\ \dot{e}_{\psi i} =& \dot{\psi}_{i-1} - (z_{23i} + \alpha_{3i}). \end{aligned} \tag{10.26}$$

The virtual control α_i is designed as follows:

$$\begin{aligned} \alpha_{1i} =& \cos(\psi_i - \varphi_i)[k_{di}e_{di}(k_{ai}^2 - e_{di}^2) + \dot{x}_{i-1}\cos\varphi_i + \dot{y}_{i-1}\sin\varphi_i], \\ \alpha_{2i} =& -\sin(\psi_i - \varphi_i)[k_{di}e_{di}(k_{ai}^2 - e_{di}^2) + \dot{x}_{i-1}\cos\varphi_i + \dot{y}_{i-1}\sin\varphi_i], \\ \alpha_{3i} =& k_{\psi i}e_{\psi i}(k_{bi}^2 - e_{\psi i}^2) + \dot{\psi}_{i-1}. \end{aligned} \tag{10.27}$$

Substituting (10.26) and (10.27) into (10.25) yields

$$\dot{V}_{1i} = -k_{di}e_{di}^2 - k_{\psi i}e_{\psi i}^2 + \Theta_{1i}, \tag{10.28}$$

where

$$\Theta_{1i} = \frac{e_{di}(-z_{21i}\cos(\psi_i-\varphi_i)+z_{22i}\sin(\psi_i-\varphi_i))}{k_{ai}^2-e_{di}^2} + \frac{e_{\psi i}(-z_{23i})}{k_{bi}^2-e_{\psi i}^2}.$$

The speed constraints of ν_i are not considered in this work, so another Lyapunov candidate function is chosen as follows:

$$V_{2i} = V_1 + \frac{1}{2} z_{2i}^T M_i z_{2i}. \tag{10.29}$$

Then the time differentiation of (10.29) can be written as

$$\begin{aligned}\dot{V}_{2i} =& -k_{di}e_{di}^2 - k_{\psi i}e_{\psi i}^2 + z_{2i}^T(\tau_i - C_i(\nu_i)\nu_i \\ & - D_i(\nu_i)\nu_i + w_i + \Delta_i - M_i\dot{\alpha}_i) + \Theta_{1i}.\end{aligned} \tag{10.30}$$

The model-based control law can be designed as:

$$\begin{aligned}\tau_i =& -K_{2i}z_{2i} + C_i(\nu_i)v_i + D_i(\nu_i)v_i - w_i \\ & + M_i\dot{\alpha}_i - \Theta_{2i} + \Delta_i,\end{aligned} \tag{10.31}$$

where

$$k_{2i} \in \mathbb{R}^{3\times 3} > 0, \tag{10.32}$$

$$\Theta_{2i} = \begin{bmatrix} \frac{e_{di}(-\cos(\psi_i-\varphi_i))}{k_{ai}^2-e_{di}^2} & \frac{e_{di}\sin(\psi_i-\varphi_i)}{k_{ai}^2-e_{di}^2} & \frac{-e_{\psi i}}{k_{bi}^2-e_{\psi i}^2}\end{bmatrix}^T. \tag{10.33}$$

Substituting (10.31) into (10.30), the equation (10.30) can be rewritten as

$$\dot{V}_{2i} = -k_{di}e_{di}^2 - k_{\psi i}e_{\psi i}^2 - k_{2i}z_{2i}^T z_{2i}. \tag{10.34}$$

Based on Lemma 10.1, we can obtain that the error e_{di} can be constrained in the range of $(-k_{ai}, k_{ai})$ and $e_{\psi i}$ could be constrained in the range of $(-k_{bi}, k_{bi})$. However, the hydrodynamic damping $D_i(\nu_i)v_i$ is a continuous nonlinear function related to the USV speed. In addition, the unmodeled dynamics of the USV and model uncertainties have not been considered above. To solve these problems, an RBF NN is used to estimate the unknown dynamics and hydrodynamic damping terms. We give

$$F_i(Z_i) = -D_i(\nu_i)\nu_i + \Delta_i, \tag{10.35}$$

where $F_i(Z_i) = [f_{1i}(Z_i), f_{2i}(Z_i), f_{3i}(Z_i)]^T \in \mathbb{R}^3$, $Z_i = \nu_i$ is the inputs of the NN, and Δ_i represents the unmodeled dynamics.

$$f_{li}(Z_i) = W_{li}^{*T} H_{li}(Z_i) + \varepsilon_{li}(Z_i), l = 1,2,3, \tag{10.36}$$

where W_{li}^* is the true constant weight value, $H_{li}(Z_i)$ is the RBF, $\varepsilon_{li}(Z_i) \leq \bar{\varepsilon}_i$ ($\bar{\varepsilon}_i > 0$ is an unknown arbitrary small constant) is the approximate error. The adaptive neural network control law is proposed as follows:

$$\begin{aligned}\tau_i =& - z_{2i}^{+T}\left(\frac{k_{di}e_{di}^2}{k_{ai}^2 - e_{di}^2} + \frac{k_{\psi i}e_{\psi i}^2}{k_{bi}^2 - e_{\psi i}^2}\right) + C_i(\nu_i) + M_i\dot{\alpha}_i \\ &- w_i - K_{2i}z_{2i} - \hat{W}_i^T H_i(Z_i) - \Theta_{2i},\end{aligned} \tag{10.37}$$

where $(\bullet)^+$ is the Moore-Penrose pseudoinverse of $(\bullet)$, $\hat{W}_i = [\hat{W}_1, \hat{W}_2, \hat{W}_3]^T$ are the weights of the RBFNN, $\hat{W}_i^T H_i(Z_i)$ is used to approximate $W_{li}^{*T} H_{li}(Z_i)$.

$$W_{li}^{*T} H_{li}(Z_i) = \hat{W}_{li}^T H_{li}(Z_i) - \varepsilon(Z_i) = -D_i(\nu_i)\nu_i + \Delta_i - \varepsilon(Z_i). \tag{10.38}$$

The adaptive update law is designed as follows:

$$\dot{\hat{W}}_i = \Gamma_i(H_i(Z_i)z_{2i} - \sigma_i|z_{2i}|\hat{W}_i),\ i = 1, 2, 3, \tag{10.39}$$

where $\Gamma_i = \Gamma_i^T > 0$ is the adaptive gain matrices and σ_i is a positive constant.

Consider the Lyapunov function candidate

$$V_{3i} = V_{2i} + \frac{1}{2}\sum_{i=1}^{3} \tilde{W}_i \Gamma_i^{-1} \tilde{W}_i, \tag{10.40}$$

where $\tilde{W}_i = \hat{W}_i - W_i^*$. Then the derivation of V_{3i} along (10.30) and (10.39) is

$$\begin{aligned}\dot{V}_{3i} =& - k_{di}e_{di}^2 - k_{\psi i}e_{\psi i}^2 + z_{2i}^T\Theta_2 + z_{2i}^T(\tau_i - C_i(\nu_i) \\ &- M_i\alpha_i - w_i + W_i^{*T} H_i(Z_i) - \varepsilon_i(Z_i)) + \sum_{i=1}^{3} \tilde{W}_i^T \Gamma_i^{-1} \dot{\hat{W}}_i.\end{aligned} \tag{10.41}$$

Substituting (10.37), (10.38) and (10.39) into (10.41), we can obtain

$$\begin{aligned}\dot{V}_{3i} =& - k_{di}e_{di}^2 - k_{\psi i}e_{\psi i}^2 - \frac{k_{di}e_{di}^2}{k_{ai}^2 - e_{di}^2} - \frac{k_{\psi i}e_{\psi i}^2}{k_{bi}^2 - e_{\psi i}^2} \\ &- z_{2i}^T K_{2i} z_{2i} - z_{2i}^T \varepsilon_i(Z_i) + \sum_{i=1}^{3} \tilde{W}_i^T \Gamma_i^{-1} \dot{\hat{W}}_i.\end{aligned} \tag{10.42}$$

Then according to the Young inequality and Lemma 10.1, we have

$$\begin{aligned}\dot{V}_3 \leq& - k_{di}\log\frac{k_{ai}^2}{k_{ai}^2 - e_{di}^2} - k_{\psi i}\log\frac{k_{bi}^2}{k_{bi}^2 - e_{\psi i}^2} \\ &- z_{2i}^T(K_{2i} - I)z_{2i} + \frac{1}{2}\|\varepsilon_i(Z_i)\|^2 \\ &+ \sum_{i=1}^{3}\frac{\sigma_i^2}{8}(\|W_i^*\|^4 + \|\tilde{W}_i\|^4 - 2\|W_i^*\|^2\|\tilde{W}_i\|^2).\end{aligned} \tag{10.43}$$

According to Lemma 10.2, we can understand that

$$\begin{aligned}\left\|\tilde{W}_i\right\| = \left\|\hat{W}_i - W_i^*\right\| \leq \left\|\hat{W}_i\right\| + \|W_i^*\| \\ \leq \frac{s_i}{\sigma_i} + \|W_i^*\| = \varpi.\end{aligned} \tag{10.44}$$

So we can obtain

$$\dot{V}_{3i} \leq -\rho V_{3i} + C, \tag{10.45}$$

where

$$\rho = \min(\min(2k_{di}, 2k_{\psi i}), \frac{2\lambda_{\min}(K_{2i} - I)}{\lambda_{\max}(M_i)}, \min(\frac{\sigma_i^2 \|W_i^*\|^2}{2\lambda_{\max}(\Gamma_i^{-1})})), \tag{10.46}$$

$$C = \frac{1}{2}\|\bar{\varepsilon}_i\|^2 + \sum_{i=1}^{3}(\frac{\sigma_i^2}{8}\|W_i^*\|^4 + \frac{\sigma_i^2}{8}\varpi^4), \tag{10.47}$$

the minimum and maximum eigenvalues of matrix $\bullet$ are denoted by $\lambda_{\min}(\bullet) and \lambda_{\max}(\bullet)$, respectively. From equation (10.45) we can obtain

$$V_{3i} \leq V_{3i}(0) + \frac{C}{\rho}. \tag{10.48}$$

According to (10.48), for e_{di} we have

$$zh \log \frac{k_{ai}^2}{k_{ai}^2 - e_{di}^2} \leq 2(V_{3i}(0) + \frac{C}{\rho}) \tag{10.49}$$

Then, the distance error

$$\|e_{di}\| \leq \sqrt{k_{ai}^2 - k_{ai}^2 e^{2(V_{3i}(0) + \frac{C}{\rho})}}.$$

Similarly, for angle error $e_{\psi i}$, we have

$$\log \frac{k_{bi}^2}{k_{bi}^2 - e_{\psi i}^2} \leq 2(V_{3i}(0) + \frac{C}{\rho}), \tag{10.50}$$

Then,

$$\|e_{\psi i}\| \leq \sqrt{k_{di}^2 - k_{di}^2 e^{(2(V_{3i}(0) + \frac{C}{\rho}))}}$$

Therefore, we can conclude that the errors e_{di} and $e_{\psi i}$ are SGUUB.

10.4 SIMULATION EXAMPLES

In this section, a string of 4 USVs are used for the numerical simulations to demonstrate the effectiveness of the proposed method. The communication relationship of the 4 USVs is shown in Fig. 10.4.

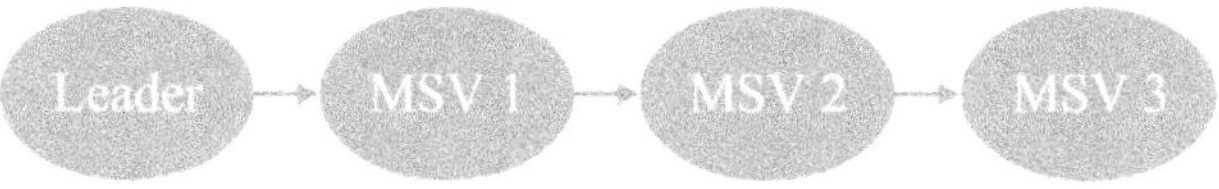

Figure 10.4 Communication graph among four USVs.

For the platoon formation control, every USV is a leader for its nearest successive USV except for the last one. So when arbitrary a USV is absence in practice, its follower should follow the virtual leader which is at the forefront and change the desired distance enough large. With this strategy, even one of USVs is absence, the formation pattern and safety can be guaranteed.

The ship model parameters used to simulate in this work are Cybership-II, which is a 1:70 scale supply vessel replica built in a marine control laboratory in the Norwegian University of Science and Technology [461].

For the sake of the comparison convenience, the desired reference trajectory is chosen similar to [90],

$$\eta_0 = \begin{cases} [1.2t, 0, 0]^T, \text{if } t \leq t_c, \\ [1.2t_c + 60\sin t', 60\,(1 - \cos t')\,, t']^T, \text{if } t > t_c, \end{cases}$$

where $t_c \geq 0$ is a time constant and $t' = 0.02(t - t_c)$. The desired distance between the follower and the leader is considered as 5m. The initial states of the USVs are given as $\eta_0 = [0, 0, 0]^T$, $\eta_1 = [0, 5.9, 0]^T$, $\eta_2 = [0, 11.8, 0]^T$, $\eta_3 = [0, 17.7, 0]^T$ and $\nu_1 = \nu_2 = \nu_3 = 0$. The time constant $t_c = 200$ s. Fig. 10.4 illustrates the validity of the position and velocity information among USVs. The maximum thrust force is 2N. In this chapter, we consider the environment disturbances as $w_i = [0.1\sin t, 0.1\sin t, 0]^T$. The minimum collision distance is 4m and the maximum connectivity distance is 6m. So the distance error should be satisfied by following constraints:

$$d_{i,col\,\min} < d_i < d_{i,con\,\max}.$$

If the inequality is satisfied during the entire process of moving, it means that the collision distance and effective connectivity distance will not be violated, so the output constraints performance can be guaranteed. In order to verify the effectiveness of the proposed controller, two cases are evaluated for Matlab simulation studies.

Case I (model – based control) : For the model-based control, the control parameters are chosen as follows: $k_{d1} = 12, k_{d2} = k_{d3} = 1, k_{\psi 1} = 1, k_{\psi 2} = k_{\psi 3} = 0.5, k_{a1} = k_{a2} = k_{a3} = 1$ and $k_{b1} = k_{b2} = k_{b3} = 0.5$. The control gains are set to $K_{2i} = diag\,[6, 6, 4]$. The parameters k_{ai} and k_{bi} are the distance constraint and angle constraint selected based on the communication equipment or other plants requirements. The value of k_{di} and $k_{\psi i}$ effect the virtual control of α_{1i}, α_{2i} and α_{3i}, if k_{di} and $k_{\psi i}$ are too large, the virtual control will

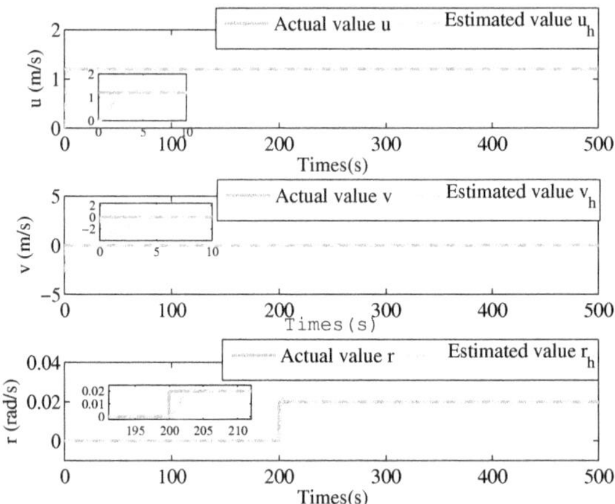

Figure 10.5 The estimate value of the finite time observer module.

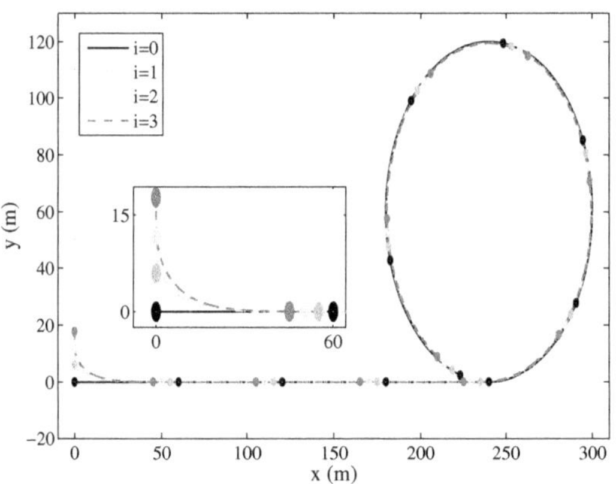

Figure 10.6 The north-east position of the 4 USVs under model-based control.

become large which may lead to the control input violating the physical limit value and affecting the stability of the system. If k_{di} and $k_{\psi i}$ are too small, the control input will not enough to drive the USV to the desired velocity.

Fig. 10.5 shows that the unmeasurable surge, sway and yaw velocity of the leader USV can be measured by the designed finite time observer with zero estimate error. From Fig. 10.6, it is obvious that the USVs can move in a desired formation pattern under the model-based method. We can observe that from Fig. 10.7, the distance between the successive vehicles is limited

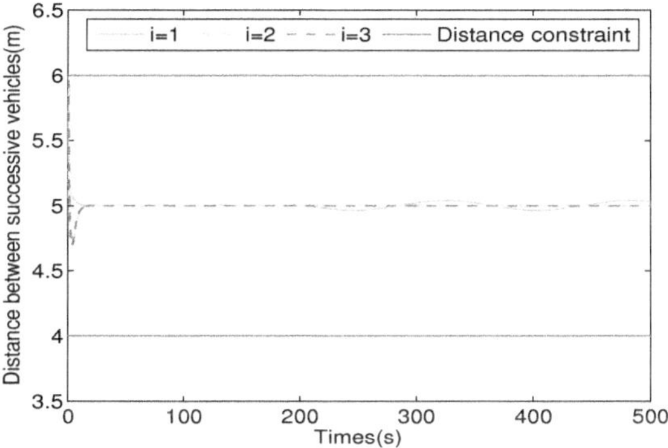

Figure 10.7 The LOS range d_i, the maximum connectivity distance and minimum collision distance under model-based control.

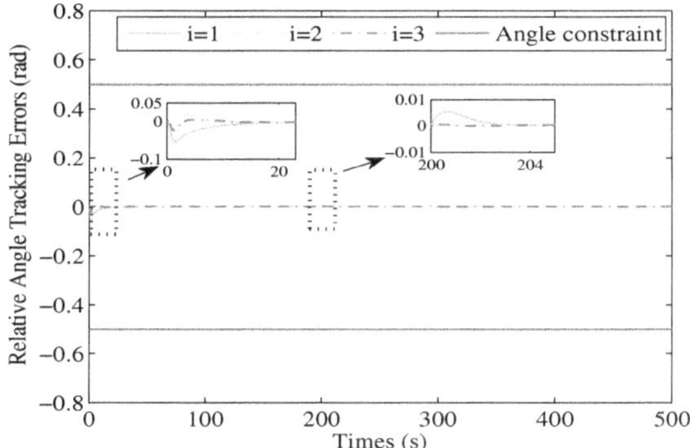

Figure 10.8 Yaw angle tracking errors along with proposed boundary.

to 4 (collision avoidance distance) to 6 (the maximum connection distance) during the whole moving process. So the distance error constraints $e_{di} < k_{ai}$ and $e_{di} > -k_{ai}$, $\forall t > 0$ and relative angle error constraints $-k_{bi} < e_{\psi i} < k_{bi}$, $\forall t > 0$ are realized. Fig. 10.7 shows yaw angle tracking errors along with proposed boundary.

Case II (*adaptive NN control*): Second, to demonstrate the control integrated with NNs, we choose the formation pattern as same as the case I. The hydrodynamic damping matrix $D_i(\nu_i)$ are unknown. Simulation parameters are chosen as follows: $k_{d1} = 11$ and $k_{d2} = k_{d3} = 3$. The NN inputs are chosen as $Z_{i1} = u_i$ and $Z_{i2} = Z_{i3} = [v_i, r_i]^T$.

The simulation results are shown in Figs. 10.9–10.11. Fig. 10.9 shows that the USVs are moving in the desired formation pattern. Figs. 10.9–10.11 also

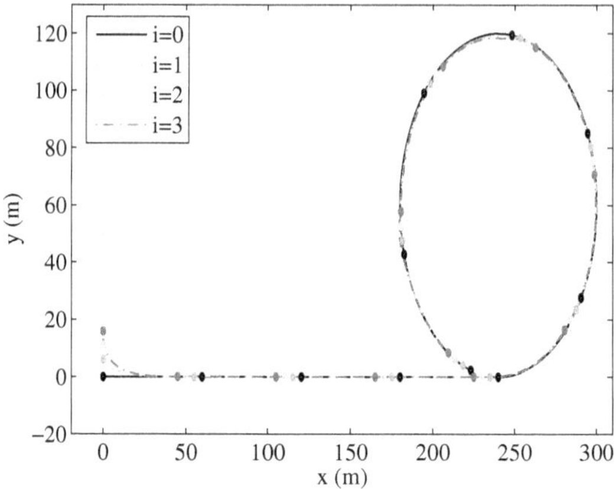

Figure 10.9 The north-east position of the 4 USVs under NN adaptive control.

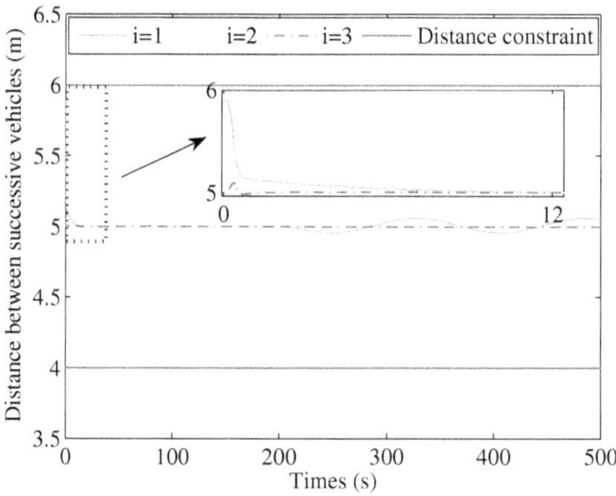

Figure 10.10 LOS range d_i and the maximum connectivity distance and minimum collision distance under NN adaptive control.

show that though the hydrodynamic damping matrix D_i is unknown to input τ_i, through NN learning LOS range d_i converges to a small neighborhood of the desired distance and the error satisfies the predefined boundary restriction, which means that during the whole process, the distance between the successive vehicles does not violate the collision and connectivity constraints. In addition, the idea of [19] is used to examine the robustness of the proposed method. To reduce the simulation time, only follower 1 is used to examine the robustness. The hydrodynamics damping term is randomly selected in the

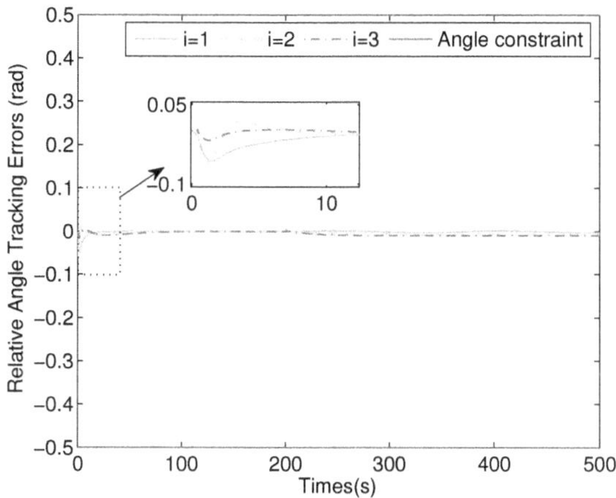

Figure 10.11 Yaw angle tracking errors along with proposed boundary.

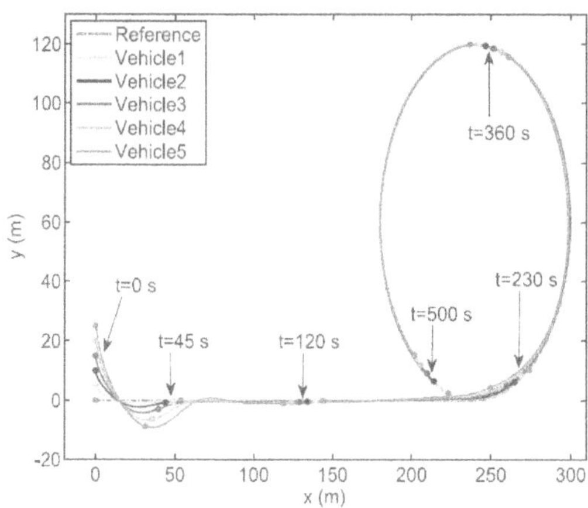

Figure 10.12 The north-east position of USVs under the method proposed in [90].

bound $[D^* - 0.5D^*, D^* + 0.5D^*]$, where D^* is equal to the value in equation (10.31), simulations are run for 100 sets of the parameters. The simulation results are shown in Figs. 10.13 and 10.14. In addition, we also add the random Gaussian noise with amplitudes of 1 cm, 10 cm, 50 cm, 1m, 1.2m and 1.3m to the trajectory of the leader, the control performance can be guaranteed below 1.2m, please see the simulation results in Figs. 10.15 and 10.16 with the amplitudes of 1m random Gaussian noise. It can be concluded that

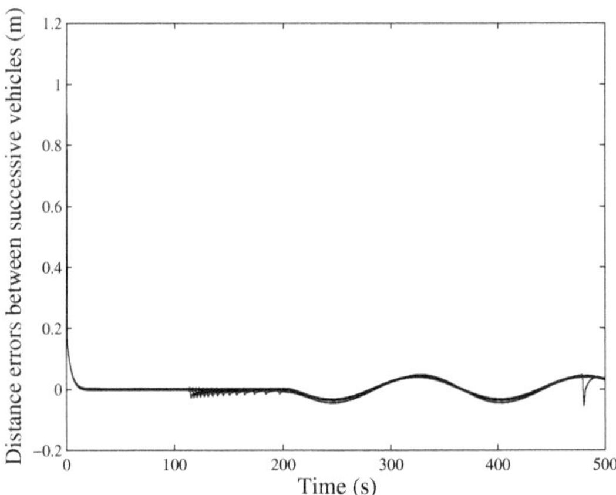

Figure 10.13 Graphs of the 100-sets simulation distance errors.

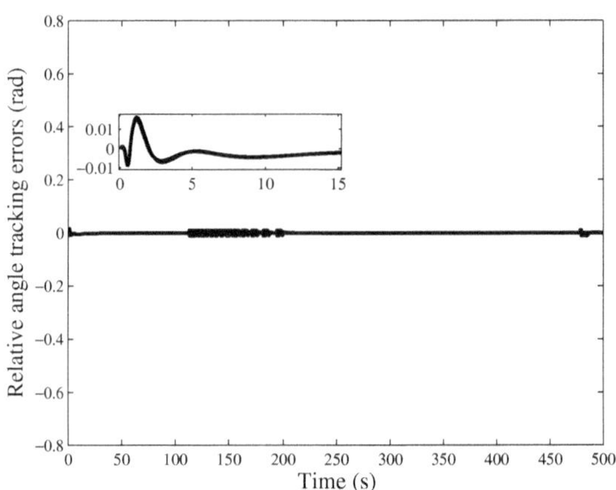

Figure 10.14 Graphs of the 100-sets relative angle tracking errors.

the proposed controller is so robust against parametric uncertainties and the USVs successfully achieve the desired formation pattern.

For the sake of the comparison convenience, the desired reference trajectory is chosen similar to that of [90]. Comparing Figs. 10.6 and 10.9 with Fig. 10.14, we can see that there is a overshoot in Fig. 10.14 which is controlled by the method in [90]. In addition, in Figs. 10.6 and 10.9, the formation pattern can be achieved near 10 meters on the x-axis, while it needs near 80 meters in [90]. So the converge performance is better than the method proposed in [90].

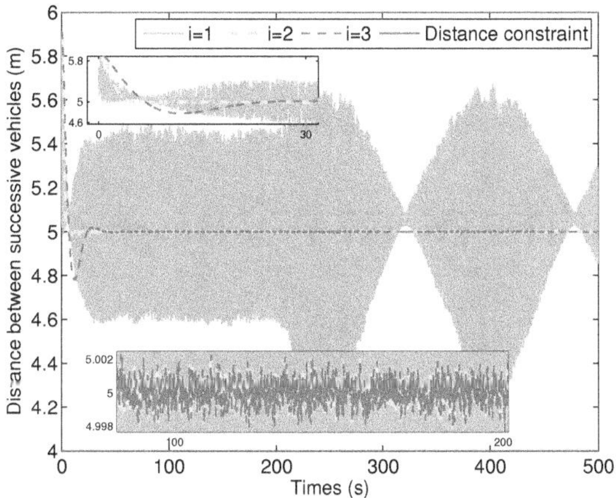

Figure 10.15 The LOS range di, the maximum connectivity distance and minimum collision distance under 1m noise.

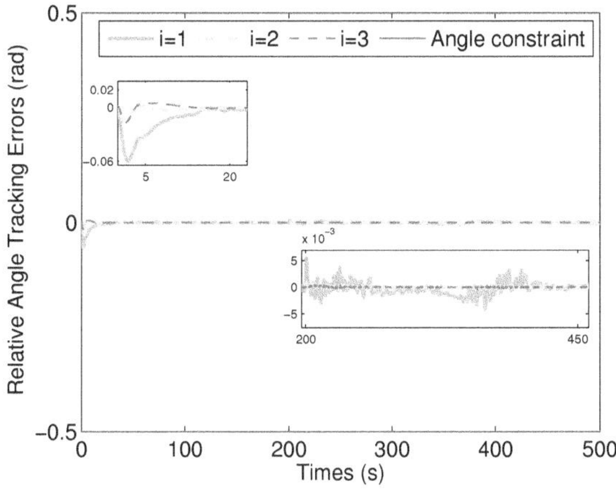

Figure 10.16 Yaw angle tracking errors under 1m noise.

10.5 CONCLUSIONS

In this chapter, the velocity-free formation control method for autonomous unmanned surface vehicles with the model uncertainties and output constraints is presented. Under the proposed model-based control, we have proven that the system is asymptotically stable and the line-of-sight range and angle errors constraint are never violated. Then under adaptive neural network control, the system is proved to be semi-globally uniformly ultimately bounded. Desired formation pattern is achieved without violation of the constraints including collision avoidance and connectivity distance. The simulation results generated confirm the performance of the proposed methods.

11 Formation Potential Field for Trajectory Tracking Control of Multi-Agents in Constrained Space

In this chapter, a formation potential field method is proposed for the multi-agents formation problem. The objective is to control a group of agents to automatically generate and maintain a specific formation while avoiding internal collisions and collisions with spatial constraints. A formation potential field is designed combining multiple local attractive potential fields with multiple local repulsive potential fields. To further relax requirements of agents' initial positions and enhance the robustness, a global attractive potential field is added outside the influence range of the local formation potential field. The optimality of the proposed scheme in formation time is analyzed as well as illustrated by contrast simulation results. Following the design of the formation potential field, two controllers are proposed accordingly to achieve a stable dynamic formation during the process of trajectory tracking, while the saturation effect of the input is taken into account. Furthermore, a collision avoidance strategy based on artificial potential field and Dirac delta function is applied to locally modify the original trajectory of the virtual leader such that agents can avoid collisions with unexpected spatial constraints while maintaining the given formation. Simulation results are presented to illustrate the performance of proposed approaches.

11.1 INTRODUCTION

Recently, research and applications on multi-agent systems (MAS) have been widely studied in various areas, including mathematics, physics, biology, sociology and engineering science (e.g., [193, 268, 541]). There are some interesting topics in MAS, such as flocking, formation and consensus, distributed problem-solving, learning control for multi-agents (e.g., [48, 54, 504, 547]). Among these topics, formation control for MAS has been applied to various applications in recent years, for example, mobile robots, unmanned air vehicles, satellites, aircraft, and spacecraft (e.g., [9, 24, 119, 345]).

Artificial potential field (APF) method provides simple and effective solutions for practical applications, such as motion planning, target tracking while also considering obstacle avoidance (e.g., [148, 199, 427]). The applications of

DOI: 10.1201/9781003298618-11

APF for obstacle avoidance was first developed by [233]. The essential is to design a potential field, which "attracts" an agent to an target, and "repulses" it from obstacles, which can be implemented quickly and provide feasible solutions without requiring many clarification. In recent years, there are many applications of APF method on formation control of MAS. [524] presented motion planning and obstacle avoidance of multi-HUG formation by combining the APF method with Kane's method, which offered the advantage that coordinated control and motion planning while in formation can be conducted synchronously. [356] proposed an approach based on potential field for the optimal path planning of a group of nonholonomic robots with coherent formation in a leader-following structure. However, in these existing research, an algorithm deciding which formation node or queue does each agent belong to is required, which involves the optimal formation problem.

A framework of general optimal formation problem was introduced by [190], where the optimal formation problem is generally divided into the minimum-time optimal formation problem and the minimax travel-distance optimal formation problem. To address this challenge, several related works have been published (e.g., [213,300,482]). These solutions have a common feature that a bijection from agents to their desired formation positions should be decided in advance. In our previous work (see [151]), the optimal bijection was obtained by minimizing the maximum distance between all agents' initial positions and their corresponding desired formation positions. However, an inevitable limitation cannot be solved by the algorithm. The potential forces generated by dynamic obstacles to avoid collisions during the process of formation tracking are ignored. If it is totaly considered, the optimal problem will be a complex dynamic problem. Furthermore, this drawback does not only exist in the mentioned approach but also do in any ones that assigning the bijection in the beginning. Besides, the optimization effect is not remarkable because most of attractive potential functions are only considered to be build in the quadratic form.

Inspired by this limitation, in this chapter, a formation potential field method is proposed to control multi-agents to automatically generate and maintain a given formation. Meanwhile, no collision among agents and with spatial constraints is also guaranteed. Each agent will automatically be attracted into an optimal desired formation position without giving specific corresponding relation between agents and their desired formation positions. Meanwhile, the attractive potential functions designed in exponential form will lead to a significant improvement in time cost.

The main contribution of this chapter is the proposal of the formation potential field method for a group of agents to automatically generate and maintain a given formation while avoiding collisions among agents. Each agent will automatically be attracted into the optimal desired formation position. The optimality compared to traditional schemes is analyzed and showed by contrast simulation results. A global potential field is designed to cover the shortage of local formation potential field. In this way, the requirement of agents' initial positions can be relaxed and the robustness can be enhanced.

11.2 SYSTEM DYNAMICS AND FORMATION MANEUVERS

Before proceeding further, the following assumptions are made in this chapter.

Assumption 19. *The influence of size and shape to control is ignored and an agent is viewed as a point.*

Assumption 20. *An agent is able to estimate its position in the world coordinate system.*

Assumption 21. *Each agent is able to broadcast information to others.*

Definition 34 (*Formation*)**.** *([151]) A formation pattern at time t is defined to be a set*

$$\mathfrak{P}(t) = \{\mathbf{x}_1^{\mathrm{d}}(t), ..., \mathbf{x}_N^{\mathrm{d}}(t)\} \tag{11.1}$$

where $\mathbf{x}_i^{\mathrm{d}}(t)$ is the desired position of agent v_i at time t, $i = 1, ..., N$ with respect to trajectory center.

Definition 35 (*Fixed Formation*)**.** *([151]) A formation pattern is defined as Fixed Formation if $\Delta\mathbf{x}_i(t) = \mathbf{x}_i^{\mathrm{d}}(t) - \mathbf{x}^{\mathrm{r}}(t), i = 1, .., N$ is time-invariant for $t \in [0, \infty)$, which indicates*

$$\begin{aligned} \Delta\dot{\mathbf{x}}_i(t) &= \dot{\mathbf{x}}_i^{\mathrm{d}}(t) - \dot{\mathbf{x}}^{\mathrm{r}}(t) = 0 \\ \Rightarrow \dot{\mathbf{x}}_i^{\mathrm{d}}(t) &= \dot{\mathbf{x}}^{\mathrm{r}}(t), \ i = 1, .., N \end{aligned} \tag{11.2}$$

In this chapter, we will take the center of agents as a virtual leader, and write this position as $\mathbf{x}_0$; take the center of the desired formation as reference, and write this position as $\mathbf{x}^{\mathrm{r}}$. We have

$$\mathbf{x}_0 = \frac{1}{N}\sum_{i=1}^{N}\mathbf{x}_i \quad \& \quad \mathbf{x}^{\mathrm{r}} = \frac{1}{N}\sum_{i=1}^{N}\mathbf{x}_i^{\mathrm{d}} \tag{11.3}$$

$\mathbf{x}_0$ is regarded as a virtual leader is because it acts as the leader for the agents' group. The formation defined in Eq. (11.1) satisfies following properties:

1. *For any $\mathbf{x}_i^{\mathrm{d}}(t) \in \mathfrak{P}(t)$, i=1,...,N, $\dot{\mathbf{x}}_i^{\mathrm{d}}(t) = \dot{\mathbf{x}}^{\mathrm{r}}(t)$ when considering fixed formation.*
2. *For any $\mathbf{x}_i^{\mathrm{d}}(t), \mathbf{x}_j^{\mathrm{d}}(t) \in \mathfrak{P}(t)$, i, j=1,...,N, $\dot{\mathbf{x}}_i^{\mathrm{d}}(t) = \dot{\mathbf{x}}_j^{\mathrm{d}}(t)$ when considering fixed formation.*

Consider a network of N agents, denoted as $v_i, i = 1, ..., N$, and $\mathbf{x}_i(t) \in \mathbb{R}^m$ is the state of agent v_i at time t. In this chapter, $\mathbf{x}_i(t)$ represents the position of an agent v_i in 2-D space, which means $m = 2$; or consider $\mathbf{x}_i(t)$ to represent the position of an agent v_i in 3-D space, which means $m = 3$. We assume N agents with the similar dynamic described by the following practical nonlinear differential equation as:

$$\dot{\mathbf{x}}_i = \mathbf{F}_i(\mathbf{x}_i) + \mathbf{G}_i(\mathbf{x}_i)\mathbf{u}_i \tag{11.4}$$

$$\mathbf{y}_i = \widetilde{\mathbf{x}}_i \tag{11.5}$$

where $i = 1, ..., N$, and $\mathbf{u}_i(t)$, $\mathbf{y}_i \in \mathbb{R}^m$ represent the control vector and output of agent v_i, respectively; $\mathbf{F}_i \in \mathbb{R}^{m\times 1}$ is a nonlinear vector; $\mathbf{G}_i \in \mathbb{R}^{m\times m}$ is a smoothly nonsingular function matrix; $\widetilde{\mathbf{x}}_i = \{\mathbf{x}_i - \mathbf{x}_p^{\mathrm{d}} : \min\{\|\mathbf{x}_i - \mathbf{x}_p^{\mathrm{d}}\|\}, p = 1, ..., N\}$, indicating the tracking error of agent v_i. The output can be interpreted as the measurement of tracking error.

In this chapter, the characteristics of the nonlinear system are not the focus. Regarding the linearity of the agents, the following assumption is made.

Assumption 22. *All agents are linear time-invariant (LTI) systems.*

Thus, $\mathbf{F}_i(\mathbf{x}_i)$ and $\mathbf{G}_i(\mathbf{x}_i)$ are considered as constant matrix and written as $\mathbf{F}_i$ and $\mathbf{G}_i$ in the following chapter.

After deciding the position of the virtual leader, the positions of all agents $\mathbf{x}_i^{\mathrm{d}}(t), i = 1, ..., N$ will be calculated relative to the virtual leader.

Control objective: Design a control strategy for a group of agents initialized on random positions to track the desired trajectory keeping the specific formation without internal collisions in non-omniscient constrained space. Consider the unknown forbidden space as Π, which means that all agents' positions satisfy the spatial constraint condition that

$$\mathbf{x}_i(t) \notin \Pi \qquad i = 1, ..., N, 0 \leq t < \infty \tag{11.6}$$

during the whole task. The control goal can be expressed mathematically as, for every $t \geq 0$,

$$\sum_{i=1}^{N} \|\widetilde{\mathbf{x}}_i(t)\| \leq \varepsilon \quad \& \quad \|\mathbf{x}_0(t) - \mathbf{x}^{\mathrm{r}}(t)\| \leq \epsilon \tag{11.7}$$

where ε and ϵ are positive constants that can be made sufficiently small when $t \to \infty$.

11.3 OPTIMAL FORMATION: FORMATION POTENTIAL FIELD

In this section, an optimal formation problem is investigated. First, the formation time cost for a single agent, e.g., agent v_i, can be expressed as $t_i^* = \min_{t_i}\{\|\widetilde{\mathbf{x}}_i(t_i)\| < \eta\}$, where η is a given constant indicating the maximum tolerable formation error for a single agent. Consider the formation task for MAS, the corresponding optimal objective function $J(\sigma)$ can be described by

$J(\sigma) = \max\{t_i^*, i = 1, ..., N\}$, representing the final formation generation time cost.

Define the network of MAS as a set marked by $\mathcal{V} = \{v_1, ..., v_N\}$. Given a specific shape of desired formation demoted by a set $\mathcal{Q} = \{\mathbf{q}_1, ..., \mathbf{q}_N\}$, the purpose of optimal formation problem is to find a bijection $\sigma : \mathcal{V} \to \mathcal{Q}$ which minimizes the formation generation time cost.

The formation potential field method will be designed in this section to dynamically generate an optimal bijection and achieve the desired formation, while avoiding internal collisions.

11.3.1 ATTRACTIVE POTENTIAL FIELD & REPULSIVE POTENTIAL FIELD

Definition 36 (*Potential Source*). *For a single potential field, there is a point* $\mathbf{x}^{\mathrm{s}}$ *to be to local extreme point of the potential field. This point* $\mathbf{x}^{\mathrm{s}}$ *is defined as the potential source. A point which is the local minima of the potential field is defined as the attractive potential source* $\mathbf{x}^{\mathrm{att}}$*; a point which is the local maxima of the potential field is defined as the repulsive potential source* $\mathbf{x}^{\mathrm{rep}}$.

Definition 37 (*Attractive Potential Field*). *A potential field* U^{att} *generated by an attractive potential source is defined as attractive potential field. For any position in a potential field, it has* $U^{\mathrm{att}}(\mathbf{x}) \geq U^{\mathrm{att}}(\mathbf{x}^{\mathrm{att}})$*, the equality holds only when* $\mathbf{x} = \mathbf{x}^{\mathrm{att}}$.

Definition 38 (*Repulsive Potential Field*). *A potential field* U^{rep} *generated by a repulsive potential source is defined as repulsive potential field. For any position in a potential field, it has* $U^{\mathrm{rep}}(\mathbf{x}) \leq U^{\mathrm{rep}}(\mathbf{x}^{\mathrm{rep}})$*, the equality holds only when* $\mathbf{x} = \mathbf{x}^{\mathrm{rep}}$.

For attractive potential field, for example, the target in a target tracking task is the attractive potential source; for repulsive potential field, for example, obstacles are considered as attractive potential sources.

11.3.2 GLOBAL TASKS AND LOCAL TASKS

Two kinds of tasks—global tasks and local tasks are defined in this study. They will be assigned different formats of potential functions for different tasks.

For global attractive tasks, for example, target tracking and path planning, the potential function is chosen to be

$$f_{\mathrm{ga}}(\mathbf{x}, \mathbf{x}^{s}) = \frac{1}{2}(\mathbf{x} - \mathbf{x}^{\mathrm{s}})^{\mathrm{T}}\mathbf{K}(\mathbf{x} - \mathbf{x}^{\mathrm{s}}) \tag{11.8}$$

where $\mathbf{K}$ is a positive definite matrix, and $\mathbf{x}^{\mathrm{s}}$ is a potential source. In this chapter, $\mathbf{K}$ is chosen to be an unit matrix ($\mathbf{K} = \mathbf{I}$, the same in the following chapter), Eq. (11.8) can be rewritten as

$$f_{\mathrm{ga}}(\mathbf{x}, \mathbf{x}^{\mathrm{s}}) = \frac{1}{2}(\mathbf{x} - \mathbf{x}^{\mathrm{s}})^{\mathrm{T}}(\mathbf{x} - \mathbf{x}^{\mathrm{s}}) \tag{11.9}$$

For local attractive tasks, for example, formation control, the potential function is chosen to be

$$f_{\text{la}}(\mathbf{x}, \mathbf{x}^{\text{s}}) = -\text{e}^{-\frac{1}{2}(\mathbf{x}-\mathbf{x}^{\text{s}})^{\text{T}}(\mathbf{x}-\mathbf{x}^{\text{s}})} \tag{11.10}$$

For local repulsive tasks, for example, obstacle avoidance, the potential function is chosen to be

$$f_{\text{lr}}(\mathbf{x}, \mathbf{x}^{\text{s}}) = \frac{1}{2(\mathbf{x}-\mathbf{x}^{\text{s}})^{\text{T}}(\mathbf{x}-\mathbf{x}^{\text{s}})} \tag{11.11}$$

The global potential function has a broader influence range on the whole task space. Its influence increases or decreases as squared when agents getting closed to the potential source. By contrast, the local potential function has a limited influence rage, emphasizing the agents near potential sources.

11.3.3 FORMATION POTENTIAL FIELD

Definition 39 (*formation potential field*). *A potential field which combines a set of attractive potential fields with repulsive potential fields to force a group of agents to generate and keep the desired formation is defined as the formation potential field.*

Two parts for the Formation Potential Filed are built to generate and keep the desired formation, one is attractive potential fields, and another one is repulsive potential fields. On the one hand, the attractive potential fields are designed to attract agents into the desired formation positions. The attractive potential sources are desired formation positions, $\mathbf{x}^{\text{att}} = \{\mathbf{x}_j^{\text{d}}, j = 1, ..., N\}$. Therefore, the attractive potential fields for agent v_i in formation potential field can be written as

$$\begin{aligned} U_i^{\text{att}} &= 1 + k_{\text{a1}} \sum_{p=1}^{N} f_{\text{la}}(\mathbf{x}_i, \mathbf{x}_p^{\text{d}}) \\ &= 1 - k_{\text{a1}} \sum_{p=1}^{N} \text{e}^{-\frac{1}{2}(\mathbf{x}_i-\mathbf{x}_p^{\text{d}})^{\text{T}}(\mathbf{x}_i-\mathbf{x}_p^{\text{d}})} \\ &= 1 - k_{\text{a1}} \sum_{p=1}^{N} \text{e}^{-\frac{1}{2}\mathbf{x}_{ip}^{\text{dT}}\mathbf{x}_{ip}^{\text{d}}} \end{aligned} \tag{11.12}$$

where $\mathbf{x}_{ip}^{\text{d}} = \mathbf{x}_i - \mathbf{x}_p^{\text{d}}$, and k_{a1} is a positive constant.

Assumption 23. *The desired formation satisfies the condition that for any position* $\mathbf{x}$ *in the task space*

$$U^{\text{att}} = 1 - k_{\text{a1}} \sum_{p=1}^{N} \text{e}^{-\frac{1}{2}(\mathbf{x}-\mathbf{x}_p^{\text{d}})^{\text{T}}(\mathbf{x}-\mathbf{x}_p^{\text{d}})} \geq 0 \tag{11.13}$$

Assumption 23 is to guarantee proper distance between any two desired formation positions.

On the other hand, repulsive potential fields are designed to avoid collisions among agents as well as guarantee that only one agent at each desired formation position and finally the desired formation can be achieved. Therefore, for agent v_i, the rest of agents are repulsive potential sources, $\mathbf{x}^{\mathrm{rep}} = \{\mathbf{x}_j, j = 1, ..., N, j \neq i\}$. Therefore, the repulsive potential fields for agent v_i in formation potential field can be written as

$$\begin{aligned} U_i^{\mathrm{rep}} &= k_{\mathrm{r1}} \sum_{q=1,q\neq i}^{N} f_{\mathrm{lr}}(\mathbf{x}_i, \mathbf{x}_q) \\ &= \frac{1}{2} k_{\mathrm{r1}} \sum_{q=1,q\neq i}^{N} \frac{1}{(\mathbf{x}_i - \mathbf{x}_q)^{\mathrm{T}} (\mathbf{x}_i - \mathbf{x}_q)} \\ &= \frac{1}{2} k_{\mathrm{r1}} \sum_{q=1,q\neq i}^{N} \frac{1}{\mathbf{x}_{iq}^{\mathrm{T}} \mathbf{x}_{iq}} \end{aligned} \tag{11.14}$$

where $\mathbf{x}_{iq} = \mathbf{x}_i - \mathbf{x}_q$, and k_{r1} is a positive constant.

To conclude, the formation potential field for agent v_i can be written as

$$\begin{aligned} U_i^{\mathrm{for}} &= U_i^{\mathrm{att}} + U_i^{\mathrm{rep}} \\ &= 1 + k_{\mathrm{a1}} \sum_{p=1}^{N} f_{\mathrm{la}}(\mathbf{x}_i, \mathbf{x}_p^{\mathrm{d}}) + k_{\mathrm{r1}} \sum_{q=1,q\neq i}^{N} f_{\mathrm{lr}}(\mathbf{x}_i, \mathbf{x}_q) \qquad (11.15) \\ &= 1 - k_{\mathrm{a1}} \sum_{p=1}^{N} \mathrm{e}^{-\frac{1}{2} \mathbf{x}_{ip}^{\mathrm{dT}} \mathbf{x}_{ip}^{\mathrm{d}}} + \frac{1}{2} k_{\mathrm{r1}} \sum_{q=1,q\neq i}^{N} \frac{1}{\mathbf{x}_{iq}^{\mathrm{T}} \mathbf{x}_{iq}} \qquad (11.16) \end{aligned}$$

Taking the formation in the form of regular triangle as example, the position of agents are initialized randomly. Accordingly, U^{att} is shown in Fig. 11.1(a) and the initial positions of three agents are also marked in the figure. U_1^{rep} is shown in Fig. 11.1(b), and U_1^{for} is shown in Fig. 11.1(c), where the position of agent v_1 is marked in the figures.

When achieving the desired formation, the agents are all at the local minima of attractive potential fields, and each local minima contains only one agent. The formation potential field for the agent v_i is turned into a single attractive potential field generated by $\mathbf{x}_i^{\mathrm{d}}$, and two repulsive potential fields generated by another two agents, while the agent is exactly at the bottom of the field. This indicates that the MAS reaches the stability.

However, the above method has some drawbacks. First, the attractive potential fields defined in Eq. (11.12) are local potential fields, therefore, the initial positions of agents have to be in the influence range of local potential fields; otherwise, the formation potential field is difficult to work successfully. Second, in the process of achieving formation or tracking in formation, if any

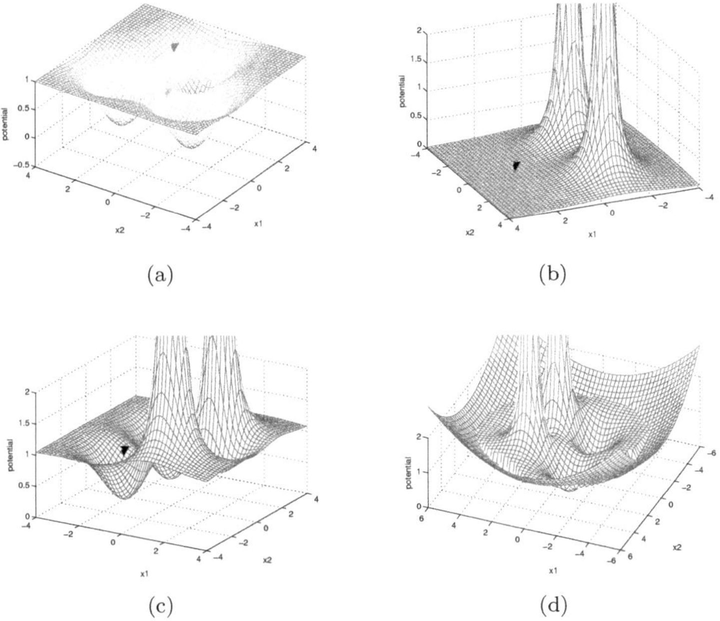

Figure 11.1 (a) Attractive potential field; (b) repulsive potential field of agent v_1; (c) formation potential field of agent v_1; (d)global formation potential field of one agent.

one agent is suddenly beyond the influence range of local potential fields, which may be caused by disturbance, the control law may fail. In order to relax requirements for initial positions of agents and improve the robustness of control approach, a global potential field is added to attract agents into the influence range of local potential field, as shown in Fig. 11.1(d). It is expressed mathematically as

$$\overline{U}_i^{\text{for}} = \begin{cases} U_i^{\text{for}} + U_i^{\text{ext}} & \text{if } \|\nabla U_i^{\text{for}}\| < \sigma \& \|\widetilde{\mathbf{x}}_i\| > \varepsilon \quad (11.17) \\ U_i^{\text{for}} & \text{otherwise} \quad (11.18) \end{cases}$$

where σ is a small constant and

$$\begin{aligned} U_i^{\text{ext}} &= k_{\text{a2}} f_{\text{ga}}(\mathbf{x}_i, \mathbf{x}^{\text{r}}) \\ &= \frac{1}{2} k_{\text{a2}} (\mathbf{x}_i - \mathbf{x}^{\text{r}})^{\text{T}} (\mathbf{x}_i - \mathbf{x}^{\text{r}}) \\ &= \frac{1}{2} k_{\text{a2}} \mathbf{x}_{i\text{r}}^{\text{T}} \mathbf{x}_{i\text{r}} \end{aligned} \quad (11.19)$$

where $\mathbf{x}_{i\text{r}} = \mathbf{x}_i - \mathbf{x}^{\text{r}}$, and k_{a2} is a positive constant.

11.3.4 PARAMETERS CHOSEN

Though the new potential functions guarantee the global minimums to be at the desired formation positions, local minima may still exist. To eliminate the local minima problem, the scaling parameters of attractive and repulsive potential functions $k_{\mathrm{a}1}$, $k_{\mathrm{r}1}$, have to be chosen properly. These two scaling parameters determine the relative intensity of attractive and repulsive force. From Fig. 11.2(a), we can see that an improper chosen of $k_{\mathrm{a}1}$, $k_{\mathrm{r}1}$ will lead to a local minima region. If the agent moves forward from behind, it will be trapped at the local minima region and cannot reach the desired formation position though there is no obstacle between the agent and the desired formation position. In the following analysis, the way to choose $k_{\mathrm{a}1}$ and $k_{\mathrm{r}1}$ to avoid the local minima will be studied.

The formation potential field designed in Section 11.3.3 can be dynamic. Specifically, the repulsive potential fields are dynamics before the desired formation is generated, and the attractive potential fields are static relative to the desired formation center. Without loss of generality, a situation where there are two agents is considered in this part, when one agent (denoted as $\mathbf{x}_q$) has reached the global minimum of an attractive potential field (denoted as $\mathbf{x}_p^{\mathrm{d}}$). The focus is to study if another agent (denoted as $\mathbf{x}_i$) will fall into the local minima. First, we have $\mathbf{x}_q = \mathbf{x}_p^{\mathrm{d}}$, therefore, we mark $\rho = \|\mathbf{x}_{ip}^{\mathrm{d}}\| = \|\mathbf{x}_{iq}\|$. The joint potential field of the attractive potential field (generated by $\mathbf{x}_p^{\mathrm{d}}$) and the repulsive potential field (generated by $\mathbf{x}_q$) is central symmetry. As the result, it is sufficient to study the 1-D potential with the $\mathbf{x}_q$ (or $\mathbf{x}_p^{\mathrm{d}}$) as the origin. Therefore, the joint potential field can be written as $U_i^{\mathrm{for}} = U_i^{\mathrm{att}} + U_i^{\mathrm{rep}} = 1 - k_{\mathrm{a}1}\mathrm{e}^{-\frac{1}{2}\rho^2} + \frac{1}{2}k_{\mathrm{r}1}\frac{1}{\rho}$. To eliminate the local minima, U_i should monotonically decrease as ρ increases, which means $\frac{\mathrm{d}U_i^{\mathrm{for}}}{\mathrm{d}\rho} < 0$ must be satisfied. As $\frac{\mathrm{d}U_i}{\mathrm{d}\rho} = k_{\mathrm{a}1}\rho\mathrm{e}^{-\frac{1}{2}\rho^2} - \frac{1}{2}k_{\mathrm{r}1}\frac{1}{\rho^2}$, to satisfy $\frac{\mathrm{d}U_i^{\mathrm{for}}}{\mathrm{d}\rho} < 0$, the chosen of $k_{\mathrm{a}1}$ and $k_{\mathrm{r}1}$ must follow that $k_{\mathrm{a}1}/k_{\mathrm{r}1} < \mathrm{e}^{\frac{1}{2}\rho^2}/2\rho^3$, which leads to that

$$\frac{k_{\mathrm{a}1}}{k_{\mathrm{r}1}} < \inf\{\frac{\mathrm{e}^{\frac{1}{2}\rho^2}}{2\rho^3}\} = \frac{\mathrm{e}^{\frac{3}{2}}}{6\sqrt{3}} = 0.4313 \tag{11.20}$$

Thus, if the chosen of $k_{\mathrm{a}1}$ and $k_{\mathrm{r}1}$ follows Eq. (11.20), there will be no free path local minima occurring. Fig. 11.2(b) show the potential field when $\frac{k_{a1}}{k_{r1}}$ is set as different values.

11.4 CONTROLLER DESIGN WITH INPUT SATURATION

In this section, a control scheme $\mathbf{u}_i$ will be designed based on the proposed formation potential field for each agent with the dynamics in Eqs. (11.4–11.5) to ensure that all agents move along a desired trajectory while generating and maintaining a specific formation without internal collisions.

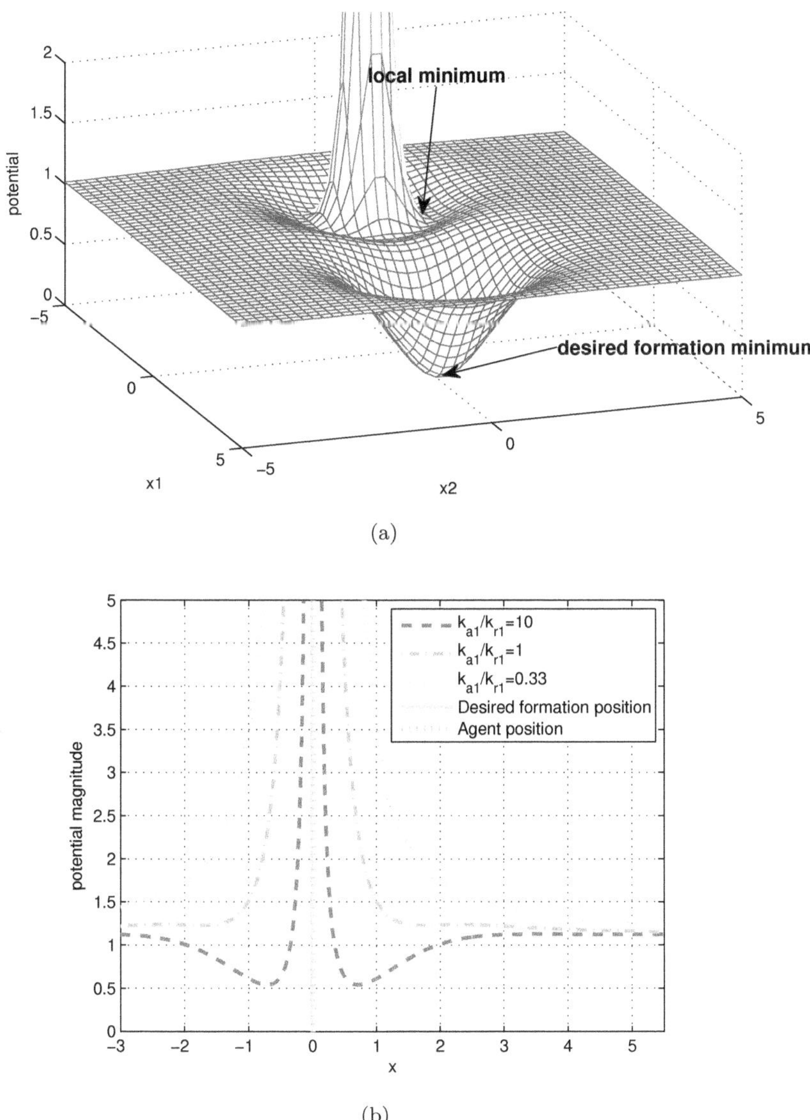

Figure 11.2 (a) Local minima region; (b) total potential field with different parameters.

In the first step, we prove the stability of local formation potential field expressed in Eq. (11.16). Define

$$\mathbf{F}_i^{\text{for}} = -k_{\text{a1}} \sum_{p=1}^{N} \frac{\partial f_{\text{la}}(\mathbf{x}_i, \mathbf{x}_p^d)}{\partial \mathbf{x}_i} - 2k_{\text{r1}} \sum_{q=1,q\neq i}^{N} \frac{\partial f_{\text{lr}}(\mathbf{x}_i, \mathbf{x}_q)}{\partial \mathbf{x}_i} \tag{11.21}$$

Choose controller $\mathbf{u}_i$ as below

$$\mathbf{u}_i = \mathbf{G}_i^+ \left(-\mathbf{F}_i + \mathbf{F}_i^{\text{for}} + \dot{\mathbf{x}}^{\text{r}}\right) \tag{11.22}$$

where $\mathbf{G}_i^+$ is the pseudoinverse of of $\mathbf{G}_i$, which is utilized here instead of inverse matrix, i.e., $\mathbf{G}_i^-$ because there is no guarantee of the non-singularity of matrix $\mathbf{G}_i$. Thus, the singular value decomposition is performed with $\mathbf{G}_i$ that

$$\mathbf{G}_i = \mathbf{U}_i \mathbf{\Sigma}_i \mathbf{V}_i^{\text{T}} = \mathbf{U}_i \begin{bmatrix} \mathbf{\Sigma}_{i0} & \mathbf{0} \\ \mathbf{0} & \mathbf{0} \end{bmatrix} \mathbf{V}_i^{\text{T}}$$

where $\mathbf{U}_i$ and $\mathbf{V}_i \in \mathbb{R}^{m\times m}$ are two orthogonal matrices, and $\mathbf{\Sigma}_{i0}$ is a diagonal matrix formed by the singular values of the matrix $\mathbf{G}_i$. We obtain

$$\mathbf{G}_i^+ = \mathbf{V}_i \mathbf{\Sigma}_i^+ \mathbf{U}_i^{\text{T}} = \mathbf{V}_i \begin{bmatrix} \mathbf{\Sigma}_{i0}^{-1} & \mathbf{0} \\ \mathbf{0} & \mathbf{0} \end{bmatrix} \mathbf{U}_i^{\text{T}}$$

where $\mathbf{\Sigma}_i^+$ is the pseudoinverse of $\mathbf{\Sigma}_i$, which is formed by replacing every non-zero diagonal entry by its reciprocal and transposing the resulting matrix ([155]).

Remark 51. *The local formation potential field is written as $U_i^{\text{for}} = U_i^{\text{att}} + U_i^{\text{rep}}, i = 1, ..., N$. Accordingly, the potential force $\mathbf{F}_i^{\text{for}}$ is given by the negative gradient of the joint potential field, $\mathbf{F}_i^{\text{for}} = \mathbf{F}_i^{\text{att}} + \mathbf{F}_i^{\text{rep}}, \quad i = 1, ..., N$ with*

$$\mathbf{F}_i^{\text{att}} = -\nabla U_i^{\text{att}} = -k_{\text{a1}} \sum_{p=1}^{N} \frac{\partial f_{\text{la}}(\mathbf{x}_i, \mathbf{x}_p^d)}{\partial \mathbf{x}_i} \tag{11.23}$$

$$\mathbf{F}_i^{\text{rep}} = -\nabla U_i^{\text{rep}} = -2k_{\text{r1}} \sum_{q=1,q\neq i}^{N} \frac{\partial f_{\text{lr}}(\mathbf{x}_i, \mathbf{x}_q)}{\partial \mathbf{x}_i} \tag{11.24}$$

The control law Eq. (11.22) is designed based on $\mathbf{F}_i^{\text{for}}$ and its motion (to be $\dot{\mathbf{x}}^{\text{r}}$) combined the system dynamics in Eq. (11.4).

Considering the presence of input constraints on $\mathbf{u}_i$, we have

$$\mathbf{S}(\mathbf{u}) = \begin{cases} \mathbf{S}_{\text{max}} \text{sign}_{\text{c}}(\mathbf{u}) & \text{if } |\mathbf{u}| \geq \mathbf{S}_{\text{max}} \quad (11.25) \\ \mathbf{u} & |\mathbf{u}| < \mathbf{S}_{\text{max}} \quad (11.26) \end{cases}$$

where $\mathbf{S}_{\max}$ is the known upper limit of input saturation constraints, and the sign function is defined in (2.28). The auxiliary design system is chosen as

$$\dot{\zeta}_i = \begin{cases} -\dfrac{\left|\Delta\mathbf{u}_i^{\mathrm{T}}\mathbf{F}_i^{\mathrm{for}}\right| + 0.5\Delta\mathbf{u}_i^{\mathrm{T}}\Delta\mathbf{u}_i}{\|\zeta_i\|^2}\zeta_i - \mathbf{K}_\zeta\zeta_i + \Delta\mathbf{u}_i & \text{if } \|\zeta_i\| \geq \mu \\ 0 & \text{if } \|\zeta_i\| < \mu \end{cases}$$

where $\Delta\mathbf{u}_i = \mathbf{G}_i(\mathbf{S}(\mathbf{u}_i) - \mathbf{u}_i)$, $\mathbf{K}_\zeta = \mathbf{K}_\zeta^{\mathrm{T}} > 0$, μ is a small positive value and $\zeta \in \mathbb{R}^n$ is the state of auxiliary design system. Thus, the control law in Eq. (11.22) can be rewritten as

$$\mathbf{u}_i = \mathbf{G}_i^{+}\left(-\mathbf{F}_i + \mathbf{F}_i^{\mathrm{for}} + \dot{\mathbf{x}}^{\mathrm{r}} - \zeta_i\right) \tag{11.27}$$

Theorem 11.1

Consider N mobile agents with similar dynamics Eqs. (11.4–11.5), Assumptions 19–23 and formation control law Eq. (11.27). The formation potential field $U^{\mathrm{for}} = \sum_{i=1}^{N} U_i^{\mathrm{for}}$ converges to zero, which gives rise to the asymptotic stability of MAS. ■

Proof of Theorem 11.1. Consider the Lyapunov candidate as

$$V = \sum_{i=1}^{N}\left[U_i^{\mathrm{for}} + \frac{1}{2}\zeta_i^T\zeta_i\right] \tag{11.28}$$

Substituting Eq. (11.15) into Eq. (11.28), we obtain

$$V = \sum_{i=1}^{N}\left[1 + k_{\mathrm{a1}}\sum_{p=1}^{N} f_{\mathrm{la}}(\mathbf{x}_i, \mathbf{x}_p^{\mathrm{d}}) + k_{\mathrm{r1}}\sum_{q=1, q\neq i}^{N} f_{\mathrm{lr}}(\mathbf{x}_i, \mathbf{x}_q) + \frac{1}{2}\zeta_i^T\zeta_i\right] \geq 0 \tag{11.29}$$

The derivative of V_1 is

$$\dot{V} = \sum_{i=1}^{N}\left[k_{\mathrm{a1}}\sum_{p=1}^{N}\dot{\mathbf{x}}_{ip}^{\mathrm{dT}}\frac{\partial f_{\mathrm{la}}(\mathbf{x}_i, \mathbf{x}_p^{\mathrm{d}})}{\partial\mathbf{x}_i} + 2k_{\mathrm{r1}}\sum_{q=1, q\neq i}^{N}\dot{\mathbf{x}}_{ir}^{\mathrm{T}}\frac{\partial f_{\mathrm{lr}}(\mathbf{x}_i, \mathbf{x}_q)}{\partial\mathbf{x}_i} + \zeta_i^T\dot{\zeta}_i\right] \tag{11.30}$$

With the help of property of formation $\dot{\mathbf{x}}_i^{\mathrm{d}}(t) = \dot{\mathbf{x}}^{\mathrm{r}}(t)$, we obtain

$$\begin{aligned}\dot{V} = \sum_{i=1}^{N}\Bigg[&-\left(\mathbf{F}_i + \mathbf{G}_i\mathbf{u}_i - \dot{\mathbf{x}}^{\mathrm{r}}\right)^{\mathrm{T}}\mathbf{F}_i^{\mathrm{for}} \\ &+\zeta_i^T\left(-\mathbf{K}_\zeta\zeta_i - \frac{\left|\Delta\mathbf{u}_i^{\mathrm{T}}\mathbf{F}_i^{\mathrm{for}}\right| + 0.5\Delta\mathbf{u}_i^{\mathrm{T}}\Delta\mathbf{u}_i}{\|\zeta_i\|^2}\zeta_i + \Delta\mathbf{u}_i\right)\Bigg]\end{aligned} \tag{11.31}$$

Substituting Eq. (11.27) into Eq. (11.31), and then we obtain

$$\dot{V} = \sum_{i=1}^{N} \Big[- \left(\mathbf{F}_i^{\text{for}} - \zeta_i + \Delta \mathbf{u}_i \right)^{\text{T}} \mathbf{F}_i^{\text{for}} - \zeta_i^T \mathbf{K}_\zeta \zeta_i \\ - \left| \Delta \mathbf{u}_i^{\text{T}} \mathbf{F}_i^{\text{for}} \right| - 0.5 \Delta \mathbf{u}_i^{\text{T}} \Delta \mathbf{u}_i + \zeta_i^{\text{T}} \Delta \mathbf{u}_i \Big] \tag{11.32}$$

Since $\zeta_i^{\text{T}} \Delta \mathbf{u}_i \leq \frac{1}{2} \zeta_i^{\text{T}} \zeta_i + \frac{1}{2} \Delta \mathbf{u}_i^{\text{T}} \Delta \mathbf{u}_i$, and $\zeta_i^{\text{T}} \mathbf{F}_i^{\text{for}} \leq \frac{1}{2} \zeta_i^{\text{T}} \zeta_i + \frac{1}{2} \mathbf{F}_i^{\text{forT}} \mathbf{F}_i^{\text{for}}$ we have

$$\dot{V} \leq \sum_{i=1}^{N} \left[-\frac{1}{2} \mathbf{F}_i^{\text{forT}} \mathbf{F}_i^{\text{for}} - \zeta_i^T \left(\mathbf{K}_\zeta - \mathbf{I} \right) \zeta_i \right] \tag{11.33}$$

Therefore, if $\mathbf{K}_\zeta \geq \mathbf{I}$, $\dot{V} < 0$, MAS will be asymptotic stable ([321]), and the desired formation tracking can be achieved. □

In the second step, considering the situation where an agent v_i is beyond the influence range of U_i^{for}, that is $\|\nabla U_i^{\text{for}}\| < \sigma \& \|\widetilde{\mathbf{x}}_i\| > \varepsilon$, we prove the stability of the global formation potential field expressed in Eq. (11.19). First, similar to Remark 51 and construction of the controller proposed in the first step, since $\overline{U}_i^{\text{for}} = U_i^{\text{for}} + U_i^{\text{ext}} = U_i^{\text{att}} + U_i^{\text{rep}} + U_i^{\text{ext}}$, denote

$$\mathbf{F}_i^{\text{ext}} = -\nabla U_i^{\text{ext}} = -k_{a2} \sum_{p=1}^{N} \frac{\partial f_{\text{ga}}(\mathbf{x}_i, \mathbf{x}^{\text{r}})}{\partial \mathbf{x}_i} \tag{11.34}$$

$$\overline{\mathbf{F}}_i^{\text{for}} = \mathbf{F}_i^{\text{att}} + \mathbf{F}_i^{\text{rep}} + \mathbf{F}_i^{\text{ext}} \tag{11.35}$$

and the controller $\mathbf{u}_i$ can be chosen as

$$\mathbf{u}_i = \mathbf{G}_i^{+} \left(-\mathbf{F}_i + \overline{\mathbf{F}}_i^{\text{for}} + \dot{\mathbf{x}}^{\text{r}} - \zeta_i \right) \tag{11.36}$$

where

$$\dot{\zeta}_i = \begin{cases} -\dfrac{\left| \Delta \mathbf{u}_i^{\text{T}} \overline{\mathbf{F}}_i^{\text{for}} \right| + 0.5 \Delta \mathbf{u}_i^{\text{T}} \Delta \mathbf{u}_i}{\|\zeta_i\|^2} \zeta_i - \mathbf{K}_\zeta \zeta_i + \Delta \mathbf{u}_i & \text{if } \|\zeta_i\| \geq \mu \\ 0 & \text{if } \|\zeta_i\| < \mu \end{cases}$$

Theorem 11.2

Consider N mobile agents with similar dynamics Eqs. (11.4–11.5), with Assumptions 19–23 and with formation control law Eq. (11.36). The whole potential function $\overline{U}^{\text{for}} = \sum_{i=1}^{N} \overline{U}_i^{\text{for}}$ converges to zero, which gives rise to the asymptotic stability of MAS. ■

Theorem 2 can be proven following the similar process as the *Proof of Theorem 11.1.*

Remark 52. *Compared to Formation Optimization Algorithm ([151]), formation potential field is a dynamic potential field, which can solve the optimal formation problem as a dynamic problem. It is able to totally consider the potential force generated by other agents and spatial constraints to avoid collisions. Furthermore, after attracting an agent into the local influence of formation potential field, the attracting force in Eq. (11.23) (applied in this chapter) is much larger than the one in Eq. (11.34) (applied in [151]), leading to a significant improvement in formation time cost.*

11.5 COLLISION AVOIDANCE SCHEME

In this section, a effective scheme will be designed to avoid collisions with spatial constraint, such as borders and obstacles. Following the proposal of the formation potential field $\overline{U}_i^{\text{for}}$, a repulsive potential field U_i^{c} generated by spatial constraints is designed to avoid moving beyond constrained space. Let $\Pi = \Pi_1 \cup \Pi_2 ... \cup \Pi_R$, where R is the number of continuous constrained space. L_r is denoted as the edge of region Π_r, $r = 1, ..., R$.

Here Dirac delta function, or simply called a δ function ([107]) is applied to gather the potential force from all points on edges of spatial constraints satisfying the following condition

$$\mathbf{x}_i - \mathbf{s} = kP_r(\mathbf{s}), \quad r = 1, ..., R \tag{11.37}$$

where k is a real number. Here, P_r is the perpendicular direction of the slope of L_r, which is also the direction of potential force, denoted as $P_r(\mathbf{s}) = \begin{bmatrix} \frac{\partial L_r}{\partial s_2} & \frac{\partial L_r}{\partial s_1} \end{bmatrix}^{\text{T}}$, where $\mathbf{s} \in L_r$, and s_1 and s_2 are the first and second dimension of $\mathbf{s}$. Then we obtain

$$U^{\text{c}}(\mathbf{x}) = \sum_{r=1}^{R} \int_{-\infty}^{+\infty} \int_{-\infty}^{+\infty} k_{cr} \frac{\delta(\|\mathbf{x} - \mathbf{s} - kP_r(\mathbf{s})\|)}{\|\mathbf{x} - \mathbf{s}\| - s_0} \mathrm{d}s_1 \mathrm{d}s_2 \tag{11.38}$$

where s_0 is a small constant regarded as safety distance to avoid collisions with constraints edges, k_{cr} is a spatial constraint potential factor, and k is a positive constant. In the following chapter, we denote "$\int_{-\infty}^{+\infty} \int_{-\infty}^{+\infty} \mathrm{d}s_1 \mathrm{d}s_2$" as "$\int \mathrm{d}\mathbf{s}$", and denote $\mathfrak{S}_i = \mathbf{x}_i - \mathbf{s} - kP_r(\mathbf{s})$ for convenience, indicating the condition to pick force points, and denote $\mathfrak{X}_i = \|\mathbf{x}_i - \mathbf{s}\| - s_0$, representing the safe distance from a certain edge of borders or obstacles. Then we have the repulsive potential force

$$\mathbf{F}^{\text{c}}(\mathbf{x}_i) = -\nabla U_i^{\text{c}} = \sum_{r=1}^{R} k_{cr} \int \int \frac{\delta(\|\mathfrak{S}_i\|)(\mathbf{x}_i - \mathbf{s})}{\mathfrak{X}_i^2 \|\mathbf{x}_i - \mathbf{s}\|} \mathrm{d}s_1 \mathrm{d}s_2 \tag{11.39}$$

The above method is introduced to model and express spatial constraints in potential functions. Please refer to our previous work ([151]) for the detailed process of modeling spatial constraints.

To guarantee the formation keeping in the process of collision avoidance, the potential field generated by constrained space is acted on the virtual leader, which is equal to locally modify the given trajectory to avoid collisions with borders and obstacles. Thus the collisions avoidance potential force on the virtual leader is written as

$$\mathbf{F}^{\mathrm{c}}(\mathbf{x}^{\mathrm{r}}) = \sum_{r=1}^{R} k_{cr} \int\int \frac{\delta(\|\mathfrak{S}\|)(\mathbf{x}^{\mathrm{r}} - \mathbf{s})}{\mathfrak{X}^2 \|\mathbf{x}^{\mathrm{r}} - \mathbf{s}\|} \mathrm{d}s_1 \mathrm{d}s_2 \tag{11.40}$$

where $\mathfrak{S}^{\mathrm{r}} = \mathbf{x}^{\mathrm{r}} - \mathbf{s} - kP_r(\mathbf{s})$ and $\mathfrak{X}^{\mathrm{r}} = \|\mathbf{x}^{\mathrm{r}} - \mathbf{s}\| - s_0 - \lambda$, and λ represents the radius of the minimum circle (in 2-D space) or sphere (in 3-D space) that is able to cover whole desired formation. By this way, the desired trajectory will be locally modified during the process of tracking to avoid collisions with unexpected spatial constraints. Denote $\widetilde{\mathbf{x}}^{\mathrm{r}} = \mathbf{x}^{\mathrm{r}*} - \mathbf{x}^{\mathrm{r}}$, indicating the bias between the modified trajectory $\mathbf{x}^{\mathrm{r}*}$ and the real desired trajectory $\mathbf{x}^{\mathrm{r}}$, the adaptive law for the modified trajectory can be written as

$$\dot{\mathbf{x}}^{\mathrm{r}*} = \dot{\mathbf{x}}^{\mathrm{r}} + \mathbf{F}^{\mathrm{c}}(\mathbf{x}^{\mathrm{r}}) - k_{\mathrm{t}}\widetilde{\mathbf{x}}^{\mathrm{r}} \tag{11.41}$$

Then the controllers in Eqs. (11.22,11.36) can be rewritten as

$$\mathbf{u}_i = \begin{cases} \mathbf{G}_i^{+}\left(-\mathbf{F}_i + \overline{\mathbf{F}}_i^{\mathrm{for}} + \dot{\mathbf{x}}^{\mathrm{r}*} - \zeta_i\right) & \text{if } \|\nabla U_i^{\mathrm{for}}\| < \sigma \& \|\widetilde{\mathbf{x}}_i\| > \varepsilon \\ \mathbf{G}_i^{+}\left(-\mathbf{F}_i + \mathbf{F}_i^{\mathrm{for}} + \dot{\mathbf{x}}^{\mathrm{r}*} - \zeta_i\right) & \text{otherwise} \end{cases} \tag{11.42}$$

11.6 SIMULATION

In this section, several examples will be presented to illustrate the effectiveness and the efficiency of the proposed approaches. Consider a group of agents with $N = 5$, and the desired formation is chosen as regular pentagon. For 2-D situation, assume N agents have the following form dynamics

$$\begin{bmatrix} \dot{x}_{i1} \\ \dot{x}_{i2} \end{bmatrix} = \begin{bmatrix} 5 \\ 10 \end{bmatrix} + \begin{bmatrix} 3 & 1 \\ 2 & 5 \end{bmatrix} \mathbf{u}_i \tag{11.43}$$

11.6.1 STATIONARY FORMATION WITH AGENTS' INITIAL POSITIONS BEING INSIDE OR OUTSIDE THE LOCAL INFLUENCE RANGE

In the first simulation, we study the effectiveness of control laws Eqs. (11.22–11.36) for stationary formation control. The agents are initialized to random positions. The stationary target is set to be $\begin{bmatrix} 0 & 0 \end{bmatrix}^{\mathrm{T}}$. Simulation results are shown in Fig. 11.4.

Fig. 11.3(a) shows that agents are able to achieve the given formation from random initial positions while avoiding collisions among agents. The curve formation generation route is to avoid internal collisions as initial distances between some agents are relative short and are mutually on the influence range

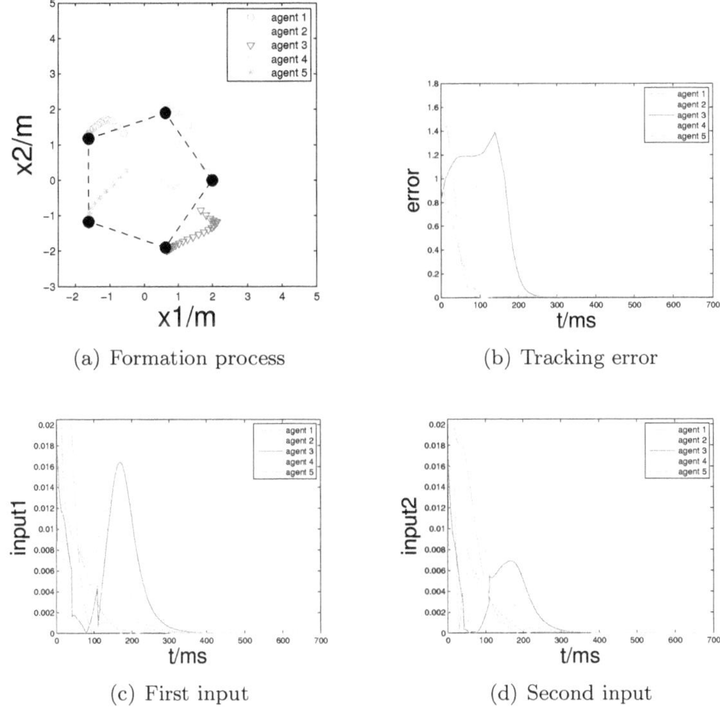

(a) Formation process (b) Tracking error

(c) First input (d) Second input

Figure 11.3 Formation results in case 1.

of local repulsive potential field. Fig. 11.3(b) shows the formation error can eventually converge to nearly zero. Figs. 11.3(c)–11.3(d) show the inputs under the condition of input saturation, where the upper limit of input saturation constraints $\mathbf{u}_{\max} = \begin{bmatrix} 20 & 20 \end{bmatrix}^{\mathrm{T}}$ m/s.

In the second simulation, we study the effectiveness of control law Eq. (11.36) for stationary formation control. The agents are initialized to random positions around a center far away from the stationary target. The center is set to be $\begin{bmatrix} 5 & 5 \end{bmatrix}^{\mathrm{T}}$. Similar simulation results are shown in Fig. 11.4.

11.6.2 OPTIMALITY STUDY

In this part, the optimality of the proposed algorithm is further investigated. There are totally three contrast simulations to be implemented. In the first one, the bijection is set randomly with the controller in [151], while in the second contrast simulation, the bijection is obtained from the *Formation Optimization Algorithm* in [151] with the same controller as in the first one. The third simulation is conducted with the formation potential field proposed in this chapter.

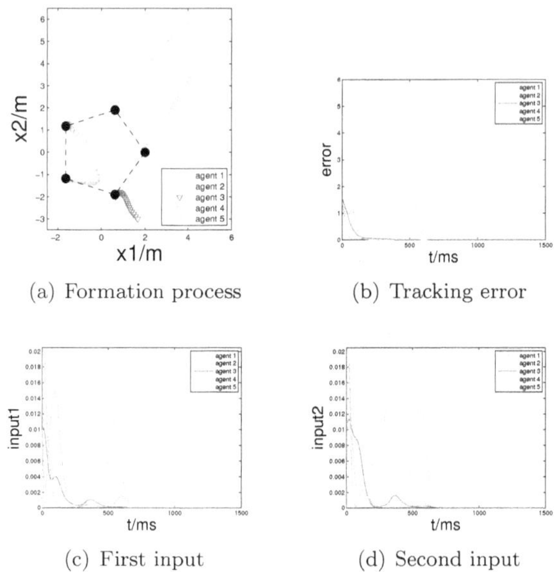

(a) Formation process (b) Tracking error

(c) First input (d) Second input

Figure 11.4 Formation results.

The initial positions all agents are around $\begin{bmatrix}3 & 3\end{bmatrix}^{\mathrm{T}}$ and kept as the same in three contrast simulations. The visual leader is at the position of $\begin{bmatrix}0 & 0\end{bmatrix}^{\mathrm{T}}$, while the desired formation is chosen as regular pentagon. η is set as 0.01 in this study. Simulation results are shown in Fig. 11.5.

Fig. 11.5 shows the performance of three contrast methods by the time cost to achieve desired formation. The optimality of the *Formation Optimization Algorithm* in time cost is not quite conspicuous; by contrast, the proposed *formation potential field* shows a remarkable improvement in time.

11.6.3 TRAJECTORY TRACKING WITH SWITCHING FORMATION SHAPES

In this simulation, the agents are initialized to random positions around the initial position of the desired trajectory, which is set to be $\mathbf{p}_0(t) = \begin{bmatrix}t & 5\sin(1.1t + \mathrm{e}^{0.3t})\end{bmatrix}^{\mathrm{T}}$. Switching signal is sent out at the time of 40s and 70s, with a formation topology set including line, pseudo V and regular pentagon. Simulation results are shown in Fig. 11.6.

We study the effectiveness of control law Eq. (11.22) and Eq. (11.36) for trajectory tracking with switching formation topology. Fig. 11.6(a) shows the tracking result of all agents, where the virtual leader can successfully track the formation center along the desired trajectory while achieving and maintaining the desired formation. Fig. 11.6(b) shows the tracking error of all agents will decrease gradually to zero with time going. A sudden increase of tracking

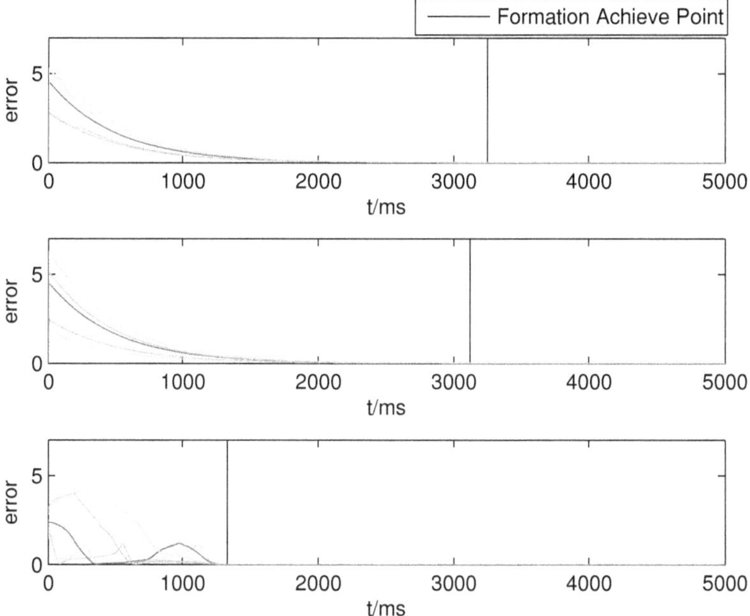

Figure 11.5 Formation errors comparison.

error happens at the moment of receiving the switching signal. Figs. 11.6(c)–11.6(d) illustrate the detailed two processes of formation switching that how MAS generate the desired formation from the previous desired formation after receiving the switching signal.

11.6.4 OBSTACLE AVOIDANCE IN 3-D SPACE

In the last simulation, the formation tracking in 3-D space with multiple spatial constraints is simulated to test the effectiveness of the spatial collisions avoidance strategy in Section 11.5 (the controllers in Eqs. (11.42–11.42)). The initial position of N agents are initialized randomly within a circle range with radius $r = 3$, and the desired trajectory is $\mathbf{p}_0(t) = \left[20\cos(2t) \;\; 7t \;\; 20\sin(2t)\right]^{\mathrm{T}}$. The initial positions of N agents are initialized randomly within a circle range with radius $r = 3m$. N agents are assumed to have the following form of dynamics

$$\begin{bmatrix} \dot{x}_{i1} \\ \dot{x}_{i2} \\ \dot{x}_{i3} \end{bmatrix} = \begin{bmatrix} 5 \\ 10 \\ 7 \end{bmatrix} + \begin{bmatrix} 3 & 1 & 2 \\ 2 & 5 & 4 \\ 1 & 3 & 8 \end{bmatrix} \mathbf{u}_i \tag{11.44}$$

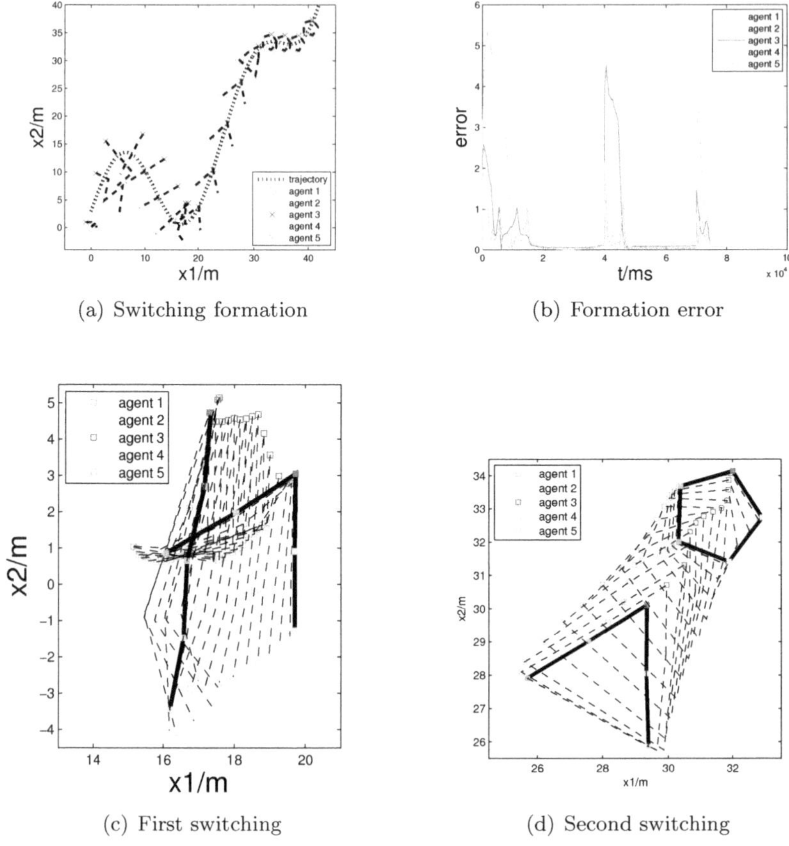

(a) Switching formation

(b) Formation error

(c) First switching

(d) Second switching

Figure 11.6 Trajectory tracking with switching formation.

A border at position $x_1 = -17$ and $x_3 = 17$ is set and a cube obstacle at position $\mathbf{x} = \begin{bmatrix}17.5 & 22.5 & 7.5\end{bmatrix}^{\mathrm{T}}$ with $5.0m$ length of edge and a sphere obstacle $\mathbf{x} = \begin{bmatrix}12 & 28 & 15\end{bmatrix}^{\mathrm{T}}$ with $4.0m$ radius. Simulation results are shown in Figs. 11.7–11.10.

Fig. 11.7 shows that all agents can successfully track the desired trajectory when there are unexpected spatial constraints, including borders and obstacles. The formation error (in Fig. 11.8) increases when an agent gets close to borders or obstacles, and decreases when being far away from borders or obstacles. Fig. 11.9 illustrates the modified trajectory of virtual leader, while Fig. 11.10 describes the whole process of how an agent avoids collisions with spatial constraints.

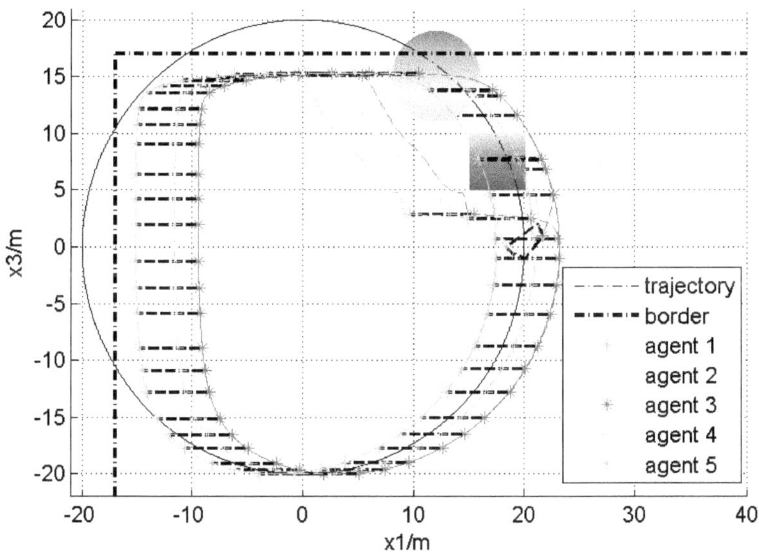

Figure 11.7 Formation tracking with spatial constraints in 3-D space.

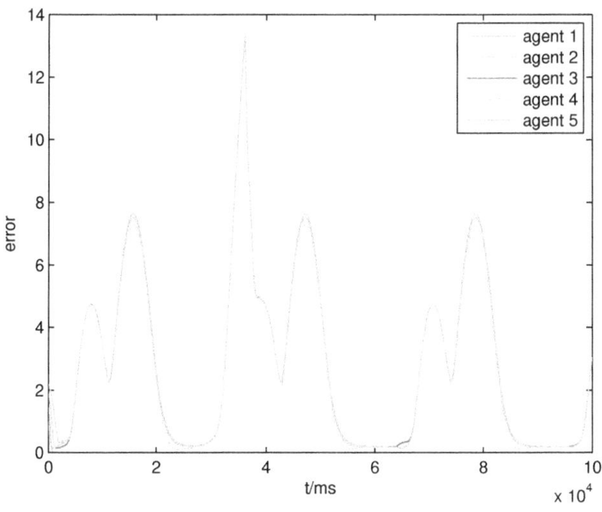

Figure 11.8 Tracking errors with spatial constraints in 3-D space.

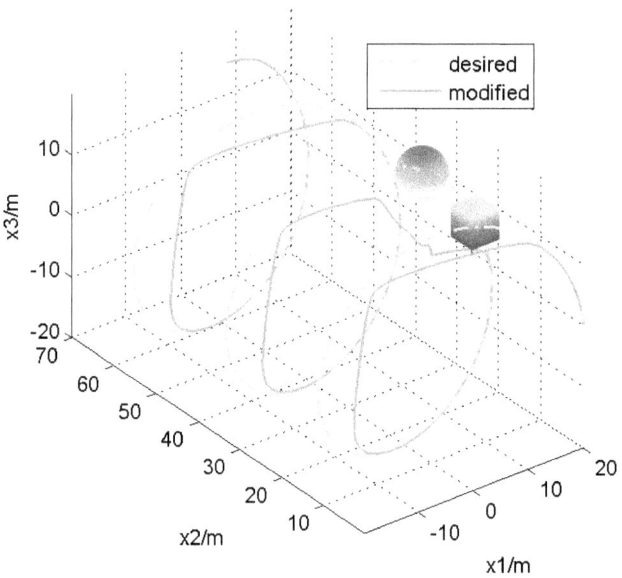

Figure 11.9 Trajectory of virtual leader.

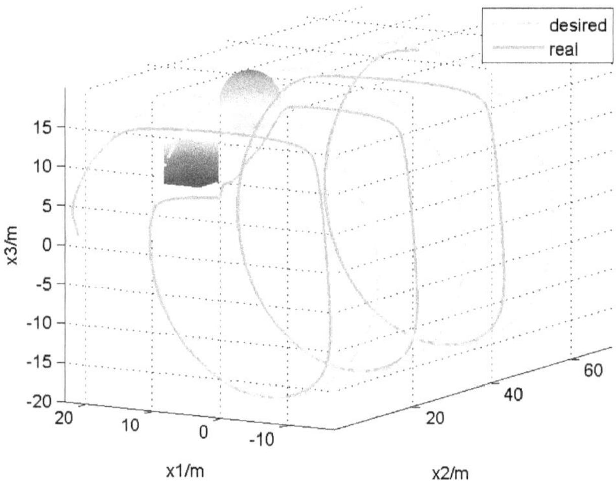

Figure 11.10 Whole process of how an agent avoids collision with spatial constraints in 3-D space.

11.7 CONCLUSION

In this chapter, the formation potential field approach has been proposed to control a group of agents to automatically achieve and maintain a given formation while avoiding internal collisions, which combines local attractive and repulsive potential fields. Furthermore, a global potential field has been added outside the influence range of local formation potential field to relax the requirements of agents' initial positions and improve robustness. Following that, two controllers with the consideration of input saturation have been proposed to achieve a stable dynamic formation and the stability of system has been proved by Lyapunov function. A collision avoidance strategy based on artificial potential field and Dirac delta function has been applied to locally modify the original trajectory of the virtual leader such that agents can avoid collisions with unexpected spatial constraints while maintaining the given formation. Simulation results have demonstrated the effectiveness and optimality of proposed approaches.

12 Formation Tracking Control of Multi-Agents in Constrained Space

In this chapter, formation tracking control based on potential field is studied. The objective is to control a group of agents to track a desired trajectory while maintaining a given formation in non-omniscient constrained space. Aiming at this purpose, the average of all agents' positions is viewed as the virtual leader, and the task is translated into that controlling the virtual leader to track the center of the desired formation. The situation where agents may not have global environmental information will lead to that they may have to avoid collisions with unexpected spatial constraints. By introducing the concepts of line source and surface source, two main types of spatial constraints: (i) borders, and (ii) obstacles are defined mathematically and unified into a common expression in an artificial potential function. In this way, the tasks of formation tracking and obstacle avoidance in a bounded environment can be solved by the control law proposed in this chapter. Furthermore, concerning about the situation of multiple spatial constraints, a Dirac delta function is introduced. Meanwhile, a formation optimal algorithm is proposed to minimize the formation generation time cost. Following that, a controller is designed, and the conditions for asymptotic stability of multi-agent systems (MAS) are proved based on the Lyapunov function. The maximum duration that the conditions cannot be satisfied is elaborated to indicate whether the tracking task should be given up. Meanwhile, the impulsive influence from spatial constraints is analyzed. Finally, simulation results are presented to illustrate the performance of proposed approaches.

12.1 INTRODUCTION

Over the past few decades, research on multi-agent systems (MAS) has spurred a broad interest in various disciplines, such as mathematics, physics, biology, sociology and engineering science [397, 541]. There are some interesting topics in MAS, including flocking, formation and consensus, distributed problem-solving, learning control for multi-agents [88, 442] and so on. Especially, formation control has been applied to various applications in the past few years, such as mobile robots [153, 356], vehicles [479, 516], satellites [345], aircrafts [15, 161], vessels [33, 115] and spacecrafts [25]. For formation control strategies to be successful applied in practical applications, many issues must be considered, including the transmission and management of shared

DOI: 10.1201/9781003298618-12

information among MAS [131], rapid collision avoidance [2, 427] and multiple constraints including both the system dynamics constraints and complex environmental constraints [333, 394]. Among these issues, the control tasks in constrained space and the collision avoidance have attracted increasing attention from different disciplines in recent research.

In [333], Mastellone et al. studied the formation control and trajectory tracking problem for a group of robots with holonomic constraints between subsystems. However, they did not consider environmental constraints, which cannot be ignored in realistic applications. In [2], Abdollahi and Rezaee proposed a technique based on the behavioral structure and designed an approach to avoid the obstacle by applying the rotational potential field. However, the method proposed had to build geometric model of obstacles, and assumed obstacle as rectangles or ellipses. Therefore, the information about obstacles had to been known in advance and the method cannot solve the problem of spatial borders. Furthermore, there was no stability analysis of the system. In [317], Lu et al. studied the control of a group of mobile agents to form a desired formation while flocking in a constrained environment. However, there was no tracking requirement in their study, and the environment was assumed to be bounded without holes. That means obstacle avoidance was not considered. To obtain the satisfying performance for the realistic control of MAS, the unknown environment modeling is extreme significant but it is still an open problem. A few studies have been done on this topics [277, 521] but are not nearly enough and there are many limitations especially on the cooperative control of MAS.

Inspired by the unexpected multiple spatial constraints in realistic task space, the problem of formation tracking control in non-omniscient constrained space is considered in this study. It has the significance meaning, because in the task of trajectory tracking, the situation where agents may not have the global environmental knowledge will lead to that they may have to avoid collisions with unexpected borders or obstacles, although the desired trajectory has been given. In addition, in spite of above existing research, there are some problems when considering spatial constraints. First, the previous studies usually considered the control in bounded space and rapidly obstacle avoidance as different problems, which renders two control schemes need to be proposed for different situations. Second, most of existing research requires the geometric information of obstacles. This limits the applicability of many approaches as the information about obstacles may not be measured in advance.

The main contribution of this work is the proposal of a novel modeling approach for spatial constraints, including borders and obstacles by introducing the concepts of line source and surface source. Dirac delta function is applied to deal with the situation of multiple complex spatial constraints and also provided a mathematical tool for analyzing the influence from suddenly detected spatial constraints to system stability. By "rapidly" we have two meanings. First, the existence of an obstacle or a border does not have to be known in advance. Second, the global geometric information of an obstacle or a border does not have to be known in advance. The control scheme can be implemented under the condition that only the local knowledge about the unexpectedly detected spatial constraints is captured. Furthermore, the conditions for asymptotic stability of MAS are studied based on the Lyapunov function, while the maximum duration that the conditions cannot be satisfied is obtained to indicate whether the formation tracking task should be given up and impulsive influence from spatial constraints is analyzed.

12.2 SYSTEM DYNAMICS AND FORMATION MANEUVERS

Before proceeding further, the following assumptions are made in this chapter.

Assumption 24. *The influence of the size and the shape of an agent to the formation tracking control are ignored, which means an agent is assumed to be of point mass.*

Assumption 25. *An agent is able to estimate its position in the world coordinate system.*

Assumption 26. *Each agent is able to broadcast information to others.*

Assumption 27. *The environment (spatial constraints) is uniform (time-invariant) in the whole task.*

Consider a network of N agents, $\mathbf{x}_i(t) \in \mathbb{R}^m$ is the position of agent i at time t. In this chapter, we consider $\mathbf{x}_i(t)$ to represent the position of an agent i in 2-D space, which means $m = 2$. We recall the following definitions in line with Definitions 34 and 35 in Chap. 11.

Definition 40 (*Formation*). *A formation pattern at time t is defined to be a set $\mathfrak{P}(t) = \{\mathbf{x}_1^{\mathrm{d}}(t), ..., \mathbf{x}_N^{\mathrm{d}}(t)\}$, where $\mathbf{x}_i^{\mathrm{d}}(t)$ is the desired position of the agent i at time t, $i = 1, ..., N$.*

In this chapter, we will take the center position of agents as a virtual leader, and write this position as $\mathbf{x}_0$; take the center of the desired formation (the trajectory) as the reference, and write this position as $\mathbf{x}^{\mathrm{r}}$. We have

$$\mathbf{x}_0 = \frac{1}{N}\sum_{i=1}^{N}\mathbf{x}_i \quad \& \quad \mathbf{x}^{\mathrm{r}} = \frac{1}{N}\sum_{i=1}^{N}\mathbf{x}_i^{\mathrm{d}} \tag{12.1}$$

Assumption 28. *There exists a constant ζ such that $0 \leq \|\dot{\mathbf{x}}^{\mathrm{r}}\| \leq \zeta < \infty$.*

Assumption 28 implies the designed trajectory is smooth.

Definition 41 (*Fixed Formation*). *A formation pattern is defined as Fixed Formation if* $\Delta \mathbf{x}_i(t) = \mathbf{x}_i^{\mathrm{d}}(t) - \mathbf{x}^{\mathrm{r}}(t), \quad i = 1,..,N$ *is time-invariant for* $t \in [0, \infty)$*, which indicates*

$$\begin{aligned} \Delta \dot{\mathbf{x}}_i(t) &= \dot{\mathbf{x}}_i^{\mathrm{d}}(t) - \dot{\mathbf{x}}^{\mathrm{r}}(t) = 0 \\ \Rightarrow \dot{\mathbf{x}}_i^{\mathrm{d}}(t) &= \dot{\mathbf{x}}^{\mathrm{r}}(t),\ i = 1,..,N \end{aligned} \tag{12.2}$$

We assume N agents with the similar dynamics described by the following nonlinear differential equation as [276, 358] did:

$$\dot{\mathbf{x}}_i = \mathbf{F}_i(\mathbf{x}_i) + \mathbf{G}_i(\mathbf{x}_i)\mathbf{u}_i \tag{12.3}$$

$$\mathbf{y}_i = \widetilde{\mathbf{x}}_i \tag{12.4}$$

where $i = 1, ..., N$, and $\mathbf{u}_i(t), \mathbf{y}_i \in \mathbb{R}^m$ represent the control vector and output of agent i; $\mathbf{F}_i(\mathbf{x}_i) \in \mathbb{R}^m$ is a smooth vector; $\mathbf{G}_i(\mathbf{x}_i) \in \mathbb{R}^{m \times m}$ is a smoothly function matrix; $\widetilde{\mathbf{x}}_i = \mathbf{x}_i - \mathbf{x}_i^{\mathrm{d}}$, representing tracking error of agent i. The output can be interpreted as the measurement of tracking error.

In this chapter, the characteristics of the nonlinear system are not the focus. Regarding the linearity of the agents, the following assumption is made.

Assumption 29. *All agents are linear time-invariant (LTI) systems.*

Thus, $\mathbf{F}_i(\mathbf{x}_i)$ and $\mathbf{G}_i(\mathbf{x}_i)$ are considered as constant matrix and written as $\mathbf{F}_i$ and $\mathbf{G}_i$ in the following chapter.

After deciding the desired position of the formation center $\mathbf{x}^r$, the desired positions of all agents $\mathbf{x}_i^d(t), i = 1, ..., N$ can be calculated relative to the formation center.

For example, assume the position of formation center is $\begin{bmatrix} x_{01}^{\mathrm{d}} & x_{02}^{\mathrm{d}} \end{bmatrix}^T$, for the first example in Fig. 12.1, the desired position of agent i is $\begin{bmatrix} x_{01}^{\mathrm{d}} + R\mathrm{cos}\alpha_i & x_{02}^{\mathrm{d}} + R\mathrm{sin}\alpha_i \end{bmatrix}^T, i = 1, ..., 5$; for the second example, the four agents' positions are $\begin{bmatrix} x_{01}^{\mathrm{d}} & x_{02}^{\mathrm{d}} + d_1 \end{bmatrix}^T$, $\begin{bmatrix} x_{01}^{\mathrm{d}} & x_{02}^{\mathrm{d}} - d_2 \end{bmatrix}^T$, $\begin{bmatrix} x_{01}^{\mathrm{d}} & x_{02}^{\mathrm{d}} + d_3 \end{bmatrix}^T$ and $\begin{bmatrix} x_{01}^{\mathrm{d}} & x_{02}^{\mathrm{d}} - d_4 \end{bmatrix}^T$, respectively.

Remark 53. *The virtual leader can be chosen as center of mass, center of gravity, center of pressure of the system according to the specified application, including properties of MAS and environmental conditions. It is not the topic of this chapter, we do not detail the method of deciding the position of virtual leader.*

Remark 54. *It is also feasible to take any one of agents as the leader. In this way, the desired positions of other agents* $\mathbf{x}_i^{\mathrm{d}}(t), i = 1, ..., N$ *can be calculated relative to the position of this leader agent.*

Control objective: Design a controller to control a group of agents initialized on random positions to track the desired trajectory in formation in non-omniscient constrained space. In other words, the center of agents and center

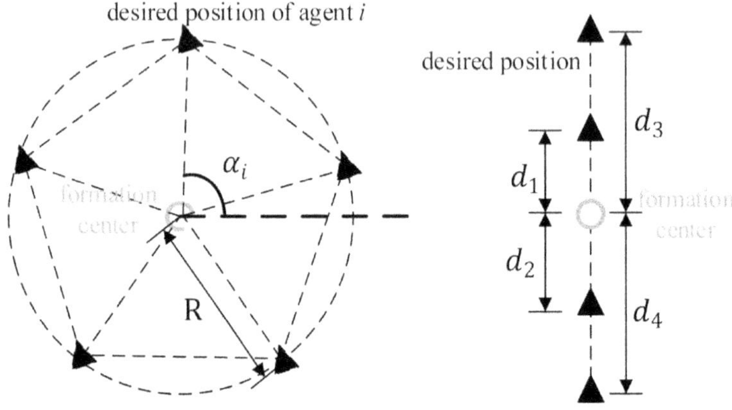

Figure 12.1 Two examples of formation.

of desired formation should coincide in the final. In the meantime, consider the unknown forbidden space as Π, which means that all agents' positions satisfy the spatial constraint condition that

$$\mathbf{x}_i(t) \notin \Pi \qquad i = 1, ..., N, 0 \leq t < \infty \tag{12.5}$$

during the whole task. The control goal can be expressed mathematically as, for every $t \geq 0$,

$$\sum_{i=1}^{N} \|\widetilde{\mathbf{x}}_i(t)\| \leq \varepsilon \quad \& \quad \|\mathbf{x}_0(t) - \mathbf{x}^{\mathrm{r}}(t)\| \leq \epsilon \tag{12.6}$$

where ε and ϵ are positive constants.

Case 1: when $\Pi = \emptyset$, ε and ϵ can be made sufficiently small when $t \to \infty$, that

$$\lim_{t\to\infty} \sum_{i=1}^{N} \|\widetilde{\mathbf{x}}_i(t)\| \to 0 \quad \& \quad \lim_{t\to\infty} \|\mathbf{x}_0(t) - \mathbf{x}^{\mathrm{r}}(t)\| \to 0 \tag{12.7}$$

Case 2: when $\Pi \neq \emptyset$,

$$\varepsilon \quad \propto \quad \frac{1}{\mathrm{d}(\mathbf{x}_i, \Pi)}, \qquad \text{and} \quad \varepsilon \leq \varepsilon_{max} \tag{12.8}$$

$$\epsilon \quad \propto \quad \frac{1}{\mathrm{d}(\mathbf{x}_i, \Pi)}, \qquad \text{and} \quad \epsilon \leq \epsilon_{max} \tag{12.9}$$

where $\mathrm{d}(\mathbf{x}_i, \Pi) = \inf_{\mathbf{s}\in\Pi} \rho(\mathbf{x}_i, \mathbf{s})$, and $\mathbf{s} \in \Pi$. $\rho(\cdot,\cdot)$ is defined as Euclidean distance in this chapter, and ε_{max}, ϵ_{max} indicate tolerable upper limits for the formation error.

12.3 ARTIFICIAL POTENTIAL FUNCTION

12.3.1 JOINT POTENTIAL FUNCTION

A moving formation is designed in this chapter since we want the virtual leader and all agents to follow a desired path. Notice that the input $\mathbf{u}_i(t)$ of the agent system can be viewed as the implementation of force created by a special force field. A method of constructing the potential of the force field is developed. This artificially created force field ensures motion of the system in accordance with the desired control goal.

We design a control law for a group of agents to track the given trajectory while keeping the desired formation. In the same time, there are spatial constraints which need to be avoided. The artificial potential field for agent i can be written as $U_i^{\mathrm{art}}(\mathbf{x}) = U_i^{\mathrm{t}}(\mathbf{x}) + U_i^{\mathrm{a}}(\mathbf{x}) + U_i^{\mathrm{c}}(\mathbf{x}), \quad i = 1,,,N$ where $U_i^{\mathrm{t}}(\mathbf{x})$ is designed to track the desired trajectory, $U_i^{\mathrm{a}}(\mathbf{x})$ is designed to avoid collisions with other agents, and $U_i^{\mathrm{c}}(\mathbf{x})$ is designed to avoid moving beyond constrained space. The joint effect of $U_i^{\mathrm{t}}(\mathbf{x})$, $U_i^{\mathrm{a}}(\mathbf{x})$ and $U_i^{\mathrm{c}}(\mathbf{x})$ is $U_i^{\mathrm{art}}(\mathbf{x})$, which is designed for the agent i to track the given trajectory in constrained space, while avoiding collisions with other agents.

Definition 42 (*Point Potential Source*)**.** *For a single potential field, there is a point $\mathbf{x}^{\mathrm{s}}$ to be to the local extreme point of the potential field. This point $\mathbf{x}^{\mathrm{s}}$ is defined as point potential source. A point which is the local minimal point of the potential field is defined as attractive point potential source $\mathbf{x}^{\mathrm{att}}$; a point which is the local maximal point of the potential field is defined as repulsive point potential source $\mathbf{x}^{\mathrm{rep}}$.*

First, as $U_i^{\mathrm{t}}(\mathbf{x})$ is designed to track the virtual leader along the given trajectory, $\mathbf{x}_i^{\mathrm{d}}$ can be viewed as the attractive point potential source for $U_i^{\mathrm{t}}(\mathbf{x})$, which is defined as

$$U_i^{\mathrm{t}} = \frac{1}{2}k_{\mathrm{t}}(\mathbf{x}_i - \mathbf{x}_i^{\mathrm{d}})^{\mathrm{T}}(\mathbf{x}_i - \mathbf{x}_i^{\mathrm{d}}) = \frac{1}{2}k_{\mathrm{t}}\widetilde{\mathbf{x}}_i^{\mathrm{T}}\widetilde{\mathbf{x}}_i \tag{12.10}$$

where k_{t} is a positive constant.

Second, as $U_i^{\mathrm{a}}(\mathbf{x})$ is designed for agent i to avoid collisions with other agents, $\mathbf{x}_q(q \neq i)$ can be viewed as repulsive point potential source for $U_i^{\mathrm{a}}(\mathbf{x})$, which is defined as

$$U_i^{\mathrm{a}} = \frac{1}{2}k_{\mathrm{a}} \sum_{q=1,q\neq i}^{N} \frac{1}{\mathbf{x}_{iq}^{\mathrm{T}}\mathbf{x}_{iq}} \tag{12.11}$$

where $\mathbf{x}_{iq} = \mathbf{x}_i - \mathbf{x}_q$, and k_{a} is a positive constant.

Third, as it is unreasonable to assume spatial constraints to be of point mass, we will model them in detail in the next subsection.

12.3.2 MODEL SPATIAL CONSTRAINTS

Let $\Pi = \Pi_1 \cup \Pi_2 ... \cup \Pi_R$, where R is the number of continuous constrained space, as showed in Fig. 12.2(a), and let L_r be the edge of region Π_r, $r = 1, ..., R$. The spatial constraints are mainly divided into two types—obstacles and borders.

Definition 43 (Border). *When $L_r(\mathbf{s})$ is an open curve, or $\int \Pi_r(\mathbf{s})\mathrm{ds} \to \infty$, $L_r(\mathbf{s})$ is called border (e.g., Π_1 and Π_2 in Fig. 12.2(a)).*

Definition 44 (Obstacle). *When $L_r(\mathbf{x})$ is a closed curve, and there exist a constant β that $\int \Pi_r(\mathbf{x})\mathrm{d}\mathbf{x} < \beta$, $\Pi_r(\mathbf{x})$ is called obstacle (e.g., Π_3 and Π_4 in Fig. 12.2(a)).*

The concepts of line source (in 2-D space) and surface source (in 3-D space) are introduced to model these spatial constraints whose size and shape cannot be ignored.

Definition 45 (*Line/Surface Potential Source*). *For a single potential field, in which the set of the local extreme points is a line/surface, this line/surface is defined as line/surface potential source.*

Assume that the edge of a spatial constraint has the like charges with the agent (viewed as a point source), making the edge form a line potential source or a surface potential source. When the agent gets close to the edge of spatial constraints, there is a repulsive potential force generated by the edge of spatial constraints. Similar to Coulomb's law [96], we build the following potential function, when just considering a single spatial constraint in 2-D space

$$U_r(\mathbf{x}) = \int_{L_r} \frac{k_\mathrm{c}}{d^2}\mathrm{dl}, \qquad r = 1, ..., R \tag{12.12}$$

where k_c is a positive constant, and d is distance from the agent's position $\mathbf{x}$ to dl.

To simplify the model of spatial constraints, firstly, we simplify an edge into a straight line, as shown in Fig. 12.2(b), then Eq. (12.12) can be rewritten as

$$\begin{aligned} U_r(\mathbf{x}) &= \int_0^{L_{r1}} \frac{k_\mathrm{c}}{d^2+l^2}\mathrm{dl} + \int_0^{L_{r2}} \frac{k_\mathrm{c}}{d^2+l^2}\mathrm{dl} \\ &= \frac{k_c}{d}(\alpha_{r1} + \alpha_{r2}) \end{aligned} \tag{12.13}$$

where the meaning of d, L_{r1}, L_{r2}, α_{r1} and α_{r2} are marked in Fig. 12.2(c). Detailedly, d denotes the perpendicular distance from the agent to the border of the spatial constraint, L_{r1} and L_{r2} denote the length of first and second half the border, α_{r1} and α_{r2} denote included angles between the perpendicular direction and directions from the agent to two vertexes. We can see how the

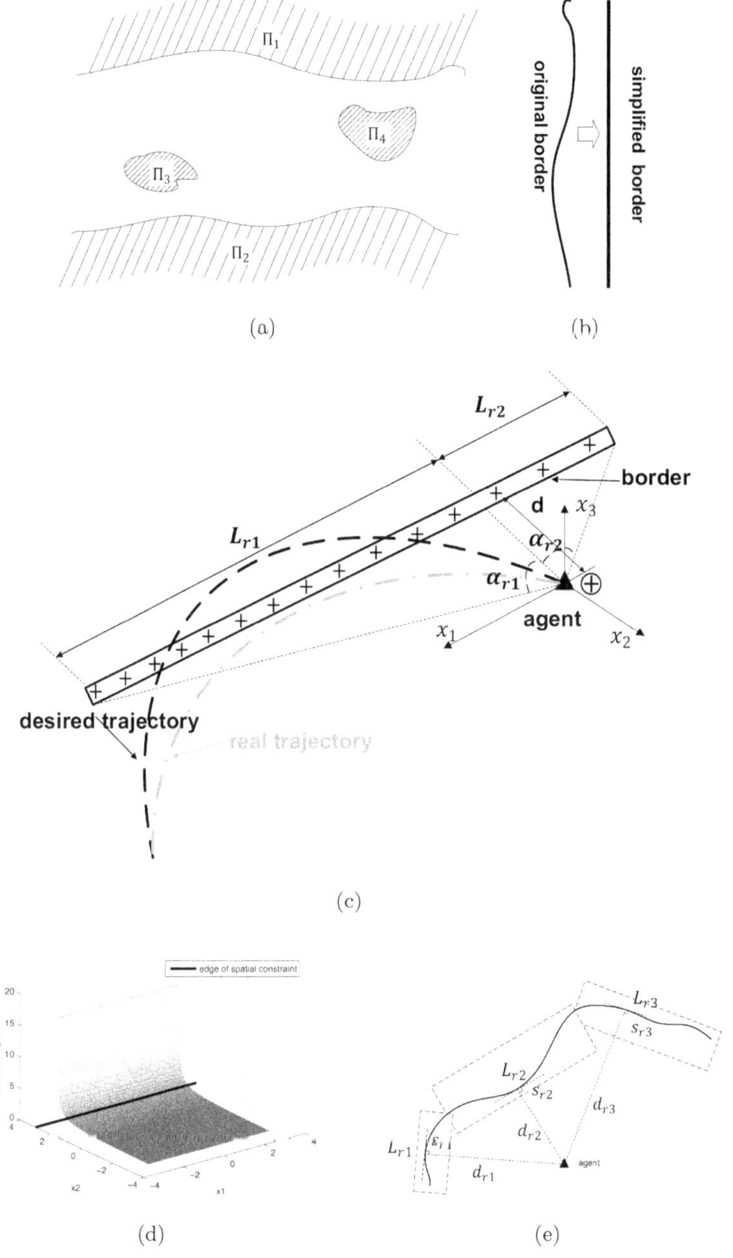

Figure 12.2 (a) Multiple spatial constraints; (b) simplify the edge into a straight line; (c) potential field calculation of a straight line; (d) potential filed of a border; (e) segmentation of irregular edges of spatial constraints.

original trajectory is influenced by the spacial constraints' potential fields. Mark $\omega_r^{\mathrm{c}} = \alpha_{r1} + \alpha_{r2}$ to indicate the relationship between agent and the spatial constraint, called spatial constraint potential factor, and we obtain

$$U_r(\mathbf{x}) = k_c \omega_r^{\mathrm{c}} \frac{1}{d} \tag{12.14}$$

It is worth noting that, ω_r^{c} should be time-varying, especially in the terms of the relationship between an agent and an obstacle; however, in this chapter, we consider it as a constant for a special spatial constraint for the convenience of analysis.

Especially, when $L_r = L_{r1} + L_{r2} \gg d \Rightarrow \omega_r^{\mathrm{c}} \to \pi$, we obtain $U_r(\mathbf{x}) = \frac{k_c \pi}{d}$, where the potential filed is illustrated in Fig. 12.2(d), and this also indicates that $\omega_r^{\mathrm{c}} < \pi$. When $L_r = L_{r1} + L_{r2} \ll d \Rightarrow \omega_r^{\mathrm{c}} \approx \frac{L_r}{d}$, we obtain $U_r(\mathbf{x}) = \frac{k_c L_r}{d^2}$.

Second, concerning irregular edges of spatial constraints as shown in Fig. 12.2(e), we divide the edge into a set of approximate straight lines ($\hat{L}_{rq}, r = 1, .., R, q = 1, ..., Q$, where q is the number of the approximate straight lines). The perpendicular direction of the slope of $\hat{L}_{rq}$ is the direction of potential force, denoted as $P_r(\mathbf{s}) = \begin{bmatrix} \frac{\partial \hat{L}_{rq}}{\partial s_2} & \frac{\partial \hat{L}_{rq}}{\partial s_1} \end{bmatrix}^T$, where $\mathbf{s} \in \hat{L}_{rq}$. The combined potential field is

$$\begin{aligned} U(\mathbf{x}) &= \sum_{r=1}^{R} \sum_{q=1}^{Q} U_{rq} \\ &= \sum_{r=1}^{R} \sum_{q=1}^{Q} k_c \omega_{rq}^{\mathrm{c}} \frac{1}{d_{rq}} = \sum_{r=1}^{R} \sum_{q=1}^{Q} k_c \omega_{rq}^{\mathrm{c}} \frac{1}{\|\mathbf{x} - \mathbf{s}_{rq}^*\|} \end{aligned} \tag{12.15}$$

where d_{rq} is distance between $\mathbf{x}$ and $\hat{L}_{rq}$, ω_{rq}^{c} is the spatial constraint potential factor related to an agent and q_{th} approximate straight line of r_{th} spatial constraints, $\mathbf{s}_{rq}^* \in \hat{L}_{rq}$ and should satisfy the following condition

$$\mathbf{x}_i - \mathbf{s}_{rq}^* = k P_{rq}(\mathbf{s}_{rq}^*), \; r = 1, ..., R, \; q = 1, ..., Q \tag{12.16}$$

where k is a real number, P_{rq} is the perpendicular direction of slope of $\hat{L}_{rq}$. This condition is used to guarantee the potential direction is perpendicular to the slope direction of $\hat{L}_{rq}(\mathbf{x})$.

Here we introduce Dirac delta function, or simply called a δ function [107] to gather the potential force from all points on edges of spatial constraints satisfying the following condition

$$\mathbf{x}_i - \mathbf{s} = k P_r(\mathbf{s}), \quad r = 1, ..., R \tag{12.17}$$

then we obtain

$$U(\mathbf{x}) = \sum_{r=1}^{R} \int_{-\infty}^{+\infty} \int_{-\infty}^{+\infty} k_c \omega_r^{\mathrm{c}} \frac{\delta(\|\mathbf{x} - \mathbf{s} - k P_r(\mathbf{s})\|)}{\|\mathbf{x} - \mathbf{s}\| - s_0} \mathrm{d}s_1 \mathrm{d}s_2 \tag{12.18}$$

where s_0 is a small constant regarded as safety distance to avoid collisions with constraints edges, and s_1 and s_2 are the first and second dimension of integral term $\mathbf{s}$. In the following chapter, we denote "$\int_{-\infty}^{+\infty}$" as "$\int$", and denote $\mathfrak{S}_i = \mathbf{x}_i - \mathbf{s} - kP_r(\mathbf{s})$ for convenience, indicating the condition to pick force points, and denote $\mathfrak{X}_i = \|\mathbf{x}_i - \mathbf{s}\| - s_0$, representing the safe distance from a certain edge of borders or obstacles.

To conclude, U_i^c can be defined as

$$U_i^c = \sum_{r=1}^{R} \int\int k_{cr} \frac{\delta(\|\mathfrak{S}_i\|)}{\mathfrak{X}_i} \mathrm{d}s_1 \mathrm{d}s_2 \tag{12.19}$$

where $k_{cr} = k_c \omega_r^c$.

Remark 55. *The influence of the size and the shape of an agent to the formation tracking control is ignored in this chapter as mentioned in Assumption 1, but it only means we do not focus on modeling the shape of an agent. However, this is unrealistic in real applications, which may also limit the application of the proposed solution. A s_0 is set in spatial collision avoidance potential function (in Eq. (12.18)) for two reasons: first, to decrease the influence of the size and the shape of an agent; second, to decrease the influence of environmental uncertainties and improve the robustness of control approach. Therefore, $s_0 = \mathfrak{R} + \Delta$, with $\mathfrak{R}$ the maximum radius of agent and Δ a safety margin.*

Remark 56. *The proposed approach to model the spatial constraints in task space can work successfully in both 2-D and 3-D situations. The line potential source in Definition 6 is used to model spatial constraints in 2-D space as described above, while the surface potential source will be used to model spatial constraints in 3-D space. As these two are parallel problem, we do not study the 3-D problem in details in this chapter to avoid unnecessarily increase the chapter's length.*

Remark 57. *The approach to model the spatial constraints in task space proposed in this chapter is applicable to both the control of individual agent and MAS.*

12.4 OPTIMAL FORMATION

In this section, a stationary formation problem or called formation generation problem is considered, which means $\dot{\mathbf{x}}_i^d(t) = \mathbf{0}, i = 1, ..., N$. In this case, the *Formation* in Eq. (11.1) can be represented as $\mathfrak{P}(0)$ or $\mathfrak{P}_0$. Given a specific shape of desired formation $\mathfrak{Q} = \{\mathbf{q}_1, ..., \mathbf{q}_N\}$. In the previous sections, we all assume the bijection from $\mathfrak{P}_0$ to $\mathfrak{Q}$ has been given, although it is not, which can be regarded as an optimal formation problem. The purpose of optimal formation problem is to find a bijection $\sigma : \mathfrak{P}_0 \rightarrow \mathfrak{Q}$ which minimizes the formation generation time cost. First, the single formation generation time cost t_i^* for the agent i can be expressed as $t_i^* = \min_{t_i}\{\|\widetilde{\mathbf{x}}_i(t_i)\| < \eta\}$, where η is a given constant indicating the maximum tolerable formation error for a single agent. The corresponding objective function $J(\sigma)$ can be described by $J(\sigma) = \max\{t_i^*, i = 1, ..., N\}$, representing the final formation generation time cost. Larger distance between an agent's initial position and its corresponding desired formation position will lead to more time cost to achieve desired formation. As described in the objective function, the maximum distance between all agents' initial positions and their corresponding desired formation positions should be minimized. The proposed formation optimal algorithm is summarized in Algorithm 1.

Algorithm 1 Formation Optimization Algorithm

Input: $\mathfrak{P}_0$, $\mathfrak{Q}$, $\mathbf{x}_i(0), i = 1, ..., N$
Output: $\sigma : \mathfrak{P}_0 \to \mathfrak{Q}$

Calculate distance matrix $\mathbf{D}$, where $\mathbf{D}(i,j) = \|\mathbf{q}_i - \mathbf{x}_j(0)\|$
Obtain ordered distance matrix $\mathbf{D}_2$ and order index matrix $\mathbf{I}$ that $\forall j$, $\mathbf{D}_2(i,j) \le \mathbf{D}_2(i^{'},j)$ and $\mathbf{D}(\mathbf{I}(i,j),j) \le \mathbf{D}(\mathbf{I}(i^{'},j),j)$, if and only if $i \le i^{'}$
$\boldsymbol{\Omega} \leftarrow \varnothing$
for $m \leftarrow N, 1$ **do**
  for $n \leftarrow 1, N$ **do**
    if $n \notin \Omega$ **then**
      obtain set $\mathbf{I}_n = \{j|\mathbf{I}(m,j) = n\}$
      count $\leftarrow$ the size of set $\mathbf{I}_n$
      if *count* > 1 **then**
        $j^* \leftarrow \arg\min_j\{\mathbf{D}_2(m,j)|j \in \mathbf{I}_n\}$
        add mapping $\sigma : \mathbf{x}_{j^*}^d \to \mathbf{q}_n$
        $\boldsymbol{\Omega} \leftarrow \boldsymbol{\Omega} \cup \{n\}$
        for $p \leftarrow 1, N$ **do**
          $\mathbf{D}(n,p) \leftarrow -1$
          $\mathbf{D}(p,j^*) \leftarrow -1$
        end for
        Reobtain ordered distance matrix $\mathbf{D}_2$ and order index matrix $\mathbf{I}$ that $\forall j$, $\mathbf{D}_2(i,j) \le \mathbf{D}_2(i^{'},j)$ and $\mathbf{D}(\mathbf{I}(i,j),j) \le \mathbf{D}(\mathbf{I}(i^{'},j),j)$, if and only if $i \le i^{'}$
      end if
    end if
  end for
end for
for $i \notin \boldsymbol{\Omega}$ **do**
  $j^* \leftarrow \arg\min_j\{\mathbf{D}(j,i)|\mathbf{D}(j,i) \ne -1\}$
  Add mapping $\sigma : \mathbf{x}_{j^*}^d \to \mathbf{q}_i$
end for

Nevertheless, there exists a drawback in the proposed algorithm. The potential force generated by other agents to avoid collisions among agents in the process of achieving formation is ignored, but this solution also considers that the potential force applied to avoid collisions among agents is a mutual force between any two agents. If it is totaly considered, the optimal problem will be too complex, as the collision avoidance among agents is a dynamic problem. Besides, the spatial constraints are also ignored, which indicates that the relative positions are only factors influencing the optimal process.

Remark 58. *Considering the trade-off between the average formation generation time cost and the final formation generation time cost, the objective function may also be described as* $J(\sigma) = \frac{1}{N}\sum_{i=1}^{N} t_i^* + \lambda \max\{t_i^*, i = 1, ..., N\}$, *where* λ *is the weighing coefficient. It will not be discussed in this chapter as the formation optimal problem is not the focus of this chapter.*

12.5 FORMATION CONTROL AND STABILITY ANALYSIS

In this section, we will design a control scheme $\mathbf{u}_i$ based on the proposed potential functions for each agent i with dynamics Eqs. (12.3–12.4) to ensure all agents move along a desired trajectory in constrained space while generating and maintaining a specific formation, without collisions among agents. Choose controller $\mathbf{u}_i$ as follows

$$\begin{aligned}\mathbf{u}_i &= \mathbf{G}_i^+ \Bigg[-\mathbf{F}_i + \dot{\mathbf{x}}^{\mathrm{r}} - k_t \widetilde{\mathbf{x}}_i + 2k_a \sum_{q=1, q\neq i}^{N} \frac{\mathbf{x}_{iq}}{\|\mathbf{x}_{iq}\|^4} \\ &\quad + \sum_{r=1}^{R} k_{cr} \int\!\!\int \frac{\delta(\|\mathfrak{S}_i\|)(\mathbf{x}_i - \mathbf{s})}{\mathfrak{X}_i^2 \|\mathbf{x}_i - \mathbf{s}\|} \mathrm{d}s_1 \mathrm{d}s_2 \Bigg] \end{aligned} \tag{12.20}$$

where $\mathbf{G}_i^+$ is the pseudoinverse of of $\mathbf{G}_i$, which is utilized here instead of inverse matrix, i.e., $\mathbf{G}_i^-$ because there is no guarantee of the non-singularity of matrix $\mathbf{G}_i$. Thus, the singular value decomposition is performed with $\mathbf{G}_i$ that

$$\mathbf{G}_i = \mathbf{U}_i \mathbf{\Sigma}_i \mathbf{V}_i^{\mathrm{T}} = \mathbf{U}_i \begin{bmatrix} \mathbf{\Sigma}_{i0} & \mathbf{0} \\ \mathbf{0} & \mathbf{0} \end{bmatrix} \mathbf{V}_i^{\mathrm{T}} \tag{12.21}$$

where $\mathbf{U}_i$ and $\mathbf{V}_i \in \mathbb{R}^{m\times m}$ are two orthogonal matrices, and $\mathbf{\Sigma}_{i0}$ is a diagonal matrix formed by the singular values of the matrix $\mathbf{G}_i$. We have

$$\mathbf{G}_i^+ = \mathbf{V}_i \mathbf{\Sigma}_i^+ \mathbf{U}_i^{\mathrm{T}} = \mathbf{V}_i \begin{bmatrix} \mathbf{\Sigma}_{i0}^{-1} & \mathbf{0} \\ \mathbf{0} & \mathbf{0} \end{bmatrix} \mathbf{U}_i^{\mathrm{T}} \tag{12.22}$$

where $\mathbf{\Sigma}_i^+$ is the pseudoinverse of $\mathbf{\Sigma}_i$, which is formed by replacing every non-zero diagonal entry by its reciprocal and transposing the resulting matrix [155].

Remark 59. *The joint potential field is written as* $U_i^{\mathrm{art}} = U_i^{\mathrm{t}} + U_i^{\mathrm{a}} + U_i^{\mathrm{c}}, i = 1, ..., N$ *which has been studied in detail in Section 3.1. Accordingly, the control force* F_i^{art} *is then given by the negative gradient of the joint potential,* $F_i^{\mathrm{art}} = F_i^{\mathrm{t}} + F_i^{\mathrm{a}} + F_i^{\mathrm{c}}, \quad i = 1, ..., N$ *with*

$$F_i^{\mathrm{t}} = -\nabla U_i^{\mathrm{t}} = -k_{\mathrm{t}}\widetilde{\mathbf{x}}_i \tag{12.23}$$

$$F_i^{\mathrm{a}} = -\nabla U_i^{\mathrm{a}} = 2k_{\mathrm{a}} \sum_{q=1, q\neq i}^{N} \frac{\mathbf{x}_{iq}}{\|\mathbf{x}_{iq}\|^4} \tag{12.24}$$

$$F_i^{\mathrm{c}} = -\nabla U_i^{\mathrm{c}} = \sum_{r=1}^{R} k_{\mathrm{c}r} \int\int \frac{\delta(\|\mathfrak{S}_i\|)(\mathbf{x}_i - \mathbf{s})}{\mathfrak{X}_i^2 \|\mathbf{x}_i - \mathbf{s}\|} \mathrm{d}s_1 \mathrm{d}s_2 \tag{12.25}$$

The control law Eq. (12.20) is designed based on F_i^{art} *and its motion (to be* $\dot{\mathbf{x}}^{\mathrm{r}}$*) combined the system dynamics Eq. (12.3).*

Theorem 12.1

Consider N mobile agents with similar dynamics Eqs. (12.3–12.4), moving in constrained space, with Assumptions 24–27 and with formation control law Eq. (12.20). The whole potential function U^{art} converges to zero, which gives rise to the asymptotic stability of MAS without considering impulsive influence from spatial constraints, if any one of the following conditions is satisfied:

1.

$$\int\int \delta(\|\mathfrak{S}_i\|)(\mathbf{x}_i - \mathbf{s})^T \dot{\mathbf{x}}^{\mathrm{r}} \mathrm{d}s_1 \mathrm{d}s_2 \geq 0 \tag{12.26}$$

2.

$$\begin{aligned} \|\dot{\mathbf{x}}^{\mathrm{r}}\| \quad &\leq \left\| \sum_{r=1}^{R} k_{\mathrm{c}r} \int\int \frac{\delta(\|\mathfrak{S}_i\|)(\mathbf{x}_i - \mathbf{s})}{\mathfrak{X}_i^2 \|\mathbf{x}_i - \mathbf{s}\|} \mathrm{d}s_1 \mathrm{d}s_2 \right\| \\ &\leq \left\| \frac{1}{2} k_{\mathrm{t}} \widetilde{\mathbf{x}}_i - k_{\mathrm{a}} \sum_{q=1, q\neq i}^{N} \frac{\mathbf{x}_{iq}}{\|\mathbf{x}_{iq}\|^4} \right\|. \end{aligned} \tag{12.27}$$

■

Proof of Theorem 12.1. The Lyapunov candidate can be written as

$$V = \sum_{i=1}^{N} (U_i^{\mathrm{t}} + U_i^{\mathrm{a}} + U_i^{\mathrm{c}}) \tag{12.28}$$

Substituting Eqs. (12.10, 12.11, 12.19) into Eq. (12.28), we obtain

$$V = \sum_{i=1}^{N}\left[\frac{1}{2}k_{\mathrm{t}}\widetilde{\mathbf{x}}_i^{\mathrm{T}}\widetilde{\mathbf{x}}_i + \frac{1}{2}k_{\mathrm{a}}\sum_{q=1,q\neq i}^{N}\frac{1}{\mathbf{x}_{iq}^{\mathrm{T}}\mathbf{x}_{iq}} + \sum_{r=1}^{R}k_{cr}\int\int\frac{\delta(\|\mathfrak{S}_i\|)}{\mathfrak{X}_i}\mathrm{d}s_1\mathrm{d}s_2\right] \tag{12.29}$$

The derivative of V is

$$\begin{aligned}\dot{V} = \sum_{i=1}^{N}\Bigg[& k_{\mathrm{t}}\widetilde{\mathbf{x}}_i^{\mathrm{T}}\dot{\widetilde{\mathbf{x}}}_i - k_{\mathrm{a}}\sum_{q=1,q\neq i}^{N}\frac{\mathbf{x}_{iq}^{\mathrm{T}}(\dot{\mathbf{x}}_i-\dot{\mathbf{x}}_q)}{\|\mathbf{x}_{iq}\|^4} \\ & -\sum_{r=1}^{R}\int\int k_{cr}\frac{\delta(\|\mathfrak{S}_i\|)(\mathbf{x}_i-\mathbf{s})^{\mathrm{T}}(\dot{\mathbf{x}}_i-\dot{\mathbf{s}})}{\mathfrak{X}_i^2\|\mathbf{x}_i-\mathbf{s}\|}\mathrm{d}s_1\mathrm{d}s_2 \\ & +\sum_{r=1}^{R}k_{cr}\int\int\frac{\dot{\delta}(\|\mathfrak{S}_i\|)}{\mathfrak{X}_i}\mathrm{d}s_1\mathrm{d}s_2\Bigg]\end{aligned} \tag{12.30}$$

We divide $\dot{V}$ into two parts and study them respectively in three steps as below. Only the first two steps are the proof of Theorem 12.1. Step 3 is to consider impulsive influence from spatial constraints for formation tracking control of MAS.

In the first step, we study the first three terms of $\dot{V}$, and mark them as $\dot{V}_1$, which is to study the stability of MAS without considering impulsive influence from spatial constraints.

$$\begin{aligned}\dot{V}_1 = & \textstyle\sum_{i=1}^{N}\left[k_{\mathrm{t}}\widetilde{\mathbf{x}}_i^{\mathrm{T}}\dot{\widetilde{\mathbf{x}}}_i - k_{\mathrm{a}}\sum_{q=1,q\neq i}^{N}\frac{\mathbf{x}_{iq}^{\mathrm{T}}(\dot{\mathbf{x}}_i-\dot{\mathbf{x}}_q)}{\|\mathbf{x}_{iq}\|^4}\right. \\ & \left.-\textstyle\sum_{r=1}^{R}\int\int k_{cr}\frac{\delta(\|\mathfrak{S}_i\|)(\mathbf{x}_i-\mathbf{s})^{\mathrm{T}}(\dot{\mathbf{x}}_i-\dot{\mathbf{s}})}{\mathfrak{X}_i^2\|\mathbf{x}_i-\mathbf{s}\|}\mathrm{d}s_1\mathrm{d}s_2\right]\end{aligned} \tag{12.31}$$

In this chapter, we consider the spatial constraints that is time invariant, which means $\dot{\mathbf{s}} = 0$. Meanwhile, by utilizing the property of fixed formation in Eq. (12.2), Eq. (12.31) can be simplified as

$$\begin{aligned}\dot{V}_1 = \sum_{i=1}^{N}\Bigg[\Bigg(& k_{\mathrm{t}}\widetilde{\mathbf{x}}_i^{\mathrm{T}} - 2k_{\mathrm{a}}\sum_{q=1,q\neq i}^{N}\frac{\mathbf{x}_{iq}^{\mathrm{T}}}{\|\mathbf{x}_{iq}\|^4} \\ & -\sum_{r=1}^{R}k_{cr}\int\int\frac{\delta(\|\mathfrak{S}_i\|)(\mathbf{x}_i-\mathbf{s})^{\mathrm{T}}}{\mathfrak{X}_i^2\|\mathbf{x}_i-\mathbf{s}\|}\mathrm{d}s_1\mathrm{d}s_2\Bigg)(\mathbf{F}_i+\mathbf{G}_i\mathbf{u}_i-\dot{\mathbf{x}}^{\mathrm{r}}) \\ & -\sum_{r=1}^{R}k_{cr}\int\int\frac{\delta(\|\mathfrak{S}_i\|)(\mathbf{x}_i-\mathbf{s})^{\mathrm{T}}\dot{\mathbf{x}}^{\mathrm{r}}}{\mathfrak{X}_i^2\|\mathbf{x}_i-\mathbf{s}\|}\mathrm{d}s_1\mathrm{d}s_2\Bigg]\end{aligned} \tag{12.32}$$

Substituting Eq. (12.20) into Eq. (12.32), we have

$$\begin{aligned}\dot{V}_1 &= \sum_{i=1}^{N}\Bigg[-\Bigg\|k_\mathrm{t}\widetilde{\mathbf{x}}_i - 2k_\mathrm{a}\sum_{q=1,q\neq i}^{N}\frac{\mathbf{x}_{iq}}{\|\mathbf{x}_{iq}\|^4}\\ &\quad -\sum_{r=1}^{R}k_{cr}\int\int\frac{\delta(\|\mathfrak{S}_i\|)(\mathbf{x}_i-\mathbf{s})}{\mathfrak{X}_i^2\|\mathbf{x}_i-\mathbf{s}\|}\mathrm{d}s_1\mathrm{d}s_2\Bigg\|^2\\ &\quad -\sum_{r=1}^{R}k_{cr}\int\int\frac{\delta(\|\mathfrak{S}_i\|)(\mathbf{x}_i-\mathbf{s})^\mathrm{T}\dot{\mathbf{x}}^\mathrm{r}}{\mathfrak{X}_i^2\|\mathbf{x}_i-\mathbf{s}\|}\mathrm{d}s_1\mathrm{d}s_2\Bigg]\end{aligned} \tag{12.33}$$

Case 1: It follows from Eq. (12.26) that

$$\begin{aligned}\dot{V}_1 &\leq -\sum_{i=1}^{N}\Bigg\|k_\mathrm{t}\widetilde{\mathbf{x}}_i - 2k_\mathrm{a}\sum_{q=1,q\neq i}^{N}\frac{\mathbf{x}_{iq}}{\|\mathbf{x}_{iq}\|^4}\\ &\quad -\sum_{r=1}^{R}k_{cr}\int\int\frac{\delta(\|\mathfrak{S}_i\|)(\mathbf{x}_i-\mathbf{s})}{\mathfrak{X}_i^2\|\mathbf{x}_i-\mathbf{s}\|}\mathrm{d}s_1\mathrm{d}s_2\Bigg\|^2\end{aligned} \tag{12.34}$$

Case 2: It follows the second part of Eq. (12.27), which is

$$\begin{aligned}&\Bigg\|\sum_{r=1}^{R}k_{cr}\int\int\frac{\delta(\|\mathfrak{S}_i\|)(\mathbf{x}_i-\mathbf{s})}{\mathfrak{X}_i^2\|\mathbf{x}_i-\mathbf{s}\|}\mathrm{d}s_1\mathrm{d}s_2\Bigg\|\\ \leq\ &\Bigg\|\frac{1}{2}k_\mathrm{t}\widetilde{\mathbf{x}}_i - k_\mathrm{a}\sum_{q=1,q\neq i}^{N}\frac{\mathbf{x}_{iq}}{\|\mathbf{x}_{iq}\|^4}\Bigg\|\end{aligned} \tag{12.35}$$

that

$$\begin{aligned}\dot{V}_1 &\leq \sum_{i=1}^{N}\Bigg[-\Bigg\|\sum_{r=1}^{R}k_{cr}\int\int\frac{\delta(\|\mathfrak{S}_i\|)(\mathbf{x}_i-\mathbf{s})}{\mathfrak{X}_i^2\|\mathbf{x}_i-\mathbf{s}\|}\mathrm{d}s_1\mathrm{d}s_2\Bigg\|^2\\ &\quad -\sum_{r=1}^{R}k_{cr}\int\int\frac{\delta(\|\mathfrak{S}_i\|)(\mathbf{x}_i-\mathbf{s})^\mathrm{T}\dot{\mathbf{x}}^\mathrm{r}}{\mathfrak{X}_i^2\|\mathbf{x}_i-\mathbf{s}\|}\mathrm{d}s_1\mathrm{d}s_2\Bigg]\end{aligned} \tag{12.36}$$

Furthermore, it follows the first part of Eq. (12.27), which is

$$\|\dot{\mathbf{x}}^\mathrm{r}\| \leq \Bigg\|\sum_{r=1}^{R}k_{cr}\int\int\frac{\delta(\|\mathfrak{S}_i\|)(\mathbf{x}_i-\mathbf{s})}{\mathfrak{X}_i^2\|\mathbf{x}_i-\mathbf{s}\|}\mathrm{d}s_1\mathrm{d}s_2\Bigg\| \tag{12.37}$$

that

$$\left\| \sum_{r=1}^{R} k_{cr} \int\int \frac{\delta(\|\mathfrak{S}_i\|)(\mathbf{x}_i - \mathbf{s})}{\mathfrak{X}_i^2 \|\mathbf{x}_i - \mathbf{s}\|} \mathrm{ds}_1 \mathrm{ds}_2 \right\|^2 + \int\int \frac{\delta(\|\mathfrak{S}_i\|)(\mathbf{x}_i - \mathbf{s})^{\mathrm{T}} \dot{\mathbf{x}}^{\mathrm{r}}}{\mathfrak{X}_i^2 \|\mathbf{x}_i - \mathbf{s}\|} \mathrm{ds}_1 \mathrm{ds}_2 \geq 0 \tag{12.38}$$

therefore, $\dot{V}_1 \leq 0$.

To conclude, if any one of conditions in Eqs. (12.26–12.27) holds, the MAS are asymptotic stable [321], and the desired formation tracking along the given trajectory in constrained space can be achieved.

It is worth noting that ω_{r}^{c} is assumed to be constant in this chapter, nevertheless it should be time varying. As $k_{cr} = k_c \omega_r^c$, this simplification indicates that a part of the time derivative of V, or $\dot{V}$ is ignored, which leaves a gap for the conclusion of this chapter for further discussion. Simply speaking, if Eq. (12.26) is satisfied, there is no need to extract further conclusion; otherwise, the condition expressed as Eq. (12.27) should be stricter.

In the second step, we prove that no collision occurs between any agents if any one of conditions in Eqs. (12.26–12.27) holds without considering impulsive influence from spatial constraints.

From the first step of proof, we known that $\dot{V}_1 \leq 0$. Integrating both sides in the interval $[0, t], \forall t > 0$, we obtain that $V_1(t) \leq V_1(0) = V(0)$. With the definition of V in Eq. (12.28), we have $U^{\mathrm{a}}(t) + U^{\mathrm{c}}(t) = \sum_{i=1}^{N} [U_i^{\mathrm{a}}(t) + U_i^{\mathrm{c}}(t)] \leq V(0)$ which also indicates $U^{\mathrm{a}}(t) \leq V(0)$ & $U^{\mathrm{c}}(t) \leq V(0)$.

According to the definition of U^{a} in Eq. (12.11) and the definition of U^{c} in Eq. (12.19), the boundedness of U^{a} and U^{c} means there is no collision among any agents or between an agent and spatial constraints for all $t > 0$. □

Remark 60. *Following Theorem 12.1, we further elaborate the bound of the input signal in Eq. (12.20).*

$$\begin{aligned} \|\mathbf{u}_i\| &= \left\| \mathbf{G}_i^{+} \left[-\mathbf{F}_i + \dot{\mathbf{x}}^{\mathrm{r}} - k_t \widetilde{\mathbf{x}}_i + 2k_a \sum_{q=1, q\neq i}^{N} \frac{\mathbf{x}_{iq}}{\|\mathbf{x}_{iq}\|^4} \right.\right. \\ &\quad \left.\left. + \sum_{r=1}^{R} k_{cr} \int\int \frac{\delta(\|\mathfrak{S}_i\|)(\mathbf{x}_i - \mathbf{s})}{\mathfrak{X}_i^2 \|\mathbf{x}_i - \mathbf{s}\|} \mathrm{ds}_1 \mathrm{ds}_2 \right] \right\| \\ &\leq \left\| \mathbf{G}_i^{+} \right\| \left[\|\mathbf{F}_i\| + \|\dot{\mathbf{x}}^{\mathrm{r}}\| + k_{\mathrm{t}} \|\widetilde{\mathbf{x}}_i\| + 2k_a \sum_{q=1, q\neq i}^{N} \frac{1}{\|\mathbf{x}_{iq}\|^3} \right. \\ &\quad \left. + k_{\mathrm{c}} \sum_{r=1}^{R} \omega_r^{\mathrm{c}} \int\int \frac{\delta(\|\mathfrak{S}_i\|)}{\mathfrak{X}_i^2} \mathrm{ds}_1 \mathrm{ds}_2 \right] \end{aligned} \tag{12.39}$$

It holds that $\dot{V} \leq 0 \Rightarrow V(t) \leq V(0), 0 \leq t < \infty \Rightarrow \|\widetilde{\mathbf{x}}_i\| \leq \sqrt{2V(0)/k_{\rm t}}$, *combining Assumption 28 and* $\omega_{\rm r}^{\rm c} < \pi$, *we have*

$$\begin{aligned}\|\mathbf{u}_i\| \leq \; & \|\mathbf{G}_i^+\| \Big[\|\mathbf{F}_i\| + \zeta + \sqrt{2k_{\rm t}V(0)} + 2k_a(N-1) \\ & + k_c \pi R_i\Big] \end{aligned} \tag{12.40}$$

following the conditions that $\|\mathbf{x}_{iq}\| \geq 1$ *and* $\|\mathfrak{X}_i\| \geq 1$. *We call this mode as normal tracking mode. Otherwise, if* $\|\mathbf{x}_{iq}\| < 1$ *or* $\|\mathfrak{X}_i\| < 1$, *MAS are in emergency avoidance mode, in which case the control input may exceed the bound in Eq. (12.40) to guarantee no collision.*

The physical interpretations on the stability conditions in Eq. (12.26–12.27) are necessary for better understanding. For the first condition in Eq. (12.26), the physical meaning is that if MAS are far away from spatial constraints, the stability of MAS can be obtained. Otherwise, if Eq. (12.26) cannot be satisfied, which means MAS are getting close to spatial constraints, we further consider the meaning of Eq. (12.27), where the first item is the reference tracking velocity, the second item is the repulsive force to avoid collisions with spatial constrains or called the effect of spatial constraints, while the third item represents the joint of repulsive force to avoid internal collisions and attractive force to keep the formation (representing the magnitude of the reference trajectory derivative). Eq. (12.27) indicates one factor leading to the stability of MAS is that the joint of repulsive force to avoid internal collisions and attractive force to keep the formation is greater than the force to avoid collisions with spatial constrains, which is greater than the desired tracking velocity.

Although we have proved the conditions leading to the asymptotic stability of MAS, the maximum continuous duration that the conditions cannot be satisfied may have more realistic meaning in a task. Therefore, reconsidering Eq. (12.33), we obtain

$$\begin{aligned}\dot{V}_1 \leq \; & \sum_{i=1}^{N} \Bigg[\|k_{\rm t}\widetilde{\mathbf{x}}_i\|^2 + 4k_{\rm a}^2 \sum_{q=1, q\neq i}^{N} \frac{1}{\|\mathbf{x}_{iq}\|^6} \\ & + \sum_{r=1}^{R} k_{cr}^2 \int\!\!\int \frac{\delta(\|\mathfrak{S}_i\|)}{\mathfrak{X}_i^4} {\rm d}s_1 {\rm d}s_2 \\ & + \sum_{r=1}^{R} k_{cr} \int\!\!\int \frac{\delta(\|\mathfrak{S}_i\|)\|\dot{\mathbf{x}}^{\rm r}\|}{\mathfrak{X}_i^2} {\rm d}s_1 {\rm d}s_2 \Bigg] \end{aligned} \tag{12.41}$$

if the conditions that $\mathfrak{X}_i \geq 1$, $\|\mathbf{x}_{iq}\| \geq 1$ are satisfied, combining Assumption 28 and $\omega_r^{\mathrm{c}} < \pi$, we have

$$\begin{aligned}\dot{V}_1 &\leq kV + \sum_{i=1}^{N}\sum_{r=1}^{R} k_{cr} \int\int \frac{\delta(\|\mathfrak{S}_i\|)\|\dot{\mathbf{x}}^{\mathrm{r}}\|}{\mathfrak{X}_i^2}\mathrm{d}s_1\mathrm{d}s_2 \\ &\leq kV + C \end{aligned} \tag{12.42}$$

where $C = k_c\pi\zeta\sum_{i=1}^{N} R_i$, R_i is the number of solutions to $\|\mathfrak{S}_i\| = 0$, and $k = \max\{2k_{\mathrm{t}}, 8k_{\mathrm{a}}, \{k_{cr}\}\}$. Without considering impulsive influence from spatial constraints, Eq. (12.42) can be written as $\dot{V} \leq kV + C$. Multiplying both sides by e^{-kt} yields

$$\frac{\mathrm{d}}{\mathrm{d}t}\left(Ve^{-kt}\right) \leq Ce^{-kt} \tag{12.43}$$

Integrating Eq. (12.43) over $[t_0, t]$, where t_0 is the latest start time that any one of conditions in Eqs. (12.26–12.27) cannot be satisfied, we have $V(t) \leq V(t_0)e^{kt-kt_0} - \frac{1}{k}C(1 - e^{kt-kt_0})$. Denoting $\Delta t = t - t_0$ yields $V(t) \leq \left(V(t_0) + \frac{1}{k}C\right)e^{k\Delta t} - \frac{1}{k}C$. We set a tolerable upper limit for Lyapunov function as $V_{\max}$

$$\begin{aligned} &\left(V(t_0) + \frac{1}{k}C\right)e^{k\Delta t} - \frac{1}{k}C \leq V_{\max} \\ \Rightarrow \quad &\Delta t \leq \frac{1}{k}\log\left(\frac{V_{\max} + \frac{1}{k}C}{V(t_0) + \frac{1}{k}C}\right) \end{aligned} \tag{12.44}$$

therefore the maximum time that the conditions cannot be satisfied can be set as $t_{\max} = \frac{1}{k}\log\left(\left(V_{\max} + \frac{1}{k}C\right)/\left(V(t_0) + \frac{1}{k}C\right)\right)$. This may be a reference in realistic tasks to indicate whether the tracking task should be given up. The basic assumption of this study is that the designed trajectory does not have too much serious conflict with spatial constraints. However, if the duration that the conditions cannot be satisfied exceeds $t_{\max}$, which may be caused by various uncertainties or the given trajectory is unreasonable, the formation tracking task may be given up.

In the third step, we study the last term of $\dot{V}$, and mark them as $\dot{V}_2$.

$$\dot{V}_2 = \sum_{i=1}^{N}\sum_{r=1}^{R} k_{cr} \int\int \frac{\dot{\delta}(\|\mathfrak{S}_i\|)}{\mathfrak{X}_i}\mathrm{d}s_1\mathrm{d}s_2 \tag{12.45}$$

when consider a single agent, we mark

$$\dot{V}_{i2} = \sum_{r=1}^{R} k_{cr} \int\int \frac{\dot{\delta}(\|\mathfrak{S}_i\|)}{\mathfrak{X}_i}\mathrm{d}s_1\mathrm{d}s_2 \tag{12.46}$$

We divide the task time into a set of slots: $[t_0, t_1)$,$[t_1, t_2)$,...
...,$[t_{k-1}, t_k)$, $k \to \infty$. Assume that in any time interval $(t_p, t_{p+1}), i = 0, 1, ...,$

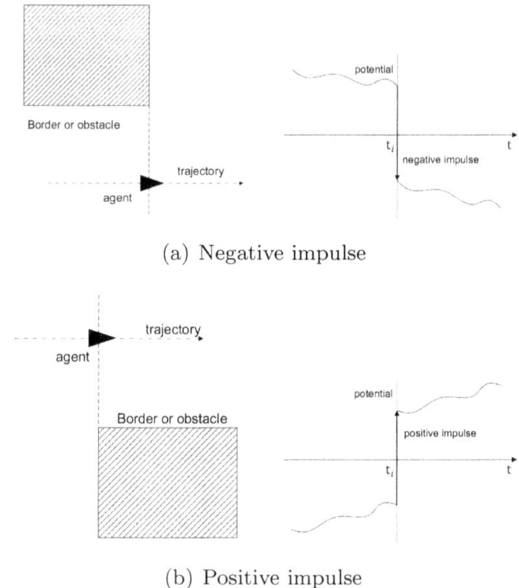

(a) Negative impulse

(b) Positive impulse

Figure 12.3 Illustration of two cases.

$\dot{V}_{i2} = 0$, which means there is no impulsive variation to the solution of Eq. (12.17), the result has been studied in Theorem 12.1. At the moment $t_p, i = 1, 2..., \dot{V}_{i2} \neq 0$, which means there is an impulse to the system. This situation is studied under the following two cases.

Case 1: $\dot{V}_{i2} < 0$, there is a negative impulse to the system, for example, an agent is leaving a constraint potential field generating by a border or obstacle, as shown in Fig. 12.3(a). In this case, V decreases, and MAS tend to be more stable.

Case 2: $\dot{V}_{i2} > 0$, there is a positive impulse to MAS, for example, an agent is accessing a new constraint potential field generating by a border or an obstacle, as shown in Fig. 12.3(b). Without loss of generality, we assume, there is just a single positive impulse to a specific agent, and we have $V_i(t_p^+) = V_i(t_p^-) + \frac{k_c}{\mathfrak{X}_i}$ where

$$V_i = \frac{1}{2}k_t\widetilde{\mathbf{x}}_i^{\mathrm{T}}\widetilde{\mathbf{x}}_i + \frac{1}{2}k_a \sum_{q=1, q\neq i}^{N} \frac{1}{\mathbf{x}_{iq}^{\mathrm{T}}\mathbf{x}_{iq}} + \sum_{r=1}^{R} k_{cr} \int\int \frac{\delta(\|\mathfrak{S}_i\|)}{\mathfrak{X}_i} \mathrm{d}s_1 \mathrm{d}s_2 \qquad (12.47)$$

according to the definition of V in Eq. (12.29). Let

$$h = \frac{V_i(t_p^+)}{V_i(t_p^-)} = 1 + \frac{k_c}{V_i(t_p^-)\mathfrak{X}_i} \tag{12.48}$$

Concerning the meaning of V_{i1} and V_{i2}, we can know that $V_i(t_p^-) = V_{i1}(t_p)$, which is the part without the impulsive influence from spatial constraints, then we also have

$$h = \frac{V_i(t_p^+)}{V_{i1}(t_p)} = 1 + \frac{k_c}{V_{i1}(t_p)\mathfrak{X}_i} \tag{12.49}$$

then we obtain $V_i(t_p^+) = h \cdot V_i(t_p^-) = h \cdot V_{i1}(t_p)$, where $V_{i1}(t_p)$ has been studied in Theorem 12.1. From the above analysis, we could know the earlier we detect a spatial constraint, the smaller influence it would give rise to stability of MAS.

Remark 61. *When there is no spatial constraint, the control law in Eq. (12.20) can be rewritten as*

$$\mathbf{u}_i = \mathbf{G}_i^+ \left[-\mathbf{F}_i + \dot{\mathbf{x}}^{\mathrm{r}} - k_{\mathrm{t}}\widetilde{\mathbf{x}}_i + 2k_{\mathrm{a}} \sum_{q=1, q\neq i}^{N} \frac{\mathbf{x}_{iq}}{\|\mathbf{x}_{iq}\|^4} \right] \tag{12.50}$$

Eq. (12.36) can be written as

$$\begin{aligned} \dot{V}_1 \;\; = \;\; & -\sum_{i=1}^{N} \Bigg\| k_{\mathrm{t}}\widetilde{\mathbf{x}}_i - 2k_{\mathrm{a}} \sum_{q=1, q\neq i}^{N} \frac{\mathbf{x}_{iq}}{\|\mathbf{x}_{iq}\|^4} - \\ & \sum_{r=1}^{R} k_{cr} \int\!\!\int \frac{\delta(\|\mathfrak{S}_i\|)(\mathbf{x}_i - \mathbf{s})}{\mathfrak{X}_i^2 \|\mathbf{x}_i - \mathbf{s}\|} \mathrm{d}s_1 \mathrm{d}s_2 \Bigg\|^2 \leq 0 \end{aligned} \tag{12.51}$$

and we also have $V = V_1$. Integrating both sides of Eq. (12.51) in the interval $[0, t], \forall t > 0$, we obtain that $V(t) \leq V(0)$. With the definition of V in Eq. (12.28), we have $U_i^{\mathrm{a}}(t) \leq V(0)$, i.e., there are no collisions among any agents for all $t > 0$.

12.6 SIMULATION

In this section, four simulation examples are presented to illustrate the effectiveness of the methods proposed in this chapter. We consider a group of agents with $N = 5$, and the desired formation is chosen as regular pentagon. We set the safety distance $s_0 = 0.1m$. We assume N agents have the following form of dynamics

$$\begin{bmatrix} \dot{x}_{i1} \\ \dot{x}_{i2} \end{bmatrix} = \begin{bmatrix} 5 \\ 10 \end{bmatrix} + \begin{bmatrix} 3 & 1 \\ 2 & 5 \end{bmatrix} \mathbf{u}_i \tag{12.52}$$

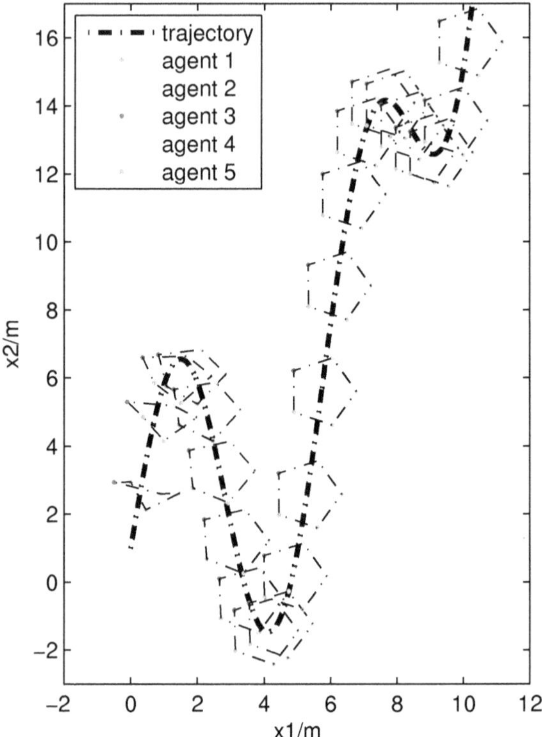

Figure 12.4 Formation tracking in free space.

12.6.1 FORMATION TRACKING IN FREE SPACE

In the first simulation, the initial positions of N agents are initialized randomly within a circle range with radius $r = 1m$. The desired trajectory is set as $\mathbf{p}_0(t) = \begin{bmatrix} t & 5\sin(1.1t + e^{0.3t}) \end{bmatrix}^T$. Simulation results are shown in Fig. 12.4–12.5.

In this simulation, we study the effectiveness of control law Eq. (12.50) for formation tracking of multi-agents in free space. Fig. 12.4 shows the tracking result of all agents, which can successfully track the desired trajectory generating and maintaining the desired formation while avoiding collisions among agents. Fig. 12.5 shows the tracking error of all agents will decrease gradually to zeros with time going.

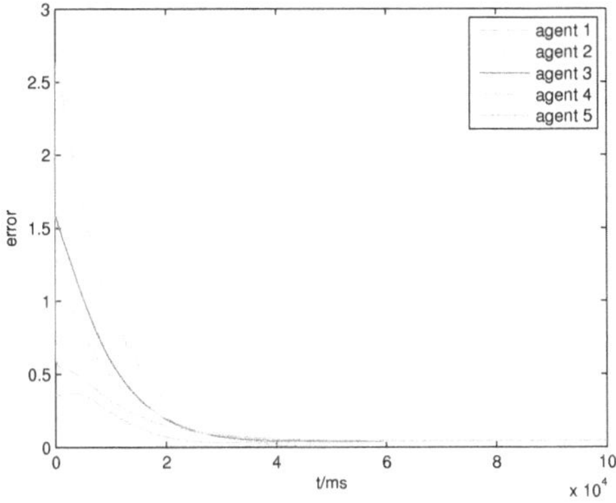

Figure 12.5 Tracking errors in free space.

12.6.2 FORMATION TRACKING IN 2-D MULTIPLE CONSTRAINED SPACE WITH SENSOR UNCERTAINTIES

In the second simulation, we simulate a similar scenario as Fig. 12.2(a) with multiple spatial constraints. The initial position of N agents are initialized randomly within a circle range with radius $r = 2$m, and the desired trajectory is $\mathbf{p}_0(t) = \begin{bmatrix} t & 20\sin(0.5t) \end{bmatrix}^T$. A border at position $x_2 = -19$ and $x_2 = 17$ is set and a square obstacle at position $\mathbf{x} = \begin{bmatrix} 13.25 & 5.25 \end{bmatrix}^T$ with 1.5m length of edge and $\mathbf{x} = \begin{bmatrix} 19.5 & -5 \end{bmatrix}^T$ with 2.0m length of edge.

To test the robustness of proposed method to sensor uncertainties, a set of white gaussian noise series with the variance $\sigma = 0.1$ are added to the estimated positions of spatial constraints [106–206]. Referring to the *Scanning Laser Range Finder UTM-30LX/LN Specification*, $\sigma < 10$mm when detection range is between 0.1 and 10m and $\sigma < 30$mm when detection range is between 10 - 30m. Thus, setting σ as 0.1 is sufficient to simulate the real scenario. Simulation results are shown in Figs. 12.6–12.8.

Fig. 12.6 shows that all agents can successfully track the desired trajectory when there are multiple borders or obstacles. The formation error (in Fig. 12.7) increases when an agent gets close to borders or obstacles, and decreases when being far away from borders or obstacles. Fig. 12.8 describes the whole process of how an agent avoids spatial constraints. All results also indicate the sensor uncertainties have little influence on the formation tracking and obstacle avoidance.

12.6.3 FORMATION TRACKING IN 3-D MULTIPLE CONSTRAINED SPACE

A similar simulation is also done in 3-D environment to show the proposed solution in this chapter also works in a 3-D problem, where the desired

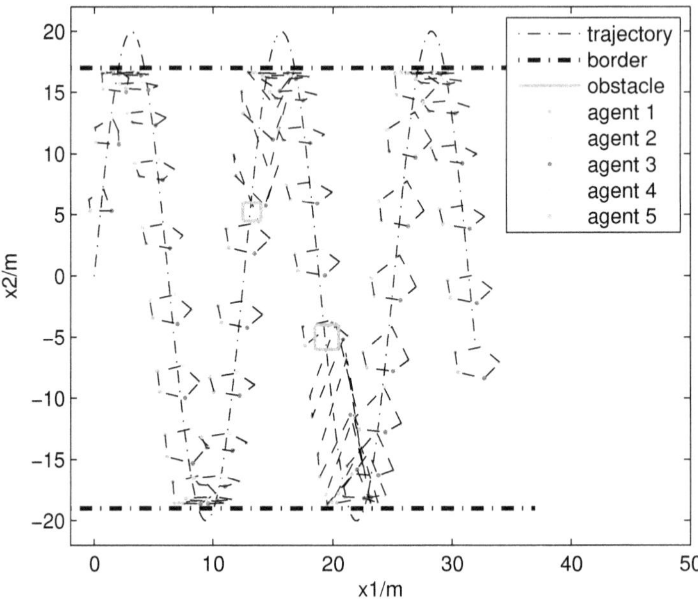

Figure 12.6 Formation tracking with multiple borders or obstacles in 2-D space.

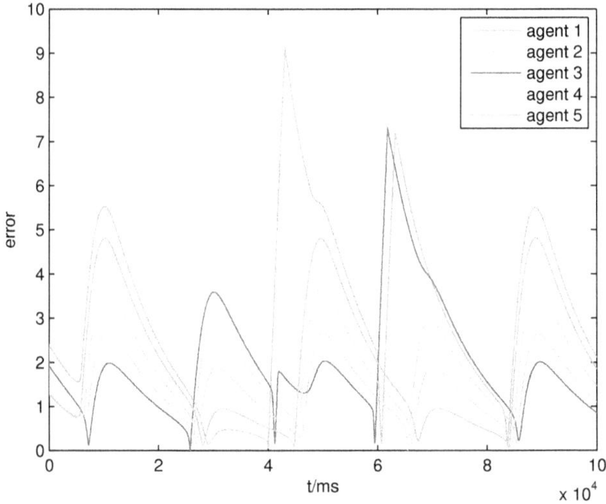

Figure 12.7 Tracking errors with multiple borders or obstacles in 2-D space.

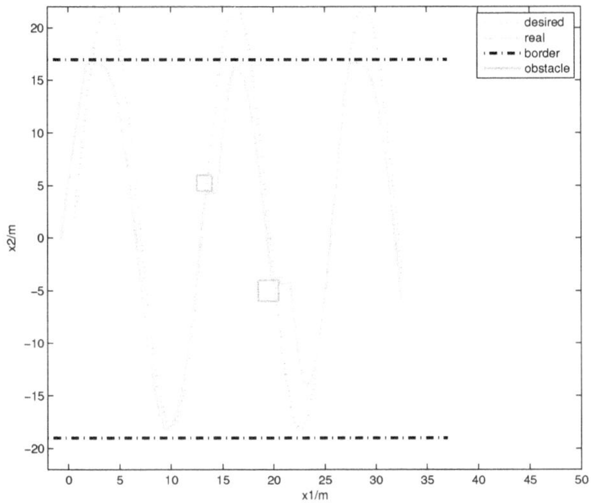

Figure 12.8 Whole process of how an agent avoids collision with multiple borders or obstacles in 2-D space.

trajectory is set to be $\mathbf{p}_0(t) = \begin{bmatrix}20\cos(2t) & 7t & 20\sin(2t)\end{bmatrix}^{\mathrm{T}}$. The initial positions of N agents are initialized randomly within a circle range with radius $r = 3m$. N agents are assumed to have the following form of dynamics

$$\begin{bmatrix}\dot{x}_{i1}\\ \dot{x}_{i2}\\ \dot{x}_{i3}\end{bmatrix} = \begin{bmatrix}5\\ 10\\ 7\end{bmatrix} + \begin{bmatrix}3 & 1 & 2\\ 2 & 5 & 4\\ 1 & 3 & 8\end{bmatrix}\mathbf{u}_i \tag{12.53}$$

A border at position $x_1 = -17$ and $x_3 = 17$ is set and a cube obstacle at position $\mathbf{x} = \begin{bmatrix}17.5 & 22.5 & 7.5\end{bmatrix}^{\mathrm{T}}$ with $5.0m$ length of edge and a sphere obstacle $\mathbf{x} = \begin{bmatrix}12 & 28 & 15\end{bmatrix}^{\mathrm{T}}$ with $4.0m$ radius. Simulation results are shown in Figs. 12.9–12.13, which are quite similar to the simulation results in 2-D environment.

Fig. 12.9 shows that all agents can successfully track the desired trajectory when there are multiple borders or obstacles. The formation error (in Fig. 12.11) increases when an agent gets close to borders or obstacles, and decreases when being far away from borders or obstacles. Fig. 12.13 describes the whole process of how an agent avoids spatial constraints with different perspectives.

12.6.4 OPTIMAL FORMATION

In this simulation, we further investigate the effectiveness of *Formation Optimization Algorithm* proposed in Section 11.4. The initial positions of all

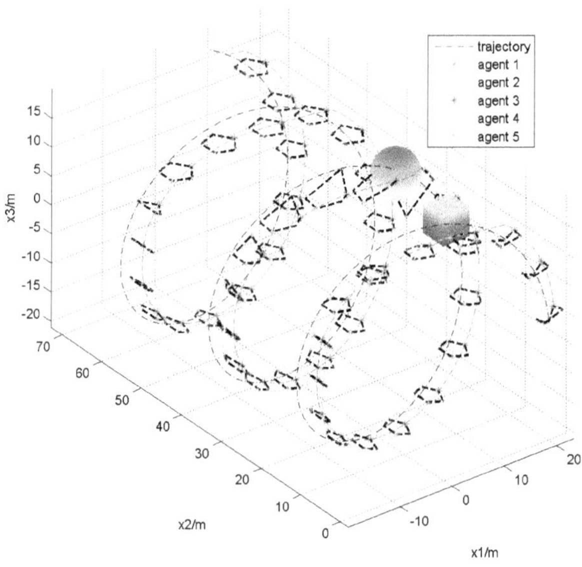

Figure 12.9 Formation tracking with multiple borders or obstacles in 3-D space.

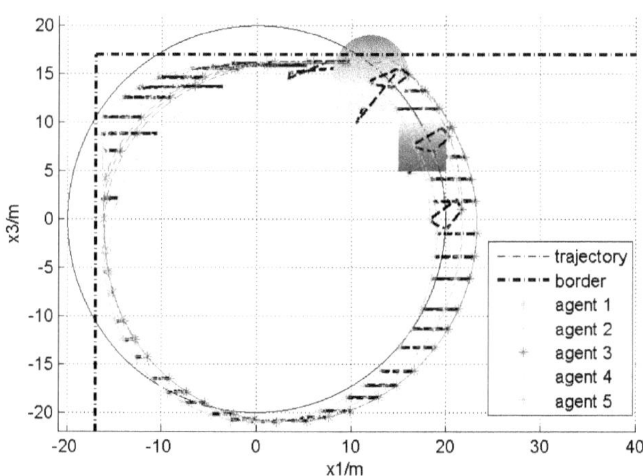

Figure 12.10 Formation tracking with multiple borders or obstacles in 3-D space.

agents are around $[3 \quad 3]^{\mathrm{T}}$. The visual leader is at the position of $[0 \quad 0]^{\mathrm{T}}$, while the desired formation is chosen as regular pentagon, and η is set as 0.01. Simulation results are shown in Figs. 12.14–12.15.

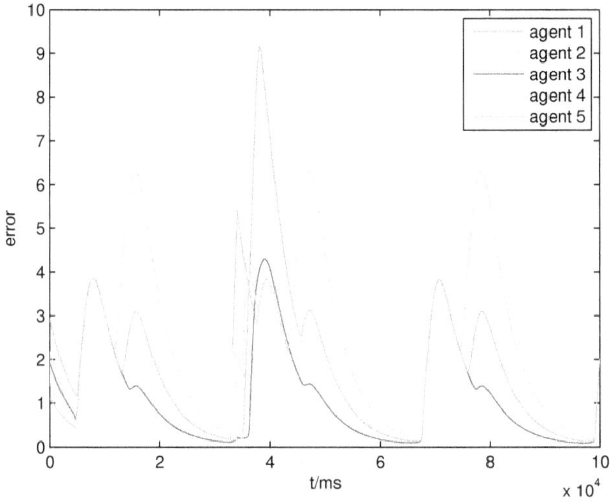

Figure 12.11 Tracking errors with multiple borders or obstacles in 3-D space.

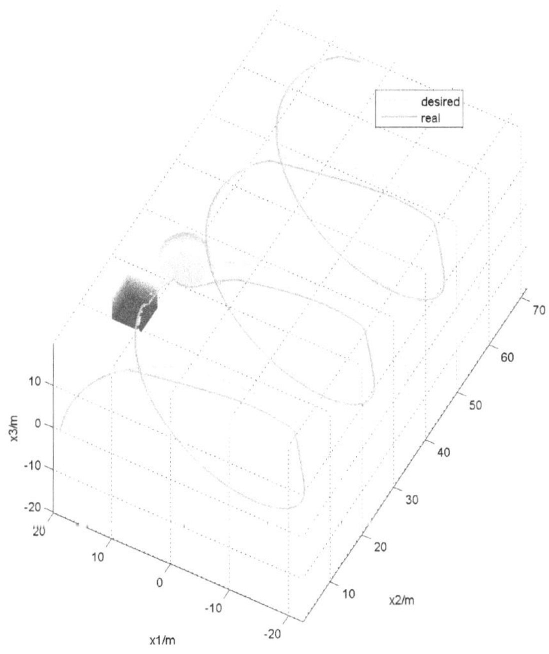

Figure 12.12 Whole process of how an agent avoids collision with multiple borders or obstacles in 3-D space.

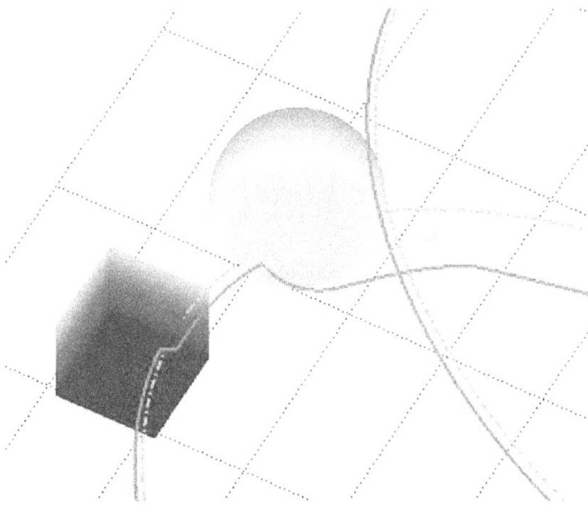

Figure 12.13 Details of how an agent avoids collision with multiple borders or obstacles in 3-D space. The green illustrates the desired trajectory while the red illustrates the real trajectory.

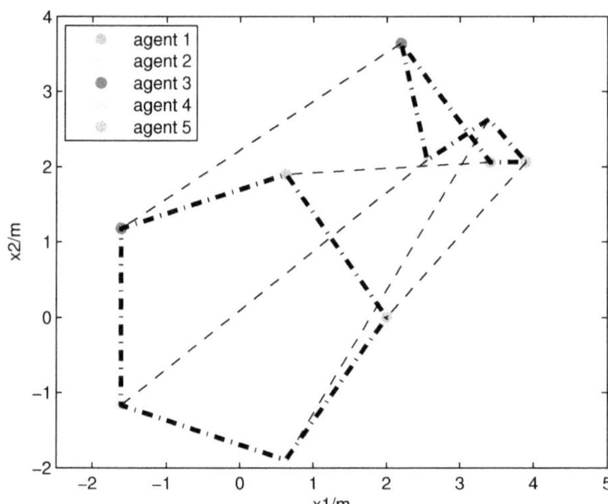

Figure 12.14 Optimal Formation Generation: initial positions and final formation positions.

Fig. 12.14 shows the optimal formation generation process and the optimal bijection relationship $\sigma : \mathfrak{P}_0 \rightarrow \mathfrak{Q}$. From the figure we could see that the desired formation position at $\begin{bmatrix}-1.6180 & -1.1756\end{bmatrix}^{\mathrm{T}}$ is the farthest one from all agents' initial positions. Therefore, agent 2, which is relatively close this desired formation position, is assigned to this formation position, reducing the

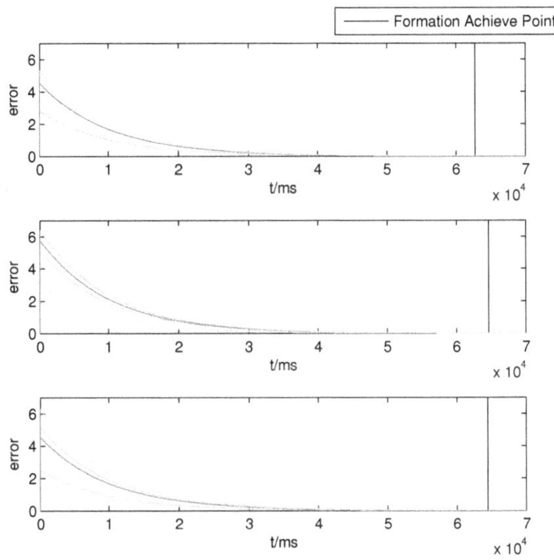

Figure 12.15 Formation Errors Comparison: the first one is obtained from the optimal formation process, while the second and third one is obtained from two random bijection.

possible maximum single formation time cost. Fig. 12.15 shows the time cost of the optimal bijection relationship and two random bijection relationships. It indicates the optimality of proposed *Formation Optimization Algorithm* in time, although the optimality is not quite conspicuous, as we only present a simple case in this simulation.

12.7 CONCLUSION

In this chapter, we have proposed a control law based on artificial potential field method for formation tracking control of multi-agents in constrained space. A novel modeling approach for spatial constraints provides a tool for tracking control or path planning in non-omniscient constrained space has been designed. Specifically, two spatial constraints: (i) borders, and (ii) obstacles have been unified into an unification expression by introducing the concepts of line source and surface source. A Dirac delta function has been introduced concerning about the situation of multiple spatial constraints. In this way, formation tracking in a bounded space and rapidly obstacle avoidance have been both solved by the method proposed in this chapter. Besides, an optimization algorithm has been designed to minimize the formation time cost. The conditions for asymptotic stability of multi-agent systems have been studied based on the Lyapunov function. Meanwhile, the maximum continuous duration that the conditions cannot be satisfied has been elaborated to indicate whether the tracking task should be given up. Simulation results have demonstrated that all agents can successfully track the given trajectory in desired formation in constrained space while avoid collisions with spatial constraints and among agents.

13 Neural-Network-Based Switching Formation Tracking Control of Multi-Agents with Uncertainties in Constrained Space

In this chapter, we present a novel approach for tracking control with switching formation in non-omniscient constrained space for multi-agent systems (MAS). The introduction of switching formation results from the situation where MAS is maneuvering in restricted path which is often the case for real world application. The preplanned trajectory may be inaccurate due to the lack of sufficient environmental information. In this case, agents may have to rapidly avoid collisions with unexpected obstacles or even switch the formation to guarantee the passability. A concept of avoidless disturbance is proposed. To solve the undesirable chattering on resulting trajectory caused by avoidless disturbance and the existing adaptation algorithm for neural network weights, a local path replanning approach is designed such that the potential force generated by avoidless disturbance is acted on the original desired trajectory outputting the locally replanned path for an agent. A neural-network-based controller is designed and the performance is validated using Lyapunov functions. Simulations are carried out to illustrate the effectiveness of proposed strategies.

13.1 INTRODUCTION

Formation control has spurred a broad interest and a large amount of results have been published over the past few years [112,118,356,379,471]. The control objective of formation is to stabilize the relative positions/distances among agents to given values. Formation control finds wide applications including various fields, such as cooperative transportation [272], satellite flying [73], unmanned aerial vehicle (UAV) [119, 408], teaming of multi-robotics [64, 487] and so on.

DOI: 10.1201/9781003298618-13

The formation tasks considering spatial constraints have attracted increasing attention from different disciplines in recent research and many achievements have been published [2,317,333]. As a matter of fact, in realistic applications, the environment is commonly restricted from flight paths of aircrafts to roads. The influences of the boundaries will destroy the formation and lead to uncertainties of positions of agents. In [369], Panagou and Kumar investigated the cooperative motion coordination of leader-follower formations moving in known restricted space applying a dipolar vector field for formation maintenance and collision avoidance. However, the appearance of unexpected obstacles presents more challenges to the success of formation control. In [524], Yang et al. studied a motion planning with formation scheme in an unknown obstacle environment for multiple hybrid-driven underwater gliders based on the artificial potential field (APF) and Kane's method. Given the task of formation tracking in unknown obstacle environment, the strategy to balance the completion of tacking and collision avoidance with unexpected obstacles needs to be solved. Furthermore, the realistic application of approaches will be limited without the consideration of system uncertainties, which exist commonly in real world.

Considering difficulties on modeling the dynamics of an agent's system, neural network (NN) is usually applied for approximating bounded and continuous functions [51, 53, 278, 343, 553, 555]. In [71], An NN-based adaptive approach was proposed for the leader-following control of MAS, where NN was explored to approximate an agent's uncertain dynamics. Similarly, in [568], Chebyshev neural networks were used to approximate unknown nonlinear function in the agent's dynamics. In the most of these existing research [71, 192], approximation error and disturbance are considered as the state error sources. In other words, the control objective is to track a given state/output trajectory and eliminate tracking error. Nevertheless, an unknown spatially constrained environment where agents do not have global knowledge about the environment will lead to that they have to avoid collisions with unexpected obstacles. It means that spacial constraints are also the approximation error sources in this situation. Under the condition that obstacles have collisions with the given trajectory, and the state error should not be eliminated to zero. In this way, traditional adaptation laws such as σ-modification, ϵ-modification and projection algorithm will result in an undesirable chattering on resulting trajectory.

APF method provides simple and effective solutions for practical applications of MAS, such as motion planning, target tracking and obstacle avoidance [199, 331, 427]. The applications of APF for obstacle avoidance was first developed by Khatib [233]. In [149], Ge and Fua represented formation structures in queues and formation vertices, as well as introduced a new concept of artificial potential trenches to effectively achieve the formation of a group of robots. In [356], Noormohammadi Asl et al. solved the optimal path planning of a group of nonholonomic robots with coherent formation in a leader-following structure based on potential field. In the most of existing research, the potential field is mostly designed to affect the agent's motion directly.

Inspired by the multiple unexpected spatial constraints in practical situation, formation control of MAS with system uncertainties in non-omniscient constrained space is studied in this chapter. Spatial constraints are modeled and expressed in a potential function. NN is explored to approximate the uncertain dynamics of MAS.

The main contribution of this work is the proposal of a solution for the adoption of the NN-based controller for trajectory tacking with avoidless external disturbance, e.g., potential force generated by unexpected spatial constraints. Traditional NN-based approach methods (see [71, 192]) consider the external disturbances such as the unstructured unmodeled dynamics and the noises, and design approaches to eliminate the effect of the disturbance. However, the approaches will result in the undesirable chattering on the trajectory. Aim at an effective solution, a local path replanning approach is proposed to act the effect of the avoidless external disturbance (e.g., potential force generated by spatial constraints) on the original trajectory rather than on the agent's motion directly. Based on the proposed local path replanning strategy, an NN-based controller is designed for tracking problem in restricted path. In addition, the passability of a restricted path is mathematically defined, while the dissatisfaction will trigger the formation switching to achieve the rapid collision avoidance with spatial constraints.

13.2 MATHEMATICAL MODELING

Before proceeding further, the following assumptions are made in this chapter.

Assumption 30. *The influence of the size and the shape of an agent to the formation tracking control is ignored, which means an agent is assumed to be of point mass.*

Assumption 31. *Each agent is able to broadcast information to others.*

Assumption 32. *An agent is able to estimate its position in the world coordinate system.*

Assumption 33. *The environment (spatial constraints) is uniform (time-invariant) in the whole task.*

13.2.1 PASSABILITY OF RESTRICTED PATH

Define a set of formation candidates as $\mathcal{T} = \{\mathcal{T}_1, ..., \mathcal{T}_Q\}$ where Q is the number of formation candidates ordered by preference and $\mathcal{T}_q$ is the qth ($q = 1, ..., Q$) predefined formation pattern in the candidates. $\mathcal{T}_1$ is the first choice of formation. In the constrained space, formation switching may be executed for obstacle avoidance as the first choice of desired formation may be impassable for the restricted path.

Denoting the restricted path as $\mathcal{P}$, we can make definitions as below about the passability of restricted path for MAS.

Definition 46 (*Rigidly-Passable*). *A restricted path is called rigidly-passable for MAS if the system can get through it with fixed formation.*

Definition 47 (*Flexibly-Passable*). *A restricted path is called flexibly-passable for MAS if the system can pass through it with switching formation.*

Define $\xi_{T_q}(t)$ as a parameter indicating the spatial condition needed for MAS with formation $\mathcal{T}_q$. Similarly, $\xi_{\mathcal{P}}(t)$ is defined for path $\mathcal{P}$ indicating the spatial condition available. Define $\varepsilon_q(t)$ as the passability index for $\mathcal{T}_q$ that $\varepsilon_q(t) \triangleq \frac{\xi_{T_q}(t)}{\xi_{\mathcal{P}}(t)}$ and we can conclude the restricted path for MAS with formation $\mathcal{T}_1$ is rigidly-passable if $\varepsilon_1(t) < 1$ all the time. The path $\mathcal{P}$ is called flexibly-passable if $\varepsilon_q(t) < 1$ can be guaranteed with switching formation $\mathcal{T}_q$, $q = 1, ..., Q$.

When MAS detects that the path is not rigidly-passable, the switching scheme is to find the first formation pattern $\mathcal{T}_q$, $q = 1, ..., Q$ in formation candidates in $\mathcal{T}$ satisfying $\varepsilon_q(t) < 1$. That means $\varepsilon_q(t) \geq 1$ is an event to trigger the formation switch for collision avoidance.

13.2.2 SYSTEM DYNAMICS & FORMATION MANEUVERS

Consider a network of MAS $\mathcal{V} = \{v_1, ..., v_N\}$, where $v_i, i = 1, ..., N$ represents i^{th} agent in the network and N is the number of agents. Denote $x_i(t) \in \mathbb{R}^m$ as the position of agent v_i at time t, where $m = 2$ (2D space) or $m = 3$ (3D space).

The center of agents is defined as the virtual leader v_0, whose position is written as x_0 that

$$x_0 = \frac{1}{N}\sum_{i=1}^{N} x_i. \tag{13.1}$$

Define the qth desired formation pattern in candidates $\mathcal{T}_q(t)$ at time t to be a set $\mathcal{T}_q(t) = \left\{x_1^{\mathrm{d}}(t), ..., x_N^{\mathrm{d}}(t)\right\}$, where $x_i^{\mathrm{d}}(t)$ is the desired position of the agent v_i. The center of the desired formation is taken as the formation reference, whose position is written as x^{r} that

$$x^{\mathrm{r}} = \frac{1}{N}\sum_{i=1}^{N} x_i^{\mathrm{d}}. \tag{13.2}$$

The desired positions of all agents $x_i^{\mathrm{d}}(t)$ $(i = 1, ..., N)$ can be calculated relative to the formation reference x^{r}.

We assume N agents with the similar dynamic described by the following practical nonlinear differential equation as

$$\dot{x}_i = f_i(x_i) + u_i \tag{13.3}$$

where $u_i(t)$ represents the control input vector of agent v_i and $f_i : \mathbb{R}^m \to \mathbb{R}^m$ is unknown nonlinear dynamics of agent v_i. Denote $\widetilde{x}_i \triangleq x_i - x_i^{\mathrm{d}}$ as the formation tracking error vector of agent v_i.

The control objective is to design a controller for MAS with uncertain dynamics initialized on random positions to track a given trajectory in formation in non-omniscient constrained space. Denoting the forbidden space as a set Π, agents' positions must satisfy the spatial constraint condition that

$$x_i(t) \notin \Pi \qquad i = 1, ..., N, \quad 0 \leq t < \infty \tag{13.4}$$

during the whole task. The control goal can be expressed mathematically as for every $t \geq 0$,

$$\sum_{i=1}^{N} \|\widetilde{x}_i(t)\| \leq \varrho, \tag{13.5}$$

where ϱ is a positive constant.
Case 1: when $\Pi = \emptyset$, ϱ can be made sufficiently small when $t \to \infty$, that

$$\lim_{t\to\infty} \sum_{i=1}^{N} \|\widetilde{x}_i(t)\| \to 0 \tag{13.6}$$

Case 2: when $\Pi \neq \emptyset$,

$$\varrho \propto \frac{1}{\rho(x_i, \Pi)}, \qquad \text{and} \quad \varrho \leq \varrho_{\max}, \tag{13.7}$$

where $\rho(x_i, \Pi) = \inf_{s\in\Pi} \rho(x_i, s)$, and $s \in \Pi$. $\rho(\cdot,\cdot)$ is defined as Euclidean distance in this chapter, and $\varrho_{\max}$ indicate tolerable upper limits for the formation error.

13.3 CONTROL DESIGN

13.3.1 ARTIFICIAL POTENTIAL FIELD

A dynamic artificial potential field will be proposed to generate a specific potential force for MAS to track a given trajectory without collisions. The control input $u_i(t)$ in Eq. (13.3) will be formulated as the implementation of the potential force, which enables motion of MAS to achieve the control objective.

Considering the control goal that MAS tracks the given trajectory while keeping the desired formation as well as avoiding internal collisions and collisions with spatial constraints, the artificial potential field for agent v_i is designed accordingly as

$$U^{\mathrm{art}}(x_i) = U^{\mathrm{t}}(x_i) + U^{\mathrm{a}}(x_i) + U^{\mathrm{c}}(x_i), \quad i = 1, ..., N \tag{13.8}$$

where $U^{\mathrm{t}}(x_i)$ is proposed for trajectory tracking, $U^{\mathrm{a}}(x_i)$ is considered to avoid collisions with other agents. Therefore, they can be defined as

$$U^{\mathrm{t}}(x_i) = \frac{1}{2}k_{\mathrm{t}}(x_i - x_i^{\mathrm{d}})^{\mathrm{T}}(x_i - x_i^{\mathrm{d}}) = \frac{1}{2}k_{\mathrm{t}}\widetilde{x}_i^{\mathrm{T}}\widetilde{x}_i \tag{13.9}$$

$$\begin{aligned} U^{\mathrm{a}}(x_i) &= \frac{1}{2}k_{\mathrm{a}} \sum_{j=1, j\neq i}^{N} \frac{1}{(x_i - x_j)^{\mathrm{T}}(x_i - x_j)} \\ &= \frac{1}{2}k_{\mathrm{a}} \sum_{j=1, j\neq i}^{N} \frac{1}{x_{ij}^{\mathrm{T}} x_{ij}} \end{aligned} \tag{13.10}$$

where $k_{\mathrm{t}}, k_{\mathrm{a}} > 0$ are two scalars, and $x_{ij} = x_i - x_j$.

Besides, $U^{\mathrm{c}}(x_i)$ is designed to avoid collisions with environmental obstacles. Let $\Pi = \Pi_1 ... \cup \Pi_R$, where Π_r $(r = 1, ..., R)$ is defined as a continuous constrained space, and R is the total number of continuous constrained space. Denote L_r as the edge of region Π_r.

Then Dirac delta function, or simply called δ function [107] is applied to integrate the potential force from all points on edges of spatial constraints satisfying the following condition

$$x_i - s = kP_r(s), \quad r = 1, ..., R \tag{13.11}$$

where k is a real number. Here, P_r is the perpendicular direction of the slope of L_r, which is also the direction of potential force, denoted as $P_r(s) = \begin{bmatrix} \frac{\partial L_r}{\partial s_2} & \frac{\partial L_r}{\partial s_1} \end{bmatrix}^{\mathrm{T}}$, where $s \in L_r$, and s_1 and s_2 are the first and second dimension of s. Then we obtain

$$U^{\mathrm{c}}(x_i) = \sum_{r=1}^{R} \int_{-\infty}^{+\infty} \int_{-\infty}^{+\infty} k_{cr} \frac{\delta(\|x_i - s - kP_r(s)\|)}{\|x_i - s\| - s_0} \mathrm{d}s_1 \mathrm{d}s_2 \tag{13.12}$$

where $s_0 > 0$ is set as safety distance for emergent collision avoidance, $k_{cr} > 0$. In the following chapter, some simplified denotations will be used for convenience and readability, e.g., defining "$\int_{-\infty}^{+\infty} \int_{-\infty}^{+\infty} \mathrm{d}s_1 \mathrm{d}s_2$" as "$\int \mathrm{d}s$", $\mathfrak{S}_{ir} = x_i - s - kP_r(s)$, indicating the condition to pick force points, and $\mathfrak{X}_i = \|x_i - s\| - s_0$, representing the safe distance from a certain edge of obstacles.

Here we provide a method to model and express spatial constraints in potential functions, mainly including borders and obstacles in a common way. Please refer to [151] for the detailed process of modeling spatial constraints. The joint effect of $U^{\mathrm{t}}(x_i)$, $U^{\mathrm{a}}(x_i)$ and $U^{\mathrm{c}}(x_i)$ is $U^{\mathrm{art}}(x_i)$, which is artificially designed for the desired control goal.

13.3.2 LOCAL PATH REPLANNING

This subsection is to address the undesirable chattering rendered by current adaptation laws for NN-based controller such as σ-modification and projection algorithm.. This is because the original adaptation law used to update the weight matrixes for NNs aims to approximate the original desired trajectory while the potential field generated by spatial constraints may have the exactly opposite effect for collision avoidance.

To solve this issue, a local path replanning algorithm is proposed, in which the potential forces generated by obstacles are acted on the desired trajectory rather than directly on the agent itself, such that a replanned desired trajectory can be generated on-line. Defining these potential sources (e.g., spatial constraints or other agents) as avoidless disturbance, the potential field is accordingly denoted as $b_k, k = 1, ..., N_b$, where N_b is the total number of avoidless disturbance. In this chapter, $N_b = 2$ indicating the repulsive force generated by the spatial constraints and other agents respectively, as below

$$b_1 = \sum_{r=1}^{R} \int k_{cr} \frac{\delta(\|\mathfrak{S}_{ir}\|)}{\mathfrak{X}_i} \mathrm{ds} \tag{13.13}$$

$$b_2 = \frac{1}{2} k_a \sum_{j=1, j\neq i}^{N} \frac{1}{x_{ij}^{\mathrm{T}} x_{ij}} \tag{13.14}$$

Define a modifying vector z_i for each agent as the updating component for the desired trajectory that

$$z_i = -\sum_{k=1}^{N_b} \frac{\partial b_k}{\partial x_i} - k_{\mathrm{t}}(x_i^{\mathrm{d}*} - x_i^{\mathrm{d}}) \tag{13.15}$$

where $x_i^{\mathrm{d}*}$ is the modified desired trajectory. The first term of Eq. (13.15) is made up by the gradient of potential field generated by avoidless disturbance, i.e.,

$$\begin{aligned} \frac{\partial b_1}{\partial x_i} &= -f_i^{\mathrm{c}} = \nabla U^{\mathrm{c}} \\ &= -\sum_{r=1}^{R} k_{cr} \int\int \frac{\delta(\|\mathfrak{S}_i\|)(x_i - s)}{\mathfrak{X}_i^2 \|x_i - \varepsilon\|} \mathrm{ds}_1 \mathrm{ds}_2 \end{aligned} \tag{13.16}$$

$$\frac{\partial b_2}{\partial x_i} = -f_i^{\mathrm{a}} = \nabla U^{\mathrm{a}} = -2k_{\mathrm{a}} \sum_{j=1, j\neq i}^{N} \frac{x_{ij}}{\|x_{ij}\|^4} \tag{13.17}$$

while the second term of Eq. (13.15) is the state feedback term. The desired path $x_i^{\mathrm{d}*}$ is on-line replanned as

$$\dot{x}_i^{\mathrm{d}*}(t) = \dot{x}_i^{\mathrm{d}}(t) + z_i(t) \tag{13.18}$$

The modified desired position $x_i^{\mathrm{d}*}$ at time t can be written as

$$x_i^{\mathrm{d}*}(t) = x_i^{\mathrm{d}}(0) + \int_0^t \dot{x}_i^{\mathrm{d}*}(\tau)\mathrm{d}\tau \tag{13.19}$$

or

$$x_i^{\mathrm{d}*}(t) = x_i^{\mathrm{d}}(t_0) + \int_{t_0}^t \dot{x}_i^{\mathrm{d}*}(\tau)\mathrm{d}\tau \tag{13.20}$$

where $t \geq t_0 > 0$. A visual tracking error $\widetilde{x}_i^*$ is also defined to measure the tracking error of the modified trajectory $x_i^{\mathrm{d}*}$ as

$$\widetilde{x}_i^* \triangleq x_i - x_i^{\mathrm{d}*} \tag{13.21}$$

Meanwhile, let

$$\widetilde{x}_i^{\mathrm{d}} \triangleq x_i^{\mathrm{d}*} - x_i^{\mathrm{d}} \tag{13.22}$$

indicating the bias between the modified trajectory $x_i^{\mathrm{d}*}$ and the original desired trajectory x_i^{d}, whose derivative can be written as

$$\dot{\widetilde{x}}_i^{\mathrm{d}} = \dot{x}_i^{\mathrm{d}*} - \dot{x}_i^{\mathrm{d}} = z_i \tag{13.23}$$

By Eq. (13.22), Eq. (13.15) can be rewritten as

$$z_i = f_i^{\mathrm{c}} + f_i^{\mathrm{a}} - k_{\mathrm{t}}\widetilde{x}_i^{\mathrm{d}} \tag{13.24}$$

13.3.3 NN-BASED CONTROLLER

As analyzed in Section 13.3.2, the potential fields generated by environmental obstacles and other agents for collision avoidance are worked on the desired trajectory to replan the trajectory $x_i^{\mathrm{d}*}$. Thus, the attractive potential source x_i^{d} is replaced by $x_i^{\mathrm{d}*}$. The attractive potential force f_i^{t} is redesigned as

$$f_i^{\mathrm{t}} = -k_{\mathrm{t}}\widetilde{x}_i^* \tag{13.25}$$

The control scheme u_i is chosen as

$$u_i = -f_i(x_i) - k_{\mathrm{t}}\widetilde{x}_i^* + \dot{x}_i^{\mathrm{d}*} \tag{13.26}$$

based on the designed potential force f_i^{t} and its motion (to be $\dot{x}_i^{\mathrm{d}*}$) combining the system dynamics in Eq. (13.3).

Note that the functions $f_i(x_i)$ is an unknown component in system dynamics, the desired controller cannot be implemented in practice. Hence, the radial basis function neural network (RBFNN) [174] is employed to approximate $f_i(x_i) : \mathbb{R}^m \rightarrow \mathbb{R}^m$ as follows

$$f_i(x_i) = W_i^{\mathrm{T}} S(x_i) \tag{13.27}$$

where weight matrix $W_i \in \mathbb{R}^{l\times m}$, l denotes the number of neurons, and $S(x_i) = [s_1(x_i), ..., s_l(x_i)]^{\mathrm{T}}$ with

$$s_j(x_i) = \exp[\frac{-(x_i-\mu_j)^{\mathrm{T}}(x_i-\mu_j)}{\sigma_j^2}], \quad j = 1, ..., l \tag{13.28}$$

and $\mu_j \in \mathbb{R}^m$ is the center of receptive field and $(\sqrt{2}\sigma_j)/2$ is the width of the Gaussian function. For any given positive constant d_0, there exists the weight matrix W^* such that

$$f(x_i) = W^{*\mathrm{T}} S(x_i) + d_i \tag{13.29}$$

giving a sufficiently large l, where d_i is the bounded function approximation error satisfying $\|d_l\| < d_0$ in Ω_Z. The optimal matrix W_i^* can be theoretically derived from

$$W_i^* = \arg \min_{W_i\in\mathbb{R}^{l\times m}} \{ \sup_{x_i\in\Omega_Z} \|f_i(x_i) - W_i^{\mathrm{T}} S(x_i)\|\} \tag{13.30}$$

For real application, its estimation $\hat{W}_i$ is used for the practical function approximation. The estimation of $f(x_i)$ is given by

$$\hat{f}_i(x_i) = \hat{W}_i^{\mathrm{T}} S(x_i). \tag{13.31}$$

The updating law for the RBFNN weight matrix $\hat{W}_i$ is derived based on the projection algorithm as Eqs. (13.32–13.33), where $W_i^{\max}$ is a given positive constant for limiting the NN weight matrixes $\hat{W}_i$. $W_i^{\max}$ is selected to satisfy $\mathrm{Tr}(W_i^{*\mathrm{T}} W_i^*) \leq W_i^{\max}$, and $\Gamma_i > 0$ is a adaption gain.

$$\dot{\hat{W}}_i = \begin{cases} \Gamma_i S_i \widetilde{x}_i^{*\mathrm{T}} & (13.32) \\ \text{if } \mathrm{Tr}(\hat{W}_i^{\mathrm{T}}\hat{W}_i) < W_i^{\max} \text{or if } \mathrm{Tr}(\hat{W}_i^{\mathrm{T}}\hat{W}_i) = W_i^{\max} \text{ and } \widetilde{x}_i^{*\mathrm{T}}\hat{W}_i^{\mathrm{T}} S_i < 0 & \\ \Gamma_i \left(S_i \widetilde{x}_i^{*\mathrm{T}} - \dfrac{\widetilde{x}_i^{*\mathrm{T}}\hat{W}_i^{\mathrm{T}} S_i}{\mathrm{Tr}(\hat{W}_i^{\mathrm{T}}\hat{W}_i)} \hat{W}_i \right) & (13.33) \\ \text{if } \mathrm{Tr}(\hat{W}_i^{\mathrm{T}}\hat{W}_i) = W_i^{\max} \text{ and } \widetilde{x}_i^{*\mathrm{T}}\hat{W}_i^{\mathrm{T}} S_i \geq 0 & \end{cases}$$

To conclude, an NN-based controller is proposed as follows:

$$\begin{aligned} u_i &= u_{i1} + u_{i2} + u_{i3} + u_{i4} && (13.34) \\ u_{i1} &= -\hat{W}_i^{\mathrm{T}} S_i(x_i) && (13.35) \\ u_{i2} &= f_i^{\mathrm{t}*} = -k_{\mathrm{t}} \widetilde{x}_i^* && (13.36) \\ u_{i3} &= \dot{x}_i^{\mathrm{d}*} && (13.37) \\ u_{i4} &= -\hat{d}_i && (13.38) \end{aligned}$$

where $\hat{d}_i$ is the estimation of d_i, and its estimation error is $\widetilde{d}_i = \hat{d}_i - d_i$ with the adaptation scheme given as

$$\dot{\hat{d}}_i = \beta_i \widetilde{x}_i^* \tag{13.39}$$

where β_i is a positive scalar.

13.3.4 STABILITY ANALYSIS

Theorem 13.1

Consider N mobile agents with similar dynamics Eq. (13.3) under Assumption 30–33, moving in constrained space, with formation control laws Eqs. (13.34–13.38), local path updating law Eq. (13.24), and NN weight updating laws Eqs. (13.32–13.33). Denote ϕ_i as

$$\begin{aligned}\phi_i &= -f_i^{\mathrm{aT}}\left(\widetilde{W}_i^{\mathrm{T}} S_i(x_i) - \widetilde{d}_i + f_i^{\mathrm{t}*} - \widetilde{x}_i^{\mathrm{d}} + z_i\right) \\ &\quad -f_i^{\mathrm{cT}}\left(\widetilde{W}_i^{\mathrm{T}} S_i(x_i) - \widetilde{d}_i + f_i^{\mathrm{t}*} - \widetilde{x}_i^{\mathrm{d}} + \dot{x}_i^{\mathrm{d}*}\right) \end{aligned} \tag{13.40}$$

For bounded initial conditions without considering impulsive influence from spatial constraints, we have that the output tracking error $\widetilde{x}_i$ converges to a small neighborhood around zero by appropriately choosing the design parameters if any one of the following conditions is satisfied:

$$\phi_i \leq 0 \tag{13.41}$$

or

$$\|\widetilde{x}_i^*\| \geq \sqrt{\frac{\phi_i}{k_{\mathrm{t}}}}, \quad \text{if} \quad \phi_i > 0 \tag{13.42}$$

or

$$\|\widetilde{x}_i^{\mathrm{d}}\| \geq \sqrt{\frac{\phi_i}{k_{\mathrm{t}}}}, \quad \text{if} \quad \phi_i > 0 \tag{13.43}$$

■

Proof of Theorem 13.1. Consider the following Lyapunov candidate

$$\begin{aligned} V &= \sum_{i=1}^{N}\left[\frac{1}{2}\widetilde{x}_i^{*\mathrm{T}}\widetilde{x}_i^* + \sum_{r=1}^{R}\int k_{cr}\frac{\delta(\|\mathfrak{S}_i\|)}{\mathfrak{X}_i}\mathrm{ds}\right. \\ &\quad + \frac{1}{2}k_{\mathrm{a}}\sum_{j=1,j\neq i}^{N}\frac{1}{x_{ij}^{\mathrm{T}}x_{ij}} + \frac{1}{2}\mathrm{Tr}\left(\frac{1}{\Gamma_i}\widetilde{W}_i^{\mathrm{T}}\widetilde{W}_i\right) \\ &\quad \left. + \frac{1}{2}\widetilde{x}_i^{\mathrm{dT}}\widetilde{x}_i^{\mathrm{d}} + \frac{1}{2\beta_i}\widetilde{d}_i^{\mathrm{T}}\widetilde{d}_i\right]. \end{aligned} \tag{13.44}$$

The derivative of V is

$$\begin{aligned}\dot{V} &= \sum_{i=1}^{N}\Bigg[\widetilde{x}_i^{*\mathrm{T}}\dot{\widetilde{x}}_i^{*} - \frac{1}{2}\sum_{j=1,j\neq i}^{N} f_{ij}^{\mathrm{aT}}(\dot{x}_i-\dot{x}_j) \\ &\quad +\widetilde{x}_i^{\mathrm{dT}}\dot{\widetilde{x}}_i^{\mathrm{d}} - f_i^{\mathrm{cT}}(\dot{x}_i-\dot{s}) - \mathrm{Tr}(\frac{1}{\Gamma_i}\widetilde{W}_i^{\mathrm{T}}\dot{W}_i) \\ &\quad +\frac{1}{\beta_i}\widetilde{d}_i^{\mathrm{T}}\dot{\widetilde{d}}_i + \sum_{r=1}^{R}\int k_{cr}\frac{\dot{\delta}(\|\mathfrak{S}_i\|)}{\mathfrak{X}_i^2}\mathrm{ds}\Bigg] \qquad (13.45)\end{aligned}$$

where $\dot{s}=0$ based on Assumption 33 and we define $\widetilde{W}_i = W_i^* - \hat{W}_i$. It will be divided into two parts and studied in the following steps.

In the first step, we study the first six terms of $\dot{V}$ and mark them as $\dot{V}_1$ that

$$\begin{aligned}\dot{V}_1 &= \sum_{i=1}^{N}\Bigg[\widetilde{x}_i^{*\mathrm{T}}\dot{\widetilde{x}}_i^{*} + \widetilde{x}_i^{\mathrm{dT}}\dot{\widetilde{x}}_i^{\mathrm{d}} - \frac{1}{2}\sum_{j=1,j\neq i}^{N} f_{ij}^{\mathrm{aT}}(\dot{x}_i-\dot{x}_j) \\ &\quad - f_i^{\mathrm{cT}}\dot{x}_i - \mathrm{Tr}(\frac{1}{\Gamma_i}\widetilde{W}_i^{\mathrm{T}}\dot{\hat{W}}_i) + \frac{1}{\beta_i}\widetilde{d}_i^{\mathrm{T}}\dot{\widetilde{d}}_i\Bigg] \qquad (13.46)\end{aligned}$$

The derivative of $\widetilde{x}_i^*$ in Eq. (13.21) can be written as

$$\begin{aligned}\dot{\widetilde{x}}_i^{*} &= f_i(x_i) + u_i - \dot{x}_i^{\mathrm{d}*} \\ &= W_i^{*\mathrm{T}}S_i(x_i) + d_i + u_i - \dot{x}_i^{\mathrm{d}*} \qquad (13.47)\end{aligned}$$

where

$$d_i = f_i(x_i) - W_i^{*\mathrm{T}}S_i(x_i) \qquad (13.48)$$

Substituting Eqs. (13.34–13.38) into Eq. (13.47) yields

$$\dot{\widetilde{x}}_i^{*} = \widetilde{W}_i^{\mathrm{T}}S_i(x_i) - k_{\mathrm{t}}\widetilde{x}_i^{*} - \widetilde{d}_i \qquad (13.49)$$

Lemma 13.1

For any i ($i = 1, ..., N$) and any $j \neq i$, denoting

$$f_{ij}^{\mathrm{a}} \triangleq 2k_{\mathrm{a}}\frac{x_{ij}}{\|x_{ij}\|^4} \qquad (13.50)$$

it holds

$$\sum_{i=1}^{N}\sum_{j=1,j\neq i}^{N} f_{ij}^{\mathrm{aT}}(\dot{x}_i-\dot{x}_j) = 2\sum_{i=1}^{N} f_i^{\mathrm{aT}}(\dot{x}_i-\dot{x}^{\mathrm{r}}) \qquad (13.51)$$

■

Proof of Lemma 13.1. Refer to APPENDIX for the Proof of Lemma 13.1. □

With the help of Lemma 13.1 and substituting Eqs. (13.23–13.24, 13.39, 13.49) into Eq. (13.46), we obtain

$$\begin{aligned}\dot{V}_1 \;=\; & \sum_{i=1}^{N}\Big[-k_t\|\widetilde{x}_i^*\|^2 + \widetilde{x}_i^{*\mathrm{T}}\widetilde{W}_i^{\mathrm{T}}S_i(x_i) \\ & -f_i^{\mathrm{cT}}(f_i+u_i) - f_i^{\mathrm{aT}}(f_i+u_i-\dot{x}^{\mathrm{r}}) \\ & +\widetilde{x}_i^{\mathrm{dT}}\left(f_i^{\mathrm{c}}+f_i^{\mathrm{a}}-k_{\mathrm{t}}\widetilde{x}_i^{\mathrm{d}}\right) - \mathrm{Tr}\big(\frac{1}{\Gamma_i}\widetilde{W}_i^{\mathrm{T}}\dot{\hat{W}}_i\big)\Big]\end{aligned} \tag{13.52}$$

Substituting Eqs. (13.34–13.38) into Eq. (13.52) outputs

$$\begin{aligned}\dot{V}_1 = & \sum_{i=1}^{N}\Big[-k_{\mathrm{t}}\left(\|\widetilde{x}_i^*\|^2+\|\widetilde{x}_i^{\mathrm{d}}\|^2\right) \\ & -\mathrm{Tr}\left(\widetilde{W}_i^{\mathrm{T}}\left(\frac{1}{\Gamma_i}\dot{\hat{W}}_i - S_i(x_i)\widetilde{x}_i^{*\mathrm{T}}\right)\right) \\ & -f_i^{\mathrm{aT}}\left(\widetilde{W}_i^{\mathrm{T}}S_i(x_i)-\widetilde{d}_i+f_i^{\mathrm{t}*}-\widetilde{x}_i^{\mathrm{d}}+z_i\right) \\ & -f_i^{\mathrm{cT}}\left(\widetilde{W}_i^{\mathrm{T}}S_i(x_i)-\widetilde{d}_i+f_i^{\mathrm{t}*}-\widetilde{x}_i^{\mathrm{d}}+\dot{x}_i^{\mathrm{d}*}\right)\Big]\end{aligned} \tag{13.53}$$

Lemma 13.2

Updating NN weights by Eqs. (13.32–13.33), it holds that

$$\mathrm{Tr}\left[\widetilde{W}_i^{\mathrm{T}}\left(\frac{1}{\Gamma_i}\dot{\hat{W}}_i - S_i(x_i)\widetilde{x}_i^{*\mathrm{T}}\right)\right] \geq 0 \tag{13.54}$$

■

Proof of Lemma 13.2. Refer to APPENDIX for the Proof of Lemma 13.2. □

Based on Lemma 13.2, we have

$$\begin{aligned}\dot{V}_1 \;\leq\; & \sum_{i=1}^{N}\Big[-k_{\mathrm{t}}\left(\|\widetilde{x}_i^*\|^2+\|\widetilde{x}_i^{\mathrm{d}}\|^2\right) \\ & -f_i^{\mathrm{aT}}\left(\widetilde{W}_i^{\mathrm{T}}S_i(x_i)-\widetilde{d}_i+f_i^{\mathrm{t}*}-\widetilde{x}_i^{\mathrm{d}}+z_i\right)\end{aligned}$$

$$-f_i^{\mathrm{cT}}\left(\widetilde{W}_i^{\mathrm{T}}S_i(x_i)-\widetilde{d}_i+f_i^{\mathrm{t}*}-\widetilde{x}_i^{\mathrm{d}}+\dot{x}_i^{\mathrm{d}*}\right)\Big]$$

$$=\sum_{i=1}^{N}\left[-k_{\mathrm{t}}\|\widetilde{x}_i^*\|^2-k_{\mathrm{t}}\|\widetilde{x}_i^{\mathrm{d}}\|^2+\phi_i\right] \tag{13.55}$$

From Eq. (13.55), we know $\dot{V}_1$ is negative as long as any one of the conditions in Eqs. (13.41–13.43) is satisfied, such that the MAS can be asymptotic stable, and the desired formation tracking can be achieved.

In the second step, we prove that no internal collision occurs if any one of conditions in Eqs. (13.41–13.43) holds without considering impulsive influence from spatial constraints.

From the above proof process, we known that $\dot{V}_1\leq 0$. Integrating both sides in the interval $[0,t],\forall t>0$, we obtain that $V_1(t)\leq V_1(0)=V(0)$. With the definition of V in Eq. (13.44), we have

$$\sum_{i=1}^{N}[U^{\mathrm{a}}(t)+U^{\mathrm{c}}(t)]\leq V(0) \tag{13.56}$$

which also indicates $\sum_{i=1}^{N}U^{\mathrm{a}}(t)\leq V(0)$ & $\sum_{i=1}^{N}U^{\mathrm{c}}(t)\leq V(0)$. According to the definition of U^{a} and U^{c} in Eqs. (13.10,13.12), the boundedness of $\sum_{i=1}^{N}U^{\mathrm{a}}$ and $\sum_{i=1}^{N}U^{\mathrm{c}}$ indicates no collision happens for all $t>0$. □

In the third step, we study the last term of $\dot{V}$ in Eq. (13.45). It is to analyze the impulsive influence from spatial constraints. A similar process have been reported in [151]. Thus the conclusion will not be elaborated in this chapter.

13.4 SIMULATIONS

In this section, we present three simulation instances to validate the performance of proposed method. Considering 5 agents randomly initialized within a circle range with radius $R=0.8m$, the safety distance is set as $s_0=0.1m$. We assume 5 agents have the following form of dynamics

$$\begin{bmatrix}\dot{x}_{i1}\\ \dot{x}_{i2}\end{bmatrix}=\begin{bmatrix}\sin(0.5x_{i1}(t))\\ \cos(0.1x_{i2}^2(t))\end{bmatrix}+\begin{bmatrix}x_{i1}\sin(0.2x_{i1}^2(t)) & x_{i1}\sin(0.1x_{i2}(t))\\ x_{i2}\cos(0.4x_{i2}(t)) & x_{i2}^2\sin(0.3x_{i1}(t))\end{bmatrix}u_i$$

The configuration of RBFNN is the same for all five agents. The number of neurons is chosen to be 25. The centers of RBFFNN activation functions $s(\cdot)$ are distributed evenly in 0.5 in the range $[-2,2]\times[-2,2]$. The initial RBFNN weight matrix $\hat{W}_i(0)$ is chosen to be a zero matrix.

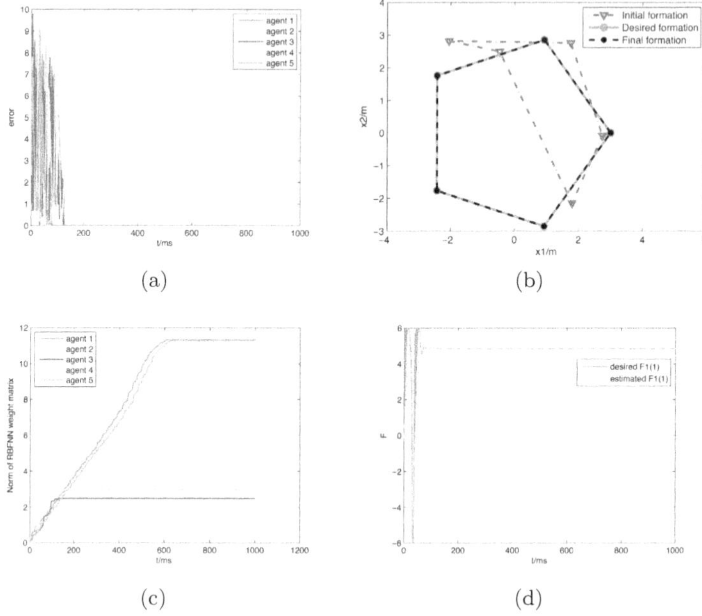

Figure 13.1 (a) Formation errors; (b) formation generation; (c) norm of RBFNN weigh matrix; (d) f_1 and its estimation.

13.4.1 FORMATION GENERATION

In the first simulation, we test the performance of proposed controller for formation generation when the desired formation is statical ($\dot{x}^{\mathrm{r}} = 0$), and also study the effectiveness of NN approximation. Simulation results are shown in Fig. 13.1.

The desired formation center is set as the origin. Fig. 13.1(a) describes that the formation errors oscillate in the beginning, but gradually decrease to zero. Fig. 13.1(b) plots the initial and final positions of 5 agents and shows the successful generation of the given formation. Fig. 13.1(c) indicates the norm of NN weight matrix converges to a constant. In Fig. 13.1(d), the successful approximation result of the first dimension of unknown function f_1, which equals to $\hat{f}_1 + \hat{d}_1$ is demonstrated . The rest of four have the similar performance, and there is no need to show each of them.

13.4.2 FORMATION TRACKING IN 2-D ENVIRONMENT WITH SPATIAL CONSTRAINTS

In the second simulation, we simulate a scenario with multiple obstacles with the desired trajectory to be set as $x^{\mathrm{r}}(t) = \begin{bmatrix} t & 20\sin(0.5t) \end{bmatrix}^{\mathrm{T}}$. Two border at positions $x_2 = -19$ and $x_2 = 17$ and two square obstacles at positions

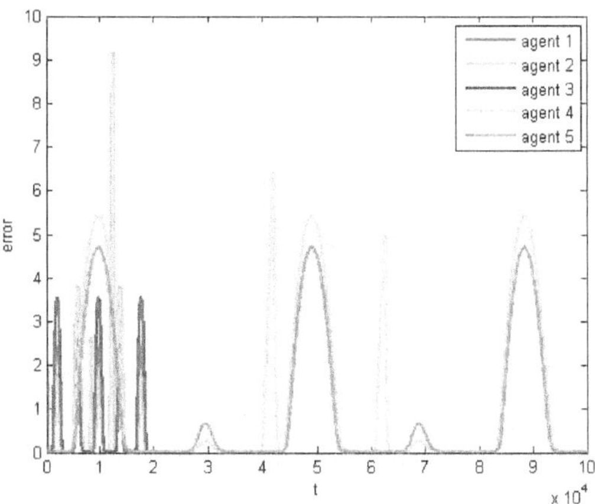

Figure 13.2 Tracking errors.

$x = \begin{bmatrix}13.25 & 5.25\end{bmatrix}^{\mathrm{T}}$ with 1.5m length of edge and $x = \begin{bmatrix}19.5 & -5\end{bmatrix}^{\mathrm{T}}$ with 2.0m length of edge are set. Simulation results are presented in Figs. 13.2-13.4.

Fig. 13.3 demonstrates the successful tracking of MAS in constrained space. The tracking error plotted in Fig. 13.2 increases when an agent gets close to obstacles, and converges to zero when being far away from obstacles. The details of collision avoidance are showed in Fig. 13.4.

As a contrast, Fig. 13.5 illustrates the tracking results of v_4 using NN-based controller, written as

$$u_i = -\hat{W}_i^{\mathrm{T}} S_i(x_i) - k_{\mathrm{t}}\widetilde{x}_i + \dot{x}_i^{\mathrm{d}} - \hat{d}_i \tag{13.57}$$

and PI controller ($K_p = 0.5$ and $K_i = 0.5$) without employing the local path replanning approach, where the line at $x_2 = -1$ is a environmental border. The result shows the undesirable chattering happens on the trajectory when the agent gets close to the spatial constraints, which verifies the significance and effectiveness of the proposed local path replanning scheme for tracking control with avoidless external disturbance. In addition, it can be observed that the NN-based controller has comparable performance in tracking control with PI controller.

13.4.3 SWITCHING FORMATION IN 3-D CONSTRAINED SPACE

In the last simulation, the formation tracking in 3-D restricted space is simulated where the path is not rigidly-passable thus formation switching is implemented such that MAS can follow the given trajectory while avoid collisions with obstacles. The desired trajectory is set as $x^{\mathrm{r}}(t) = \begin{bmatrix}t & 0 & 0\end{bmatrix}^{\mathrm{T}}$. We define a set of desired formation candidates with three members as $\mathcal{T} = \{\mathcal{T}_1, \mathcal{T}_2, \mathcal{T}_3\}$, shown in Fig. 13.6. Simulation results are shown in Figs. 13.7–13.9.

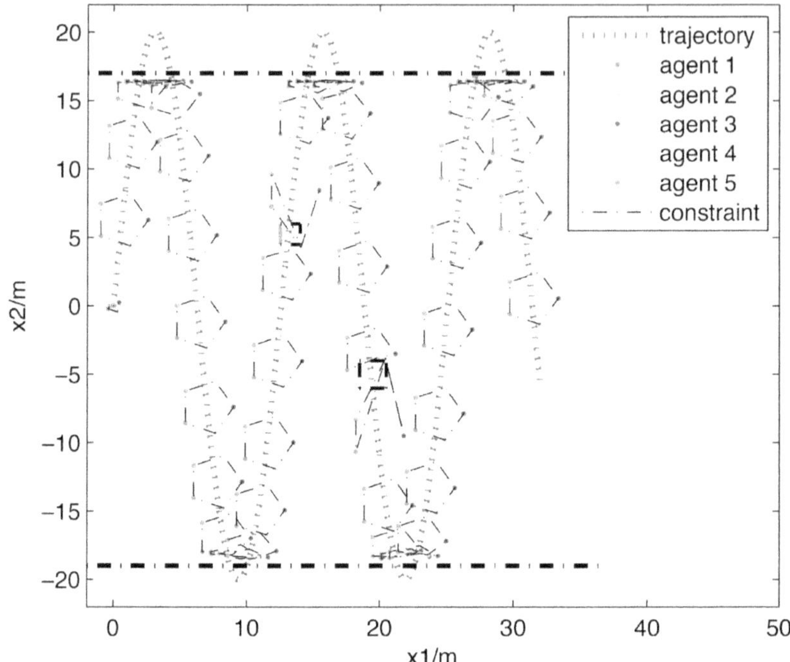

Figure 13.3 Formation tracking in constrained space.

The path is not rigidly-passable for the original formation, thus formation switching is carried for MAS to flexibly pass the path. Figs. 13.8 and 13.9 indicate that all agents can successfully track the desired trajectory while avoid collisions with obstacles. Fig. 13.7 plots the convergency of formation errors. In the second phase of task, there is a constant formation error due to the repulsive force of the spatial constraints.

13.5 CONCLUSION

In this chapter, formation tracking for MAS has been studied to control multiple agents under dynamic uncertainties to track a predefined trajectory with switching formation in non-omniscient constrained environment. The potential functions used to track the desired trajectory and avoid collisions have been established. Formation switching strategy has been considered to guarantee the passability of restricted path. The concept of avoidless disturbance has been proposed. To solve the undesirable chattering on the trajectory caused by avoidless disturbance and the existing adaptation algorithm, a local path replanning approach has been designed to act potential field generated by avoidless disturbance on the original desired trajectory rather than the agent's motion directly. An neural network-based controller has been designed and the

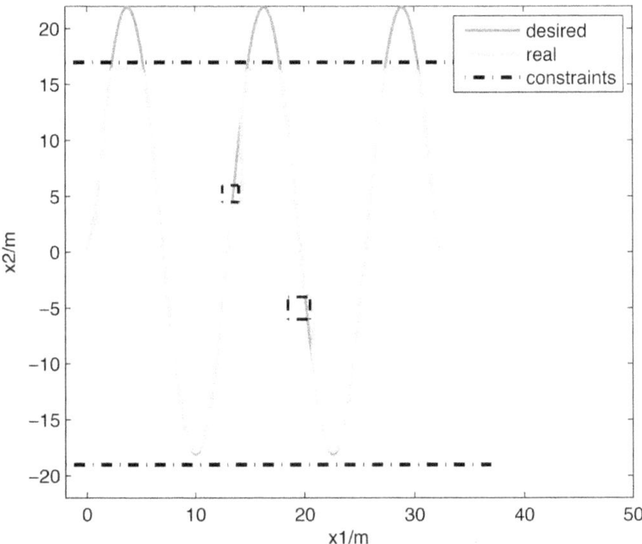

Figure 13.4 Process of how an agent avoids collision with spatial constraints.

stability has been analyzed using the Lyapunov functions. Simulations have been executed to demonstrate the effectiveness of proposed approaches. The contrast simulation studies have been carried out to verify the significance of the local path replanning method to the stability of the system with avoidless external disturbance. In addition, the outcome has proved the proposed neural network-based controller has comparable performance with the conventional PI controller.

Compared to [71, 192], the effect of external disturbance has been further investigated in this study for the case of collision avoidance in non-omniscient constrained space. Concerning the disturbance that should not be avoided, e.g., the influence of obstacles, we have proposes a strategy acting the influence of disturbance on the trajectory to realize the local path replanning rather than designing a controller to eliminate the effect of the disturbance in approximation, which will result in undesirable tracking chattering. The improvement enables the methods to be applied to realistic tasks of formation tracking in constrained space. Compared to [151], the design of neural network-based controller has addressed the issue of unknown or uncertain system dynamics of an agent, which exists commonly in real systems. The outcome will be more practical for various applications.

For the proposed schemes to be successful applied in practical applications, many issues must be further considered, such as communication limitations and communication delay [406, 492], which will be addressed in our future study.

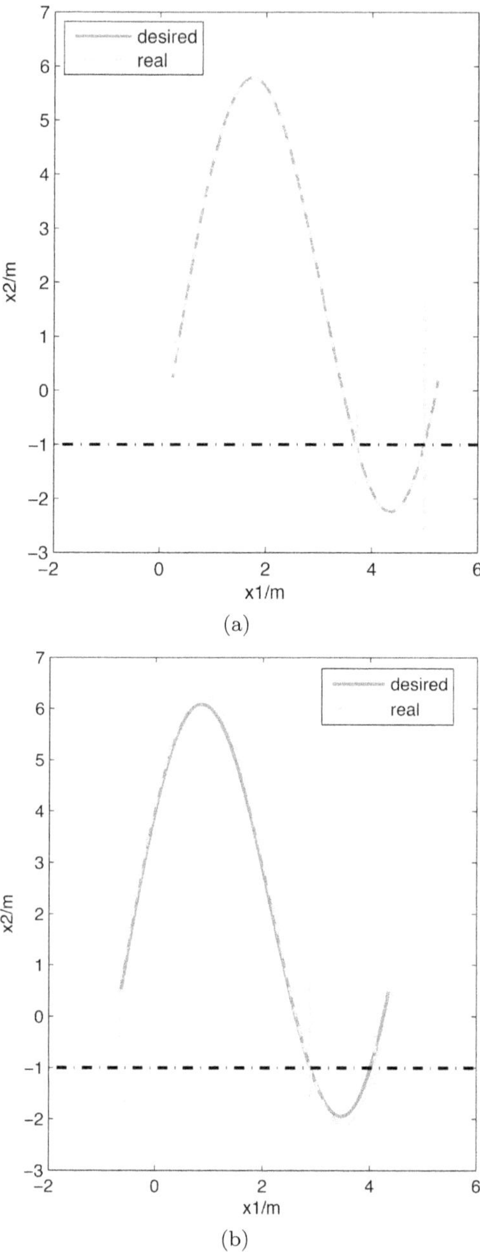

Figure 13.5 Undesirable tracking chattering in constrained space: (a) NN-based controller with σ-modification adaptation; (b) PI controller.

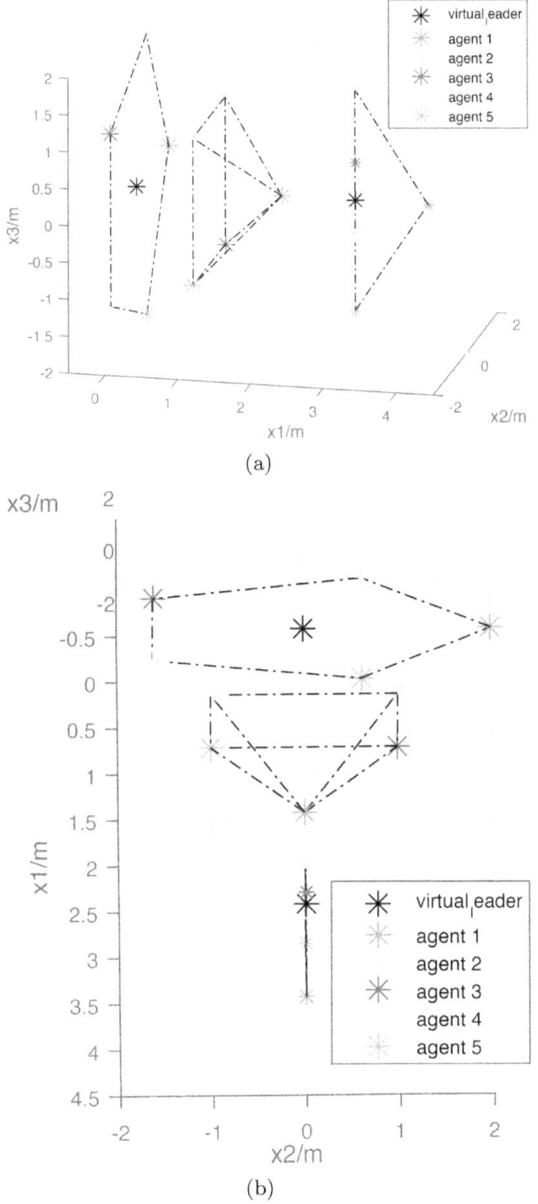

Figure 13.6 Formation candidates.

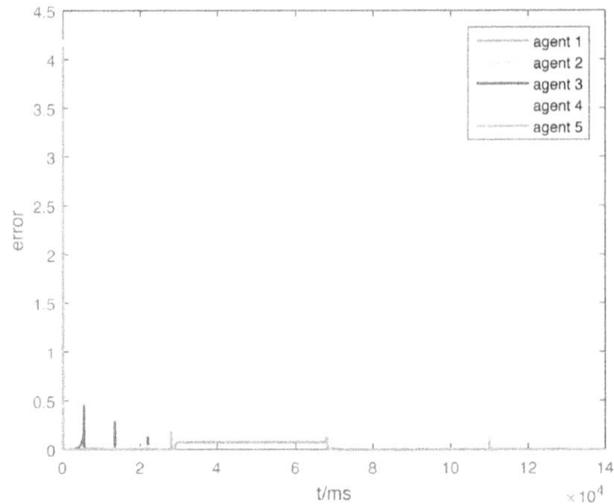

Figure 13.7 Formation errors.

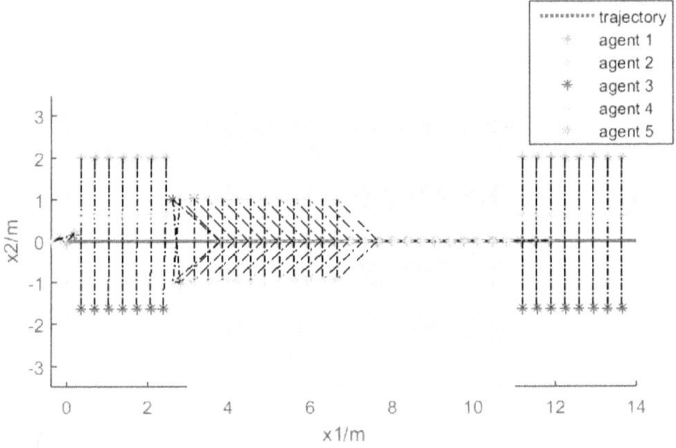

Figure 13.8 Switching formation tracking in restricted path, top view.

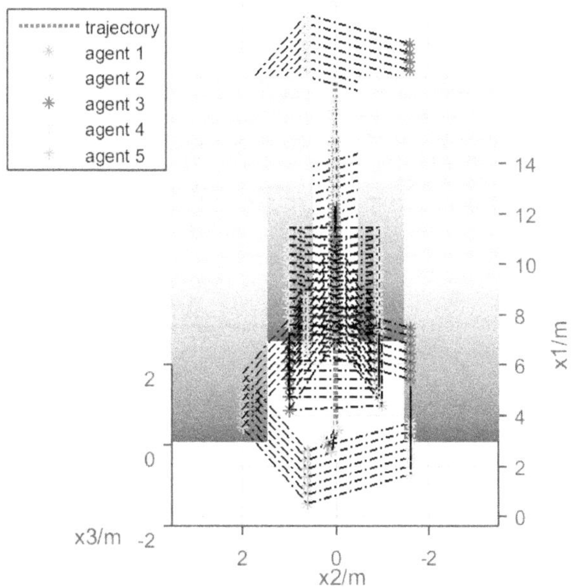

Figure 13.9 Switching formation tracking in restricted path, side view.

13.6 APPENDIX

Proof of Lemma 13.1.

$$
\begin{aligned}
& \sum_{i=1}^{N} \sum_{j=1, j \neq i}^{N} f_{ij}^{\mathrm{aT}} (\dot{x}_i - \dot{x}_j) \\
= \; & 2k_{\mathrm{a}} \sum_{i=1}^{N} \sum_{j=1, j \neq i}^{N} \frac{x_{ij}^{\mathrm{T}} (\dot{x}_i - \dot{x}_j)}{\|x_{ij}\|^4} \\
= \; & 2k_{\mathrm{a}} \sum_{i=1}^{N} \sum_{j=1, j \neq i}^{N} \frac{x_{ij}^{\mathrm{T}} (\dot{x}_i - \dot{x}^{\mathrm{r}} + \dot{x}^{\mathrm{r}} - \dot{x}_j)}{\|x_{ij}\|^4} \\
= \; & 2k_{\mathrm{a}} \sum_{i=1}^{N} \sum_{j-1, j \neq i}^{N} \frac{x_{ij}^{\mathrm{T}} (\dot{x}_i - \dot{x}^{\mathrm{r}})}{\|x_{ij}\|^4} + \frac{x_{ji}^{\mathrm{T}} (\dot{x}_j - \dot{x}^{\mathrm{r}})}{\|x_{ji}\|^4} \\
= \; & \underbrace{2k_{\mathrm{a}} \frac{x_{12}^{\mathrm{T}} (\dot{x}_1 - \dot{x}^{\mathrm{r}})}{\|x_{12}\|^4} + 2k_{\mathrm{a}} \frac{x_{21}^{\mathrm{T}} (\dot{x}_{21} - \dot{x}^{\mathrm{r}})}{\|x_{21}\|^4}}_{i=1, \quad j=2} + \cdots
\end{aligned}
$$

$$
\begin{aligned}
&\cdots + \underbrace{2k_{\mathrm{a}}\frac{x_{qp}^{\mathrm{T}}(\dot{x}_q - \dot{x}^{\mathrm{r}})}{\|x_{qp}\|^4} + 2k_{\mathrm{a}}\frac{x_{pq}^{\mathrm{T}}(\dot{x}_p - \dot{x}^{\mathrm{r}})}{\|x_{pq}\|^4}}_{i=q,\quad j=p} + \cdots \\
&\cdots + \underbrace{2k_{\mathrm{a}}\frac{x_{pq}^{\mathrm{T}}(\dot{x}_p - \dot{x}^{\mathrm{r}})}{\|x_{pq}\|^4} + 2k_{\mathrm{a}}\frac{x_{qp}^{\mathrm{T}}(\dot{x}_q - \dot{x}^{\mathrm{r}})}{\|x_{qp}\|^4}}_{i=p,\quad j=q} + \cdots \\
+ &\underbrace{2k_{\mathrm{a}}\frac{x_{N(N-1)}^{\mathrm{T}}(\dot{x}_N - \dot{x}^{\mathrm{r}})}{\|x_{N(N-1)}\|^4} + 2k_{\mathrm{a}}\frac{x_{(N-1)N}^{\mathrm{T}}(\dot{x}_{(N-1)N} - \dot{x}^{\mathrm{r}})}{\|x_{(N-1)N}\|^4}}_{i=N,\quad j=N-1} \\
= &\ 4k_{\mathrm{a}}\frac{x_{12}^{\mathrm{T}}(\dot{x}_1 - \dot{x}^{\mathrm{r}})}{\|x_{12}\|^4}\cdots + 4k_{\mathrm{a}}\frac{x_{qp}^{\mathrm{T}}(\dot{x}_q - \dot{x}^{\mathrm{r}})}{\|x_{qp}\|^4} + \cdots \\
&\cdots + 4k_{\mathrm{a}}\frac{x_{pq}^{\mathrm{T}}(\dot{x}_p - \dot{x}^{\mathrm{r}})}{\|x_{pq}\|^4} + \cdots + 4k_{\mathrm{a}}\frac{x_{N(N-1)}^{\mathrm{T}}(\dot{x}_N - \dot{x}^{\mathrm{r}})}{\|x_{N(N-1)}\|^4} \\
= &\ 4k_{\mathrm{a}}\sum_{i=1}^{N}\sum_{j=1, j\neq i}^{N}\frac{x_{ij}^{\mathrm{T}}(\dot{x}_i - \dot{x}^{\mathrm{r}})}{\|x_{ij}\|^4} \\
= &\ 2\sum_{i=1}^{N} f_i^{\mathrm{aT}}(\dot{x}_i - \dot{x}^{\mathrm{r}})
\end{aligned}
$$

□

Proof of Lemma 13.2. There must exist an upper bound $W_i^{\max}$ such that $\mathrm{Tr}(W_i^{*\mathrm{T}}W_i^*) \leq W_i^{\max}$ as W_i^* is a constant. Choose a sufficient large value for $W_i^{\max}$. Assuming the initial NN weight matrix $\hat{W}_i(0)$ satisfies that

$$
\mathrm{Tr}(\hat{W}_i^{\mathrm{T}}(0)\hat{W}_i(0)) \leq W_i^{\max} \tag{13.58}
$$

and the NN weight matrix updating law is implemented by Eqs. (13.32–13.33) then we have

$$
\forall t \geq 0 \quad \mathrm{Tr}(\hat{W}_i^{\mathrm{T}}(t)\hat{W}_i(t)) \leq W_i^{\max}, \quad i = 1, ..., N. \tag{13.59}
$$

The following cases are considered:

1) For $\dot{\hat{W}}_i = \Gamma_i S_i \widetilde{x}_i^{*\mathrm{T}}$, then

$$
\mathrm{Tr}[\widetilde{W}_i^{\mathrm{T}}(\frac{1}{\Gamma_i}\dot{\hat{W}}_i - S_i(x_i)\widetilde{x}_i^{*\mathrm{T}})] = 0
$$

2) For $\dot{\hat{W}}_i = \Gamma_i S_i \widetilde{x}_i^{*\mathrm{T}} - \Gamma_i \frac{\widetilde{x}_i^{*\mathrm{T}}\hat{W}_i^{\mathrm{T}}S_i}{\mathrm{Tr}(\hat{W}_i^{\mathrm{T}}\hat{W}_i)}\hat{W}_i$, then

$$
\mathrm{Tr}(\hat{W}_i^{\mathrm{T}}\hat{W}_i) = W_i^{\max} \quad \text{and} \quad \widetilde{x}_i^{*\mathrm{T}}\hat{W}_i^{\mathrm{T}}S_i \geq 0
$$

Therefore,

$$\begin{aligned} & \mathrm{Tr}[\widetilde{W}_i^{\mathrm{T}}(\frac{1}{\Gamma_i}\dot{\hat{W}}_i - S_i(x_i)\widetilde{x}_i^{*\mathrm{T}})] \\ = & -\frac{\widetilde{x}_i^{*\mathrm{T}}\hat{W}_i^{\mathrm{T}}S_i}{\mathrm{Tr}(\hat{W}_i^{\mathrm{T}}\hat{W}_i)}\mathrm{Tr}(\widetilde{W}_i^{\mathrm{T}}\hat{W}_i) \end{aligned} \tag{13.60}$$

It is noted that

$$\begin{aligned} \mathrm{Tr}(\widetilde{W}_i^{\mathrm{T}}W_i) &= \mathrm{Tr}(\widetilde{W}_i^{\mathrm{T}}W_i^*) - \mathrm{Tr}(\widetilde{W}_i^{\mathrm{T}}\widetilde{W}_i) \\ &= \frac{1}{2}\mathrm{Tr}(W_i^{*\mathrm{T}}W_i^*) - \frac{1}{2}\mathrm{Tr}(\widetilde{W}_i^{\mathrm{T}}\widetilde{W}_i) \\ &\quad -\frac{1}{2}\mathrm{Tr}(\hat{W}_i^{\mathrm{T}}\hat{W}_i) \leq 0 \end{aligned} \tag{13.61}$$

where the facts $\mathrm{Tr}(\hat{W}_i^{\mathrm{T}}\hat{W}_i) = W_i^{\max} \geq \mathrm{Tr}(W_i^{*\mathrm{T}}W_i^*)$ and $\mathrm{Tr}(\widetilde{W}_i^{\mathrm{T}}\widetilde{W}_i) \geq 0$ have been used. Then it is easy to prove that

$$\mathrm{Tr}\left[\widetilde{W}_i^{\mathrm{T}}(\frac{1}{\Gamma_i}\dot{\hat{W}}_i - S_i(x_i)\widetilde{x}_i^{*\mathrm{T}})\right] \geq 0. \tag{13.62}$$

Therefore, in both cases, the following condition holds

$$\mathrm{Tr}\left[\widetilde{W}_i^{\mathrm{T}}(\frac{1}{\Gamma_i}\dot{\hat{W}}_i - S_i(x_i)\widetilde{x}_i^{*\mathrm{T}})\right] \geq 0. \tag{13.63}$$

□

14 Role Switching for Tracking Control of Multi-Agents in Constrained Space

In this chapter, formation tracking control is studied for multi-agent systems (MAS) with communication limitations. The objective is to control a group of agents to track a desired trajectory while maintaining a given formation in non-omniscient constrained space. In the situation where agents do not have global environmental information, they have to avoid collisions with unexpected spatial constraints. A coordination mechanism is proposed for the leader and followers to avoid collisions together while maintaining the given formation by locally adjusting the predefined trajectory. A formation scaling factor is introduced to scale up or scale down the given formation size in the case that the region is impassable for MAS with the original formation size. Controllers for the leader and followers are designed and the adaptation law is developed to update the formation scaling factor. The conditions for asymptotic stability of MAS are discussed based on the Lyapunov theory. The maximum duration for which the conditions cannot be satisfied is used to indicate whether the tracking task should be given up. Simulation results are presented to illustrate the performance of proposed approaches.

14.1 INTRODUCTION

In recent years, there has been a tremendous interest in cooperative control of multi-agent systems (MAS) [193, 268, 498, 541]. Among all the topics in this field, formation control has attracted considerable attention from many researchers. Formation control is defined as the coordination of a group of robots that enter into and maintain a formation with specified geometrical shapes. Potential application areas of formation control include cooperative tasks such as exploring, surveillance, search and rescue, transporting of large objects and control of arrays of satellites [12, 73, 221, 227].

Leader-following formation control of multi-agent systems (MAS) have been widely studied in recent year [89, 127, 450], where the leader tracks a predefined path and the follower maintains a desired geometric configuration with the leader. The fact that only a single group leader is involved in the team implies that the leader-follower approach is simple to implement and understand, and the requirement on communication bandwidth is reduced. This is, however, a single point of massive failure type system because the loss of the group leader causes the entire group to fail. Another issue with

DOI: 10.1201/9781003298618-14

the typical leader-follower approach is the lack of inter-vehicle information feedback throughout the group.

The importance of adjustable leader/follower roles for shared control has been emphasized in a recent review in the field of human-robot interaction [207], and there are several works in this direction [208,334,335,347,414]. Such an human-robot system is formulated as a two-agent system with one leader and one follower. Similarly, in displacement-based control [85,283,399] of MAS, when given a task of trajectory tracking, only the leader has the knowledge of desired trajectory while the followers are aware of the displacements with respect to the leader to achieve the desired formation. In this structure, since only the followers track the leader and there is no feedback from the followers to the leader, if a follower fails to follow properly, no mechanism can guarantee the formation keeping. It is prone to failure especially in dynamic and uncertain environments. Furthermore, the positions of other followers have no influence on the motion of the leader agent, so the formation can become disjoint and followers can be left behind if they are not able to track the motion of the leader accurately.

These issues challenge the success of formation control especially in unknown spacial constrained environment, which have attracted increasing attention from different disciplines in recent research. In [2], Abdollahi and Rezaee proposed a technique based on the behavioral structure and designed an approach to avoid the obstacle by applying the rotational potential field. In this study, agents were designed to avoid collisions with obstacles individually, which means the coordination is broken temporarily during the process of collision avoidance. If any one of agents fail to follow, the original desired formation has to be given up. The situation where agents may not have global environmental information will lead to that they may have to avoid collisions with unexpected spatial constraints. The introduction of role switching for MAS is significant which provide a coordination mechanism to avoid collisions with spacial constraints. In [202], a scaling matrix was introduced to scale up or scale down the specific geometric formation shape depending on the given task in bounded region. Some limitations existed on the approach. Two objective functions have to be built respectively for outer sub-region and inner sub-region but no details are provided on construction method. The simulation examples actually assumed the formation center should be approximately the center of constrained region and in the mean time approximated the bounded region as ellipse, which is obviously unrealistic. In [317], Lu et al. studied the control of a group of mobile agents to form a desired formation while flocking in a constrained environment. However, there was no tracking requirement in their work, which means the role of leader and follower has little meaning for the task. The study of role-switching strategy for formation tracking in non-omniscient environment remains a significant topic.

Inspired by the unexpected multiple spatial constraints in bounded task space, the problem of coordination formation tracking control of MAS with limited communication in non-omniscient constrained space is studied in this chapter. The leader has the information of the trajectory and will be treated as the reference for formation tracking. When any one or more followers detect the existence of environmental obstacles, the MAS will convert to the obstacle avoidance mode. In this situation, followers in the influence range of the obstacles and the leader will switch into the role of coordinator, which will be influenced both by the potential filed generated by the given trajectory or spatial constraints and the neighbor agents' positions to achieve obstacle avoidance while keeping formation in the same time. The size of formation shape may be scale up or scale down in the case that the path is impassable for the MAS with original size.

The main contributions are as below:

A role-switching strategy is proposed for MAS to formation tracking in constrained space with limited communications. A novel role 'coordinator' is introduced as the coordination mechanism between the leader and followers to avoid collision with obstacles while maintaining given formation.

An adaptation law for formation scaling factor is designed such that the size of the specific geometric formation shape can be scale up or scale down according to various spatial constraints.

The conditions for asymptotic stability of MAS are studied based on the Lyapunov function, while the maximum duration that the conditions cannot be satisfied is obtained to indicate whether the formation tracking task should be given up.

14.2 SYSTEM DYNAMICS AND FORMATION MANEUVERS

Before proceeding further, the following assumptions are made in this chapter.

Assumption 34. *The influence of the size and the shape of an agent to the formation tracking control are ignored, which means an agent is assumed to be of point mass.*

Assumption 35. *An agent is able to estimate its position in the world coordinate system.*

Assumption 36. *The environment (spatial constraints) is uniform (time-invariant) in the whole task.*

Assumption 37. *Without loss of generality, it is assumed that the first agent v_1 is the pre-nominated leader for trajectory tracking, who knows the desired reference $\mathbf{y}^d(t)$ while other agents are unaware of the desired reference but know the agent is the leader.*

Assumption 38. *There exists a constant* ζ_1 *such that* $0 \le \|\dot{\mathbf{y}}^{\mathrm{d}}\| \le \zeta_1 < \infty$.

Assumption 38 implies the designed trajectory is smooth. Consider a network of N agents, $\mathbf{x}_i(t) \in \mathbb{R}^m$ is the position of agent i at time t. In this chapter, we consider $\mathbf{x}_i(t)$ to represent the position of an agent i in 2-D or 3-D space, which means $m = 2$ or $m = 3$.

Definition 48 (*Formation*). *A formation pattern at time t is defined to be a set*

$$\mathfrak{P}(t) = \{\mathbf{x}_1^{\mathrm{d}}(t), ..., \mathbf{x}_N^{\mathrm{d}}(t)\} \tag{14.1}$$

where $\mathbf{x}_i^{\mathrm{d}}(t)$ *is the desired position of the agent* i *at time* t, $i = 1, ..., N$.

Denote δ_i as the specific displacement between $\mathbf{x}_i^{\mathrm{d}}$ and $\mathbf{x}_1$ that

$$\mathbf{x}_i^{\mathrm{d}} = \mathbf{x}_1 + \delta_i \tag{14.2}$$

and the desired displacement between $\mathbf{x}_i$ and $\mathbf{x}_j$ is written as δ_{ij}, we have

$$\mathbf{x}_i^{\mathrm{d}} - \mathbf{x}_j^{\mathrm{d}} = \delta_i - \delta_j = \delta_{ij} \tag{14.3}$$

We assume N agents with the similar dynamics described by the following linear equation

$$\dot{\mathbf{x}}_i = \mathbf{u}_i \tag{14.4}$$

where $\mathbf{u}_i(t)$ represent the control vector agent v_i. The trajectory tracking error for the leader v_1 is marked as

$$\varepsilon_1 = \mathbf{x}_1 - \mathbf{y}^{\mathrm{d}} \tag{14.5}$$

while the formation error for followers $v_i, i = 2, ..., N$ is written as

$$\varepsilon_i = \mathbf{x}_i - \mathbf{x}_i^{\mathrm{d}} = \mathbf{x}_i - \mathbf{x}_1 - \delta_i \tag{14.6}$$

Control objective: Design a controller to control a group of agents initialized on random positions to track the desired trajectory in formation in non-omniscient constrained space. In other words, the center of agents and center of desired formation should coincide in the final. In the meantime, consider the unknown forbidden space as Π, which means that all agents' positions satisfy the spatial constraint condition that

$$\mathbf{x}_i(t) \notin \Pi \qquad i = 1, ..., N, 0 \le t < \infty \tag{14.7}$$

during the whole task. The control goal can be expressed mathematically as, every $t \ge 0$,

$$\sum_{i=1}^{N} \|\varepsilon_i(t)\| \le \epsilon, i = 1, ..., N \tag{14.8}$$

where ε and ϵ are positive constants.

Case 1: when $\Pi = \emptyset$, ε and ϵ can be made sufficiently small when $t \to \infty$, that

$$\sum_{i=1}^{N} \|\varepsilon_i(t)\| \leq 0, i = 1, ..., N \tag{14.9}$$

Case 2: when $\Pi \neq \emptyset$,

$$\epsilon \propto \frac{1}{\mathrm{d}(\mathbf{x}_i, \Pi)}, \qquad \text{and} \quad \epsilon \leq \epsilon_{max} \tag{14.10}$$

where $\mathrm{d}(\mathbf{x}_i, \Pi) = \inf_{\mathbf{s} \in \Pi} \rho(\mathbf{x}_i, \mathbf{s})$, and $\mathbf{s} \in \Pi$. $\rho(\cdot, \cdot)$ is defined as Euclidean distance in this chapter, and ϵ_{max} indicate tolerable upper limits for the formation error.

14.3 GRAPH THEORY

A team of agents interacts with each other via communication or sensing networks to achieve collective objectives. It is convenient to model the information exchanges among agents by undirected graphs. An undirected graph $\mathcal{G}$ is a pair $(\mathcal{V}, \mathcal{E})$, where $\mathcal{V} = \{v_1, ..., v_N\}$ is a nonempty finite node set and $\mathcal{E} \subseteq \mathcal{V} \times \mathcal{V}$ is an edge set of ordered pairs of nodes, called edges. The adjacency matrix $\mathcal{A} = [a_{ij}] \in \mathbb{R}^{N \times N}$ associated with indirected graph $\mathcal{G}$ is defined such that $a_{ii} = 0$, $a_{ij} = a_{ji} = 1$ if $(v_i, v_j) \in \mathcal{E}$ (or $(v_j, v_i) \in \mathcal{E}$) and $a_{ij} = 0$ otherwise. The Laplacian matrix $\mathcal{L} = [\mathcal{L}_{ij}] \in \mathbb{R}^{N \times N}$ of graph $\mathcal{G}$ is defined as $\mathcal{L}_{ii} = \sum_{j \neq i} a_{ij}$ and $\mathcal{L}_{ij} = -a_{ij}$, $i \neq j$. The Laplacian matrix can be written into a compact form as $\mathcal{L} = \mathcal{D} - \mathcal{A}$, where $\mathcal{D} = \mathrm{diag}(d_1, ..., d_N)$ is the degree matrix with d_i as the in-degree of the i-th node.

Assumption 39. *The subgraph $\mathcal{G}_s$ associated with the followers is undirected and in the graph $\mathcal{G}$ the leader has directed paths to all followers. (Equivalently, $\mathcal{G}$ contains a directed spanning tree with the leader as the root.)*

14.4 ROLE SWITCHING FOR MULTI-AGENTS COORDINATION

To proceed the role-switching strategy in this study, the following definitions of roles in the task of formation tracking in constrained space are made first.

Definition 49 (*Leader*)**.** *In MAS, the agent which has the knowledge of the desired trajectory or only be influenced by the environment is defied as the Leader.*

Definition 50 (*Coordinator*)**.** *In MAS, the agent whose motion is decided both by environment and neighbor positions is defined as the Coordinator.*

Definition 51 (*Follower*)**.** *In MAS, the agent only influenced by neighbor motions is defined as the Follower.*

14.4.1 TRAJECTORY TRACKING & FORMATION FOLLOWING

Based on Assumption 37, the first agent knows the reference trajectory $\mathbf{y}^d(t)$, so the control for first agent can be directly designed by using the certainly equivalence principal to track $\mathbf{y}^d(t)$. The controller is given as followers:

$$\mathbf{u}_1 = -k_t\varepsilon_1 + \dot{\mathbf{y}}^{\mathrm{d}} \tag{14.11}$$

On the other hand, followers $v_i, i = 2, ..., N$, have no information about the desired trajectory, however, track the leader in formation with the detection of neighbors' positions. The controller is proposed as

$$\begin{aligned} \mathbf{u}_i = & - k_n \frac{b_l}{d_i}(\mathbf{x}_i - \mathbf{x}_1 - \delta_i) - k_n \sum_{j=2}^{N} \frac{a_{ij}}{d_i}(\mathbf{x}_i - \mathbf{x}_j - \delta_{ij}) \\ & - k_g g\left(\frac{b_i}{d_i}(\mathbf{x}_i - \mathbf{x}_1 - \delta_i) + \sum_{j=2}^{N} \frac{a_{ij}}{d_i}(\mathbf{x}_i - \mathbf{x}_j - \delta_{ij})\right) \end{aligned} \tag{14.12}$$

where $k_n > 0$ and $k_g > 0 \in \mathbb{R}$ are two constants, and $g(\cdot)$ is a nonlinear (norm-normalized) sign function as defined in (2.32) such that for $\omega \in \mathbb{R}^m$,

$$g(\omega) = \begin{cases} \frac{\omega}{\|\omega\|} & \text{if } \|\omega\| \neq 0 \quad (14.13) \\ 0 & \text{if } \|\omega\| = 0 \quad (14.14) \end{cases}$$

14.4.2 SPATIAL COLLISION AVOIDANCE SCHEME

$U_i^{\mathrm{c}}(\mathbf{x})$ is designed to avoid moving beyond constrained space. It is unreasonable to assume spatial constraints to be of point mass. Let $\Pi = \Pi_1 \cup \Pi_2 ... \cup \Pi_R$, where R is the number of continuous constrained space, for example, a border or an obstacle. L_r is denoted as the edge of region Π_r, $r = 1, ..., R$.

Here we introduce Dirac delta function, or simply called a δ function [107] to gather the potential force from all points on edges of spatial constraints satisfying the following condition

$$\mathbf{x}_i - \mathbf{s} = kP_r(\mathbf{s}), \quad r = 1, ..., R \tag{14.15}$$

where k is a real number. Here, P_r is the perpendicular direction of the slope of L_r, which is also the direction of potential force, denoted as $P_r(\mathbf{s}) = \begin{bmatrix} \frac{\partial L_r}{\partial s_2} & \frac{\partial L_r}{\partial s_1} \end{bmatrix}^{\mathrm{T}}$, where $\mathbf{s} \in L_r$, and s_1 and s_2 are the first and second dimension of $\mathbf{s}$. Then we obtain

$$U^{\mathrm{c}}(\mathbf{x}) = \sum_{r=1}^{R} \int_{-\infty}^{+\infty} \int_{-\infty}^{+\infty} \frac{\delta(\|\mathbf{x} - \mathbf{s} - kP_r(\mathbf{s})\|)}{\|\mathbf{x} - \mathbf{s}\| - s_0} \mathrm{d}s_1 \mathrm{d}s_2 \tag{14.16}$$

where s_0 is a small constant regarded as safety distance to avoid collisions with constraints edges, k_{cr} is a spatial constraint potential factor, and k is a positive constant. In the following chapter, we denote "$\int_{-\infty}^{+\infty}\int_{-\infty}^{+\infty} \mathrm{ds}_1\mathrm{ds}_2$" as "$\int \mathrm{ds}$", and denote $\mathfrak{S}_i = \mathbf{x}_i - \mathbf{s} - kP_r(\mathbf{s})$ for convenience, indicating the condition to pick force points, and denote $\mathfrak{X}_i = \|\mathbf{x}_i - \mathbf{s}\| - s_0$, representing the safe distance from a certain edge of borders or obstacles.

$$\mathbf{F}_i^{\mathrm{c}} = -\nabla U_i^{\mathrm{c}} = \sum_{r=1}^{R} \int\int \frac{\delta(\|\mathfrak{S}_i\|)(\mathbf{x}_i - \mathbf{s})}{\mathfrak{X}_i^2\|\mathbf{x}_i - \mathbf{s}\|} \mathrm{ds}_1\mathrm{ds}_2 \tag{14.17}$$

Here we provide a method to model and express spatial constraints in potential functions, mainly including borders and obstacles in a common way. Please refer to [151] for the detailed process of modeling spatial constraints.

Any agent satisfying the condition that $U_i^c > 0$ has the potential to be nominated as the temporary leader. The controller of the leader and followers satisfying Eq. (14.15) is switched into:

$$\begin{aligned} \mathbf{u}_1 =& k_c\mathbf{F}_1^{\mathrm{c}} + \dot{\mathbf{y}}^{\mathrm{d}} - k_t\varepsilon_1 - k_n \sum_{i=2}^{N} \frac{b_j}{d_1}(\mathbf{x}_1 - \mathbf{x}_i + \delta_i) \\ &\cdot \mathrm{sign_c}\left(\sum_{i=2}^{N} \frac{b_i}{d_1}(\mathbf{x}_1 - \mathbf{x}_i + \delta_i)^{\mathrm{T}}(k_c\mathbf{F}_1^{\mathrm{c}} - k_t\varepsilon_1)\right) \end{aligned} \tag{14.18}$$

$$\begin{aligned} \mathbf{u}_i =& k_c\mathbf{F}_i^{\mathrm{c}} - k_n\frac{b_i}{d_i}(\mathbf{x}_i - \mathbf{x}_1 - \delta_{ij}) - k_n\sum_{j=2}^{N}\frac{a_{ij}}{d_i}(\mathbf{x}_i - \mathbf{x}_j - \delta_{ij}) \\ &- k_g g\left(\frac{b_i}{d_i}(\mathbf{x}_i - \mathbf{x}_1 - \delta_{ij}) + \sum_{j=2}^{N}\frac{a_{ij}}{d_i}(\mathbf{x}_i - \mathbf{x}_j - \delta_{ij})\right) \\ & i = 2, ..., N \end{aligned} \tag{14.19}$$

where $k_c > 0$ is a constant and $\mathrm{sign_c}(\cdot)$ is the classical sign function as defined in (2.28).

In this situation, the leader downgrades to the coordinator to receive the information of spatial constraints detected by other agents then locally adjust the original trajectory to achieve collision avoidance. Meanwhile, the original followers $\{v_i|\mathbf{F}_i^{\mathrm{c}} \neq 0, i = 2, ..., N\}$ upgrade to the coordinator to spread the influence of spatial constraints to MAS communication network such that the given formation can be achieved while avoid collisions with obstacles.

14.4.3 FORMATION SCALING

The agent who is not pre-nominated leader or does not satisfies Eqs. (14.11–14.47) will be the formation follower in the task. The controller for follower is designed as below:

$$\begin{aligned}\mathbf{u}_1 &= k_c\mathbf{F}_1^c + \dot{\mathbf{y}}^d - k_t\varepsilon_1 - k_n\sum_{i=2}^{N}\frac{b_j}{d_1}(\mathbf{x}_1 - \mathbf{x}_i + \lambda\delta_i)\,\mathrm{sign_c} \\ &\cdot\left(\sum_{i=2}^{N}\frac{b_i}{d_1}(\mathbf{x}_1 - \mathbf{x}_i + \lambda\delta_i)^{\mathrm{T}}\left(k_c\mathbf{F}_1^c - k_t\varepsilon_1\right)\right)\end{aligned} \tag{14.20}$$

$$\begin{aligned}\mathbf{u}_i &= k_c\mathbf{F}_i^c - k_n\frac{b_i}{d_i}(\mathbf{x}_i - \mathbf{x}_1 - \lambda\delta_{ij}) - k_n\sum_{j=2}^{N}\frac{a_{ij}}{d_i}(\mathbf{x}_i - \mathbf{x}_j - \lambda\delta_{ij}) \\ &-k_g g\left(\frac{b_i}{d_i}(\mathbf{x}_i - \mathbf{x}_1 - \lambda\delta_{ij}) + \sum_{j=2}^{N}\frac{a_{ij}}{d_i}(\mathbf{x}_i - \mathbf{x}_j - \lambda\delta_{ij})\right)\end{aligned} \tag{14.21}$$

A time-varying scaling factor $\lambda(t)$ is introduced to adjust the size of the given formation geometric shape initialized as 1. The adaptation law will be studied in detail in the next section.

Define $\hat{\varepsilon}_i$ as the scaling formation error for agent $v_i, i = 2, ..., N$, such that

$$\hat{\varepsilon}_i = \mathbf{x}_i - \mathbf{x}_1 - \lambda\delta_i \tag{14.22}$$

Two examples that the formation scaling is required to enable MAS to pass the restricted region are described in Fig. 14.1. In the first case (see Fig. 14.1(a)), the size of formation should be scaled up; while in the second case (see Fig. 14.1(b)), the size of formation should be scaled down. We will elaborate the criteria in detail in the subsequent section.

14.5 CONTROL PERFORMANCE ANALYSIS

In this section, we will prove the effectiveness of the proposed strategies in several cases.

Case 1: Trajectory tracking and formation following in free space

---

Theorem 14.1

Consider N mobile agents with similar dynamics Eq. (14.4), moving in free space, with Assumptions 34–39. There is one leader knowing the information of the given trajectory with the control law Eq. (14.11) and the rest of followers knowing the information of displacement to achieve the desired geometric formation with the control law Eq. (14.12). The asymptotic stability of MAS without considering impulsive influence from spatial constraints can be acquired.

■

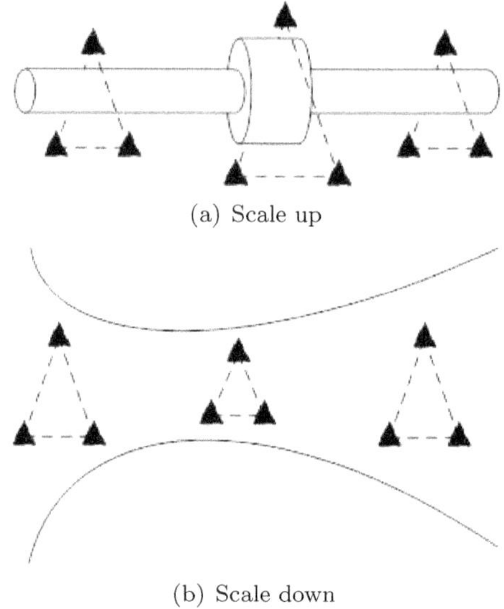

(a) Scale up

(b) Scale down

Figure 14.1 Formation scaling examples: (a) scale up; (b) scale down.

Proof of Theorem 14.1. **Step 1**: Leader tracking

Denote the tracking error of the leader v_1 as below

$$\varepsilon_1 = \mathbf{x}_1 - \mathbf{y}^{\mathrm{d}} \tag{14.23}$$

whose time derivative can be written as

$$\begin{aligned} \dot{\varepsilon}_1 &= \dot{\mathbf{x}}_1 - \dot{\mathbf{y}}^{\mathrm{d}} && (14.24) \\ &= \mathbf{u}_1 - \dot{\mathbf{y}}^{\mathrm{d}} && (14.25) \end{aligned}$$

Substituting Eq. (14.11) into Eq. (14.25), we obtain

$$\dot{\varepsilon}_1 = -k_t \varepsilon_1 \tag{14.26}$$

The Lyapunov candidate for leader tracking can be chosen as

$$V_1 = \frac{1}{2} \varepsilon_1^{\mathrm{T}} \varepsilon_1 \tag{14.27}$$

The derivative of V_1 is

$$\dot{V}_1 = \dot{\varepsilon}_1^{\mathrm{T}} \varepsilon_1 \tag{14.28}$$

Substituting Eq. (14.25) into Eq. (14.28), we have

$$\dot{V}_1 = -k_t \varepsilon_1^{\mathrm{T}} \varepsilon_1 \leq 0 \tag{14.29}$$

Therefore, the leader is able to track the desired trajectory in free space.

Step 2: Follower in formation Without considering the flexible formation adaptation in free space, the control law for followers in Eq. (14.21) can be rewritten as

$$\mathbf{u}_i = -k_n \frac{b_i}{d_i}\varepsilon_i - k_n \sum_{j=2}^{N} \frac{a_{ij}}{d_i}(\varepsilon_i - \varepsilon_j) - k_g g\left(\frac{b_i}{d_i}\varepsilon_i + \sum_{j=2}^{N} \frac{a_{ij}}{d_i}(\varepsilon_i - \varepsilon_j)\right) \quad (14.30)$$

Mark the formation following error as

$$\varepsilon_i = \mathbf{x}_i - \mathbf{x}_1 - \delta_i \quad (14.31)$$

whose deviation is

$$\begin{aligned} \dot{\varepsilon}_i &= \dot{\mathbf{x}}_i - \dot{\mathbf{x}}_1 &(14.32)\\ &= \mathbf{u}_i - \dot{\mathbf{x}}_1 &(14.33) \end{aligned}$$

Substituting Eq. (14.30) into Eq. (14.33), we obtain

$$\begin{aligned} \dot{\varepsilon}_i =& -k_n \frac{b_i}{d_i}\varepsilon_i - k_n \sum_{j=2}^{N} \frac{a_{ij}}{d_i}(\varepsilon_i - \varepsilon_j) \\ & - k_g g\left(\frac{b_i}{d_i}\varepsilon_i + \sum_{j=2}^{N} \frac{a_{ij}}{d_i}(\varepsilon_i - \varepsilon_j)\right) - \dot{\mathbf{x}}_1 \\ =& -k_n \frac{b_i}{d_i}\varepsilon_i - k_n \sum_{j=2}^{N} \frac{a_{ij}}{d_i}(\varepsilon_i - \varepsilon_j) \\ & - k_g g\left(\frac{b_i}{d_i}\varepsilon_i + \sum_{j=2}^{N} \frac{a_{ij}}{d_i}(\varepsilon_i - \varepsilon_j)\right) + k_{\mathrm{t}}\varepsilon_1 - \dot{\mathbf{y}}^{\mathrm{d}} \quad (14.34) \end{aligned}$$

Let $\mathbf{\Xi} = [\varepsilon_2, \varepsilon_3, ..., \varepsilon_N]^{\mathrm{T}}$, and

$$\mathbf{\Upsilon} = \begin{bmatrix} \frac{1}{d_2} & 0 & \cdots & 0 \\ 0 & \frac{1}{d_3} & \cdots & 0 \\ \vdots & \vdots & \ddots & \vdots \\ 0 & 0 & \cdots & \frac{1}{d_N} \end{bmatrix} \quad (14.35)$$

The deviation of $\mathbf{\Xi}$ can be obtained as

$$\begin{aligned} \dot{\mathbf{\Xi}} =& -k_n(\mathbf{\Upsilon}\mathcal{L} \otimes \mathbf{I}_m)\mathbf{\Xi} - k_n(\mathbf{\Upsilon}\mathbf{B} \otimes \mathbf{I}_m)\mathbf{\Xi} - k_g\mathbf{I}_{m(N-1)}\mathbf{G}(\mathbf{\Xi}) - (\mathbf{1} \otimes \mathbf{I}_m)\dot{\mathbf{x}}_1 \\ =& -k_n\left[\mathbf{\Upsilon}(\mathcal{L} + \mathbf{B}) \otimes \mathbf{I}_m\right]\mathbf{\Xi} - k_g\mathbf{I}_{m(N-1)}\mathbf{G}(\mathbf{\Xi}) - (\mathbf{1} \otimes \mathbf{I}_m)\dot{\mathbf{x}}_1 \\ =& -k_n\left[\mathbf{\Upsilon}(\mathcal{L} + \mathbf{B}) \otimes \mathbf{I}_m\right]\mathbf{\Xi} - k_g\mathbf{I}_{m(N-1)}\mathbf{G}(\mathbf{\Xi}) \\ & + k_t(\mathbf{1} \otimes \mathbf{I}_m)\varepsilon_1 - (\mathbf{1} \otimes \mathbf{I}_m)\dot{\mathbf{y}}^{\mathrm{d}} \quad (14.36) \end{aligned}$$

where

$$\mathbf{G}(\Xi) \triangleq \begin{bmatrix} g\left(\frac{1}{d_2}b_2\varepsilon_2 + \frac{1}{d_2}\sum_{j=2}^{N}\mathcal{L}_{2j}\varepsilon_j\right) \\ \vdots \\ g\left(\frac{1}{d_N}b_N\varepsilon_N + \frac{1}{d_N}\sum_{j=2}^{N}\mathcal{L}_{Nj}\varepsilon_j\right) \end{bmatrix} = \begin{bmatrix} g\left(\sum_{j=2}^{N}\mathcal{H}_{j2}\varepsilon_j\right) \\ \vdots \\ g\left(\sum_{j=2}^{N}\mathcal{H}_{jN}\varepsilon_j\right) \end{bmatrix} \tag{14.37}$$

Choose the Lyapunov candidate for followers as

$$V_2 = \frac{1}{2}\Xi^{\mathrm{T}}\left[\boldsymbol{\Upsilon}(\mathcal{L}+\mathbf{B})\otimes\mathbf{I}_m\right]\Xi \tag{14.38}$$

Define

$$\begin{aligned} \mathbf{H} &= \boldsymbol{\Upsilon}(\mathcal{L}+\mathbf{B}) \\ &= \boldsymbol{\Upsilon}(\mathbf{D}-\mathbf{A}+\mathbf{B}) \\ &= \mathbf{I}_N + \boldsymbol{\Upsilon}(\mathbf{B}-\mathbf{A}) \qquad (14.39) \\ H_{ij} &= \mathrm{I}_{ij} - \frac{a_{ij}}{d_i} + \frac{b_{ij}}{d_i} \qquad (14.40) \\ \sum_{j=2}^{N} H_{ij} &= \sum_{j=2}^{N}\left(\mathrm{I}_{ij} - \frac{a_{ij}}{d_i} + \frac{b_{ij}}{d_i}\right) = 2\frac{b_i}{d_i} \qquad (14.41) \end{aligned}$$

The derivative of V_2 can be represented as

$$\begin{aligned} \dot{V}_2 =& \Xi^{\mathrm{T}}\left(\mathbf{H}\otimes\mathbf{I}_m\right)\dot{\Xi} \\ =& -k_n\Xi^{\mathrm{T}}\left(\mathbf{H}^2\otimes\mathbf{I}_m\right)\Xi - k_g\Xi^{\mathrm{T}}\left[\mathbf{H}\otimes\mathbf{I}_m\right]\mathbf{G}(\Xi) \\ &+ k_t\Xi^{\mathrm{T}}\left(\mathbf{H1}\otimes\mathbf{I}_m\right)\varepsilon_1 - \Xi^{\mathrm{T}}\left[\mathbf{H1}\otimes\mathbf{I}_m\right]\dot{\mathbf{y}}^{\mathrm{d}} \\ \leq& -k_n\Xi^{\mathrm{T}}\left(\mathbf{H}^2\otimes\mathbf{I}_m\right)\Xi - k_g\Xi^{\mathrm{T}}\left(\mathbf{H}\otimes\mathbf{I}_m\right)\mathbf{G}(\Xi) + \frac{k_t(N-1)}{2}\varepsilon_1^{\mathrm{T}}\varepsilon_1 \\ &+ \frac{k_t}{2}\Xi^{\mathrm{T}}\left(\mathbf{H}^2\otimes\mathbf{I}_m\right)\Xi - \sum_{i=2}^{N}\sum_{j=2}^{N}H_{ij}\varepsilon_i^{\mathrm{T}}\dot{\mathbf{y}}^{\mathrm{d}} \\ \leq& -\left(k_n - \frac{k_t}{2}\right)\Xi^{\mathrm{T}}\left(\mathbf{H}^2\otimes\mathbf{I}_m\right)\Xi - k_g\Xi^{\mathrm{T}}\left(\mathbf{H}\otimes\mathbf{I}_m\right)\mathbf{G}(\Xi) \\ &+ \frac{k_t(N-1)}{2}\varepsilon_1^{\mathrm{T}}\varepsilon_1 + \zeta_1\sum_{i=2}^{N}\sum_{j=2}^{N}\|H_{ij}\varepsilon_i\| \end{aligned} \tag{14.42}$$

where we have

$$
\begin{aligned}
&\Xi^{\mathrm{T}}\left[\boldsymbol{\Upsilon}(\mathcal{L}+\mathbf{B})\otimes\mathbf{I}_m\right]\mathbf{G}(\Xi)\\
&=\left[\sum_{j=2}^{N}H_{j2}\varepsilon_j^{\mathrm{T}}\quad\cdots\quad\sum_{j=2}^{N}H_{jN}\varepsilon_j^{\mathrm{T}}\right]\times\begin{bmatrix}\frac{\sum_{j=2}^{N}\mathcal{H}_{j2}\varepsilon_j}{\left\|\sum_{j=2}^{N}\mathcal{H}_{j2}\varepsilon_j\right\|}\\ \vdots\\ \frac{\sum_{j=2}^{N}\mathcal{H}_{jN}\varepsilon_j}{\left\|\sum_{j=2}^{N}\mathcal{H}_{jN}\varepsilon_j\right\|}\end{bmatrix}\\
&=\sum_{i=2}^{N}\sum_{j=2}^{N}\|H_{ij}\varepsilon_i\|
\end{aligned}
\tag{14.43}
$$

It indicates that

$$
\begin{aligned}
\dot{V}_2\leq&-k_n\Xi^{\mathrm{T}}\left[\boldsymbol{\Upsilon}(\mathcal{L}+\mathbf{B})\otimes\mathbf{I}_m\right]^2\Xi\\
&-(k_g-\zeta_1)\sum_{i=2}^{N}\sum_{j=2}^{N}\|H_{ij}\varepsilon_i\|+\frac{k_t(N-1)}{2}\varepsilon_1^{\mathrm{T}}\varepsilon_1
\end{aligned}
\tag{14.44}
$$

Under the condition that $c_2\geq\zeta$, we will obtain that $\dot{V}_2<0$, which proves that the desired formation can be achieved by all followers in free space. □

Case 2: Trajectory tracking and formation following in constrained space

---

Theorem 14.2

Consider N mobile agents with similar dynamics Eq. (14.4), moving in constrained space, with Assumptions 34–39. There is one leader knowing the information of the given trajectory with the control law Eq. (14.18) and the rest of followers knowing the information of displacement to achieve the desired geometric formation with the control law Eq. (14.21). The asymptotic stability of MAS without considering impulsive influence from spatial constraints can be acquired if the following conditions are satisfied:

1.
$$\mathbf{F}_1^{\mathrm{c}\mathrm{T}}\dot{\mathbf{y}}^{\mathrm{d}}>0 \tag{14.45}$$

2.
$$\hat{\Xi}^{\mathrm{T}}(\mathbf{B1}\otimes\mathbf{I}_m)\left(k_c\mathbf{F}_1^{\mathrm{c}}-k_t\varepsilon_1\right)\geq0 \tag{14.46}$$

■

Proof of Theorem 14.2. Let $\hat{\boldsymbol{\Xi}} = [\hat{\varepsilon}_2, \hat{\varepsilon}_3, ..., \hat{\varepsilon}_N]^{\mathrm{T}}$. The control for the leader in Eq. (14.18) can be rewritten as

$$\mathbf{u}_1 = \dot{\mathbf{y}}^{\mathrm{d}} - k_t\varepsilon_1 + k_c\mathbf{F}_1^{\mathrm{c}} + k_n \sum_{j=2}^{N} \frac{b_j}{d_1}\hat{\varepsilon}_j \mathrm{sign}_{\mathrm{c}} \left(\sum_{j=2}^{N} b_j \hat{\varepsilon}_j^{\mathrm{T}} \left(k_c\mathbf{F}_1^{\mathrm{c}} - k_t\varepsilon_1\right) \right) \quad (14.47)$$

Choose the Lyapunov candidate for the leader as below

$$V_3 = \frac{1}{2} k_t \varepsilon_1^{\mathrm{T}} \varepsilon_1 + U_1^{\mathrm{c}} \quad (14.48)$$

The time deviation for V_3 can be written as

$$\begin{aligned} \dot{V}_3 &= k_t\varepsilon_1^{\mathrm{T}}\dot{\varepsilon}_1 - k_c\mathbf{F}_1^{\mathrm{cT}}\dot{\mathbf{x}}_1 & (14.49)\\ &= k_t\varepsilon_1^{\mathrm{T}}\left(\dot{\mathbf{x}}_1 - \dot{\mathbf{y}}^{\mathrm{d}}\right) - k_c\mathbf{F}_1^{\mathrm{cT}}\dot{\mathbf{x}}_1 & (14.50)\\ &= \left(k_t\varepsilon_1 - k_c\mathbf{F}_1^{\mathrm{c}}\right)^{\mathrm{T}}\dot{\mathbf{x}}_1 - k_t\varepsilon_1^{\mathrm{T}}\dot{\mathbf{y}}^{\mathrm{d}} & (14.51)\\ &= \left(k_t\varepsilon_1 - k_c\mathbf{F}_1^{\mathrm{c}}\right)^{\mathrm{T}}\mathbf{u}_1 - k_t\varepsilon_1^{\mathrm{T}}\dot{\mathbf{y}}^{\mathrm{d}} & (14.52) \end{aligned}$$

Substituting Eq. (14.47) into Eq. (14.52), we have

$$\dot{V}_3 = -\left\|k_t\varepsilon_1 - k_c\mathbf{F}_1^{\mathrm{c}}\right\|^2 - k_n \left\| \hat{\boldsymbol{\Xi}}^{\mathrm{T}}\left(\frac{1}{d_1}\mathbf{B}\mathbf{1} \otimes \mathbf{I}_m\right)\left(k_c\mathbf{F}_1^{\mathrm{c}} - k_t\varepsilon_1\right) \right\| - \mathbf{F}_1^{\mathrm{cT}}\dot{\mathbf{y}}^{\mathrm{d}} \quad (14.53)$$

If the condition in Eq. (14.45) is satisfied, we have $\dot{V}_3 < 0$, then the asymptotic stability can be easily obtained [321]. The boundary of ε_1 and $\mathbf{F}_1^{\mathrm{c}}$ can be guaranteed and we may assume that $0 \leq \|\varepsilon_1\| \leq \phi_e < \infty$ and $0 \leq \|\mathbf{F}_1^{\mathrm{c}}\| \leq \phi_{f1} < \infty$.

Otherwise, if the condition in Eq. (14.45) is not satisfied, which means $\mathbf{F}_1^{\mathrm{cT}}\dot{\mathbf{y}}^{\mathrm{d}} < 0$, indicating that the leader agent is getting close to an obstacle. In this case, we may set a tolerable upper limit $t_{\max}$ for the continuous duration that $\mathbf{F}_1^{\mathrm{cT}}\dot{\mathbf{y}}^{\mathrm{d}} < 0$. This may be a reference in realistic tasks to indicate whether the tracking task should be given up or the desired trajectory should be re-planned. The basic assumption of this chapter is that the designed trajectory for the leader does not have too much serious conflict with spatial constraints. However, if the duration that the condition that $\mathbf{F}_1^{\mathrm{cT}}\dot{\mathbf{y}}^{\mathrm{d}} \geq 0$ cannot be satisfied exceeds $t_{\max}$, which may be caused by various uncertainties or the given trajectory is unreasonable, the formation tracking task may be given up or the desired should be redesigned.

On the other hand, the controllers for followers in Eq. (14.21) can be rewritten as

$$\mathbf{u}_i = -k_n\frac{b_i}{d_i}\hat{\varepsilon}_i - k_n\sum_{j=2}^{N}\frac{a_i}{d_i}\hat{\varepsilon}_{ij} + k_c\mathbf{F}_i^{\mathrm{c}} - k_g g\left(\frac{b_i}{d_i}\hat{\varepsilon}_i + \sum_{j=2}^{N}\frac{a_{ij}}{d_i}\hat{\varepsilon}_{ij}\right) \quad (14.54)$$

Thus the deviation of the scaling formation following errors for followers $v_i, i = 2, ..., N$ can be represented as

$$\begin{aligned}
\dot{\hat{\varepsilon}}_i =& \dot{\mathbf{x}}_i - \dot{\mathbf{x}}_1 - \dot{\lambda}\delta_i \\
=& \mathbf{u}_i - \mathbf{u}_1 - \dot{\lambda}\delta_i \\
=& - k_n \frac{b_i}{d_i}\hat{\varepsilon}_i - k_n \sum_{j=2}^{N} \frac{a_{ij}}{d_i}\hat{\varepsilon}_{ij} + k_c \mathbf{F}_i^{\mathrm{c}} - k_g g \left(\frac{b_i}{d_i}\hat{\varepsilon}_i + \sum_{j=2}^{N} \frac{a_{ij}}{d_i}\hat{\varepsilon}_{ij} \right) - \mathbf{u}_1 - \dot{\lambda}\delta_i \\
=& - k_n \frac{b_i}{d_i}\hat{\varepsilon}_i - k_n \sum_{j=2}^{N} \frac{a_{ij}}{d_i}\hat{\varepsilon}_{ij} + k_t \varepsilon_1 - \dot{\mathbf{y}}^{\mathrm{d}} + k_c \mathbf{F}_i^{\mathrm{c}} - k_c \mathbf{F}_1^{\mathrm{c}} \\
& - k_g g \left(\frac{b_i}{d_i}\hat{\varepsilon}_i + \sum_{j=2}^{N} \frac{a_{ij}}{d_i}\hat{\varepsilon}_{ij} \right) \\
& - k_n \sum_{j=2}^{N} \frac{b_j}{d_1}\hat{\varepsilon}_j \mathrm{sign} \left(\sum_{j=2}^{N} \frac{b_j}{d_1}\hat{\varepsilon}_j^{\mathrm{T}} \left(k_c \mathbf{F}_1^{\mathrm{c}} - k_t \varepsilon_1 \right) \right) - \dot{\lambda}\delta_i
\end{aligned} \tag{14.55}$$

Denote $\mathbf{B}_2$ as $\mathbf{B}_{2ij} = b_j$ and $\mathbf{F}^{\mathrm{c}} = \left[\mathbf{F}_2^{\mathrm{c}}, \mathbf{F}_3^{\mathrm{c}}, ..., \mathbf{F}_N^{\mathrm{c}}\right]^{\mathrm{T}}$, we obtain

$$\begin{aligned}
\dot{\hat{\boldsymbol{\Xi}}} =& - k_n \left(\boldsymbol{\Upsilon}\mathbf{B} \otimes \mathbf{I}_m \right) \hat{\boldsymbol{\Xi}} - k_n \left(\boldsymbol{\Upsilon}\mathcal{L} \otimes \mathbf{I}_m \right) \hat{\boldsymbol{\Xi}} + k_c \mathbf{I}_{m(N-1)} \mathbf{F}^{\mathrm{c}} \\
& - k_g \mathbf{I}_{m(N-1)} \mathbf{G}(\hat{\boldsymbol{\Xi}}) - (\mathbf{1} \otimes \mathbf{I}_m)\mathbf{u}_1 - \dot{\lambda}\boldsymbol{\Delta} \\
=& - k_n \left(\mathbf{H} \otimes \mathbf{I}_m \right) \hat{\boldsymbol{\Xi}} + k_c \mathbf{I}_{m(N-1)} \mathbf{F}^{\mathrm{c}} - k_g \mathbf{I}_{m(N-1)} \mathbf{G}(\hat{\boldsymbol{\Xi}}) - (\mathbf{1} \otimes \mathbf{I}_m)\mathbf{u}_1 - \dot{\lambda}\boldsymbol{\Delta}
\end{aligned} \tag{14.56}$$

where $\Delta = \left[\delta_2, \delta_3, ..., \delta_N\right]^{\mathrm{T}}$. Substituting Eq. (14.47) into Eq. (14.56), we have

$$\begin{aligned}
\dot{\hat{\boldsymbol{\Xi}}} =& - k_n \left(\mathbf{H} \otimes \mathbf{I}_m \right) \hat{\boldsymbol{\Xi}} + k_t \left(\mathbf{1} \otimes \mathbf{I}_m \right) \varepsilon_1 - \left(\mathbf{1} \otimes \mathbf{I}_m \right) \left(\dot{\mathbf{y}}^{\mathrm{d}} + k_c \mathbf{F}_1^{\mathrm{c}} \right) \\
& + k_c \mathbf{I}_{m(N-1)} \mathbf{F}^{\mathrm{c}} - k_g \mathbf{I}_{m(N-1)} \mathbf{G}(\boldsymbol{\Xi}) \\
& - \frac{k_n}{d_1} \left(\mathbf{B}_2 \otimes \mathbf{I}_m \right) \hat{\boldsymbol{\Xi}} \mathrm{sign}_{\mathrm{c}} \left(\hat{\boldsymbol{\Xi}}^{\mathrm{T}} \left(\frac{1}{d_1} \mathbf{B1} \otimes \mathbf{I}_m \right) \left(k_c \mathbf{F}_1^{\mathrm{c}} - k_t \varepsilon_1 \right) \right) - \dot{\lambda}\boldsymbol{\Delta}
\end{aligned} \tag{14.57}$$

Choose the Lyapunov candidate for followers as below

$$V_4 = \frac{1}{2} \hat{\boldsymbol{\Xi}}^{\mathrm{T}} \left(\mathbf{H} \otimes \mathbf{I}_m \right) \hat{\boldsymbol{\Xi}} + \frac{1}{k_c} \sum_{i=2}^{N} U_i^{\mathrm{c}} \tag{14.58}$$

The time deviation for V_4 can be written as

$$\dot{V}_4 = \hat{\boldsymbol{\Xi}}^{\mathrm{T}} \left(\mathbf{H} \otimes \mathbf{I}_m \right) \dot{\hat{\boldsymbol{\Xi}}} - \sum_{i=2}^{N} \mathbf{F}_i^{\mathrm{cT}} \dot{\mathbf{x}}_i \tag{14.59}$$

with

$$\begin{aligned}
&\hat{\Xi}^{\mathrm{T}}\left(\mathbf{H}\otimes\mathbf{I}_m\right)\dot{\hat{\Xi}} \\
=&-k_n\hat{\Xi}^{\mathrm{T}}\left(\mathbf{H}^2\otimes\mathbf{I}_m\right)\hat{\Xi}+k_c\Xi^{\mathrm{T}}\left(\mathbf{H}\otimes\mathbf{I}_m\right)\mathbf{F}^{\mathrm{c}} \\
&-k_g\Xi^{\mathrm{T}}\left(\mathbf{H}\otimes\mathbf{I}_m\right)\mathbf{G}(\Xi)-\dot{\lambda}\hat{\Xi}^{\mathrm{T}}\left(\mathbf{H}\otimes\mathbf{I}_m\right)\boldsymbol{\Delta} \\
&-\hat{\Xi}^{\mathrm{T}}\left(\mathbf{H1}\otimes\mathbf{I}_m\right)\left(\dot{\mathbf{y}}^{\mathrm{d}}-k_t\varepsilon_1+k_c\mathbf{F}_1^{\mathrm{c}}\right) \\
&-\frac{k_n}{d_1}\hat{\Xi}^{\mathrm{T}}\left(\mathbf{HB}_2\otimes\mathbf{I}_m\right)\hat{\Xi}\mathrm{sign}_{\mathrm{c}}\left(\hat{\Xi}^{\mathrm{T}}(\mathbf{B1}\otimes\mathbf{I}_m)\left(k_c\mathbf{F}_1^{\mathrm{c}}-k_t\varepsilon_1\right)\right)
\end{aligned} \tag{14.60}$$

and

$$\begin{aligned}
-\sum_{i=2}^{N}\mathbf{F}_i^{\mathrm{cT}}\dot{\mathbf{x}}_i=&-\sum_{i=2}^{N}\mathbf{F}_i^{\mathrm{cT}}\left[k_c\mathbf{F}_i^{\mathrm{c}}-k_n\frac{b_i}{d_i}\hat{\varepsilon}_i-k_n\sum_{j=2}^{N}\frac{a_{ij}}{d_i}\hat{\varepsilon}_{ij}\right. \\
&\left.-k_g g\left(\frac{b_i}{d_i}\hat{\varepsilon}_i+\sum_{j=2}^{N}\frac{a_{ij}}{d_i}\hat{\varepsilon}_{ij}-k_c\mathbf{F}_i^{\mathrm{c}}\right)\right] \\
=&-k_c\sum_{i=2}^{N}\|\mathbf{F}_i^{\mathrm{c}}\|^2+k_n\mathbf{F}^{\mathrm{cT}}(\mathbf{H}\otimes\mathbf{I}_m)\hat{\Xi}+k_g\mathbf{F}^{\mathrm{cT}}\mathbf{G}(\hat{\Xi})
\end{aligned} \tag{14.61}$$

deducing that

$$\begin{aligned}
\dot{V}_4=&-k_n\hat{\Xi}^{\mathrm{T}}\left(\mathbf{H}^2\otimes\mathbf{I}_m\right)\hat{\Xi}-k_c\sum_{i=2}^{N}\|\mathbf{F}_i^{\mathrm{c}}\|^2 \\
&-\frac{k_n}{d_1}\hat{\Xi}^{\mathrm{T}}\left(\mathbf{HB}_2\otimes\mathbf{I}_m\right)\hat{\Xi}\mathrm{sign}_{\mathrm{c}}\left(\hat{\Xi}^{\mathrm{T}}(\mathbf{B1}\otimes\mathbf{I}_m)\left(k_c\mathbf{F}_1^{\mathrm{c}}-k_t\varepsilon_1\right)\right) \\
&-\hat{\Xi}^{\mathrm{T}}\left(\mathbf{H1}\otimes\mathbf{I}_m\right)\left(\dot{\mathbf{y}}^{\mathrm{d}}+k_c\mathbf{F}_1^{\mathrm{c}}\right)+k_t\hat{\Xi}^{\mathrm{T}}\left(\mathbf{H1}\otimes\mathbf{I}_m\right)\varepsilon_1 \\
&+\left(k_c+k_n\right)\hat{\Xi}^{\mathrm{T}}\left(\mathbf{H}\otimes\mathbf{I}_m\right)\mathbf{F}^{\mathrm{c}} \\
&-k_g\hat{\Xi}^{\mathrm{T}}\left(\mathbf{H}\otimes\mathbf{I}_m\right)\mathbf{G}(\hat{\Xi})+k_g\mathbf{F}^{\mathrm{cT}}\mathbf{G}(\hat{\Xi})-\dot{\lambda}\hat{\Xi}^{\mathrm{T}}\left(\mathbf{H}\otimes\mathbf{I}_m\right)\boldsymbol{\Delta}
\end{aligned} \tag{14.62}$$

Assumption 40. *There exists a constant ζ_2 such that $0\le\left|\dot{\lambda}\right|\le\zeta_2<\infty$.*

Given a desired formation, $\|\Delta\|$ is a constant. Define $\zeta_3=\zeta_1+\zeta_2\|\Delta\|$, we have

$$\begin{aligned}
\dot{V}_4\le&-k_n\hat{\Xi}^{\mathrm{T}}\left(\mathbf{H}^2\otimes\mathbf{I}_m\right)\hat{\Xi}-k_c\sum_{i=2}^{N}\|\mathbf{F}_i^{\mathrm{c}}\|^2 \\
&-\frac{k_n}{d_1}\hat{\Xi}^{\mathrm{T}}\left(\mathbf{HB}_2\otimes\mathbf{I}_m\right)\Xi\mathrm{sign}_{\mathrm{c}}\left(\hat{\Xi}^{\mathrm{T}}(\mathbf{B1}\otimes\mathbf{I}_m)\left(k_c\mathbf{F}_1^{\mathrm{c}}-k_t\varepsilon_1\right)\right)
\end{aligned}$$

$$+\zeta_3\sum_{i=2}^{N}\sum_{j=2}^{N}\|H_{ij}\varepsilon_i\| - k_c\hat{\boldsymbol{\Xi}}^{\mathrm{T}}(\mathbf{H1}\otimes\mathbf{I}_m)\mathbf{F}_1^{\mathrm{c}} + k_t\hat{\boldsymbol{\Xi}}^{\mathrm{T}}(\mathbf{H1}\otimes\mathbf{I}_m)\varepsilon_1$$
$$+(k_c+k_n)\hat{\boldsymbol{\Xi}}^{\mathrm{T}}(\mathbf{H}\otimes\mathbf{I}_m)\mathbf{F}^{\mathrm{c}}$$
$$-k_g\hat{\boldsymbol{\Xi}}^{\mathrm{T}}(\mathbf{H}\otimes\mathbf{I}_m)\mathbf{G}(\hat{\boldsymbol{\Xi}}) + k_g\mathbf{F}^{\mathrm{cT}}\mathbf{G}(\hat{\boldsymbol{\Xi}}) \tag{14.63}$$

Following the same process as shown in Eq. (14.43), we could obtain that

$$-\boldsymbol{\Xi}^{\mathrm{T}}(\mathbf{H1}\otimes\mathbf{I}_m)\dot{\mathbf{y}}^{\mathrm{d}} - \dot{\lambda}\hat{\boldsymbol{\Xi}}^{\mathrm{T}}(\mathbf{H}\otimes\mathbf{I}_m)\boldsymbol{\Delta} - k_g\boldsymbol{\Xi}^{\mathrm{T}}(\mathbf{H}\otimes\mathbf{I}_m)\mathbf{G}(\boldsymbol{\Xi})$$
$$= -(k_g-\zeta_3)\sum_{i=2}^{N}\sum_{j=2}^{N}\|H_{ij}\varepsilon_i\| \tag{14.64}$$

Guarantee that $k_g \geq \zeta_3$, we have

$$-\boldsymbol{\Xi}^{\mathrm{T}}(\mathbf{H1}\otimes\mathbf{I}_m)\dot{\mathbf{y}}^{\mathrm{d}} - \dot{\lambda}\hat{\boldsymbol{\Xi}}^{\mathrm{T}}(\mathbf{H}\otimes\mathbf{I}_m)\boldsymbol{\Delta} - k_g\boldsymbol{\Xi}^{\mathrm{T}}(\mathbf{H}\otimes\mathbf{I}_m)\mathbf{G}(\boldsymbol{\Xi}) \leq 0 \tag{14.65}$$

Therefore

$$\begin{aligned}\dot{V}_4 \leq\ & -k_n\hat{\boldsymbol{\Xi}}^{\mathrm{T}}\left(\mathbf{H}^2\otimes\mathbf{I}_m\right)\hat{\boldsymbol{\Xi}} - k_c\sum_{i=2}^{N}\|\mathbf{F}_i^{\mathrm{c}}\|^2\\ & -\frac{k_n}{d_1}\hat{\boldsymbol{\Xi}}^{\mathrm{T}}(\mathbf{HB}_2\otimes\mathbf{I}_m)\boldsymbol{\Xi}\mathrm{sign}_{\mathrm{c}}\left(\hat{\boldsymbol{\Xi}}^{\mathrm{T}}(\mathbf{B1}\otimes\mathbf{I}_m)(k_c\mathbf{F}_1^{\mathrm{c}} - k_t\varepsilon_1)\right)\\ & -k_c\hat{\boldsymbol{\Xi}}^{\mathrm{T}}(\mathbf{H1}\otimes\mathbf{I}_m)\mathbf{F}_1^{\mathrm{c}} + k_t\hat{\boldsymbol{\Xi}}^{\mathrm{T}}(\mathbf{H1}\otimes\mathbf{I}_m)\varepsilon_1\\ & +(k_c+k_n)\hat{\boldsymbol{\Xi}}^{\mathrm{T}}(\mathbf{H}\otimes\mathbf{I}_m)\mathbf{F}^{\mathrm{c}} + k_g\mathbf{F}^{\mathrm{cT}}\mathbf{G}(\hat{\boldsymbol{\Xi}})\end{aligned} \tag{14.66}$$

Considering that $\|\mathbf{G}(\hat{\boldsymbol{\Xi}})\| = \sqrt{N-1}$ we can obtain

$$\begin{aligned}\dot{V}_4 \leq\ & -k_n\hat{\boldsymbol{\Xi}}^{\mathrm{T}}\left(\mathbf{H}^2\otimes\mathbf{I}_m\right)\hat{\boldsymbol{\Xi}} - k_c\sum_{i=2}^{N}\|\mathbf{F}_i^{\mathrm{c}}\|^2\\ & -\frac{k_n}{d_1}\hat{\boldsymbol{\Xi}}^{\mathrm{T}}(\mathbf{HB}_2\otimes\mathbf{I}_m)\boldsymbol{\Xi}\mathrm{sign}_{\mathrm{c}}\left(\hat{\boldsymbol{\Xi}}^{\mathrm{T}}(\mathbf{B1}\otimes\mathbf{I}_m)(k_c\mathbf{F}_1^{\mathrm{c}} - k_t\varepsilon_1)\right)\\ & -k_c\hat{\boldsymbol{\Xi}}^{\mathrm{T}}(\mathbf{H1}\otimes\mathbf{I}_m)\mathbf{F}_1^{\mathrm{c}} + k_t\hat{\boldsymbol{\Xi}}^{\mathrm{T}}(\mathbf{H1}\otimes\mathbf{I}_m)\varepsilon_1\\ & +(k_c+k_n)\hat{\boldsymbol{\Xi}}^{\mathrm{T}}(\mathbf{H}\otimes\mathbf{I}_m)\mathbf{F}^{\mathrm{c}} + k_g\sqrt{N-1}\sum_{i=2}^{N}\|\mathbf{F}_i^{\mathrm{c}}\|\end{aligned} \tag{14.67}$$

Under the condition we have proved above that $\|\varepsilon_1\| \leq \phi_e$ and $\|\mathbf{F}_1^{\mathrm{c}}\| \leq \phi_{f1}$, we have

$$k_n\hat{\boldsymbol{\Xi}}^{\mathrm{T}}(\mathbf{H}\otimes\mathbf{I}_m)\mathbf{F}^{\mathrm{c}} - \frac{k_n}{2}\hat{\boldsymbol{\Xi}}^{\mathrm{T}}\left(\mathbf{H}^2\otimes\mathbf{I}_m\right)\hat{\boldsymbol{\Xi}} - \frac{k_n}{2}\sum_{i=2}^{N}\|\mathbf{F}_i^{\mathrm{c}}\|^2 \leq 0 \tag{14.68}$$

$$\hat{\boldsymbol{\Xi}}^{\mathrm{T}}(\mathbf{H1}\otimes\mathbf{I}_m)(k_t\varepsilon_1 - k_c\mathbf{F}_1^{\mathrm{c}}) - \varrho\hat{\boldsymbol{\Xi}}^{\mathrm{T}}\left(\mathbf{H}^2\otimes\mathbf{I}_m\right)\hat{\boldsymbol{\Xi}} - \psi_1 \leq 0 \tag{14.69}$$

$$k_g\sqrt{N-1}\sum_{i=2}^{N}\|\mathbf{F}_i^{\mathrm{c}}\| - \varrho\sum_{i=2}^{N}\|\mathbf{F}_i^{\mathrm{c}}\|^2 - \psi_2 \leq 0 \tag{14.70}$$

where $\varrho > 0$, $\psi_1 > 0$ and $\psi_2 > 0$ are three positive constants. The smaller value of ϱ will lead to larger value of ψ_1 and ψ_2. We should guarantee that $\varrho < k_n$ and $\varrho < k_c - \frac{k_n}{2}$. Define $\psi = \psi_1 + \psi_2$ we obtain

$$\begin{aligned}\dot{V}_4 \leq & -\frac{k_n}{2}\hat{\boldsymbol{\Xi}}^{\mathrm{T}}\left(\mathbf{H}^2 \otimes \mathbf{I}_m\right)\hat{\boldsymbol{\Xi}} - \left(k_c - \frac{k_n}{2}\right)\sum_{i=2}^{N}\|\mathbf{F}_i^{\mathrm{c}}\|^2 \\ & -\frac{k_n}{d_1}\hat{\boldsymbol{\Xi}}^{\mathrm{T}}\left(\mathbf{H}\mathbf{B}_2 \otimes \mathbf{I}_m\right)\boldsymbol{\Xi}\mathrm{sign}_{\mathrm{c}}\left(\hat{\boldsymbol{\Xi}}^{\mathrm{T}}(\mathbf{B1} \otimes \mathbf{I}_m)\left(k_c\mathbf{F}_1^{\mathrm{c}} - k_t\varepsilon_1\right)\right) \\ & +k_c\boldsymbol{\Xi}^{\mathrm{T}}\left(\mathbf{H} \otimes \mathbf{I}_m\right)\left[\mathbf{F}^{\mathrm{c}} - \left(\mathbf{1} \otimes \mathbf{I}_m\right)\mathbf{F}_1^{\mathrm{c}}\right] \\ & +k_c(1-\lambda)\boldsymbol{\Delta}^{\mathrm{T}}\left(\mathbf{H} \otimes \mathbf{I}_m\right)\left[\mathbf{F}^{\mathrm{c}} - \left(\mathbf{1} \otimes \mathbf{I}_m\right)\mathbf{F}_1^{\mathrm{c}}\right] + \psi \end{aligned} \tag{14.71}$$

If the condition in Eq. (14.46) is satisfied, we have

$$\begin{aligned}\dot{V}_4 \leq & -\frac{k_n}{2}\hat{\boldsymbol{\Xi}}^{\mathrm{T}}\left(\mathbf{H}^2 \otimes \mathbf{I}_m\right)\hat{\boldsymbol{\Xi}} - \left(k_c - \frac{k_n}{2}\right)\sum_{i=2}^{N}\|\mathbf{F}_i^{\mathrm{c}}\|^2 - \frac{k_n}{d_1}\hat{\boldsymbol{\Xi}}^{\mathrm{T}}\left(\mathbf{H}\mathbf{B}_2 \otimes \mathbf{I}_m\right)\boldsymbol{\Xi} \\ & +k_c\boldsymbol{\Xi}^{\mathrm{T}}\left(\mathbf{H} \otimes \mathbf{I}_m\right)\left[\mathbf{F}^{\mathrm{c}} - \left(\mathbf{1} \otimes \mathbf{I}_m\right)\mathbf{F}_1^{\mathrm{c}}\right] \\ & +k_c(1-\lambda)\boldsymbol{\Delta}^{\mathrm{T}}\left(\mathbf{H} \otimes \mathbf{I}_m\right)\left[\mathbf{F}^{\mathrm{c}} - \left(\mathbf{1} \otimes \mathbf{I}_m\right)\mathbf{F}_1^{\mathrm{c}}\right] + \psi \end{aligned} \tag{14.72}$$

When $\hat{\boldsymbol{\Xi}}^{\mathrm{T}}\left(\mathbf{H} \otimes \mathbf{I}_m\right)\left[\mathbf{F}^{\mathrm{c}} - \left(\mathbf{1} \otimes \mathbf{I}_m\right)\mathbf{F}_1^{\mathrm{c}}\right] \leq 0$, we set $\dot{\lambda} = k_\lambda/\lambda$, there is

$$\begin{aligned}\dot{V}_4 \leq & -\frac{k_n}{2}\hat{\boldsymbol{\Xi}}^{\mathrm{T}}\left(\mathbf{H}^2 \otimes \mathbf{I}_m\right)\hat{\boldsymbol{\Xi}} - \left(k_c - \frac{k_n}{2}\right)\sum_{i=2}^{N}\|\mathbf{F}_i^{\mathrm{c}}\|^2 - \frac{k_n}{d_1}\hat{\boldsymbol{\Xi}}^{\mathrm{T}}\left(\mathbf{H}\mathbf{B}_2 \otimes \mathbf{I}_m\right)\boldsymbol{\Xi} \\ & +k_c(1-\lambda)\boldsymbol{\Delta}^{\mathrm{T}}\left(\mathbf{H} \otimes \mathbf{I}_m\right)\left[\mathbf{F}^{\mathrm{c}} - \left(\mathbf{1} \otimes \mathbf{I}_m\right)\mathbf{F}_1^{\mathrm{c}}\right] + \psi \end{aligned} \tag{14.73}$$

In this situation, $\lim_{t\to\infty} k_c(1-\lambda)\boldsymbol{\Delta}^{\mathrm{T}}\left(\mathbf{H} \otimes \mathbf{I}_m\right)\left[\mathbf{F}^{\mathrm{c}} - \left(\mathbf{1} \otimes \mathbf{I}_m\right)\mathbf{F}_1^{\mathrm{c}}\right] = 0$.

When $\hat{\boldsymbol{\Xi}}^{\mathrm{T}}\left(\mathbf{H} \otimes \mathbf{I}_m\right)\left[\mathbf{F}^{\mathrm{c}} - \left(\mathbf{1} \otimes \mathbf{I}_m\right)\mathbf{F}_1^{\mathrm{c}}\right] > 0$, the adaptation of λ should guarantee that $\lambda > 1$ when $\boldsymbol{\Delta}^{\mathrm{T}}\left(\mathbf{H} \otimes \mathbf{I}_m\right)\left[\mathbf{F}^{\mathrm{c}} - \left(\mathbf{1} \otimes \mathbf{I}_m\right)\mathbf{F}_1^{\mathrm{c}}\right] > 0$ and $\lambda < 1$ when $\boldsymbol{\Delta}^{\mathrm{T}}\left(\mathbf{H} \otimes \mathbf{I}_m\right)\left[\mathbf{F}^{\mathrm{c}} - \left(\mathbf{1} \otimes \mathbf{I}_m\right)\mathbf{F}_1^{\mathrm{c}}\right] < 0$. Meanwhile, a constant $\tau > 0$ is set to avoid over-scaling. Therefore, we design that

$$\dot{\lambda} = \begin{cases} \alpha\mathrm{sign}_{\mathrm{c}}\left(\boldsymbol{\Delta}^{\mathrm{T}}\left(\mathbf{H} \otimes \mathbf{I}_m\right)\left[\mathbf{F}^{\mathrm{c}} - \left(\mathbf{1} \otimes \mathbf{I}_m\right)\mathbf{F}_1^{\mathrm{c}}\right]\right) & (14.74) \\ \text{if} \quad \hat{\boldsymbol{\Xi}}^{\mathrm{T}}\left(\mathbf{H} \otimes \mathbf{I}_m\right)\left[\mathbf{F}^{\mathrm{c}} - \left(\mathbf{1} \otimes \mathbf{I}_m\right)\mathbf{F}_1^{\mathrm{c}}\right] > \tau \quad \& \quad \lambda \geq \underline{\lambda} > 0; & \\ k_\lambda(1-\lambda) & (14.75) \\ \text{if} \quad \hat{\boldsymbol{\Xi}}^{\mathrm{T}}\left(\mathbf{H} \otimes \mathbf{I}_m\right)\left[\mathbf{F}^{\mathrm{c}} - \left(\mathbf{1} \otimes \mathbf{I}_m\right)\mathbf{F}_1^{\mathrm{c}}\right] \leq \tau, & \end{cases}$$

where α and k_λ are two positive constants. To avoid internal collisions among agents, we should set a lower limit $0 < \underline{\lambda} < 1$ for the scaling factor, such that the adjustment of λ should satisfy that $\lambda \geq \underline{\lambda} > 0$. If this condition condition cannot be satisfied, that means the path is passable with current formation shape. A possible solution is to switch into another formation shape or give

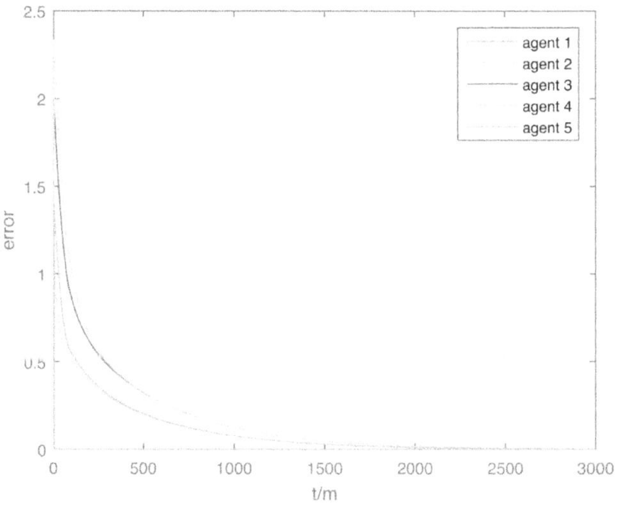

Figure 14.2 Formation error.

up the formation task and enter emergency avoidance mode. Then we have

$$\begin{aligned}\dot{V}_4 \leq & -cV_4+\psi \\ & +k_c \boldsymbol{\Xi}^{\mathrm{T}}\left(\mathbf{H} \otimes \mathbf{I}_m\right)\left[\mathbf{F}^{\mathrm{c}}-\left(\mathbf{1} \otimes \mathbf{I}_m\right) \mathbf{F}_1^{\mathrm{c}}\right] \\ & +k_c(1-\lambda) \boldsymbol{\Delta}^{\mathrm{T}}\left(\mathbf{H} \otimes \mathbf{I}_m\right)\left[\mathbf{F}^{\mathrm{c}}-\left(\mathbf{1} \otimes \mathbf{I}_m\right) \mathbf{F}_1^{\mathrm{c}}\right]\end{aligned} \tag{14.76}$$

By introducing of size scaling adaptation factor λ,

$$k_c(1-\lambda) \boldsymbol{\Delta}^{\mathrm{T}}\left(\mathbf{H} \otimes \mathbf{I}_m\right)\left[\mathbf{F}^{\mathrm{c}}-\left(\mathbf{1} \otimes \mathbf{I}_m\right) \mathbf{F}_1^{\mathrm{c}}\right] \tag{14.77}$$

will limit the norm of $k_c \boldsymbol{\Xi}^{\mathrm{T}}\left(\mathbf{H} \otimes \mathbf{I}_m\right)\left[\mathbf{F}^{\mathrm{c}}-\left(\mathbf{1} \otimes \mathbf{I}_m\right) \mathbf{F}_1^{\mathrm{c}}\right]$ to a certain extent. □

14.6 SIMULATION STUDIES

In this section, three simulation examples are presented to illustrate the effectiveness of the methods proposed in this chapter. We consider a group of agents with $N=5$, and the desired formation is chosen as regular pentagon. We set the safety distance $s_0=0.1m$.

14.6.1 FORMATION TRACKING IN 2-D FREE SPACE

In the first simulation, the initial positions of N agents are initialized randomly within a circle range with radius $r=2m$. we study the effectiveness of control

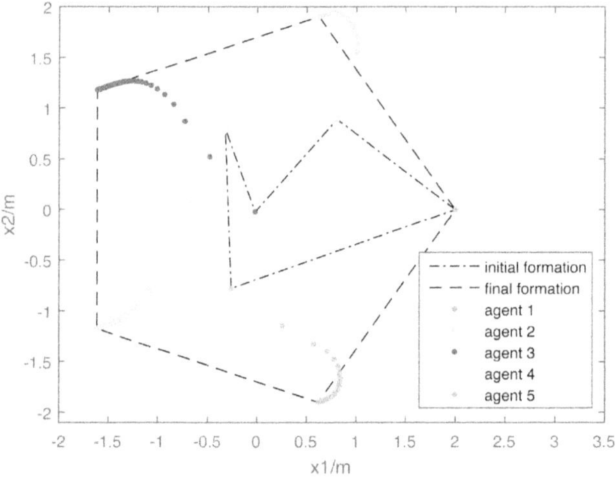

Figure 14.3 Formation generation.

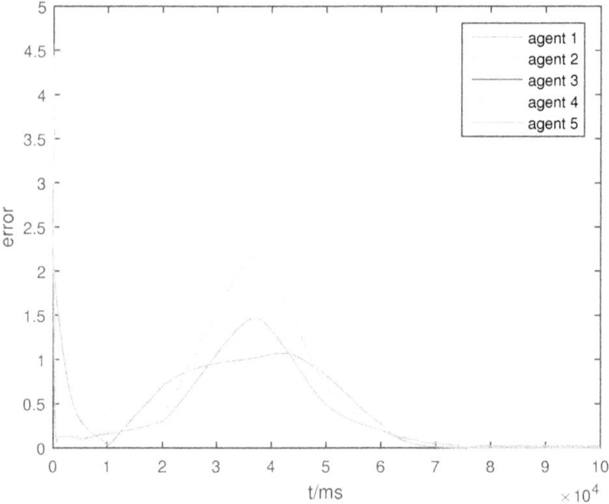

Figure 14.4 Formation tracking error.

laws Eqs. (14.11–14.12) for stationary formation control. Simulation results are shown in Figs. 14.2–14.3.

Fig. 14.3 shows that agents are able to achieve the given formation from random initial positions. Fig. 14.2 shows the tracking error of all agents will decrease gradually to zeros with time going.

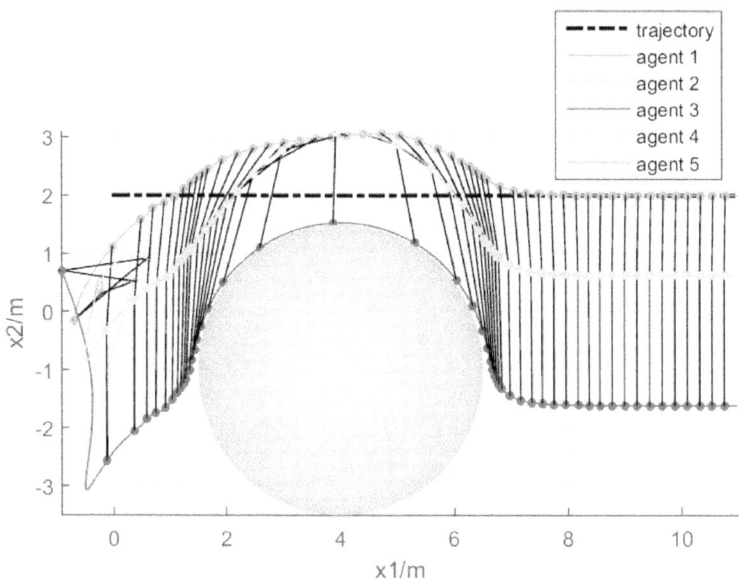

Figure 14.5 Formation tracking.

14.6.2 FORMATION TRACKING IN SIMPLE 3-D CONSTRAINED SPACE

In the second simulation, the performance of control laws Eqs. (14.18–14.19) is tested. Two sphere obstacle center in $\mathbf{x} = \begin{bmatrix} 4 & -1 & 1.7 \end{bmatrix}^{\mathrm{T}}$ and $\mathbf{x} = \begin{bmatrix} 4 & -1 & -1.7 \end{bmatrix}^{\mathrm{T}}$ with 2.5m radius. The desired trajectory for the leader is set as $\mathbf{y}^{\mathrm{d}}(t) = \begin{bmatrix} 1.1t & 2 & 0 \end{bmatrix}^{T}$. Simulation results are shown in Figs. 14.4–14.6.

Figs. 14.5 and 14.6 shows that all agents can successfully track the desired trajectory when there are multiple borders or obstacles. The formation error (in Fig. 14.4) increases when an agent gets close to borders or obstacles, and decreases when being far away from borders or obstacles.

14.6.3 FORMATION TRACKING IN 3-D MULTIPLE CONSTRAINED SPACE WITH FORMATION SCALING

In the third simulation, the effectiveness of control laws Eqs. (14.20–14.21) is studied. As mentioned in Section 4.3, two kind of restricted environments are built. We build a restricted region with two cuboid obstacles. The desired trajectory for the leader is set as $\mathbf{y}^{\mathrm{d}}(t) = \begin{bmatrix} t & 2 & 0 \end{bmatrix}^{T}$. Simulation results are shown in Figs. 14.7–14.14.

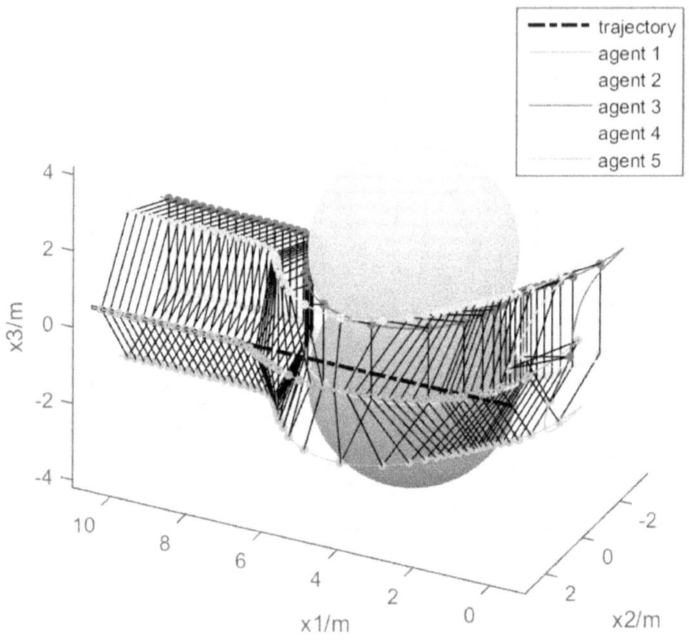

Figure 14.6 Formation tracking.

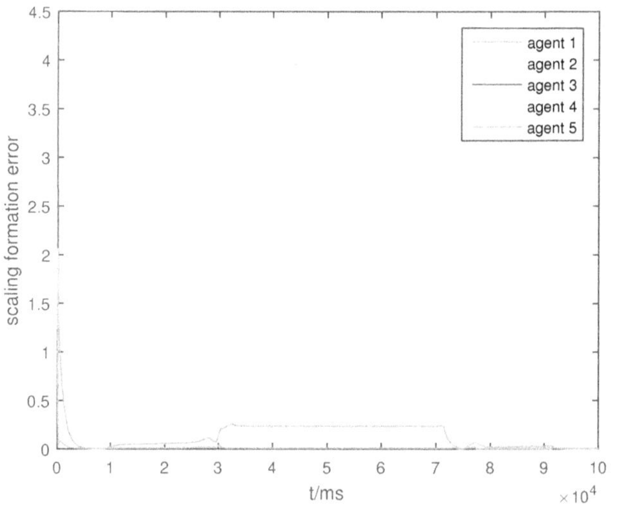

Figure 14.7 Formation tracking error.

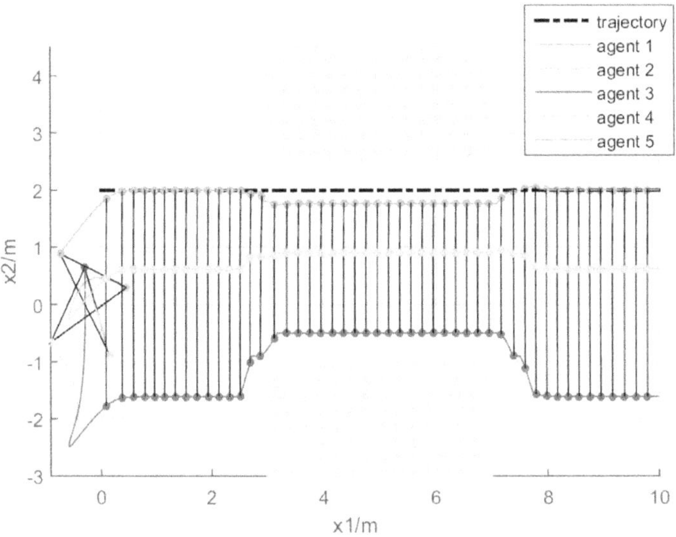

Figure 14.8 Formation tracking with size adaptation in constrained space.

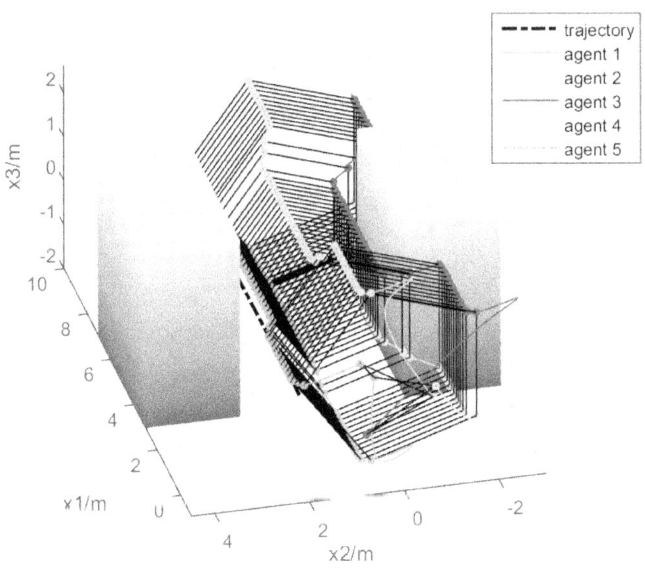

Figure 14.9 Formation tracking with size adaptation in constrained space.

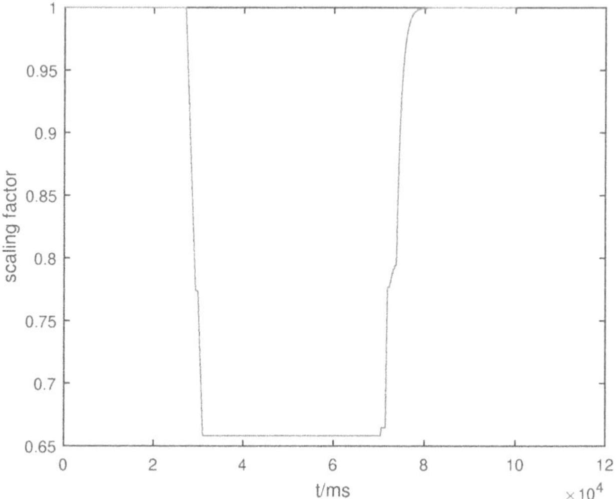

Figure 14.10 Scaling factor.

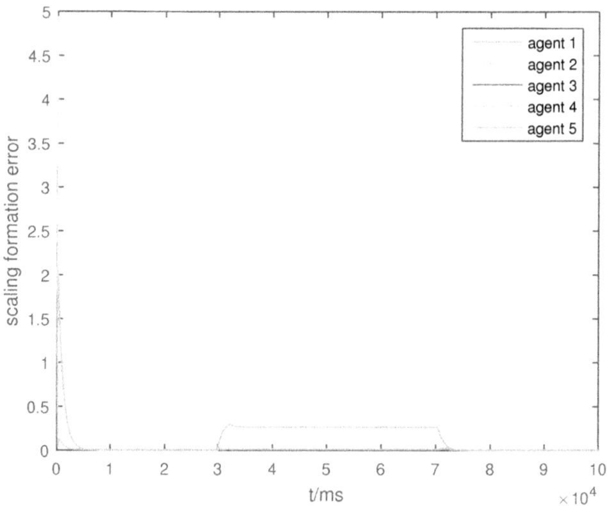

Figure 14.11 Formation tracking error.

Figs. 14.7 and 14.11 show the scaling formation errors for agents $v_i, i = 2, .., 5$ converge to zero gradually. The tracking error for the agent v_1 decreases to zero in the first phase but increases to a constant when entering the influence range of the obstacle and decreases to zero once again after leaving

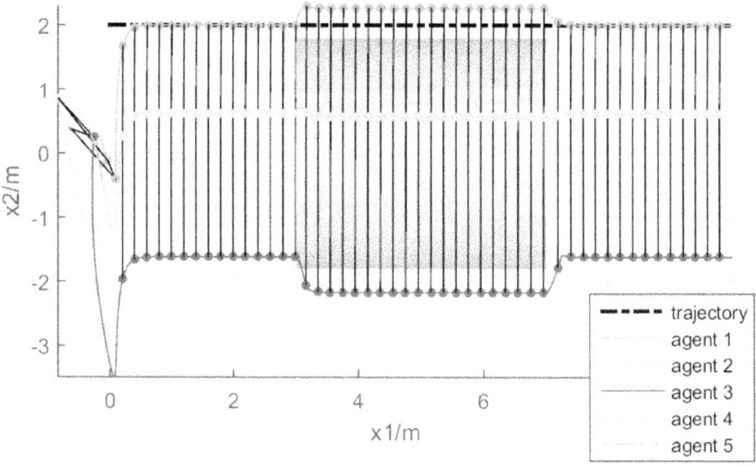

Figure 14.12 Formation tracking with size adaptation in constrained space.

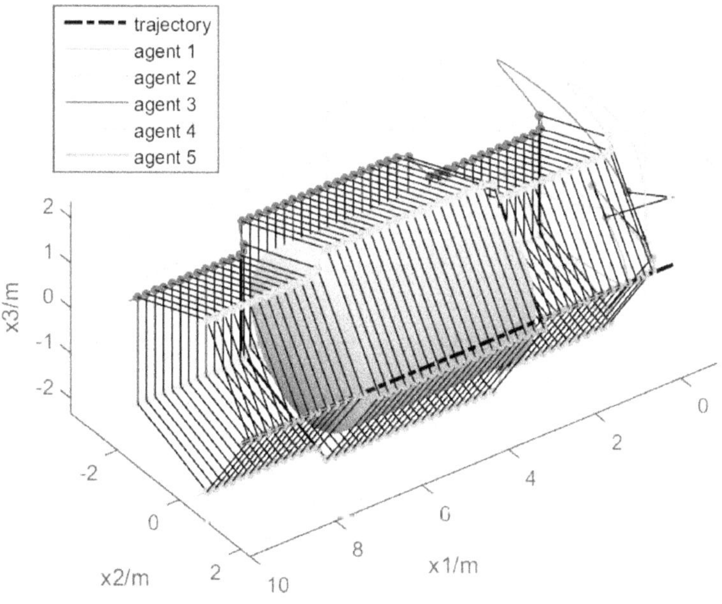

Figure 14.13 Formation tracking with size adaptation in constrained space.

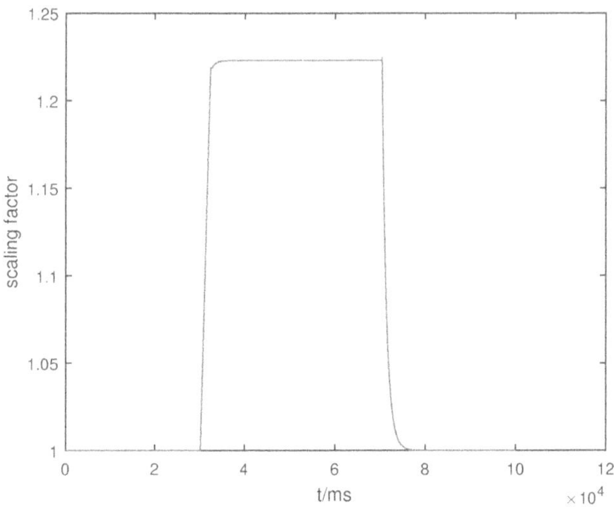

Figure 14.14 Scaling factor.

the influence range of the obstacle. Figs. 14.8, 14.9, 14.12, and 14.13 indicate that all agents can successfully track the desired trajectory while maintaining the predefined formation. The size of formation shape is scaled down in the first situation and scaled up in the second situation to enable all agents keep safe distance from the obstacles when entering the influence range of spatial constraints. Figs. 14.10 and 14.14 describe the variation tendency of formation scaling factors in two cases.

14.7 CONCLUSION

In this chapter, formation tracking control of multi-agent systems with communication limitations in constrained space has been studied based on predefined displacement. A role of coordinator has been introduced to as a feedback and coordination approach to guarantee obstacle avoidance as well as formation keeping. Controllers have been designed according to the local-exchange information and locally detected environmental constraints. A scaling factor has been proposed to scale up or scale down the given formation size in the case that the restricted path is impassable for the desired formation and the adaptation law is deduced. Meanwhile, the conditions for asymptotic stability of multi-agent systems have been proved based on the Lyapunov function. The maximum duration that the conditions cannot be satisfied has been elaborated to indicate whether the tracking task should be given up. Finally, simulation results have been presented to illustrate the performance of proposed approaches.

15 Vision-Based Leader-Follower Formation Control of Multi-Agents with Visibility Constraints

In this chapter, vision-based tracking is studied for the leader-follower formation control of multi-agent systems (MAS) under visibility constraints without communication among agents. A kinect is the sole sensor installed on each agent to estimate the pose including relative orientations and relative positions of the local leader for the formation control on the basis of visual tracking of the local leader. A dipolar vector field is establish to generate and maintain desired formation given the desired displacement on positions between agents. The visibility constraints of kinect are converted into inputs constrains acquired the pose information of the local leader, and an auxiliary system is designed to analyze the effort of visibility constraints. Combining the state of the auxiliary system with the vector filed, the proposed control protocol is able to achieve the formation tracking, which do not require global or local communication among agents. To enable the global leader move in the obstacle environment without collisions, a speech navigation system is designed to translate speech signals into control inputs for the global leader as a motion planner. Simulation results on Robot Operating System (ROS) using Gazebo simulator as well as experimental results on Turtlebot 2 are presented to illustrate the performance.

15.1 INTRODUCTION

Vision-based control has attracted increasing attention in recent years, and it is not restricted to formation control, such as collision avoidance [323] and target tracking [470]. Cameras have been strongly tied to robotic applications benefiting from their low prices and sufficient information they can capture compared with conventional sensors, e.g., laser range finders and sonars. A framework for vision-based formation control was proposed in [92]. In [369], the authors studied the formation control of multiple robots with visibility and communication constraints based on vision. In [64], the L-F formation control of mobile robots without communication was solved by using a pan-controlled camera as the sole sensor. In [328], a vision-based localization protocol for L-F formation control of unicycle robots equipped with a panoramic camera

DOI: 10.1201/9781003298618-15

was proposed, while the necessary conditions of observability was discussed. Similar schemes for vision-based formation control were also proposed in [134, 346, 467]. On the other hand, some work explores the potential of applying global vision sensor for solutions. In [316], the position and the orientation of robots were estimated by a global vision system that can cover whole area for the formation control. A similar achievement was also reported in [133]. [396] studied the effects of different visual feedbacks with multiple cameras on the performance of human operators to achieve swarm of multiple unmanned aerial vehicles.

The essential problem for the formation control using on-board cameras is to estimate the pose of the robot w.r.t. a reference frame. Despite the progresses achieved, this topic remains challenging because traditional cameras providing the view-angles to an object but not distances. Kinect, featureing both RGB camera and depth sensor, provides a solution to this problem. A few publications have reported the applications of kinect on the cooperative control of MAS [269] but far from being desired. First, the commonly estimated information via kinect is distance at most the angle of an object relative to the center of the kinect, which renders that the criteria of the formation has to be weaken (e.g., to control the follower to keep desired distance from the leader). Most of existing solutions needed the combination of multiple sensors for desired performance. From a localization viewpoint, at least three landmarks (better be static) should be equipped, which is not easily achievable in kinect-based formation without information sharing. Due to the limited range of view of kinect or specific formation shape such as pseudo V, there is normally one leader agent detected in field of vision. Meanwhile, it is practically impossible to keep the visibility of multiple visual targets in the whole procedure. Hence, the exploration of the solution for the rigid formation of MAS with the sole on-board cameras is still challenging.

The L-F formation control is commonly considered in the situation where agents are subject to limited sensing and communication and lack of localization in global coordination [66, 286]. The limited field of view challenges the success of the formation control via onboard kinect. Efforts have been made to deal with visibility of fixed landmarks for the navigation problem of mobile robots with limited sensing [76] and visual servoing control [5]. However, few studies have aimed at visibility constraints in formation control, where the objective enhances the difficulty of visibility maintenance, since both the agent itself and the visual target are mobile. To maintain the visibility especially when a specific displacement on the pose w.r.t. the mobile object (e.g., leader) has to be kept, the relative pose or motion of the mobile target should be accurately estimated visually as feedback massages. In addition, it remains unexplored that how to incorporate these feedback messages into the controller design as the bearing angular is not commonly involved in the agent's system dynamics.

Motivated by the issue of vision-based formation control, the L-F formation tracking of MAS using kinect as the sole sensor under visibility constraints without communication among agents is addressed in this chapter. The kinect has dual role in this scheme, one as the sensor to detect the pose of the leader, and another one as the orientational indicator for the follower, which replaces the usage of a gyroscopes or acceleration sensors. The body-fixed frame of the leader agent can be set up with one landmark.

The primary contribution is the proposal of a vision-based pose estimation method with RGBD images captured by the individual kinect board on the follower required for the formation control and visibility maintenance. This is difficult to be achieved by sole traditional sensors (e.g., laser rangefinder and sonar). Consequently, the formation tracking based on predefined position displacement between agents is able to be realized with consensus in motion direction provided the kinect on each agent. The visibility constraints of kinect are converted into a set of input constraints. An auxiliary design system is presented to analyze the effect of visibility constraints, and states of auxiliary design system are utilized to compensate the formation controller ensuring the formation keeping under visibility maintenance.

15.2 MATHEMATICAL MODELING

Before proceeding further, the following definitions are made in this chapter.

Definition 52 (*Global Leader*). *In MAS, the agent which has the knowledge of the desired trajectory or is assigned the task to explore the environment is defined as the Global Leader.*

Definition 53 (*Local Leader*). *In MAS, the agent which is in the range of visibility of another agent and viewed as the formation reference of this agent is defined as the Local Leader of this agent.*

15.2.1 SYSTEM DYNAMICS

Consider a network of multi-agent systems $V = \{v_1, ..., v_N\}$, where $v_i(i = 1, ..., N)$ represents i^{th} agent in the network and N is the number of agents.

Assumption 41. *Without loss of generality, it is assumed that the first agent v_1 is the global leader for the given task (e.g., trajectory tracking or path planning), who knows the desired trajectory or is equipped with more advanced sensors to explore the environment while other agents know only which agent is the leader and the desired position displacement to achieve predefined formation. The followers' objective is to keep specific geometric configuration with the leader.*

We assume N agents with similar motions w.r.t. a global frame $\mathcal{G}$ described by the following equation

$$\dot{\mathbf{q}}_{0i} = \begin{bmatrix} \cos\theta_i & 0 \\ \sin\theta_i & 0 \\ 0 & 1 \end{bmatrix} \mathbf{u}_i \Rightarrow \begin{bmatrix} \dot{x}_{0i} \\ \dot{y}_{0i} \\ \dot{\theta}_i \end{bmatrix} = \begin{bmatrix} \cos\theta_i & 0 \\ \sin\theta_i & 0 \\ 0 & 1 \end{bmatrix} \begin{bmatrix} \nu_i \\ \omega_i \end{bmatrix} \tag{15.1}$$

where $\mathbf{q}_{0i} = [x_{0i} \quad y_{0i} \quad \theta_i]^{\mathrm{T}}$ is the configuration vector of agent v_i, $\mathbf{r}_{0i} = [x_{0i} \quad y_{0i}]^{\mathrm{T}}$ represents the position of agent v_i in the world coordinate system, and θ_i is the orientation of agent v_i w.r.t. $\mathcal{G}$; $\mathbf{u}_i = [\nu_i \quad \omega_i]^{\mathrm{T}}$ is the system input (ν_i and ω_i are linear and angular velocity of agent v_i, respectively).

It is assumed that each follower agent v_i ($i = 2, .., N$) has a local leader v_j ($j \neq i$) acted as its formation reference, which may be another follower agent in the forward-looking vision range or the local leader. To mathematically express the motion of the follower v_i w.r.t. the leader' body-fixed frame $\mathcal{J}$ ($\mathcal{J} = 1, ..., N$), write the position $\mathbf{r}_i = [x_i \quad y_i]^{\mathrm{T}}$ of the follower v_i w.r.t. its local leader frame $\mathcal{J}$, given as $\mathbf{r}_i = R(-\theta_i)(\mathbf{r}_i - \mathbf{r}_j)$. The time derivative of $\mathbf{r}_i$ is

$$\dot{\mathbf{r}}_i = \dot{\mathbf{R}}(-\theta_j)(\mathbf{r}_i - \mathbf{r}_j) + \mathbf{R}(-\theta_j)(\dot{\mathbf{r}}_i - \dot{\mathbf{r}}_j) \tag{15.2}$$

where

$$\begin{aligned} \mathbf{R}(-\theta_j) &= \begin{bmatrix} \cos(-\theta_j) & -\sin(-\theta_j) \\ \sin(-\theta_j) & \cos(-\theta_j) \end{bmatrix} \\ &= \begin{bmatrix} \cos\theta_j & \sin\theta_j \\ -\sin\theta_j & \cos\theta_j \end{bmatrix} \end{aligned} \tag{15.3}$$

is the rotation matrix of the body-fixed frame $\mathcal{J}$ w.r.t. frame $\mathcal{G}$, and

$$\dot{\mathbf{R}}(\theta_j) = \begin{bmatrix} 0 & \omega_j \\ -\omega_j & 0 \end{bmatrix} \mathbf{R}(-\theta_j) \tag{15.4}$$

Substituting Eqs. (15.1, 15.3–15.4) into Eq. (15.2), we can get

$$\begin{bmatrix} \dot{x}_i \\ \dot{y}_i \end{bmatrix} = \begin{bmatrix} -1 & y_i \\ 0 & -x_i \end{bmatrix} \begin{bmatrix} \nu_j \\ \omega_j \end{bmatrix} + \begin{bmatrix} \cos(\theta_i - \theta_j) \\ \sin(\theta_i - \theta_j) \end{bmatrix} \nu_i \tag{15.5}$$

Define $\beta_i \triangleq \theta_i - \theta_j$, whose time differentiating can be written as

$$\dot{\beta}_i = \dot{\theta}_i - \dot{\theta}_j = \omega_i - \omega_j \tag{15.6}$$

Combining Eqs. (15.5–15.6) outputs the local system dynamics as

$$\begin{aligned} \begin{bmatrix} \dot{x}_i \\ \dot{y}_i \\ \dot{\beta}_i \end{bmatrix} &= \begin{bmatrix} \cos\beta_i & 0 \\ \sin\beta_i & 0 \\ 0 & 1 \end{bmatrix} \begin{bmatrix} \nu_i \\ \omega_i \end{bmatrix} + \begin{bmatrix} -1 & y_i \\ 0 & -x_i \\ 0 & -1 \end{bmatrix} \begin{bmatrix} \nu_j \\ \omega_j \end{bmatrix} \\ &= \begin{bmatrix} \cos\beta_i & 0 \\ \sin\beta_i & 0 \\ 0 & 1 \end{bmatrix} \mathbf{u}_i + \begin{bmatrix} -1 & y_i \\ 0 & -x_i \\ 0 & -1 \end{bmatrix} \mathbf{u}_j \end{aligned} \tag{15.7}$$

where $\mathbf{q}_i = [x_i \quad y_i \quad \beta_i]^{\mathrm{T}} \in \mathbb{R}^3$ is the state vector comprising the position $\mathbf{r}_i = [x_i \quad y_i]^{\mathrm{T}}$ as well as the orientation β_i of the follower v_i w.r.t. the leader frame $\mathcal{J}$, $\mathbf{u}_i = [\nu_i \quad \omega_i]^{\mathrm{T}}$ is the control inputs of the follower v_i and $\mathbf{u}_j = [\nu_j \quad \omega_j]^{\mathrm{T}}$ is the control inputs of the local leader v_j.

Assumption 42. *There exists a constant $\nu_{\max}$ such that $0 \leq |\nu_i| \leq \nu_{\max} < \infty$ and another constant $\omega_{\max}$ such that $0 \leq |\omega_i| \leq \omega_{\max} < \infty$, $i = 1, ..., N$. Each agent has priori knowledge on the velocity bounds $\nu_{\max}$ and $\omega_{\max}$.*

15.2.2 FORMATION MANEUVERS

The formation pattern at time t is defined to be a set of desired displacement $\mathcal{P}(t) = \{\delta_2(t), ..., \delta_N(t)\}$ for followers $v_i, i = 2, ..., N$ w.r.t. the local leader frame $\mathcal{J}$, where $\delta_i(t)$ is the desired position displacement of the agent v_i at time t relative to its leader v_j. Thus we have

$$\begin{bmatrix} x_i^{\mathrm{d}} \\ y_i^{\mathrm{d}} \\ \beta_i^{\mathrm{d}} \end{bmatrix} = \begin{bmatrix} \delta_{xi} \\ \delta_{yi} \\ 0 \end{bmatrix} \tag{15.8}$$

where $\mathbf{r}_i^{\mathrm{d}} = [x_i^{\mathrm{d}} \quad y_i^{\mathrm{d}}]^{\mathrm{T}}$ is the desired formation position vector, β_i^{d} is the desired tracking orientation, and $\delta_i = [\delta_{xi} \quad \delta_{yi}]^{\mathrm{T}}$ is the desired displacement of agent v_i at time t w.r.t the local leader frame $\mathcal{J}$. We set $\beta_i^{\mathrm{d}} = 0$ indicating that the consensus of MAS in motion direction is required to be achieved such that $\theta_i = \theta_j (i = 2, .., N)$ w.r.t. the global frame $\mathcal{G}$.

The formation error for the follower $v_i, i = 2, ..., N$ is written as

$$\begin{bmatrix} \varepsilon_{xi} \\ \varepsilon_{yi} \end{bmatrix} = \begin{bmatrix} x_i - \delta_{xi} \\ y_i - \delta_{yi} \end{bmatrix} \tag{15.9}$$

where $\varepsilon_i = [\varepsilon_{xi} \quad \varepsilon_{yi}]^{\mathrm{T}}$ is the vector of formation errors w.r.t. the leader frame $\mathcal{J}$.

15.2.3 VISIBILITY CONSTRAINTS

A team of agents interacts with each other via communication or sensing networks to achieve collective objectives. The information exchanges among agents via forward-looking vision can be modeled by directed graphs. A directed graph G is a pair (V, E), where $V = \{v_1, ..., v_N\}$ is a nonempty finite node set and $E \subseteq V \times V$ is an edge set of ordered pairs of nodes, called edges.

We assume that each follower is equipped with a fixed onboard kinect of limited angle of view $2\gamma_{\max} < \pi$ and all kinects equipped are holonomic. The limited sensing region is modeled as a cone of view, which essentially is an isosceles triangle in obstacle-free environments. The follower v_i is localized w.r.t. its local leader v_j, i.e., that the displacement in x-coordinate of $\mathcal{J}$, the displacement in y-coordinate of $\mathcal{J}$ and relative orientation β_i or bearing angle

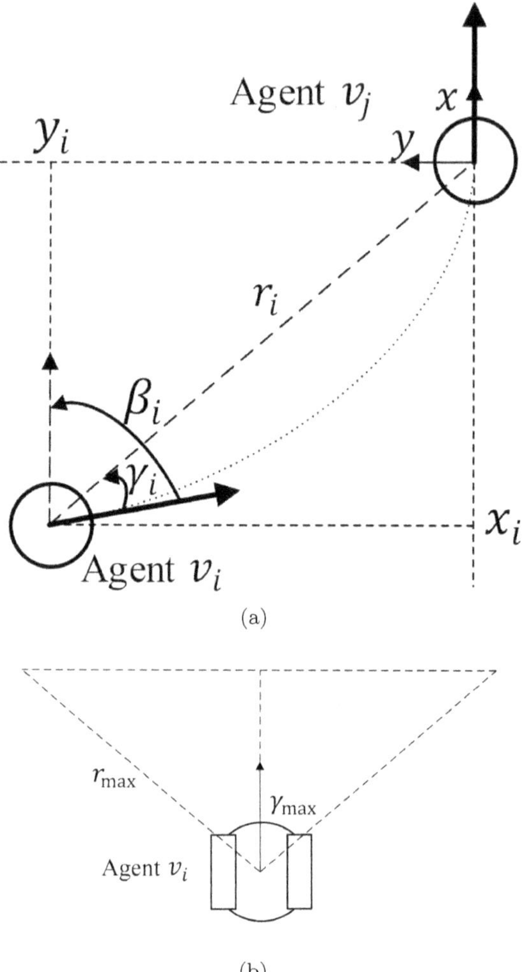

Figure 15.1 Visibility constraints: (a) vision-based localization; (b) visual detection range.

$\gamma_i \in [-\gamma_{\max}, \gamma_{\max}]$ that are measured as shown in Fig. 15.1(a). Define the relative distance as $r_i = \sqrt{x_i^2 + y_i^2}$, therefore, an agent has a limited visual detection range such that an object can only be detected if the following conditions can be satisfied that

$$|\gamma_i| \leq \gamma_{\max} \tag{15.10}$$

$$\|\mathbf{r}_i\| \leq \frac{r_{\max} \cos \gamma_{\max}}{|\cos \gamma_i|} \tag{15.11}$$

where $r_{\max}$ is the length of the equal sides of the cone of view. The available detection range for an agent is illustrated in Fig. 15.1(b).

Assumption 43. *A follower can reliably detect its local leader which lie within a limited region w.r.t. the forward-looking direction.*

Assumption 44. *The subgraph G_s associated with the followers is directed and in the graph G the leader has directed paths to all followers. (Equivalently, G contains a directed spanning tree with the leader as the root.)*

Remark 62. *There may be more than one agents in the visible range of a followers. Selected a specific agent in the filed of vision as the local leader in the beginning of the task, this role will not be changed during the whole process of the task without human intervention. Thus, the specific task for a follower is to track its local leader with specific position displacement while maintaining visibility of the local leader.*

15.3 FORMATION CONTROL

15.3.1 FORMATION FOLLOWING

In this chapter, a dipolar vector field [248] will be build for each follower to achieve desired formation w.r.t. the local leader frame $\mathcal{J}$ as Eq. (15.8). A vector field $\mathbf{F} : \mathbb{R}^2 \to \mathbb{R}^2$ can be mathematically expressed as the following form

$$\mathbf{F}(\eta) = \lambda \left(\mathbf{p}^{\mathrm{T}}\eta\right) \eta - \mathbf{p}\left(\eta^{\mathrm{T}}\eta\right) \tag{15.12}$$

where $\lambda \in \mathbb{R}$ and $\eta = [\eta_x \quad \eta_y]^{\mathrm{T}}$ is a position vector and $\mathbf{p} = [p_x \quad p_y]^{\mathrm{T}}$, $\mathbf{p} \neq 0$. The analytical formation of the vector field components F_{xi} and F_{yi} for agent v_i is

$$\begin{aligned} F_{xi} &= (\lambda - 1)p_x\varepsilon_{xi}^2 + \lambda p_y\varepsilon_{xi}\varepsilon_{yi} - p_x\varepsilon_{yi}^2 && (15.13)\\ F_{yi} &= (\lambda - 1)p_y\varepsilon_{yi}^2 + \lambda p_x\varepsilon_{xi}\varepsilon_{yi} - p_y\varepsilon_{xi}^2 && (15.14)\end{aligned}$$

All integral lines in the vector field converge to the original $(0, 0)$ parallel to the axis the $\mathbf{p}$ lies on (see Fig. 15.2). To achieve the desired formation of the follower w.r.t. the local leader frame $\mathcal{J}$ with consensus in motion direction, an attractive vector field is establish where $\lambda = 3$ and $\mathbf{p} = [1 \quad 0]^{\mathrm{T}}$ such that $\varphi_p \triangleq \operatorname{atan2}(p_y, p_x) = 0$ with $p_x = 1$ and $p_y = 0$. Meanwhile, we pick the formation error vector ε_i w.r.t. the local leader frame $\mathcal{J}$ as the position vector η. Consequently, Eqs. (15.12–15.14) can be rewritten as

$$\begin{aligned} \mathbf{F}_i(\varepsilon_i) &= 3\left(\mathbf{p}^{\mathrm{T}}\varepsilon_i\right)\varepsilon_i - \mathbf{p}\left(\varepsilon_i^{\mathrm{T}}\varepsilon_i\right) && (15.15)\\ F_{xi} &= 2\varepsilon_{xi}^2 - \varepsilon_{yi}^2 && (15.16)\\ F_{yi} &= 3\varepsilon_{xi}\varepsilon_{yi} && (15.17)\end{aligned}$$

The corresponding controller $\mathbf{u}_{0i} = [\nu_{0i} \quad \omega_{0i}]^{\mathrm{T}}$ for v_i is proposed as

$$\begin{aligned} \nu_{0i} &= -k_1 \operatorname{sign}_{\mathrm{c}}\left(\varepsilon_i^{\mathrm{T}} \begin{bmatrix}\cos\beta_i\\ \sin\beta_i\end{bmatrix}\right) \|\varepsilon_i\| - \operatorname{sign}_{\mathrm{c}}(\varepsilon_{xi})\,\nu_{\max} && (15.18)\\ \omega_{0i} &= -k_2(\beta_i - \varphi_i) + \dot{\varphi}_i && (15.19)\end{aligned}$$

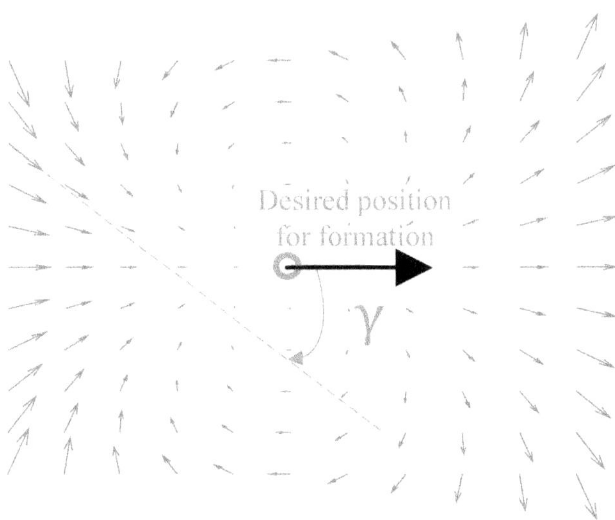

Figure 15.2 Dipolar vector field.

where ν_{0i} and ω_{0i} denote linear and angular velocity respectively, $\varphi_i \triangleq \text{atan2}(F_{yi}, F_{xi})$ is the orientation of the vector field at (x_i, y_i), and $k_1 > 0$, $k_2 > 0$ are two constant parameters. The sign function sign_c is in line with (2.28).

15.3.2 VISIBILITY MAINTENANCE

Given that the local leader is decided among the visible agents in the beginning of the task, the visibility constraints modeled in Eqs. (15.10–15.11) must be satisfied initially as

$$|\gamma_i(0)| \leq \gamma_{\max} \tag{15.20}$$

$$\|\mathbf{r}_i(0)\| \leq \frac{r_{\max}\cos\gamma_{\max}}{|\cos\gamma_i(0)|} \tag{15.21}$$

for $i = 2, ..., N$.

In the following, the visibility constraints will be divided into two sections, bearing angle constraint and distance constraint and analyze, respectively. First the bearing angle constraint can be equally written as $-\gamma_{\max} \leq \gamma_i \leq \gamma_{\max}$. Define $\alpha_i \triangleq \arctan\left(\frac{y_i}{x_i}\right)$, we have

$$\alpha_i = \arctan\left(\frac{\varepsilon_{yi} + \delta_{yi}}{\varepsilon_{xi} + \delta_{xi}}\right) \tag{15.22}$$

$$\beta_i = \alpha_i - \gamma_i. \tag{15.23}$$

as shown in Fig. 15.4. Converting the constraint on γ_i into the constraint on β_i yields $\beta_i^-(t) \leq \beta_i \leq \beta_i^+(t)$, with

$$\beta_i^-(t) = \arctan\left(\frac{\varepsilon_{yi}+\delta_{yi}}{\varepsilon_{xi}+\delta_{xi}}\right) - \gamma_{\max} \tag{15.24}$$

$$\beta_i^+(t) = \arctan\left(\frac{\varepsilon_{yi}+\delta_{yi}}{\varepsilon_{xi}+\delta_{xi}}\right) + \gamma_{\max} \tag{15.25}$$

which are time-varying state constraints. Eq. (15.20) implies the following constraints can be obtained initially that $\beta_i^-(0) \leq \beta_i(0) \leq \beta_i^+(0)$. Based on Eq. (15.20), the two-sided inequalities of the motion constraints are converted following the procedures below. Choose positive scalar k_3 and consider the scalar $\dot{\beta}_i$ regulated as

$$-k_3\left(\beta_i - \beta_i^-\right) + \dot{\beta}_i^- \leq \dot{\beta}_i \leq -k_3\left(\beta_i - \beta_i^+\right) + \dot{\beta}_i^+ \tag{15.26}$$

Considering that $\dot{\beta}_i = \omega_i - \omega_j$, the above inequality can be rewritten as the input constraints as

$$-k_3\left(\beta_i - \beta_i^-\right) + \dot{\beta}_i^- + \omega_j \leq \omega_i \leq -k_3\left(\beta_i - \beta_i^+\right) + \dot{\beta}_i^+ + \omega_j \tag{15.27}$$

where we have $\dot{\beta}_i^-(t) = \dot{\beta}_i^+(t) = \dot{\alpha}_i$ deducing the input constraints as

$$-k_3\left(\beta_i - \beta_i^-\right) + \dot{\alpha}_i + \omega_j \leq \omega_i \leq -k_3\left(\beta_i - \beta_i^+\right) + \dot{\alpha}_i + \omega_j \tag{15.28}$$

Without the estimation of ω_j, this constraint is further strengthened as

$$-k_3\left(\beta_i - \beta_i^-\right) + \dot{\alpha}_i + \omega_{\max} \leq \omega_i \leq -k_3\left(\beta_i - \beta_i^+\right) + \dot{\alpha}_i - \omega_{\max} \tag{15.29}$$

Meanwhile, it should be satisfied that $\omega_{\max} < \frac{1}{2}k_3\left(\beta_i^+ - \beta_i^-\right) = k_3\gamma_{\max}$. Considering the presence of input constraints on ω_i and defining

$$\omega_i^- = -k_3\left(\beta_i - \beta_i^-\right) + \dot{\alpha}_i + \omega_{\max} \tag{15.30}$$

$$\omega_i^+ = -k_3\left(\beta_i - \beta_i^+\right) + \dot{\alpha}_i - \omega_{\max} \tag{15.31}$$

we have the following input saturation function.

$$S(\omega_i) = \begin{cases} \omega_i^- & \text{if } \omega_i < \omega_i^- & (15.32) \\ \omega_i^+ & \text{if } \omega_i > \omega_i^+ & (15.33) \\ \omega_i & \text{otherwise} & (15.34) \end{cases}$$

Similar to [65], defining $\Delta\omega_i = S(\omega_i) - \omega_i$, the control scheme on the angular velocity Eq. (15.19) is redesigned as

$$\omega_i = -k_2\left(\beta_i - \varphi_i\right) + \dot{\varphi}_i - \zeta_{1i} \tag{15.35}$$

with the auxiliary design system chosen as

$$\dot{\zeta}_{1i} = \begin{cases} -k_4\zeta_{1i} - \frac{|(\beta_i - \varphi_i)\Delta\omega_i| + 0.5\Delta\omega_i^2}{\|\zeta_{1i}\|^2}\zeta_{1i} + \Delta\omega_i & \text{if } |\zeta_{1i}| \geq \chi_1 \quad (15.36) \\ 0 & \text{if } |\zeta_{1i}| < \chi_1 \quad (15.37) \end{cases}$$

where $k_4 > 0$, χ_1 is a small positive value and $\zeta_{1i} \in \mathbb{R}$ is the state of the auxiliary design system. Then studying the distance constraint, define

$$R_i^{\max} = \frac{r_{\max}\cos\gamma_{\max}}{|\cos\gamma_i|} \tag{15.38}$$

and we have the distance constraint as

$$r_i \triangleq \|\mathbf{r}_i\| = \sqrt{x_i^2 + y_i^2} \leq R_i^{\max} \tag{15.39}$$

The time derivative of r_i is

$$\dot{r}_i = \frac{x_i\dot{x}_i + y_i\dot{y}_i}{r_i} \tag{15.40}$$

Substituting Eq. (15.7) into Eq. (15.40) we obtain

$$\begin{aligned} \dot{r}_i &= \frac{x_i(-\nu_j + y_i\omega_j + \nu_i\cos\beta_i)}{r_i} \\ &\quad + \frac{y_i(-x_i\omega_j + \nu_i\cos\beta_i)}{r_i} \\ &= \frac{x_i\nu_i\cos\beta_i + y_i\nu_i\cos\beta_i - x_i\nu_j}{r_i} \\ &= -(\cos\alpha_i\cos\beta_i + \sin\alpha_i\cos\beta_i)\nu_i + \nu_j\cos\alpha_i \\ &= -\nu_i\cos(\alpha_i - \beta_i) + \nu_j\cos\alpha_i \\ &= -\nu_i\cos\gamma_i + \nu_j\cos\alpha_i \end{aligned} \tag{15.41}$$

Following the similar steps above, choose positive scalar k_5 and consider the scalar $\dot{r}_i$ regulated as

$$\dot{r}_i < -k_5(r_i - R_i^{\max}) + \dot{R}_i^{\max} \tag{15.42}$$

Combining Eqs. (15.41–15.42) we obtain

$$\nu_i > \frac{k_5(r_i - R_i^{\max}) - \dot{R}_i^{\max} + \nu_j\cos\alpha_i}{\cos\gamma_i} \tag{15.43}$$

Without the estimation of v_j, this constraint is further strengthened as

$$\nu_i > \frac{k_5(r_i - R_i^{\max}) - \dot{R}_i^{\max} + \nu_{\max}\cos\alpha_i}{\cos\gamma_i} \tag{15.44}$$

Considering the presence of input constraints on ν_i and defining

$$\nu_i^- = \frac{k_5 \left(r_i - R_i^{\max}\right) - \dot{R}_i^{\max} + \nu_{\max} \cos \alpha_i}{\cos \gamma_i} \tag{15.45}$$

we have

$$S(\nu_i) = \begin{cases} \nu_i^- & \text{if } \nu_i < \nu_i^- \quad (15.46) \\ \nu_i & \text{otherwise} \quad (15.47) \end{cases}$$

Defining $\Delta\nu_i = S(\nu_i) - \nu_i$, the control scheme on the linear velocity Eq. (15.18) is redesigned as

$$\nu_i = -k_1 \mathrm{sign_c}\left(\varepsilon_i^{\mathrm{T}} \begin{bmatrix} \cos \beta_i \\ \sin \beta_i \end{bmatrix}\right) \|\varepsilon_i\| - \mathrm{sign_c}\left(\varepsilon_{xi}\right) \nu_{\max} - \zeta_{2i} \tag{15.48}$$

with the auxiliary design system proposed as

$$\dot{\zeta}_{2i} = \begin{cases} -k_6 \zeta_{2i} - \dfrac{\|\varepsilon_i\| \, |\Delta\nu_i| + 0.5 \Delta\nu_i^2}{\|\zeta_{2i}\|^2} \zeta_{2i} + \Delta\nu_i & \text{if } \|\zeta_{2i}\| \geq \chi_2 \quad (15.49) \\ 0 & \text{if } \|\zeta_{2i}\| < \chi_2 \quad (15.50) \end{cases}$$

where $k_6 > 0$, χ_2 is a small positive value and $\zeta_{2i} \in \mathbb{R}$ is the state of the auxiliary design system.

15.3.3 PERFORMANCE ANALYSIS

Theorem 15.1

Consider N mobile agents with similar dynamics Eq. (15.1), Assumptions 41–44 and formation control laws Eqs. (15.32–15.35) and Eqs. (15.46–15.48), the predefined formation expressed in Eq. (15.8) can be achieved. ■

Proof of Theorem 15.1. Consider the Lyapunov candidate as

$$V_i = \frac{1}{2}\varepsilon_{xi}^2 + \frac{1}{2}\varepsilon_{yi}^2 + \frac{1}{2}\left(\beta_i - \varphi_i\right)^2 + \frac{1}{2}\zeta_{1i}^2 + \frac{1}{2}\zeta_{2i}^2 \tag{15.51}$$

for $i = 2, \ldots, N$. The derivative of V_i is

$$\dot{V}_i = \varepsilon_{xi}\dot{\varepsilon}_{xi} + \varepsilon_{yi}\dot{\varepsilon}_{yi} + \left(\beta_i - \varphi_i\right)\left(\dot{\beta}_i - \dot{\varphi}_i\right) + \zeta_{1i}\dot{\zeta}_{1i} + \zeta_{2i}\dot{\zeta}_{2i} \tag{15.52}$$

with

$$\dot{\varepsilon}_{xi} = S(\nu_i) \cos \beta_i - \nu_j + y_i \omega_j \tag{15.53}$$

$$\dot{\varepsilon}_{yi} = S(\nu_i) \sin \beta_i - x_i \omega_j \tag{15.54}$$

We obtain that

$$\begin{aligned}
\dot{V}_i &= \varepsilon_{xi}\left(S(\nu_i)\cos\beta_i - \nu_j + y_i\omega_j\right) + \zeta_{1i}\dot{\zeta}_{1i} \\
&\quad + \varepsilon_{yi}\left(S(\nu_i)\sin\beta_i - x_i\omega_j\right) + \zeta_{2i}\dot{\zeta}_{2i} \\
&\quad + \left(\beta_i - \varphi_i\right)\left(S(\omega_i) - \omega_j - \dot{\varphi}_i\right) \\
&= \begin{bmatrix}\varepsilon_{xi} & \varepsilon_{yi}\end{bmatrix}\begin{bmatrix}\cos\beta_i \\ \sin\beta_i\end{bmatrix} S(\nu_i) + \zeta_{1i}\dot{\zeta}_{1i} \\
&\quad + \left(\beta_i - \varphi_i\right)\left(S(\omega_i) - \dot{\varphi}_i\right) + \zeta_{2i}\dot{\zeta}_{2i} \\
&\quad + \begin{bmatrix}\varepsilon_{xi} & \varepsilon_{yi} & \beta_i - \varphi_i\end{bmatrix}\begin{bmatrix}-\nu_j + y_i\omega_j \\ -x_i\omega_j \\ -\omega_j\end{bmatrix}
\end{aligned} \tag{15.55}$$

Specifically, we define

$$P_1 \triangleq \left(\varepsilon_{xi}\cos\beta_i + \varepsilon_{yi}\sin\beta_i\right) S(\nu_i) + \zeta_{2i}\dot{\zeta}_{2i} \tag{15.56}$$

$$P_2 \triangleq \left(\beta_i - \varphi_i\right)\left(S(\omega_i) - \dot{\varphi}_i\right) + \zeta_{1i}\dot{\zeta}_{1i} \tag{15.57}$$

$$P_3 \triangleq \begin{bmatrix}\varepsilon_{xi} & \varepsilon_{yi} & \beta_i - \varphi_i\end{bmatrix}\begin{bmatrix}-\nu_j + y_i\omega_j \\ -x_i\omega_j \\ -\omega_j\end{bmatrix} \tag{15.58}$$

Substituting Eqs. (15.48–15.50) into Eq. (15.56) yields that

$$\begin{aligned}
P_1 &= -k_6\zeta_{2i}^2 - 0.5\Delta\nu_i^2 + \zeta_{2i}\Delta\nu_i - \|\varepsilon_i\|\,|\Delta\nu_i| \\
&\quad + \left(\varepsilon_{xi}\cos\beta_i + \varepsilon_{yi}\sin\beta_i\right)\left(\nu_i + \Delta\nu_i\right) \\
&\le -\left(k_1 - 0.5\right)\left|\frac{\varepsilon_i^{\mathrm{T}}}{\|\varepsilon_i\|}\begin{bmatrix}\cos\beta_i \\ \sin\beta_i\end{bmatrix}\right|\|\varepsilon_i\|^2 - \left(k_6 - 1\right)\zeta_{2i}^2 \\
&\quad - \mathrm{sign_c}\left(\varepsilon_{xi}\right)\nu_{\max}\left(\varepsilon_i^{\mathrm{T}}\begin{bmatrix}\cos\beta_i \\ \sin\beta_i\end{bmatrix}\right)
\end{aligned} \tag{15.59}$$

Substituting Eqs. (15.35–15.37) into Eq. (15.57) yields that

$$\begin{aligned}
P_2 &= \left(\beta_i - \varphi_i\right)\left(\omega_i + \Delta\omega_i - \dot{\varphi}_i\right) - k_4\zeta_{1i}^2 - 0.5\Delta\omega_i^2 \\
&\quad - |\left(\beta_i - \varphi_i\right)\Delta\omega_i| + \zeta_{1i}\Delta\omega_i \\
&= -k_2\left(\beta_i - \varphi_i\right)^2 + \left(\beta_i - \varphi_i\right)\left(-\zeta_{1i} + \Delta\omega_i\right) - k_4\zeta_{1i}^2 \\
&\quad - 0.5\Delta\omega_i^2 - |\left(\beta_i - \varphi_i\right)\Delta\omega_i| + \zeta_{1i}\Delta\omega_i \\
&\le -k_2\left(\beta_i - \varphi_i\right)^2 - \left(\beta_i - \varphi_i\right)\zeta_{1i} - k_4\zeta_{1i}^2 \\
&\quad - 0.5\Delta\omega_i^2 + \zeta_{1i}\Delta\omega_i \\
&\le -\left(k_2 - 0.5\right)\left(\beta_i - \varphi_i\right)^2 - \left(k_4 - 1\right)\zeta_{1i}^2
\end{aligned} \tag{15.60}$$

and

$$
\begin{aligned}
P_3 &= -\varepsilon_{xi}\nu_j + \varepsilon_{xi}y_i\omega_j - \varepsilon_{yi}x_i\omega_j - (\beta_i - \varphi_i)\,\omega_j \\
&= \begin{bmatrix}\varepsilon_{xi} & \varepsilon_{yi}\end{bmatrix}\begin{bmatrix}-\nu_j \\ 0\end{bmatrix} + \begin{bmatrix}\varepsilon_{xi} & \varepsilon_{yi}\end{bmatrix}\begin{bmatrix}y_i \\ -x_i\end{bmatrix}\omega_j \\
&\quad - (\beta_i - \varphi_i)\,\omega_j \\
&\leq \|\varepsilon_i\|\,\nu_{\max} + \begin{bmatrix}\varepsilon_{xi} & \varepsilon_{yi}\end{bmatrix}\begin{bmatrix}\varepsilon_{yi} + \delta_{yi} \\ -\varepsilon_{xi} - \delta_{xi}\end{bmatrix} \\
&\quad - (\beta_i - \varphi_i)\,\omega_j \\
&= \|\varepsilon_i\|\,\nu_{\max} + \varepsilon_i^{\mathrm{T}}\begin{bmatrix}\delta_{yi} \\ -\delta_{xi}\end{bmatrix} - (\beta_i - \varphi_i)\,\omega_j
\end{aligned} \tag{15.61}
$$

Let $\vartheta_i = \beta_i - \varphi_i$, representing the dynamics of the orientation error of v_i we have $\dot{\vartheta}_i = \dot{\beta}_i - \dot{\varphi}_i = \dot{\theta}_i - \dot{\theta}_j - \dot{\varphi}_i = \omega_i - \omega_j - \dot{\varphi}_i = -k_2(\beta_i - \varphi_i) - \omega_j = -k_2\vartheta_i - \omega_j$. If $\omega_j = 0$, it follows that $\vartheta_i \to 0$ exponentially, i.e., that $\beta_i \to \varphi_i$. However, w_j is not in general equal to zero. Assuming that ω_j is continuously differentiable and $|\dot{\omega}_j|$ is sufficiently small, ω_j can be treated as a frozen parameter, and the frozen system $0 = -k_2\vartheta_i - \omega_j$ has a continuously differentiable isolated root $\vartheta_i = -\frac{\omega_j}{k_2} = h(\omega_j)$. Therefore, P_3 can be further written as

$$
P_3 = \|\varepsilon_i\|\,\nu_1^{\max} + \varepsilon_i^{\mathrm{T}}\begin{bmatrix}\delta_{yi} \\ -\delta_{xi}\end{bmatrix} + \frac{\omega_j^2}{k_2} \tag{15.62}
$$

Eq. (15.55) can be rewritten as

$$
\begin{aligned}
\dot{V}_i &\leq -(k_1 - 0.5)\left|\frac{\varepsilon_i^{\mathrm{T}}}{\|\varepsilon_i\|}\begin{bmatrix}\cos\beta_i \\ \sin\beta_i\end{bmatrix}\right|\|\varepsilon_i\|^2 + \|\varepsilon_i\|\,\nu_{\max} \\
&\quad -(k_2 - 0.5)(\beta_i - \varphi_i)^2 - (k_4 - 1)\,\zeta_{1i}^2 \\
&\quad -(k_6 - 1)\,\zeta_{2i}^2 - \mathrm{sign_c}(\varepsilon_{xi})\,\nu_{\max}\left(\varepsilon_i^{\mathrm{T}}\begin{bmatrix}\cos\beta_i \\ \sin\beta_i\end{bmatrix}\right) \\
&\quad +\varepsilon_i^{\mathrm{T}}\begin{bmatrix}\delta_{yi} \\ -\delta_{xi}\end{bmatrix} + \frac{\omega_j^2}{k_2} \\
&\leq -(k_1 - 0.5)\left|\frac{\varepsilon_i^{\mathrm{T}}}{\|\varepsilon_i\|}\begin{bmatrix}\cos\beta_i \\ \sin\beta_i\end{bmatrix}\right|\|\varepsilon_i\|^2 - (k_4 - 1)\,\zeta_{1i}^2 \\
&\quad (k_2 - 0.5)(\beta_i - \varphi_i)^2 - \nu_{\max}\left|\varepsilon_i^{\mathrm{T}}\begin{bmatrix}\cos\beta_i \\ \sin\beta_i\end{bmatrix}\right| \\
&\quad -(k_6 - 1)\,\zeta_{2i}^2 + \|\varepsilon_i\|\,(\nu_{\max} + \|\delta_{ij}\|) + \frac{\omega_j^2}{k_2} \\
&\leq -\left[(k_1 - 0.5)\left|\frac{\varepsilon_i^{\mathrm{T}}}{\|\varepsilon_i\|}\begin{bmatrix}\cos\beta_i \\ \sin\beta_i\end{bmatrix}\right| - 0.5\right]\|\varepsilon_i\|^2
\end{aligned}
$$

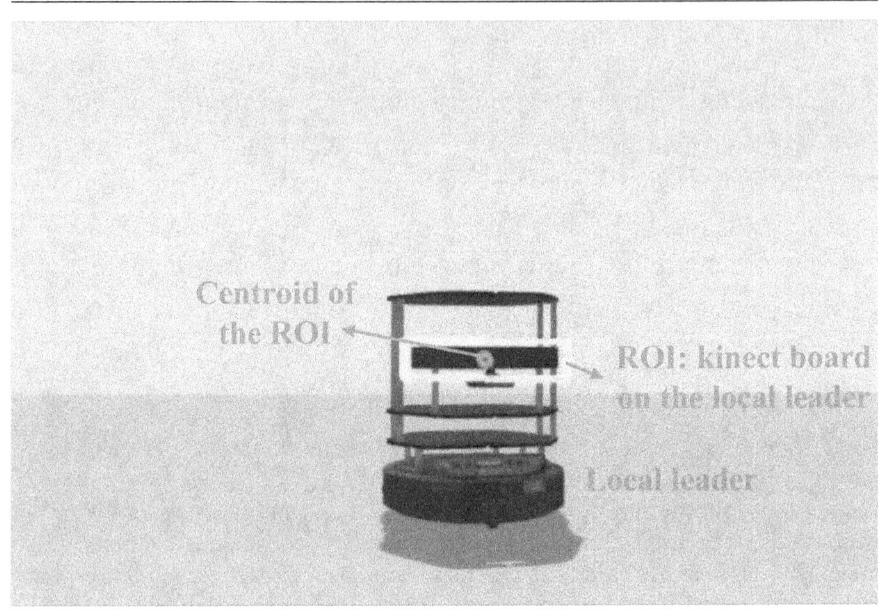

Figure 15.3 Visual tracking in Gazebo.

$$
\begin{aligned}
& -\left(k_2-0.5\right)\left(\beta_i-\varphi_i\right)^2-\left(k_4-1\right)\zeta_{1i}^2-\left(k_6-1\right)\zeta_{2i}^2 \\
& -\nu_{\max}\left|\varepsilon_i^{\mathrm{T}}\begin{bmatrix}\cos\beta_i \\ \sin\beta_i\end{bmatrix}\right|+\frac{1}{2}\left(\nu_{\max}+\left\|\delta_{ij}\right\|\right)^2+\frac{\omega_j^2}{k_2} \\
\leq\ & -k_v V_i+\phi_i
\end{aligned}
\tag{15.63}
$$

where $\phi_i=\frac{1}{2}\left(\nu_1^{\max}+\left\|\delta_{ij}\right\|\right)^2+\frac{\omega_{\max}^2}{k_2}$. Hence, by properly choosing parameters, MAS will be asymptotic stable [321], and the desired leader-follower formation tracking can be achieved under visibility constraints. □

15.4 VISION-BASED POSE ESTIMATION

In this section, a method to estimate the pose of an object using RGBD images captured by kinect will be proposed. The Microsoft Kinect combines 3D depth map information with traditional RGB color camera data. The advantages of this sensor over stereo cameras is the low computational expense in calculating depth and the accuracy of the range measurements. The CamShift algorithm [201] is applied to track a selected object, which can be implemented in real time, confirming the real time performance of the control strategy. To further obtain the required information to estimate the pose of the leader, the region of the kinect board in the leader is chosen as the region of interest (ROI) in visual tracking algorithm. Fig. 15.3 demonstrates the visual tracking outcome in Gazebo where the green line encloses the ROI and the red circle marks the

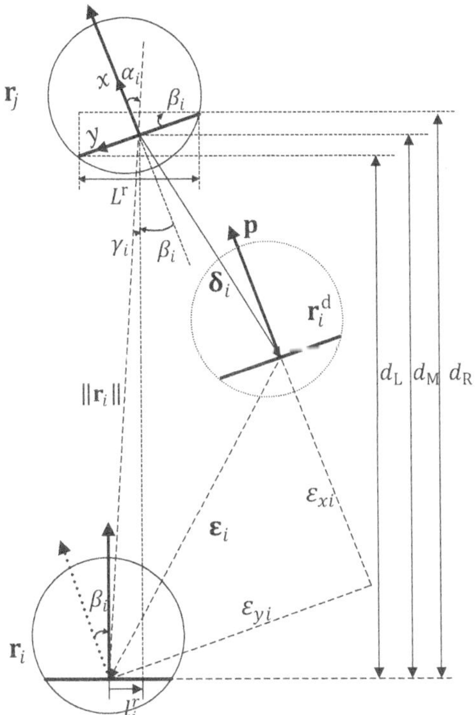

Figure 15.4 Variables in a local leader frame.

centroid of the ROI. Then the following information can be directed acquired from an RGBD image executing the object tracking algorithm:

$\mathbf{r}_L = [x_L, y_L, d_L]$: the midpoint of the left border of the region tracked on the depth image, with location $[x_L, y_L]$ and depth value d_L;
$\mathbf{r}_M = [x_M, y_M, d_M]$: the centroid of the region tracked on the depth image, with location $[x_M, y_M]$ and depth value d_M;
$\mathbf{r}_R = [x_R, y_R, d_R]$: the midpoint of the right border of the region tracked on the depth image, with location $[x_R, y_R]$ and depth value d_R.

Define L^r as the real length of the kinect of the local leader in the axis that the long side of the kinect of the follower lies on, while L^f is the length of the kinect of the local leader in the view of the follower. Above variables have been marked in Figs. 15.4 and 15.5. Then we have

$$L^f = x_R - x_L \tag{15.64}$$
$$L^r = \kappa d_M L^f \tag{15.65}$$

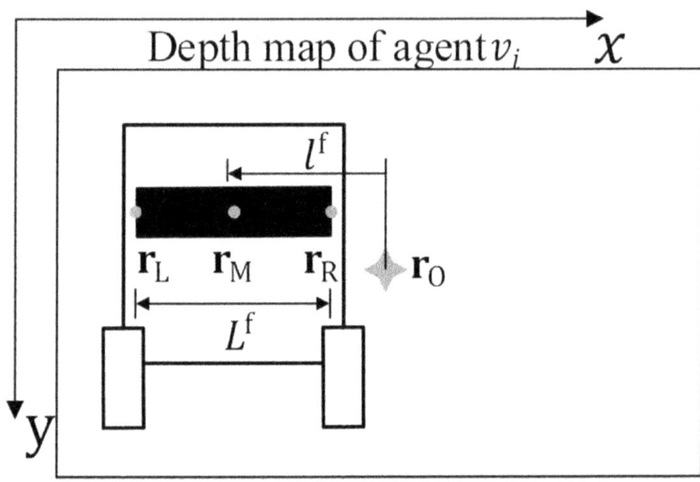

Figure 15.5 Vision map.

where $\kappa > 0$ is a scaling factor mapping lengths measured in pixel units to lengths measured in meters. The estimation of the orientation β_i w.r.t. the local leader frame $\mathcal{J}$ can be obtained as

$$\beta_i = \arctan\left(\frac{d_\mathrm{L} - d_\mathrm{R}}{L^\mathrm{r}}\right) \tag{15.66}$$

Define $r_\mathrm{O} = [x_\mathrm{O}, y_\mathrm{O}]$ as the centroid of the depth image and l^f as the horizontal offset in the depth image that

$$l^\mathrm{f} = x_\mathrm{M} - x_\mathrm{O} \tag{15.67}$$

Let l_i^r be the offset of the kinect center of the leader v_j to the kinect centroid of the follower v_i in the axis the long side of the kinect of the follower v_i lies on in real world, which can be estimated as

$$l_i^\mathrm{r} = -\kappa d_\mathrm{M} l^\mathrm{f} = d_\mathrm{M} \tan\gamma_i \tag{15.68}$$

and we could infer that

$$\gamma_i = \arctan\left(-\kappa l^\mathrm{f}\right) \tag{15.69}$$

which further yields that

$$\alpha_i = \arctan\left(-\kappa l^\mathrm{f}\right) + \arctan\left(\frac{d_\mathrm{L} - d_\mathrm{R}}{L^\mathrm{r}}\right) \tag{15.70}$$

Then the distance between the follower v_i and the local leader v_j can be got as

$$r_i = \sqrt{d_\mathrm{M}^2 + l_i^{r2}} \tag{15.71}$$

The follower v_i can be localized w.r.t. it local leader v_j as

$$x_i = -r_i \cos \alpha_i \quad (15.72)$$
$$y_i = -r_i \sin \alpha_i \quad (15.73)$$

inferring the formation error for the follower v_i as below.

$$\varepsilon_{xi} = -r_i \cos \alpha_i - \delta_{xi} \quad (15.74)$$
$$\varepsilon_{yi} = -r_i \sin \alpha_i - \delta_{yi} \quad (15.75)$$

With the help of variables acquired in this section, the control schemes proposed in Section 15.3 is able to be implemented without communication among agents or a laser range finder to determine the distance to the leader or a gyroscope to measure the orientation.

Remark 63. *This study selects the region of kinect equipped by the leader as the ROI in visual tracking algorithm. In fact, any visible region in the leader that can indicate the relative rotate angle of the leader can be chosen as the formation reference, especially a flat with distinguishing color that is easy to be visually tracked.*

Remark 64. *As the depth value of the point at the border is used in the estimation procedure, to eliminate the influence of noise, we set a small constant as the width of the edge to narrow down the tracking region ensuring that no redundant region has been tracked.*

15.5 SPEECH NAVIGATION SYSTEM.

In this section, we will study the motion plan for the global leader moving in constrained space. The leader will generally be provided with laser range finders or cameras to sense the environment. To improve the efficiency of world exploration by introducing the speech guidance of human, a speech navigation system is developed to translate speech commands into continuous control inputs for motion planning for the global leader agent. It consists of three main parts: speech recognition system, speech measurement system and control system. Fig. 15.6 shows the overall system block diagram.

Speech recognition system is developed to recognize objective contents of speech, while the speech measurement system measures subjective contents of human speech inputs. These two subsystems are jointly defined as the speech system converting audio signals into numerical signals, which is similar to the analog-to-digital converter (A/D converter). The outputs of speech recognition system and speech measurement system will be fused as results of the speech system. Following that, the control system translates fused outcomes into control inputs for the agent. These three submodules will be detailed subsequently.

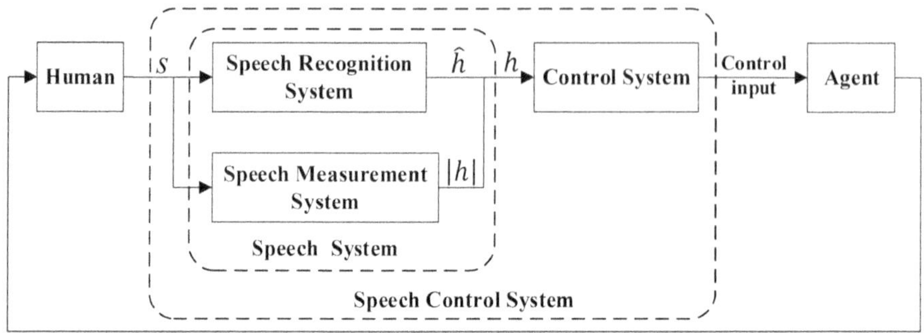

Figure 15.6 Speech navigation system diagram.

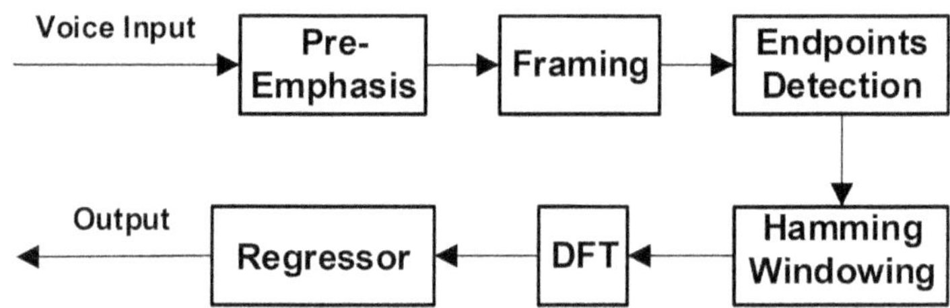

Figure 15.7 Extraction of proposed spectrum-based feature.

A. Speech Recognition System

We choose two single speech commands to be mapped into the control input ω_1 as $H = \{h^+, h^-\}$, where h^+ and h^- are corresponding to positive and negative signs of ω_1. Hence, two speech commands are needed to be recognized in the speech recognition system. In the speech recognition system, a map function $f_1 : s \to \hat{h}$ is establish, where s is a speech signal, and $\hat{h} = \pm 1$. If $s \in h^+$, then $\hat{h} = 1$; if $s \in h^-$, then $\hat{h} = -1$.

We utilize Mel-frequency cepstral coefficients (MFCC) to represent the acoustic input [95] and Dynamic Time Warping (DTW) algorithm [415] to measure similarity between two MFCC coefficients, where one is for a testing sample and another one is for a training templet. The classification result will be the target of the training template with the maximum similarity to the testing sample.

B. Speech Measurement System

In speech measurement system, a map function $f_2 : s \to |h|$ is to be establish. The control intention written as $|h|$ will be measured in this subsystem. We propose a spectrum-based feature to represent control intentions. The process of extracting spectrum-based features is explained in Fig. 15.7. Then an algorithm named *Random Forest* (RF) [34] is utilized as the regressor to map the extracted spectrum-based feature to the control magnitude $|h|$.

C. Control System

Obtaining the $\hat{h}$ from the speech recognition system and $|h|$ from the speech measurement system, the fused result of the speech system would be $g = \hat{h}\times|h|$ taken as the input of the control system. The fused result will be linearly mapped into the control input ω_1 as $\omega(t) = k_\omega h(t)$, where $k_\omega > 0$ is a scalar. To conclude, the control schemes for the global leader v_1 is proposed as

$$\nu_1 = \text{const} \leq \nu_{\max} \tag{15.76}$$

$$\omega_1 = k_\omega h \tag{15.77}$$

More details may refer to our previous work in [305]. This speech navigation system is designed assuming that the environment is unknown to MAS while can be observed by the human, which means MAS can only sense the environment by outputs of the speech control system. This assumption may be too incredible in realistic; however, it is to motivate the development of a speech-based human-robot interaction (HRI) system used jointly with other sensors (e.g., cameras and laser range finders).

15.6 SIMULATION & EXPERIMENT STUDIES

The Turtlebot 2 with Kinect for Xbox 360 is chosen as the agent in both simulation and experiment studies. The angular field of kinect view is 57° horizontally and 43° vertically [166]. In this chapter, only the horizontal view limit is considered; thus, we choose $\gamma_{\max} = 25°$ leaving a gap for safety and robustness issues. Moreover, kinect has a practical ranging limit of 3.5m distance and to further enable the visual tracking of the kinect to work during the whole task, a smaller value of distance limit should be adopted, which is set to be $r_{\max} = 2.0$m.

15.6.1 SIMULATIONS

A simulation platform for formation control of MAS is developed using Gazebo simulator on ROS implemented by Python. To study the performance of control laws Eqs. (15.32–15.35) & Eqs. (15.46–15.48) as well as the pose estimation algorithm introduced in Section 15.4, we consider a group of agents with $N = 3$ to achieve a specific formation. The simulation interface is designed as Fig. 15.8 with two subwindows illustrating the kinect view of two followers and a main window demonstrating the motion of three agents in the environment built.

In this section, we will explore two formation topologies as shown in Fig. 15.9. The results are shown in Figs. 15.10–15.12.

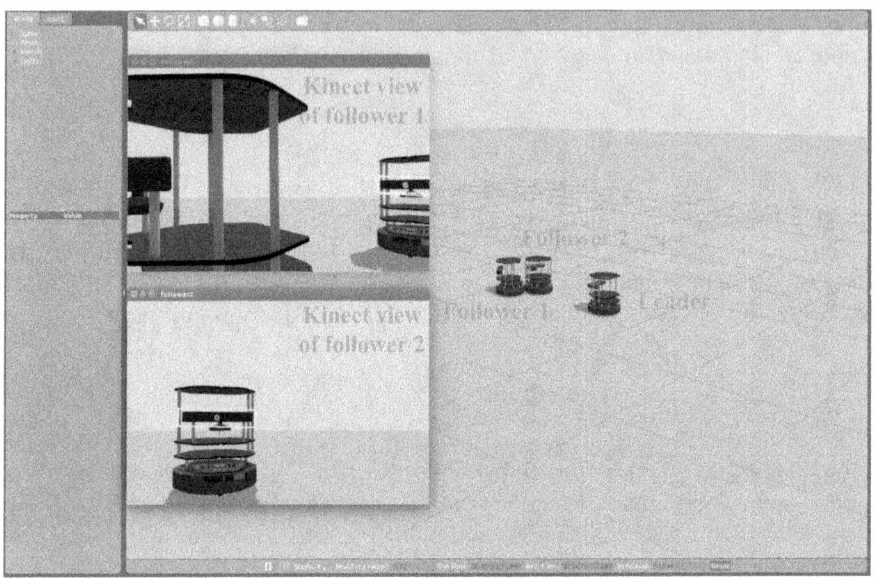

Figure 15.8 Simulation interface.

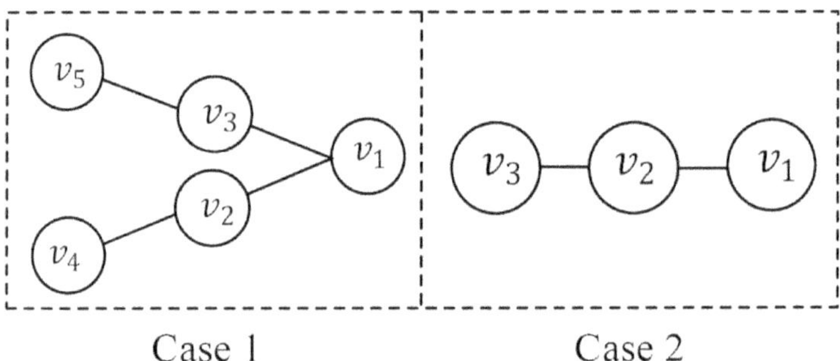

Figure 15.9 Topologies for two cases in Section A.

For the first case, MAS will generate Pseudo V formation topology tracking a straight path ($\nu_1 = 0.2$, $\omega_1 = 0$), where v_1 is the leader for v_2 and v_3. The configuration vectors for three agents in world coordinate system are initialized as $\mathbf{q}_{01} = [0 \quad 0 \quad 0]^{\mathrm{T}}$, $\mathbf{q}_{02} = [-1.5 \quad -0.5 \quad 0]^{\mathrm{T}}$ and $\mathbf{q}_{03} = [-1.7 \quad 0.6 \quad 0]^{\mathrm{T}}$ (see Fig. 15.11(c)). The desired formation position vectors for v_2 and v_3 w.r.t. the local leader frame $\mathcal{J}$ are set as $\mathbf{r}_2^{\mathrm{d}} = [-1.0 \quad -0.3]^{\mathrm{T}}$ and $\mathbf{r}_3^{\mathrm{d}} = [-1.0 \quad 0.3]^{\mathrm{T}}$.

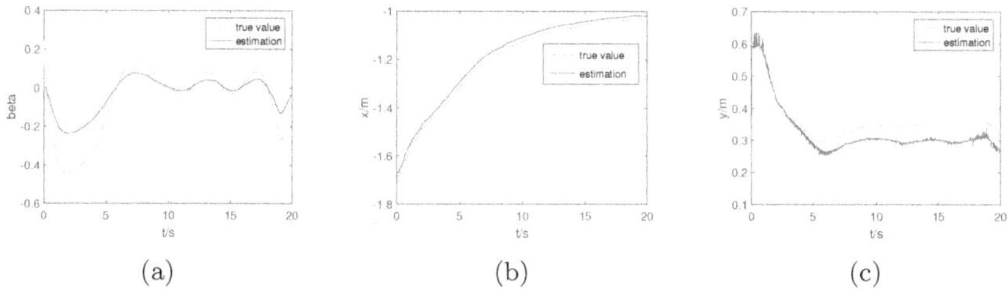

Figure 15.10 Vision-based pose estimation results: (a) the estimation of the relative orientation; (b) the estimation of the relative position in x-coordinate; (c) the estimation of the relative position in y-coordinate.

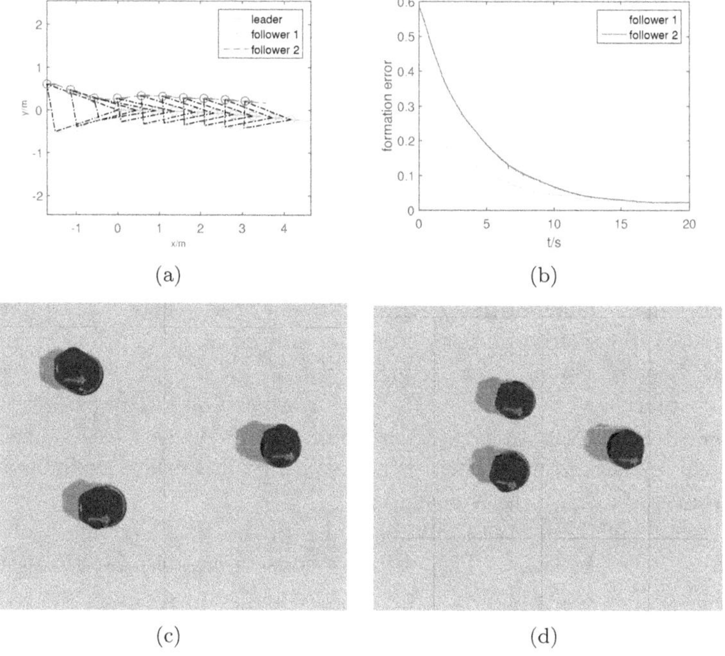

Figure 15.11 Formation tracking results for the first case in Section A: (a) trajectories for whole tacking process; (b) time history of formation errors; (c) $t = 0s$; (d) $t = 4s$.

A. Vision-based Formation tracking

First, the effectiveness of vision-based algorithm for pose estimation is validated for the formation tracking with the first topology. The estimation results for the relative orientation, relative positions in x-coordinate and y-coordinate of the leader v_1 by the follower v_3 are illustrated in Fig. 15.10, while the true values are got from *Odometry* message source of the navigation stack in Turtlebot 2 plotted in the figure for comparison.

Fig. 15.11(d) plots the positions of three agents at the end, where we could see the desired formation is realized. The whole tracking trajectories are illustrated in Fig. 15.11(a). The formation errors, which are calculated by *Odometry* message source of the navigation stack in Turtlebot 2, decrease gradually to a small value with time going as shown in Fig. 15.11(b). The main stabilization error sources for the formation tracking are: (i) the non-zero input for the leader agent; (ii) the measurement error of the kinect on the depth measurement.

Similarly, in the second one, a line formation topology will be achieved to track a circle path ($\nu_1 = 0.2$, $\omega_1 = 0.1$), where v_1 is the leader for v_2 and v_2 is the local leader for v_3. The configuration vectors for three agents in world coordinate system are initialized as $\mathbf{q}_{01} = [0 \quad 0 \quad 0]^{\mathrm{T}}$, $\mathbf{q}_{02} = [-1.5 \quad -0.3 \quad 0]^{\mathrm{T}}$ and $\mathbf{q}_{03} = [-2.7 \quad -0.5 \quad 0]^{\mathrm{T}}$. The desired formation position vectors for v_2 and v_3 w.r.t. the leader frame $\mathcal{J}$ are set as $\mathbf{r}_2^{\mathrm{d}} = [-1.0 \quad 0]^{\mathrm{T}}$ and $\mathbf{r}_3^{\mathrm{d}} = [-1.0 \quad 0]^{\mathrm{T}}$.

The whole tracking trajectories are illustrated in Fig. 15.12(a), while the beginning formation generation phase is detailed in Fig. 15.12(b). Fig. 15.12(c) plots the initialized positions and orientations of three agents. Although the initialized positions seem form a straight line in the world coordinate system, the displacement in y-coordinate w.r.t. the leader frame does not equal to zero. Fig. 15.12(c) illustrates the scene that the desired formation has been nearly achieved. Due to the non-zero angular velocity of the global leader v_1, the desired displacement on relative positions has been almost reached, although there is a small offset in motion direction of the follower w.r.t. the local leader frame, which leads to a formation of broken line in world coordinate system rather than a straight line as designed.

The success of these two simulations also prove the effectiveness of visibility maintenance algorithm. The visibility constraints may be dissatisfactory resulting from the designed dipolar vector field as indicated in Fig. 15.2 where $\gamma > \gamma_{\max}$, and the non-zero angular velocity of the leader. The proposed visibility maintenance strategies guarantee the success of visual tracking and the connection of the sensing networks.

B. Speech Navigation System

In this section, we are going to validate the effectiveness of speech navigation system for path planning in constrained space. Audio-Technica AT2020 Cardioid Condenser Studio Microphone is chosen as the audio input device for the speech control module. A simple obstacle environment is built (see Fig. 15.13) with a stop sign set at $[8 \quad 8]^{\mathrm{T}}$ viewed as the target of the task for MAS and six cubes of size $1\mathrm{m} * 1\mathrm{m} * 1\mathrm{m}$ set as obstacles. The configuration

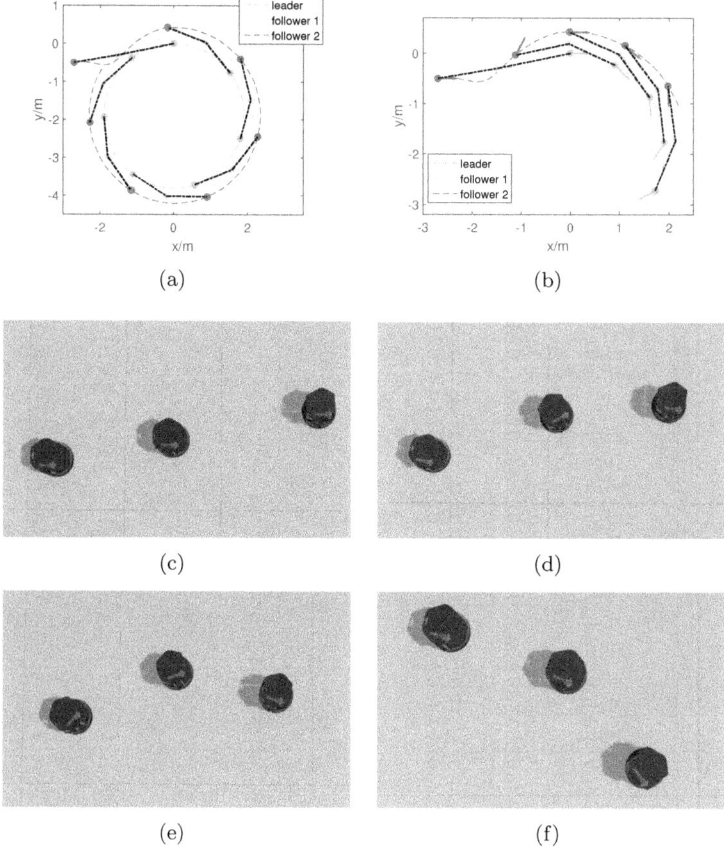

Figure 15.12 Formation tracking results for the second case in Section A: (a) trajectories for whole tacking process; (b) trajectories for the beginning formation generation process; (c) $t = 0s$; (d) $t = 2s$; (e) $t = 4s$; (f) $t = 8s$.

vectors for three agents in world coordinate system are initialized as $\mathbf{q}_{01} = [-6 \quad -7 \quad 0]^{\mathrm{T}}$, $\mathbf{q}_{02} = [-7.5 \quad -7.5 \quad 0]^{\mathrm{T}}$ and $\mathbf{q}_{03} = [-7.7 \quad -6.6 \quad 0]^{\mathrm{T}}$. The desired formation topology is chosen the same as the first case in Section A. The linear velocity of the leader v_1 is set as 0.4. The kincct board on the leader has never been used for obstacle detection in this simulation to shown how the leader explore the world under the total guidance of the speech navigation system.

Five targets $(0.2, 0.4, ..., 1)$ are assigned as outcomes of RF for model training. Training set will be collected round by round. In every round, the subject is required to express the control intensity in order from 0.2 to 1.0. There are 30 training samples collected for each target and 150 training samples in

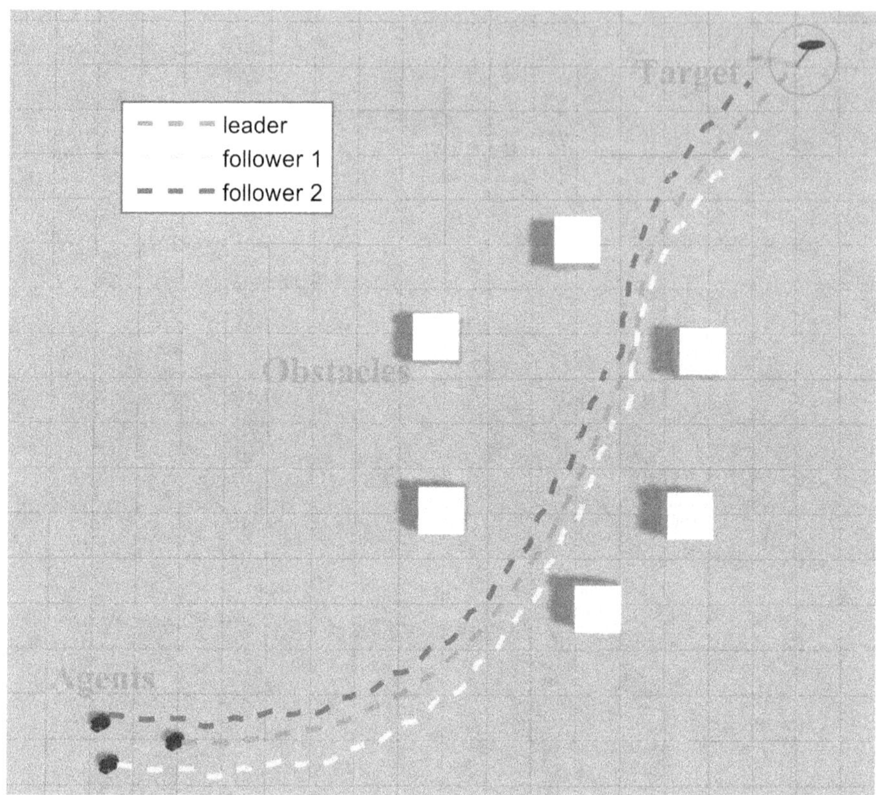

Figure 15.13 Trajectories of MAS in constrained space.

total for the speech measurement system. Meanwhile, these samples are also used to test the accuracy of the speech recognition system. We obtain 96.0% accuracy in totaly with 95.3% on "left" and 96.7% on "right".

Resulting trajectories (see Fig. 15.13) indicate that the proposed speech navigation system can efficiently convert the speech inputs into control inputs for obstacle avoidance and target tracking. Meanwhile, the given formation is achieved and maintained during the process of collision avoidance as detailed in Fig. 15.14. In Fig. 15.14(a), the desired formation has been formed basically, where the formation generation process may refer to the first case in Section A (see Fig. 15.11(a)). After that, MAS enters the collision avoidance mode with the help of the speech navigation system. Finally, the leader is able to reach the destination exactly (see Fig. 15.14(f)).

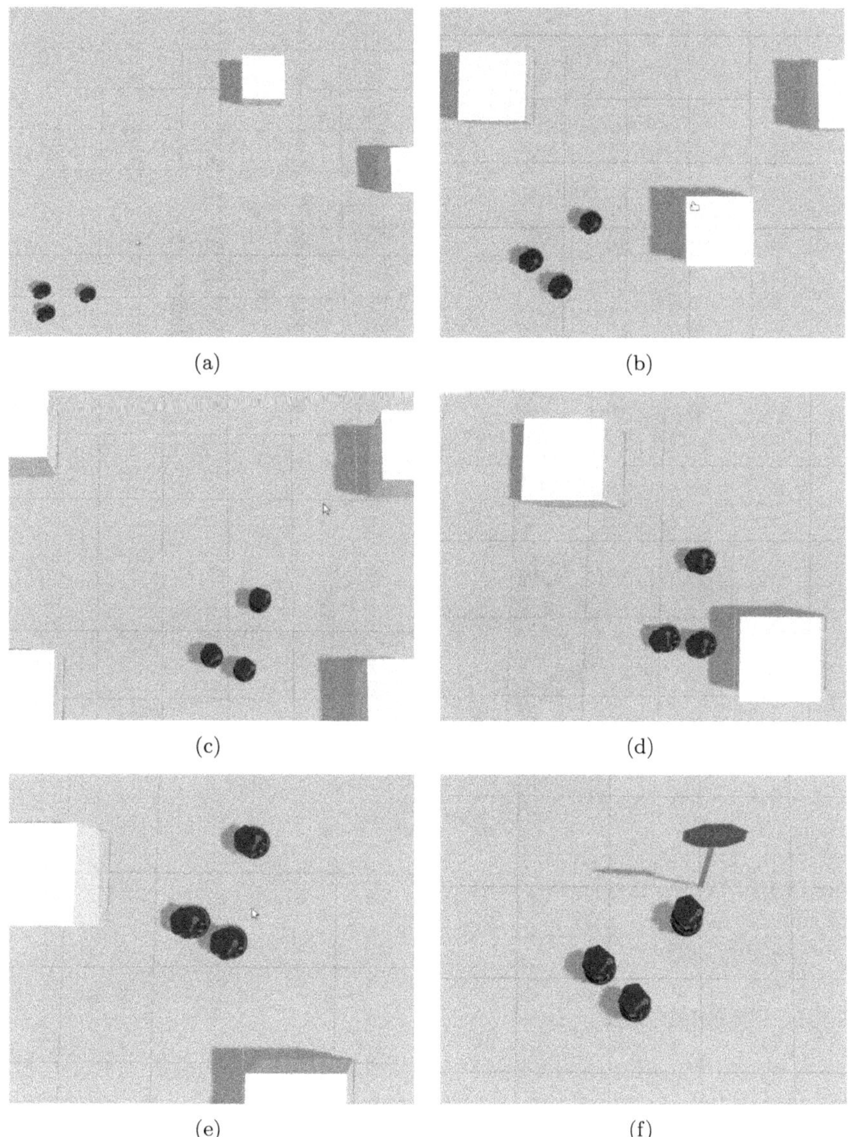

Figure 15.14 Formation tracking in constrained space with speech navigation: (a) $t = 5s$; (b) $t = 23s$; (c) $t = 34s$; (d) $t = 43s$; (e) $t = 48s$; (f) $t = 58s$.

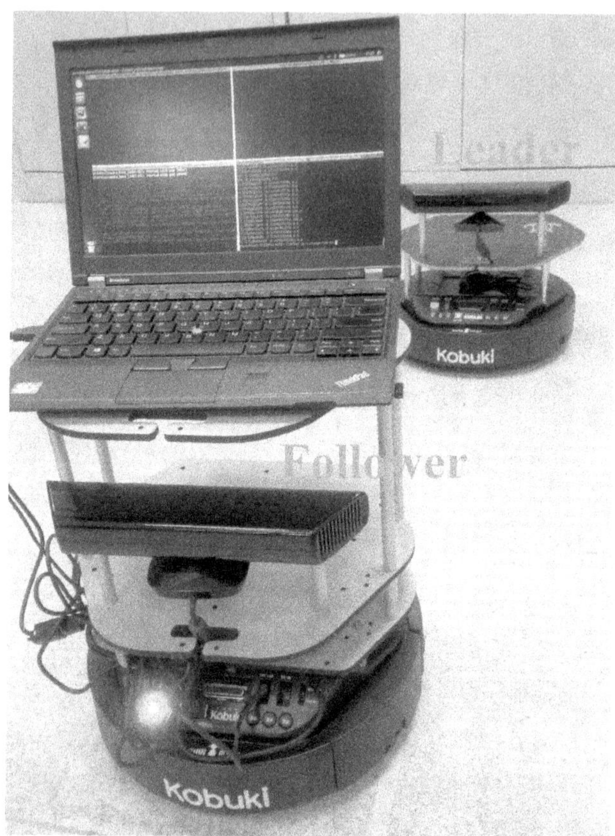

Figure 15.15 Experiment setup.

15.6.2 EXPERIMENTS

In order to test the proposed vision-based formation control strategy in a realistic scenario, a simple experiment is carried out at Robotics Research Laboratory (RRL), National University of Singapore, Singapore. The experimental setup consists of two Turtlebots acting as the leader and the follower as shown in Fig. 15.15. The follower Turtlebot consists of a laptop (Thinkpad X230, Lenovo Inc.) with ROS, while there is no laptop connected to the leader Turtlebot proving no communication between two agents. The leader is chosen to be static ($\nu_1 = 0$, $\omega_1 = 0$) in this experiment to observe the process of formation generation and formation error with zero input of the leader. The visual tracking result in real situation is demonstrated in Fig. 15.16.

Figure 15.16 Visual tracking result in experiment.

The configuration vectors for three agents in world coordinate system are initialized as $\mathbf{q}_{01} = [0 \quad 0 \quad 0]^{\mathrm{T}}$ and $\mathbf{q}_{02} = [-1.5 \quad 0.3 \quad 0]^{\mathrm{T}}$ (see Fig. 15.17(a)). The desired formation position vector for v_2 w.r.t. the leader frame is set as $\mathbf{r}_2^{\mathrm{d}} = [-0.8 \quad 0]^{\mathrm{T}}$. A switching signal will be manually sent after the first given formation has been formed, and then the desired formation position vector for v_2 w.r.t. the leader frame is set as $\mathbf{r}_2^{\mathrm{d}} = [-1.5 \quad -0.3]^{\mathrm{T}}$. Figs. 15.17(b) and 15.17(c) plot the scenes where the switching formations have been basically achieved. The trajectory of followers relative to the position of the leader and the time history of the formation error is illustrated in Fig. 15.18. The formation error decreases and remains close to zero approximately in $t \in [0, 8.4]s$. At around $t = 8.4s$, there is an impulsive increase of the formation error due to the switching signal, then the given formation is achieved again at around $t = 14s$. As expected, the zero input for the leader agent results to that the formation error is able to converge to zero closely. The slight vibration might come from the measurement error of the kinect, which caused by the shake of the Turtlebot in the experiment.

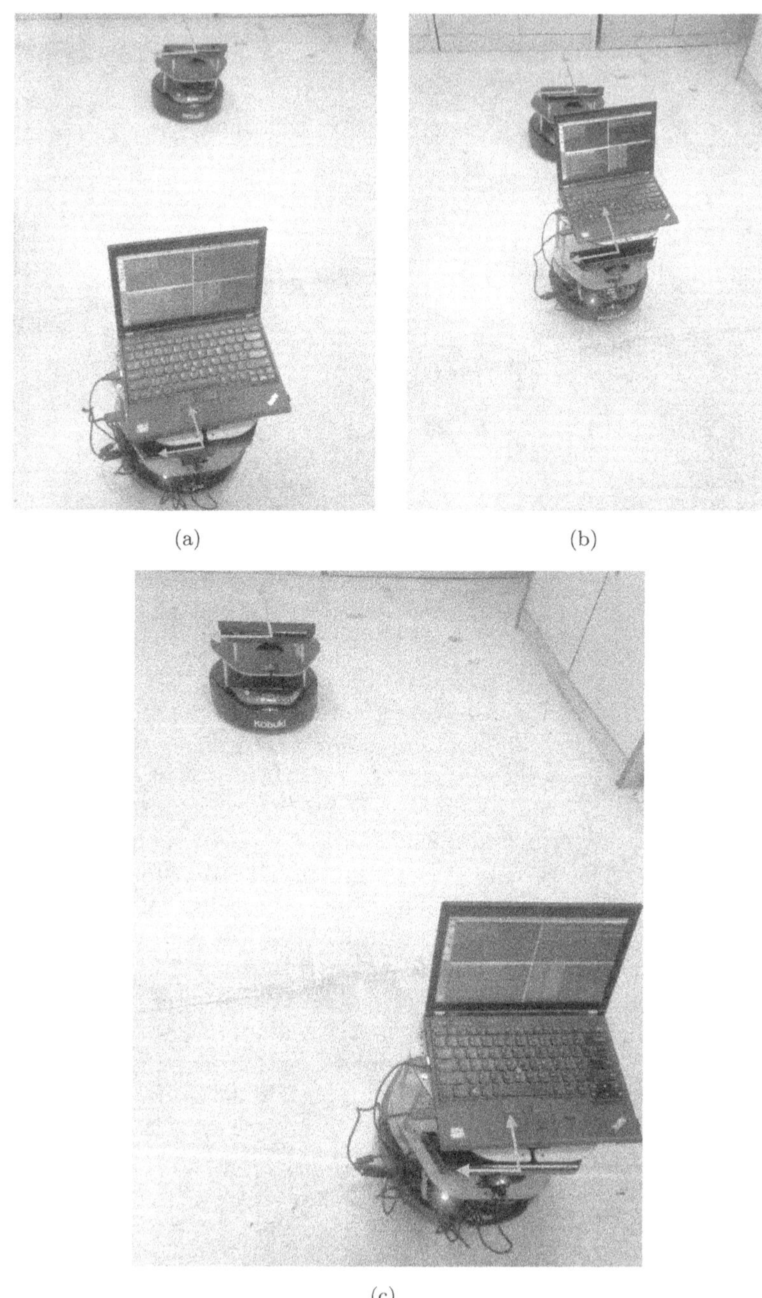

Figure 15.17 Experimental result: (a) initial positions; (b) the first desired formation has been formed; (c) the second desired formation has been formed.

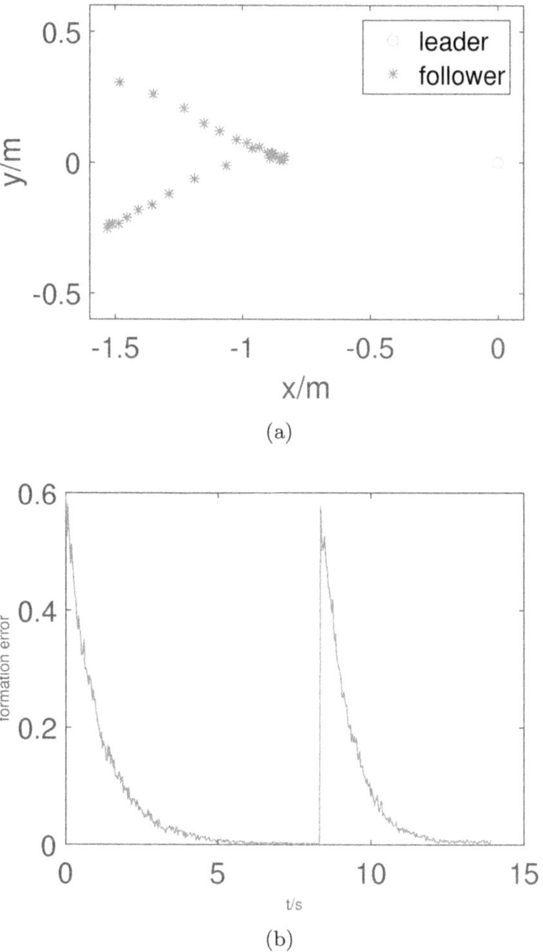

Figure 15.18 Experimental result: (a) trajectory of the agent; (b) time history of formation error.

15.7 CONCLUSION

This chapter has presented a kinect-based formation tracking control for multi-agent systems under visibility constraints. A pose estimation algorithm using RGBD images has been proposed to estimate relative orientation and relative positions of the local leader. The visual localization benefiting the success of leader-follower formation without the communication among agents. The desired formation based on the given position displacement is generated with the consensus in motion direction by the designed dipolar vector field.

The visibility constraints, resulting from the limited vision range of the kinect used as the sole sensor for formation, are converted into inputs constrains. Under the effort of the designed auxiliary system, the control protocol has been proved to achieve the formation tasks while maintaining the visibility constraints. A speech navigation system has been developed to translate speech signals into continuous control inputs for the global leader as the motion planner. Finally, simulation results using Gazebo simulator on Robot Operating System as well as experimental results on Turtlebot 2 are presented to demonstrate the performance.

16 Conclusion

A team of agents is expected to cooperate with each other to perform tasks in a constrained environment with myriad realistic applications ranging from exploration, surveillance, search, and rescue. The desirable performance requires the consideration of many issues such as mapping, obstacle avoidance, collision avoidance, communication limitations, system dynamics constraints, etc. They are also envisioned to collaborate with human beings in the foreseeable future for productivity, service, and operations with guaranteed efficiency and quality. In this book, we are attempting to discuss the formation problem under the consideration of practical applications summarized in the following sections. For each chapter, together with theoretical derivation, the simulation results are also presented to demonstrate the effectiveness of the proposed methodology.

In Chap. 3, a multilayer formation control problem has been defined with the proposal of a layered distributed finite-time estimator for agents in each layer to obtain their target positions and velocities based on the information of agents in their prior layers. A model-based control law has been designed to achieve multilayer formation, in the presence of practical issues, such as model uncertainties and loss of velocity measurements.

In Chap. 4, to address the formation control problem of flapping-wing vehicles given the model uncertainty and the measurement inaccuracy, adaptive neural networks are developed with the control algorithm. Based on the exploration of the coupling property between the translation motion and the rotational motion, an attitude estimation algorithm has been proposed, while the stability is proved together with the control scheme.

In Chap. 5, a layered affine formation problem has been solved by a fully distributed control law over a directed interaction topology for networked Euler-Lagrange systems. The method improves the feasibility as no global information is demanded. Adaptive-neural-network-based formation control has been integrated to tackle the model uncertainties by updating the norm of weight matrix targeting at less computation cost.

In Chap. 6, the satellite formation problem has been studied under a cooperative circumnavigation framework. The affine Laplacian matrix has been introduced to characterize the desired formation shape of a set of networked microsatellites. Based on the potential functions and the affine transformation, a cooperative circumnavigation control law is proposed, and we have proved that the closed-loop errors will converge to the origin exponentially.

In Chap. 7, the cooperative circumnavigation problem has been revisited for networked unmanned aerial vehicles. Benefiting from affine transformations, cooperative circumnavigation control design is structured, which can deploy unmanned aerial vehicles on spatial orbits with arbitrary radii concerning a

DOI: 10.1201/9781003298618-16

moving target in 3-D space. Moreover, global information is not needed for any follower unmanned aerial vehicles since they are completely maneuvered by the leader unmanned aerial vehicles.

In Chap. 8, we have developed the measure-theoretic approach for generating the collective behavior of flocking of planar autonomous vehicles under an all-to-all communication scheme. The measure-valued transition equation has been retrieved to induce an auxiliary one-body continuum model from the exact multi-body particle model of the collective system. Benefiting from the systematic design, the system does not rely on many mobile robots to achieve the flocking problem.

In Chap. 9, we have introduced the bioinspired neuro-dynamics to achieve the desired formation pattern without violation of the constraints including collision avoidance and connectivity distance. The observer has been designed to overcome the leader's velocity unavailable problem and decrease the communication burden.

In Chap. 10, we have proposed a velocity-free formation control method for autonomous unmanned surface vehicles with the model uncertainties being approximated by the neural networks, while the velocity of the leader is estimated by the designed reconstruction module, which will accomplish in finite time and reduce the communication burden. The output constraints are addressed by the symmetric barrier Lyapunov function method. The spatial constraints and connectivity distance constraints are also under consideration when achieving the desired formation pattern.

In Chap. 11, the formation tracking problem based on the potential field has been investigated. The contributions are not ultimately depending on the usage of the potential field. Our studies adapt the potential field protocols to the formation tracking control in complex constrained space and meanwhile under dynamic uncertainties.

In Chap. 12, a novel modeling approach for spatial constraints has been designed as a tool for trajectory tracking in non-omniscient constrained space, where the Dirac delta function has been introduced to design the artificial potential field for multiple spatial constraints.

In Chap. 13, the virtual-leader-based formation control has been studied. The model-based and the neural network-based controller have been deduced using the artificial potential field method for the formation tracking of multi-agents in constrained space. The proposed method has been proven to automatically generate and maintain a desired formation while avoiding internal collisions, which combines local attractive and repulsive potential fields.

In Chap. 14 and Chap. 15, the leader-follower-based formation control has been studied. Chap. 14 has introduced the role of "coordinator" and a coordination approach to guarantee obstacle avoidance as well as formation keeping. A scaling factor has been proposed to scale up or scale down the given formation size in the case that the restricted path is impassable for the desired formation and the adaptation law has been developed.

In Chap. 15, a visual-based pose estimation algorithm has been proposed to estimate the relative orientation and relative positions of the local leader in the leader-follower formation problem. The desired formation is generated with the consensus in motion direction by the designed dipolar vector field. The visibility constraints, resulting from the limited vision range of the camera used as the sole sensor for formation, are converted into input constraints. Under the effort of the designed auxiliary system, the control protocol has been proven to achieve the formation tasks while maintaining the visibility constraints.

References

1. Abdelkader Abdessameud and Abdelhamid Tayebi. On consensus algorithms for double-integrator dynamics without velocity measurements and with input constraints. *Systems & Control Letters*, 59(12):812–821, 2010.
2. F Abdollahi and H Rezaee. A decentralized cooperative control scheme with obstacle avoidance for a team of mobile robots. *Industrial Electronics, IEEE Transactions on*, 61(1):347–354, 2014.
3. Amit Ailon and Ilan Zohar. Control strategies for driving a group of non-holonomic kinematic mobile robots in formation along a time-parameterized path. *IEEE/ASME Transactions on Mechatronics*, 17(2):326–336, 2011.
4. M Alanyali, S Venkatesh, O Savas, and S Aeron. Distributed bayesian hypothesis testing in sensor networks. In *Proceedings of the 2004 American Control Conference*, volume 6, pages 5369–5374. IEEE, 2004.
5. Guillaume Allibert, Estelle Courtial, and François Chaumette. Predictive control for constrained image-based visual servoing. *IEEE Transactions on Robotics*, 26(5):933–939, 2010.
6. David Angeli and P-A Bliman. Extension of a result by moreau on stability of leaderless multi-agent systems. In *Proceedings of the 44th IEEE Conference on Decision and Control*, pages 759–764. IEEE, 2005.
7. Vassilis Angelopoulos. The THEMIS mission. In *The THEMIS Mission*, pages 5–34. Springer, 2009.
8. Ali Noormohammadi Asl, Mohammad Bagher Menhaj, and Atena Sajedin. Control of leader–follower formation and path planning of mobile robots using asexual reproduction optimization (aro). *Applied Soft Computing*, 14:563–576, 2014.
9. Sofía Avila-Becerril, Gerardo Espinosa-Pérez, Elena Panteley, and Romeo Ortega. Consensus control of flexible-joint robots. *International Journal of Control*, 88(6):1201–1208, 2015.
10. Reza Babazadeh and Rastko Selmic. Anoptimal displacement-based leader-follower formation control for multi-agent systems with energy consumption constraints. In *2018 26th Mediterranean Conference on Control and Automation (MED)*, pages 179–184. IEEE, 2018.
11. Reza Babazadeh and Rastko Selmic. Distance-based multiagent formation control with energy constraints using sdre. *IEEE Transactions on Aerospace and Electronic Systems*, 56(1):41–56, 2019.
12. He Bai and John T Wen. Cooperative load transport: a formation-control perspective. *Robotics, IEEE Transactions on*, 26(4):742–750, 2010.
13. Wenbin Bai, Xiande Wu, Yaen Xie, Yu Wang, Han Zhao, Kefan Chen, Yong Li, and Yong Hao. A cooperative route planning method for multi-uavs based-on the fusion of artificial potential field and b-spline interpolation. In *2018 37th Chinese Control Conference (CCC)*, pages 6733–6738. IEEE, 2018.
14. Drumi D Bainov and Pavel S Simeonov. *Integral Inequalities and Applications*, volume 57. Springer Science & Business Media, 2013.

15. Efstathios Bakolas and Panagiotis Tsiotras. Minimum-time paths for a small aircraft in the presence of regionally-varying strong winds. *AIAA Infotech@ Aerospace, Atlanta, GA*, pages 2010–3380, 2010.
16. Tucker Balch and Ronald C Arkin. Behavior-based formation control for multirobot teams. *Robotics and Automation, IEEE Transactions on*, 14(6):926–939, 1998.
17. Afshin Banazadeh and Neda Taymourtash. Adaptive attitude and position control of an insect-like flapping wing air vehicle. *Nonlinear Dynamics*, 85, 07 2016.
18. Roger R Bate, Donald D Mueller, and Jerry E White. *Fundamentals of Astrodynamics*. Courier Corporation, 1971.
19. Yazdan Batmani, Mohammad Davoodi, and Nader Meskin. Nonlinear suboptimal tracking controller design using state-dependent riccati equation technique. *IEEE Transactions on Control Systems Technology*, 1–7, 10 2016.
20. Yazdan Batmani and Shahabeddin Najafi. Event-triggered h∞ depth control of remotely operated underwater vehicles. *IEEE Transactions on Systems, Man, and Cybernetics: Systems*, 1–9, 02 2019.
21. Frank H Bauer, Kate Hartman, Jonathan P How, John Bristow, David Weidow, Franz Busse, and Others. Enabling spacecraft formation flying through spaceborne GPS and enhanced automation technologies. In *ION-GPS Conference*, volume 1, pages 369–384, 1999.
22. Dario Bauso, Laura Giarré, and Raffaele Pesenti. Attitude alignment of a team of uavs under decentralized information structure. In *Control Applications, 2003. CCA 2003. Proceedings of 2003 IEEE Conference on*, volume 1, pages 486–491. IEEE, 2003.
23. Dario Bauso, Laura Giarré, and Raffaele Pesenti. Distributed consensus protocols for coordinating buyers. In *42nd IEEE International Conference on Decision and Control (IEEE Cat. No. 03CH37475)*, volume 1, pages 588–592. IEEE, 2003.
24. Ismail Bayezit and Barıs Fidan. Distributed cohesive motion control of flight vehicle formations. *Industrial Electronics, IEEE Transactions on*, 60(12):5763–5772, 2013.
25. Randal W Beard, Jonathan Lawton, and Fred Y Hadaegh. A coordination architecture for spacecraft formation control. *Control Systems Technology, IEEE Transactions on*, 9(6):777–790, 2001.
26. Alberto Bemporad and Claudio Rocchi. Decentralized linear time-varying model predictive control of a formation of unmanned aerial vehicles. In *2011 50th IEEE Conference on Decision and Control and European Control Conference*, pages 7488–7493. IEEE, 2011.
27. Eshel Ben-Jacob, Inon Cohen, and Herbert Levine. Cooperative self-organization of microorganisms. *Advances in Physics*, 49(4):395–554, 2000.
28. Sanjay P Bhat and Dennis S Bernstein. Finite-time stability of continuous autonomous systems. *SIAM Journal on Control and Optimization*, 38(3):751–766, 2000.
29. Marco Bibuli, Yogang Singh, Sanjay Sharma, Rhondasutton Sutton, Daniel Hatton, and Asiya Khan. A two layered optimal approach towards cooperative motion planning of unmanned surface vehicles in a constrained maritime environment. *IFAC-PapersOnLine*, 51:378–383, 01 2018.

30. Patrick Billingsley. *Convergence of Probability Measures.* John Wiley & Sons, 2013.
31. Stefano Boccaletti, Ginestra Bianconi, Regino Criado, Charo I Del Genio, Jesús Gómez-Gardenes, Miguel Romance, Irene Sendina-Nadal, Zhen Wang, and Massimiliano Zanin. The structure and dynamics of multilayer networks. *Physics Reports*, 544(1):1–122, 2014.
32. Even Børhaug, Alexey Pavlov, Elena Panteley, and Kristin Y Pettersen. Straight line path following for formations of underactuated marine surface vessels. *IEEE Transactions on Control Systems Technology*, 19(3):493–506, 2010.
33. Even Børhaug, Alexey Pavlov, Elena Panteley, and Kristin Ytterstad Pettersen. Straight line path following for formations of underactuated marine surface vessels. *Control Systems Technology, IEEE Transactions on*, 19(3):493–506, 2011.
34. Leo Breiman. Random forests. *Machine Learning*, 45(1):5–32, 2001.
35. José A Canizo, José A Carrillo, and Jesús Rosado. A well-posedness theory in measures for some kinetic models of collective motion. *Mathematical Models and Methods in Applied Sciences*, 21(03):515–539, 2011.
36. Kun Cao, Dongyu Li, and Lihua Xie. Bearing-ratio-of-distance rigidity theory with application to directly similar formation control. *Automatica*, 109:108540, 2019.
37. Kun Cao, Zhirong Qiu, and Lihua Xie. Relative Docking and Formation Control via Range and Odometry Measurements. *IEEE Transactions on Control of Network Systems, in press, DOI: 10.1109/TCNS.2019.2951893*, 2019.
38. Yongcan Cao and Wei Ren. Containment control with multiple stationary or dynamic leaders under a directed interaction graph. In *Proceedings of the 48th IEEE Conference on Decision and Control (CDC) Held Jointly with 2009 28th Chinese Control Conference*, pages 3014–3019, 2009.
39. Yongcan Cao, Wei Ren, and Magnus Egerstedt. Distributed containment control with multiple stationary or dynamic leaders in fixed and switching directed networks. *Automatica*, 48(8):1586–1597, 2012.
40. Yongcan Cao, Wei Ren, and Ziyang Meng. Decentralized finite-time sliding mode estimators and their applications in decentralized finite-time formation tracking. *Systems and Control Letters*, 59(9):522–529, 2010.
41. Yongcan Cao, Wenwu Yu, Wei Ren, and Guanrong Chen. An overview of recent progress in the study of distributed multi-agent coordination. *IEEE Transactions on Industrial Informatics*, 9(1):427–438, 2012.
42. Zhiqiang Cao, Min Tan, Shuo Wang, Yong Fan, and Bin Zhang. The optimization research of formation control for multiple mobile robots. In *Proceedings of the 4th World Congress on Intelligent Control and Automation (Cat. No. 02EX527)*, volume 2, pages 1270–1274. IEEE, 2002.
43. Zhiqiang Cao, Liangjun Xie, Bin Zhang, Shuo Wang, and Min Tan. Formation constrained multi-robot system in unknown environments. In *2003 IEEE International Conference on Robotics and Automation (Cat. No. 03CH37422)*, volume 1, pages 735–740. IEEE, 2003.
44. José A Carrillo, Maria R D'Orsogna, and Vladislav Panferov. Double milling in self-propelled swarms from kinetic theory. *Kinetic & Related Models*, 2(2):363, 2009.

45. José A Carrillo, Massimo Fornasier, Jesús Rosado, and Giuseppe Toscani. Asymptotic flocking dynamics for the kinetic cucker–smale model. *SIAM Journal on Mathematical Analysis*, 42(1):218–236, 2010.
46. José A Carrillo, Massimo Fornasier, Giuseppe Toscani, and Francesco Vecil. Particle, kinetic, and hydrodynamic models of swarming. In *Mathematical Modeling of Collective Behavior in Socio-economic and Life Sciences*, pages 297–336. Springer, 2010.
47. Luis Rodolfo Garcia Carrillo, Alejandro Enrique Dzul López, R Lozano, and Claude Pégard. *Quad Rotorcraft Control.* Springer, New York, NY, USA, 2013.
48. Giovanna Castellano, Mario GCA Cimino, Anna Maria Fanelli, Beatrice Lazzerini, Francesco Marcelloni, and Maria Alessandra Torsello. A multi-agent system for enabling collaborative situation awareness via position-based stigmergy and neuro-fuzzy learning. *Neurocomputing*, 135:86–97, 2014.
49. Zeze Chang, Junjie Wang, and Zhongkui Li. Fully Distributed Event-Triggered Affine Formation Maneuver Control Over Directed Graphs. *IFAC-PapersOnLine*, 55(3):178–183, 2022.
50. Chien Chern Cheah, Saing Paul Hou, and Jean Jacques E Slotine. Region-based shape control for a swarm of robots. *Automatica*, 45(10):2406–2411, 2009.
51. LPC Chen, G Wen, Y Liu, and F Wang Adaptive consensus control for a class of nonlinear multiagent time-delay systems using neural networks. *Neural Networks and Learning Systems*, 2014.
52. C Chen, SS Xu, and Y Liang. Study of Nonlinear Integral Sliding Mode Fault-Tolerant Control. *IEEE/ASME Transactions on Mechatronics*, 21(2):1160–1168, 2016.
53. CL Philip Chen, Guo-Xing Wen, Yan-Jun Liu, and Zhi Liu. Observer-based adaptive backstepping consensus tracking control for high-order nonlinear semi-strict-feedback multiagent systems. *IEEE Transactions on Cybernetics*, 46(7):1591–1601, 2016.
54. CL Philip Chen, Guo-Xing Wen, Yan-Jun Liu, and Fei-Yue Wang. Adaptive consensus control for a class of nonlinear multiagent time-delay systems using neural networks. *Neural Networks and Learning Systems, IEEE Transactions on*, 25(6):1217–1226, 2014.
55. Jiannan Chen, Changchun Hua, and Fang Wang. Distributed adaptive containment control of uncertain quav multiagents with time-varying payloads and multiple variable constraints. *ISA Transactions*, 90, 07 2019.
56. Lepeng Chen, Rongxin Cui, Chenguang Yang, and Weisheng Yan. Adaptive neural network control of underactuated surface vessels with guaranteed transient performance: Theory and experimental results. *IEEE Transactions on Industrial Electronics*, PP:1–1, 05 2019.
57. Liangliang Chen, Chuanjiang Li, Yanchao Sun, and Guangfu Ma. Distributed finite-time tracking control for multiple uncertain Euler–Lagrange systems with input saturations and error constraints. *IET Control Theory & Applications*, 13(1):123–133, 2018.
58. Liangming Chen, Ming Cao, and Chuanjiang Li. Angle rigidity and its usage to stabilize multiagent formations in 2-d. *IEEE Transactions on Automatic Control*, 66(8):3667–3681, 2020.

59. Liangming Chen, Chuanjiang Li, Bing Xiao, and Yanning Guo. Formation-containment control of networked Euler–Lagrange systems: An event-triggered framework. *ISA Transactions*, 2018.
60. Liangming Chen, Jie Mei, Chuanjiang Li, and Guangfu Ma. Distributed Leader–Follower Affine Formation Maneuver Control for High-Order Multiagent Systems. *IEEE Transactions on Automatic Control*, 65(11):4941–4948, November 2020.
61. Liangming Chen, Zhongqi Sun, Chuanjiang Li, Baolong Zhu, and Cheng Wang. Satellite affine formation flying with obstacle avoidance. *Proceedings of the Institution of Mechanical Engineers, Part G: Journal of Aerospace Engineering*, 233(16):5992–6004, December 2019.
62. Liangming Chen, Qingkai Yang, Mingming Shi, Yanan Li, and Mir Feroskhan. Stabilizing angle rigid formations with prescribed orientation and scale. *IEEE Transactions on Industrial Electronics*, 69(11):11654–11664, 2021.
63. Mou Chen and Shuzhi Ge. Adaptive neural output feedback control of uncertain nonlinear systems with unknown hysteresis using disturbance observer. *IEEE Transactions on Industrial Electronics*, 62:1–1, 12 2015.
64. Mou Chen and Shuzhi Sam Ge. Adaptive neural output feedback control of uncertain nonlinear systems with unknown hysteresis using disturbance observer. *IEEE Transactions on Industrial Electronics*, 62(12):7706–7716, 2015.
65. Mou Chen, Shuzhi Sam Ge, and Beibei Ren. Adaptive tracking control of uncertain mimo nonlinear systems with input constraints. *Automatica*, 47(3):452–465, 2011.
66. Xiaohan Chen and Yingmin Jia. Input-constrained formation control of differential-drive mobile robots: geometric analysis and optimisation. *IET Control Theory & Applications*, 8(7):522–533, 2014.
67. Yang Quan Chen and Zhongmin Wang. Formation control: a review and a new consideration. In *2005 IEEE/RSJ International Conference on Intelligent Robots and Systems*, pages 3181–3186. IEEE, 2005.
68. Yang Quan Chen and Zhongmin Wang. Formation control: a review and a new consideration. In *2005 IEEE/RSJ International Conference on Intelligent Robots and Systems*, pages 3181–3186, 2005.
69. YongBo Chen, JianQiao Yu, XiaoLong Su, and GuanChen Luo. Path planning for multi-uav formation. *Journal of Intelligent & Robotic Systems*, 77(1):229–246, 2015.
70. Zhiyong Chen and Hai-Tao Zhang. No-beacon collective circular motion of jointly connected multi-agents. *Automatica*, 47(9):1929–1937, 2011.
71. Long Cheng, Zeng-Guang Hou, Min Tan, Yingzi Lin, and Wenjun Zhang. Neural-network-based adaptive leader-following control for multiagent systems with uncertainties. *Neural Networks, IEEE Transactions on*, 21(8):1351–1358, 2010.
72. Pakpong Chirarattananon, Yufeng Chen, E. Helbling, Kevin Ma, Richard Cheng, and Robert Wood. Dynamics and flight control of a flappingwing robotic insect in the presence of wind gusts. *Interface Focus: A Theme Supplement of Journal of the Royal Society Interface*, 7, 02 2017.
73. Hancheol Cho, Sang-Young Park, Han-Earl Park, and Kyu-Hong Choi. Analytic solution to optimal reconfigurations of satellite formation flying in

circular orbit under j_2 perturbation. *Aerospace and Electronic Systems, IEEE Transactions on*, 48(3):2180–2197, 2012.

74. Yun Ho Choi and Doik Kim. Distance-based formation control with goal assignment for global asymptotic stability of multi-robot systems. *IEEE Robotics and Automation Letters*, 6(2):2020–2027, 2021.
75. Dhrubajit Chowdhury and Hassan K Khalil. Synchronization in networks of identical linear systems with reduced information. In *2018 Annual American Control Conference (ACC)*, pages 5706–5711. IEEE, 2018.
76. Woojin Chung, Seokgyu Kim, Minki Choi, Jaesik Choi, Hoyeon Kim, Changbae Moon, and Jae-Bok Song. Safe navigation of a mobile robot considering visibility of environment. *IEEE Transactions on Industrial Electronics*, 56(10):3941–3950, 2009.
77. Andrew Clark, Basel Alomair, Linda Bushnell, and Radha Poovendran. Leader selection in multi-agent systems for smooth convergence via fast mixing. In *2012 IEEE 51st IEEE Conference on Decision and Control (CDC)*, pages 818–824. IEEE, 2012.
78. Andrew Clark, Linda Bushnell, and Radha Poovendran. Joint leader and link weight selection for fast convergence in multi-agent systems. In *2013 American Control Conference*, pages 3814–3820. IEEE, 2013.
79. Andrew Clark, Linda Bushnell, and Radha Poovendran. A supermodular optimization framework for leader selection under link noise in linear multi-agent systems. *IEEE Transactions on Automatic Control*, 59(2):283–296, 2013.
80. Julio B Clempner and Wen Yu. *New Perspectives and Applications of Modern Control Theory: In Honor of Alexander S. Poznyak*. Springer, 2017.
81. Christopher I Connolly, J Brian Burns, and Rich Weiss. Path planning using laplace's equation. In *Proceedings., IEEE International Conference on Robotics and Automation*, pages 2102–2106. IEEE, 1990.
82. Luca Consolini, Fabio Morbidi, Domenico Prattichizzo, and Mario Tosques. Leader–follower formation control of nonholonomic mobile robots with input constraints. *Automatica*, 44(5):1343–1349, 2008.
83. Samuel Coogan and Murat Arcak. Scaling the size of a formation using relative position feedback. *Automatica*, 48(10):2677–2685, 2012.
84. Jorge Cortés. Distributed algorithms for reaching consensus on general functions. *Automatica*, 44(3):726–737, 2008.
85. Jorge Cortés. Global and robust formation-shape stabilization of relative sensing networks. *Automatica*, 45(12):2754–2762, 2009.
86. Felipe Cucker and Jiu-Gang Dong. Avoiding collisions in flocks. *IEEE Transactions on Automatic Control*, 55(5):1238–1243, 2010.
87. Felipe Cucker and Steve Smale. Emergent behavior in flocks. *IEEE Transactions on Automatic Control*, 52(5):852–862, 2007.
88. R Cui, B Ren, and SS Ge. Synchronised tracking control of multi-agent system with high order dynamics. *Control Theory & Applications, IET*, 6(5):603–614, 2012.
89. Rongxin Cui, Shuzhi Ge, Bernard How, and Y.s Choo. Leader–follower formation control of underactuated autonomous underwater vehicles. *Ocean Engineering*, 37:1491–1502, 12 2010.

90. Shi-Lu Dai, Shude He, Hai Lin, and Cong Wang. Platoon formation control with prescribed performance guarantees for usvs. *IEEE Transactions on Industrial Electronics*, 65:4237–4246, 2018.
91. Robert Daily and David M Bevly. Harmonic potential field path planning for high speed vehicles. In *2008 American Control Conference*, pages 4609–4614. IEEE, 2008.
92. Aveek K Das, Rafael Fierro, Vijay Kumar, James P Ostrowski, John Spletzer, and Camillo J Taylor. A vision-based formation control framework. *IEEE Transactions on Robotics and Automation*, 18(5):813–825, 2002.
93. Madhu Sudan Das, Sourish Sanyal, and Sanjoy Mandal. Navigation of multiple robots in formative manner in an unknown environment using artificial potential field based path planning algorithm. *Ain Shams Engineering Journal*, 13(5):101675, 2022.
94. PK Das and Prabir Kumar Jena. Multi-robot path planning using improved particle swarm optimization algorithm through novel evolutionary operators. *Applied Soft Computing*, 92:106312, 2020.
95. Steven B Davis and Paul Mermelstein. Comparison of parametric representations for monosyllabic word recognition in continuously spoken sentences. *Acoustics, Speech and Signal Processing, IEEE Transactions on*, 28(4):357–366, 1980.
96. Charles Augustin de Coulomb. Troisieme mémoire sur l électricité et le magnétisme. *Histoire de l Académie Royale des Sciences*, pages 612–638, 1785.
97. Hector Garcia de Marina, Juan Jimenez Castellanos, and Weijia Yao. Leaderless collective motions in affine formation control, April 2021.
98. Hector Garcia de Marina, Zhiyong Sun, Murat Bronz, and Gautier Hattenberger. Circular formation control of fixed-wing uavs with constant speeds. In *2017 IEEE/RSJ International Conference on Intelligent Robots and Systems (IROS)*, pages 5298–5303, 2017.
99. Klaus Deimling. *Multivalued Differential Equations*. Walter de Gruyter, 2011.
100. JD DeLaurier. An aerodynamic model for flapping-wing flight. *The Aeronautical Journal (1968)*, 97(964):125–130, 1993.
101. JD DeLaurier and JM Harris. A study of mechanical flapping-wing flight. *The Aeronautical Journal (1968)*, 97(968):277–286, 1993.
102. Jaydev P Desai, James P Ostrowski, and Vijay Kumar. Modeling and control of formations of nonholonomic mobile robots. *IEEE Transactions on Robotics and Automation*, 17(6):905–908, 2001.
103. Jaydev P Desai, Jim Ostrowski, and Vijay Kumar. Controlling formations of multiple mobile robots. In *Robotics and Automation, 1998. Proceedings. 1998 IEEE International Conference on*, volume 4, pages 2864–2869. IEEE, 1998.
104. Lei Ding and Ge Guo. Formation control for ship fleet based on backstepping. *Control and Decision*, 27:299–303, 2012.
105. Wei Ding, Gangfeng Yan, and Zhiyun Lin. Collective motions and formations under pursuit strategies on directed acyclic graphs. *Automatica*, 46(1):174–181, 2010.
106. Albert Diosi. *Laser Range Finder and Advanced Sonar Based Simultaneous Localization and Mapping for Mobile Robots*. Monash University, 2005.
107. PAM Dirac. *The Principles of Quantum Mechanics*, 4th edn. Clarendon, 1958.

108. K Do. Synchronization motion tracking control of multiple underactuated ships with collision avoidance. *IEEE Transactions on Industrial Electronics*, 63:1–1, 05 2016.
109. KD Do. Formation control of underactuated ships with elliptical shape approximation and limited communication ranges. *Automatica*, 48:1380–1388, 2012.
110. Khac Duc Do. Bounded controllers for formation stabilization of mobile agents with limited sensing ranges. *Automatic Control, IEEE Transactions on*, 52(3):569–576, 2007.
111. Khac Duc Do and Jie Pan. Nonlinear formation control of unicycle-type mobile robots. *Robotics and Autonomous Systems*, 55(3):191–204, 2007.
112. Dong, Wang, Ning, Zhang, Jianliang, Wang, Wei, and Wang. Cooperative containment control of multiagent systems based on follower observers with time delay. *IEEE Transactions on Systems Man & Cybernetics Systems*, 2016.
113. Chaoyang Dong, Yang Liu, and Qing Wang. Barrier lyapunov function based adaptive finite-time control for hypersonic flight vehicles with state constraints. *ISA Transactions*, 96, 07 2019.
114. Guowei Dong, Hongyi Li, Hui Ma, and Renquan Lu. Finite-time consensus tracking neural network ftc of multi-agent systems. *IEEE Transactions on Neural Networks and Learning Systems*, 32(2):653–662, 2020.
115. W Dong and JA Farrell. Formation control of multiple underactuated surface vessels. *IET Control Theory & Applications*, 2(12):1077–1085, 2008.
116. Xiwang Dong, Liang Han, Qingdong Li, Jian Chen, and Zhang Ren. Time-varying formation tracking for second-order multi-agent systems with one leader. In *2015 Chinese Automation Congress (CAC)*, pages 1046–1051, 2015.
117. Xiwang Dong, Liang Han, Qingdong Li, Jian Chen, and Zhang Ren. Containment analysis and design for general linear multi-agent systems with time-varying delays. *Neurocomputing*, 173:2062–2068, 2016.
118. Xiwang Dong, Zongying Shi, Geng Lu, and Yisheng Zhong. Formation-containment analysis and design for high-order linear time-invariant swarm systems. *International Journal of Robust and Nonlinear Control*, 25(17):3439–3456, 2015.
119. Xiwang Dong, Bocheng Yu, Zongying Shi, and Yisheng Zhong. Time-varying formation control for unmanned aerial vehicles: Theories and applications. *Control Systems Technology, IEEE Transactions on*, 23(1):340–348, 2015.
120. Yi Dong, Youfeng Su, Yue Liu, and Shengyuan Xu. An internal model approach for multi-agent rendezvous and connectivity preservation with nonlinear dynamics. *Automatica*, 89:300–307, 2018.
121. Ali Dorri, Salil S Kanhere, and Raja Jurdak. Multi-agent systems: A survey. *IEEE Access*, 6:28573–28593, 2018.
122. M dos Santos Soares and J Vrancken. A real-life test bed for multi-agent monitoring of road network performance. *International Journal of Critical Infrastructures*, 5(4), 2009.
123. David Dovrat and Alfred Bruckstein. On gathering and control of unicycle a(ge)nts with crude sensing capabilities. *IEEE Intelligent Systems*, 32:40–46, 11 2017.
124. Haibo Du, Yigang He, and Yingying Cheng. Finite-time synchronization of a class of second-order nonlinear multi-agent systems using output

feedback control. *IEEE Transactions on Circuits and Systems I: Regular Papers*, 61(6):1778–1788, 2014.

125. Leah Edelstein-Keshet. *Mathematical Models in Biology.* SIAM, 2005.
126. Christopher Edwards and Sarah Spurgeon. *Sliding Mode Control: Theory and Applications.* CRC Press, 1998.
127. DB Edwards, TA Bean, DL Odell, and MJ Anderson. *A Leader-Follower Algorithm for Multiple AUV Formations.* IEEE, 2004.
128. Yuan Fan, Jun Chen, Cheng Song, and Yong Wang. Event-triggered coordination control for multi-agent systems with connectivity preservation. *International Journal of Control, Automation and Systems*, 18(4):966–979, 2020.
129. Zhenyuan Fan and Huiping Li. Two-layer model predictive formation control of unmanned surface vehicle. pages 6002–6007, 10 2017.
130. Kaveh Fathian, Tyler H Summers, and Nicholas R Gans. Distributed formation control and navigation of fixed-wing uavs at constant altitude. In *2018 International Conference on Unmanned Aircraft Systems (ICUAS)*, pages 300–307, 2018.
131. J Alexander Fax and Richard M Murray. Information flow and cooperative control of vehicle formations. *Automatic Control, IEEE Transactions on*, 49(9):1465–1476, 2004.
132. Zhiguang Feng and Wei Xing Zheng. On extended dissipativity of discrete-time neural networks with time delay. *IEEE Transactions on Neural Networks and Learning Systems*, 26, 02 2015.
133. ED Ferreira-Vazquez, EG Hernandez-Martinez, JJ Flores-Godoy, G Fernandez-Anaya, and P Paniagua-Contro. Distance-based formation control using angular information between robots. *Journal of Intelligent & Robotic Systems*, 83(3-4):543–560, 2016.
134. Baris Fidan, Veysel Gazi, Shaohao Zhai, Na Cen, and Engin Karatas. Single-view distance-estimation-based formation control of robotic swarms. *Industrial Electronics, IEEE Transactions on*, 60(12):5781–5791, 2013.
135. Aleksei Fedorovich Filippov. *Differential Equations with Discontinuous Right-Hand Side.* Dordrecht, The Netherlands: Kluwer, 1988.
136. Aleksei Fedorovich Filippov. *Differential Equations with Discontinuous Right-hand Sides: Control Systems*, volume 18. Springer Science & Business Media, 2013.
137. WT Fong, SK Ong, and AYC Nee. Methods for in-field user calibration of an inertial measurement unit without external equipment. *Measurement Science and Technology*, 19(8):817–822, 2008.
138. Junjie Fu and Jinzhi Wang. Adaptive coordinated tracking of multi-agent systems with quantized information. *Systems & Control Letters*, 74:115–125, 2014.
139. Ming-yu Fu, Duansong Wang, and Cheng-long Wang. Formation control for water-jet usv based on bio-inspired method. *China Ocean Engineering*, 32:117–122, 2018.
140. Cheng-Heng Fua, Shuzhi Sam Ge, Khac Duc Do, and Khiang Wee Lim. Multi-robot formations based on the queue-formation scheme with limited communications. In *Proceedings 2007 IEEE International Conference on Robotics and Automation*, pages 2385–2390. IEEE, 2007.

141. Ken-Ichi Funahashi. On the approximate realization of continuous mappings by neural networks. *Neural Networks*, 2(3):183–192, 1989.
142. Joseph Gangestad, Brian Hardy, and David Hinkley. Operations, orbit determination, and formation control of the aerocube-4 cubesats. *Small Satellite Conference*, 2013.
143. Lan Gao, Xiaofeng Liao, Huaqing Li, and Guo Chen. Event-triggered control for multi-agent systems with general directed topology and time delays. *Asian Journal of Control*, 18(3):945–953, 2016.
144. Santiago Garrido, Luis Moreno, and Pedro U Lima. Robot formation motion planning using fast marching. *Robotics and Autonomous Systems*, 59(9):675–683, 2011.
145. S Ge and Cong Wang. Adaptive neural control of uncertain mimo nonlinear systems. *Neural Networks, IEEE Transactions on*, 15:674–692, 2004.
146. SS Ge, CC Hang, TH Lee, and Tao Zhang. Stable adaptive neural network control. *Springer International*, 13, 2002.
147. SS Ge, TH Lee, and CJ Harris. *Adaptive Neural Network Control of Robotic Manipulators*. World Scientific, 1998.
148. Shuzhi S Ge and Yun J Cui. Dynamic motion planning for mobile robots using potential field method. *Autonomous Robots*, 13(3):207–222, 2002.
149. Shuzhi Sam Ge and C-H Fua. Queues and artificial potential trenches for multirobot formations. *Robotics, IEEE Transactions on*, 21(4):646–656, 2005.
150. Shuzhi Sam Ge, Tong Heng Lee, and Christopher J Harris. *Adaptive Neural Network Control of Robotic Manipulators*. World Scientific, London, UK, 1998.
151. Shuzhi Sam Ge, Xiaomei Liu, Cher Hiang Goh, and Liguang Xu. Formation tracking control of multi-agents in constrained space. *Control Systems Technology, IEEE Transactions on*, 24(3):992–1003, 2015.
152. Sheida Ghapani, Jie Mei, Wei Ren, and Yongduan Song. Fully distributed flocking with a moving leader for lagrange networks with parametric uncertainties. *Automatica*, 67, 07 2015.
153. Jawhar Ghommam, Hasan Mehrjerdi, Maarouf Saad, and Faiçal Mnif. Formation path following control of unicycle-type mobile robots. *Robotics and Autonomous Systems*, 58(5):727–736, 2010.
154. Eberhard Gill, Oliver Montenbruck, and Simone D'Amico. Autonomous formation flying for the PRISMA mission. *Journal of Spacecraft and Rockets*, 44(3):671–681, 2007.
155. G Golub and W Kahan. Calculating the singular values and pseudo-inverse of a matrix. *Journal of the Society for Industrial & Applied Mathematics*, 2(2):205–224, 1965.
156. Javier V Gómez, Alejandro Lumbier, Santiago Garrido, and Luis Moreno. Planning robot formations with fast marching square including uncertainty conditions. *Robotics and Autonomous Systems*, 61(2):137–152, 2013.
157. Tenoch Gonzalez, Jaime A Moreno, and Leonid Fridman. Variable gain super-twisting sliding mode control. *IEEE Transactions on Automatic Control*, 57(8):2100–2105, 2011.
158. Antonio González-Pardo, Pablo Varona, David Camacho, and Francisco de Borja Rodriguez Ortiz. Communication by identity discrimination in

bio-inspired multi-agent systems. *Concurrency and Computation: Practice and Experience*, 24(6):589–603, 2012.

159. Steven J Gortler, Alexander D Healy, and Dylan P Thurston. Characterizing generic global rigidity. *American Journal of Mathematics*, 132(4):897–939, 2010.
160. Stephen Grossberg. Nonlinear neural networks: Principle, mechanisms, and architectures. *Neural Networks*, 1:17–61, 1988.
161. Y Gu, B Seanor, G Campa, MR Napolitano, L Rowe, S Gururajan, and S Wan. Design and flight testing evaluation of formation control laws. *IEEE Transactions on Control Systems Technology*, 14:p.1105–1112, 2006.
162. Ge Guo, Lei Ding, and Qing-Long Han. A distributed event-triggered transmission strategy for sampled-data consensus of multi-agent systems. *Automatica*, 50(5):1489–1496, 2014.
163. Ge Guo and Zhenyu Gao. Velocity free leader-follower formation control for autonomous underwater vehicles with line-of-sight range and angle constraints. *Information Sciences*, 486, 02 2019.
164. Seung-Yeal Ha and Eitan Tadmor. From particle to kinetic and hydrodynamic descriptions of flocking. *arXiv preprint arXiv:0806.2182*, 2008.
165. Mirza Tariq Hamayun, Christopher Edwards, and Halim Alwi. Design and analysis of an integral sliding mode fault-tolerant control scheme. *IEEE Transactions on Automatic Control*, 57(7):1783–1789, 2012.
166. Jungong Han, Ling Shao, Dong Xu, and Jamie Shotton. Enhanced computer vision with microsoft kinect sensor: A review. *IEEE Transactions on Cybernetics*, 43(5):1318–1334, 2013.
167. Tingrui Han, Zhiyun Lin, Ronghao Zheng, and Minyue Fu. A barycentric coordinate-based approach to formation control under directed and switching sensing graphs. *IEEE Transactions on Cybernetics*, 48(4):1202–1215, 2017.
168. Zhimin Han, Kexin Guo, Lihua Xie, and Zhiyun Lin. Integrated relative localization & leader-follower formation control. *IEEE Transactions on Automatic Control*, 64(1):20–34, 2018.
169. Zhimin Han, Lili Wang, Zhiyun Lin, and Ronghao Zheng. Formation control with size scaling via a complex laplacian-based approach. *IEEE Transactions on Cybernetics*, 46(10):2348–2359, 2015.
170. Zhimin Han, Lili Wang, Zhiyun Lin, and Ronghao Zheng. Formation control with size scaling via a complex Laplacian-based approach. *IEEE Transactions on Cybernetics*, 46(10):2348–2359, 2016.
171. Philip Hartman. *Ordinary Differential Equations*. SIAM, 2002.
172. Mariann Hauge, Lars Landmark, Piotr Łubkowski, Marek Amanowicz, and Krzysztof Maślanka. Selected issues of qos provision in heterogenous military networks. *International Journal of Electronics and Telecommunications*, 60(1):1–7, 2014.
173. Andreas J Häusler, Alessandro Saccon, A Pedro Aguiar, John Hauser, and António M Pascoal. Cooperative motion planning for multiple autonomous marine vehicles. *IFAC Proceedings Volumes*, 45(27):244–249, 2012.
174. SS Haykin. Neural networks : A comprehensive foundation. *Neural Networks A Comprehensive Foundation*, 1994.

175. Shude He, Min Wang, Shi-Lu Dai, and Fei Luo. Leader–follower formation control of usvs with prescribed performance and collision avoidance. *IEEE Transactions on Industrial Informatics*, 15(1):572–581, 2018.
176. Shude He, Min Wang, Shi-Lu Dai, and Fei Luo. Leader-follower formation control of usvs with prescribed performance and collision avoidance. *IEEE Transactions on Industrial Informatics*, 15:572–581, 01 2019.
177. W He, YH Chen, and Z Yin. Adaptive neural network control of an uncertain robot with full-state constraints. *IEEE Transactions on Cybernetics*, 46(3):620–629, 2016.
178. Wangli He, Biao Zhang, Qing-Long Han, Feng Qian, Jürgen Kurths, and Jinde Cao. Leader-following consensus of nonlinear multiagent systems with stochastic sampling. *IEEE Transactions on Cybernetics*, 47(2):327–338, 2016.
179. Wei He, Yuhao Chen, and Zhao Yin. Adaptive neural network control of an uncertain robot with full-state constraints. *IEEE Transactions on Cybernetics*, 46(3):620–629, 2016.
180. Wei He, Shuzhi Sam Ge, Yanan Li, Effie Chew, and Yee Sien Ng. Neural network control of a rehabilitation robot by state and output feedback. *Journal of Intelligent & Robotic Systems*, 80(1):15–31, 2015.
181. Wei He, Xinxing Mu, Liang Zhang, and Yao Zou. Modeling and Trajectory Tracking Control for Flapping-Wing Micro Aerial Vehicles. *IEEE/CAA Journal of Automatica Sinica, in press, DOI: 10.1109/JAS.2020.1003417*, 2020.
182. Wei He, Tingting Wang, Xiuyu He, Lung-Jieh Yang, and Okyay Kaynak. Dynamical Modeling and Boundary Vibration Control of a Rigid-Flexible Wing System. *IEEE/ASME Transactions on Mechatronics, in press, DOI:10.1109/TMECH.2020.2987963*, 2020.
183. Wei He, Zichen Yan, Changyin Sun, and Yunan Chen. Adaptive neural network control of a flapping wing micro aerial vehicle with disturbance observer. *IEEE Transactions on Cybernetics*, 47(10):3452–3465, 2017.
184. Wei He, Zhao Yin, and Changyin Sun. Adaptive neural network control of a marine vessel with constraints using the asymmetric barrier lyapunov function. *IEEE Transactions on Cybernetics*, 47:1–11, 05 2016.
185. Sonia Hernandez and Derek A Paley. Three-dimensional motion coordination in a spatiotemporal flowfield. *IEEE Transactions on Automatic Control*, 55(12):2805–2810, 2010.
186. Janning F Herrmann, Florian Kretschmer, Stephanie Hoeppener, Christiane Höppener, and Ulrich S Schubert. Ordered arrangement and optical properties of silica-stabilized gold nanoparticle–PNIPAM core–satellite clusters for sensitive Raman detection. *Small*, 13(39):1701095, 2017.
187. Steven Ho, Hany Nassef, Nick Pornsinsirirak, Yu-Chong Tai, and Chih-Ming Ho. Unsteady aerodynamics and flow control for flapping wing flyers. *Progress in Aerospace Sciences*, 39(8):635–681, 2003.
188. A Hodgkin and A Huxley. A quantitative description of membrane current and its application to conduction and excitation in nerve. *Bulletin of Mathematical Biology*, 52:25–71, 01 1990.
189. Yiguang Hong, Guanrong Chen, and Linda Bushnell. Distributed observers design for leader-following control of multi-agent networks. *Automatica*, 44(3):846–850, 2008.

190. Ma Hongbin, Wang Meiling, Jia Zhenchao, and Yang Chenguang. A new framework of optimal multi-robot formation problem. In *Control Conference (CCC), 2011 30th Chinese*, pages 4139–4144. IEEE, 2011.
191. Roberto Horowitz and Pravin Varaiya. Control design of an automated highway system. *Proceedings of the IEEE*, 88:913 – 925, 08 2000.
192. Zeng-Guang Hou, Long Cheng, and Min Tan. Decentralized robust adaptive control for the multiagent system consensus problem using neural networks. *Systems, Man, and Cybernetics, Part B: Cybernetics, IEEE Transactions on*, 39(3):636–647, 2009.
193. Jiangping Hu and Wei Xing Zheng. Emergent collective behaviors on coopetition networks. *Physics Letters A*, 378(26-27):1787–1796, 2014.
194. Qinglei Hu, Hongyang Dong, Youmin Zhang, and Guangfu Ma. Tracking control of spacecraft formation flying with collision avoidance. *Aerospace Science and Technology*, 42:353–364, 2015.
195. Qinglei Hu, Bing Xiao, and Peng Shi. Tracking control of uncertain euler–lagrange systems with finite-time convergence. *International Journal of Robust and Nonlinear Control*, 25, 10 2014.
196. Qinglei Hu and Jian Zhang. Relative position finite-time coordinated tracking control of spacecraft formation without velocity measurements. *ISA Transactions*, 54:60–74, 2015.
197. Qinglei Hu, Jian Zhang, and Youmin Zhang. Velocity-free attitude coordinated tracking control for spacecraft formation flying. *ISA Transactions*, 73, 12 2017.
198. Huang Huang, Changbin Yu, and Qinghe Wu. Autonomous scale control of multiagent formations with only shape constraints. *International Journal of Robust and Nonlinear Control*, 23(7):765–791, 2013.
199. L Huang. Velocity planning for a mobile robot to track a moving target a potential field approach. *Robotics and Autonomous Systems*, 57(1):55–63, 2009.
200. Yi Huang and Ziyang Meng. Bearing-based distributed formation control of multiple vertical take-off and landing uavs. *IEEE Transactions on Control of Network Systems*, 8(3):1281–1292, 2021.
201. C Intel. Open source computer vision library reference manual. *Intel Corporation Microprocessor Research Labs*, 2001.
202. ZH Ismail, N Sarman, and MW Dunnigan. Dynamic region boundary-based control scheme for multiple autonomous underwater vehicles. In *OCEANS, 2012-Yeosu*, pages 1–6. IEEE, 2012.
203. A Jadbabaie, Jie Lin, and AS Morse. Coordination of groups of mobile autonomous agents using nearest neighbor rules. *IEEE Transactions on Automatic Control*, 48(6):988–1001, 2003.
204. Ali Jadbabaie, Jie Lin, and A Stephen Morse. Coordination of groups of mobile autonomous agents using nearest neighbor rules. *Automatic Control, IEEE Transactions on*, 48(6):988–1001, 2003.
205. Ali Jadbabaie, Nader Motee, and Mauricio Barahona. On the stability of the kuramoto model of coupled nonlinear oscillators. In *Proceedings of the 2004 American Control Conference*, volume 5, pages 4296–4301. IEEE, 2004.

206. S Jain, S Nandy, G Chakraborty, CS Kumar, and SN Shome. Error modeling of laser range finder for robotic application using time domain technique. In *IEEE International Conference on Signal Processing*, 2011.
207. Nathanaël Jarrassé, Vittorio Sanguineti, and Etienne Burdet. Slaves no longer: review on role assignment for human–robot joint motor action. *Adaptive Behavior*, 22(1):70–82, 2014.
208. Nathanal Jarrassé, T Charalambous, and E Burdet. A framework to describe, analyze and generate interactive motor behaviors. *PLoS ONE*, 7(11):e49945, 2012.
209. Harold Jeffreys, Bertha Jeffreys, and Bertha Swirles. *Methods of Mathematical Physics*. Cambridge University Press, 1999.
210. M Ji, G Ferrari-Trecate, M Egerstedt, and A Buffa. Containment Control in Mobile Networks. *IEEE Transactions on Automatic Control*, 53(8):1972–1975, 2008.
211. Meng Ji and Magnus Egerstedt. A graph-theoretic characterization of controllability for multi-agent systems. In *2007 American Control Conference*, pages 4588–4593. IEEE, 2007.
212. Yanan Ji, Zhijian Ji, Yungang Liu, and Chong Lin. Target controllability of multi-agent systems. In *2022 IEEE 11th Data Driven Control and Learning Systems Conference (DDCLS)*, pages 1377–1382. IEEE, 2022.
213. Zhenchao Jia, Hongbin Ma, Chenguang Yang, and Meiling Wang. Three-robot minimax travel-distance optimal formation. In *Decision and Control and European Control Conference (CDC-ECC), 2011 50th IEEE Conference on*, pages 7641–7646. IEEE, 2011.
214. Fangcui Jiang, Long Wang, and Guangming Xie. Consensus of high-order dynamic multi-agent systems with switching topology and time-varying delays. *Journal of Control Theory and Applications*, 8(1):52–60, 2010.
215. Lei Jiao, Zhihong Peng, Lele Xi, Shuxin Ding, and Jinqiang Cui. Multi-agent coverage path planning via proximity interaction and cooperation. *IEEE Sensors Journal*, 22(6):6196–6207, 2022.
216. M Jin, SH Kang, PH Chang, and J Lee. Robust control of robot manipulators using inclusive and enhanced time delay control. *IEEE/ASME Transactions on Mechatronics*, 22(5):2141–2152, 2017.
217. Xu Jin. Fault tolerant finite-time leader-follower formation control for autonomous surface vessels with LOS range and angle constraints. *Automatica*, 68:228–236, 2016.
218. Gangshan Jing and Long Wang. Multiagent flocking with angle-based formation shape control. *IEEE Transactions on Automatic Control*, 65(2):817–823, 2019.
219. Wang Jinhuan and Cheng Daizhan. Consensus of multi-agent systems with higher order dynamics. In *2007 Chinese Control Conference*, pages 761–765. IEEE, 2007.
220. Eric W Justh and PS Krishnaprasad. Equilibria and steering laws for planar formations. *Systems & Control Letters*, 52(1):25–38, 2004.
221. Wang Kai and Huang Xian Lin. The optimal target assignment in the cooperative surveillance task of satellite formation flying. In *Control Conference (CCC), 2014 33rd Chinese*, pages 7493–7497. IEEE, 2014.

222. Rahul Kala. Multi-robot path planning using co-evolutionary genetic programming. *Expert Systems with Applications*, 39(3):3817–3831, 2012.
223. Arti Kalra, SG Anavatti, and R Padhi. Aggressive formation flying of fixed-wing uavs with differential geometric guidance. *Unmanned Systems*, 05:97–113, 2017.
224. Vikram Kapila, Andrew G Sparks, James M Buffington, and Qiguo Yan. Spacecraft formation flying: Dynamics and control. *Journal of Guidance, Control, and Dynamics*, 23(3):561–564, 2000.
225. Matthew Keennon, Karl Klingebiel, and Henry Won. Development of the nano hummingbird: A tailless flapping wing micro air vehicle. In *50th AIAA Aerospace Sciences Meeting Including the New Horizons Forum and Aerospace Exposition*, 01 2012.
226. Farid Kendoul. Survey of advances in guidance, navigation, and control of unmanned rotorcraft systems. *Journal of Field Robotics*, 29(2):315–378, 2012.
227. Amirreza M Khaleghi, Dong Xu, Sara Minaeian, Mingyang Li, Yifei Yuan, Jian Liu, Young-Jun Son, Christopher Vo, and Jyh-Ming Lien. A dddams-based uav and ugv team formation approach for surveillance and crowd control. In *Proceedings of the 2014 Winter Simulation Conference*, pages 2907–2918. IEEE Press, 2014.
228. HK Khalil. *Nonlinear Systems* 2nd ed. 1996.
229. HK Khalil. *Nonlinear Systems*, Third Edition. Upper Saddle River, NJ: Prentice-Hall, Inc., 2002.
230. Hassan K Khalil. *Noninear Systems*. Prentice-Hall, New Jersey, 2(5):1–5, 1996.
231. Hassan K Khalil and JW Grizzle. *Nonlinear Systems*. Prentice Hall, New Jersey, 1996.
232. Oussama Khatib. Real-time obstacle avoidance for manipulators and mobile robots. In *Proceedings. 1985 IEEE International Conference on Robotics and Automation*, volume 2, pages 500–505. IEEE, 1985.
233. Oussama Khatib. Real-time obstacle avoidance for manipulators and mobile robots. *The International Journal of Robotics Research*, 5(1):90–98, 1986.
234. Saman Khodaverdian. On the synchronization of linear heterogeneous multi-agent systems in cycle-free communication networks. In *Proceedings of 2014 International Conference on Modelling, Identification & Control*, pages 190–195. IEEE, 2014.
235. CW Kilmister and C Cercignani. The boltzmann equation and its applications. *The Mathematical Gazette*, 73(463):61, 1989.
236. Jin-Oh Kim and Pradeep Khosla. Real-time obstacle avoidance using harmonic potential functions. *IEEE Transactions on Robotics and Automation*, 1992.
237. Tae-Hyoung Kim and Toshiharu Sugie. Cooperative control for target-capturing task based on a cyclic pursuit strategy. *Automatica*, 43(8):1426–1431, 2007.
238. Yongho Kim and Eric T Matson. A realistic decision making for task allocation in heterogeneous multi-agent systems. *Procedia Computer Science*, 94:386–391, 2016.

239. Michael Kirschner, Oliver Montenbruck, and Srinivas Bettadpur. Flight dynamics aspects of the GRACE formation flying. In *2nd International Workshop on Satellite Constellations and Formation Flying*, pages 19–20, 2001.
240. MD Kirszbraun. Über die zusammenziehende und lipschitzsche transformationen. *Fundamenta Mathematicae*, 22:77–108, 1934.
241. Linghuan Kong, Wei He, Chenguang Yang, Guang Li, and Zhengqiang Zhang. Adaptive fuzzy control for a marine vessel with time-varying constraints. *IET Control Theory & Applications*, 12, 03 2018.
242. Linghuan Kong, Wei He, Chenguang Yang, Zhijun Li, and Changyin Sun. Adaptive fuzzy control for coordinated multiple robots with constraint using impedance learning. *IEEE Transactions on Cybernetics*, 49(8):3052–3063, 2019.
243. Richard E Korf. Real-time heuristic search. *Artificial Intelligence*, 42(2-3):189–211, 1990.
244. Ajay D Kshemkalyani and Mukesh Singhal. *Distributed Computing: Principles, Algorithms, and Systems.* Cambridge University Press, 2011.
245. Miroslav Kubat. Neural networks: A comprehensive foundation by simon haykin, macmillan, 1994, isbn 0-02-352781-7. *Knowledge Engineering Review*, 13(4):409–412, 1999.
246. Miroslav Kubat. Neural networks: a comprehensive foundation by Simon Haykin, Macmillan, 1994, ISBN 0-02-352781-7. *Knowledge Engineering Review*, 13(4):409–412, 1999.
247. Jean-Michel Lasry and Pierre-Louis Lions. Mean field games. *Japanese Journal of Mathematics*, 2(1):229–260, 2007.
248. Steven M LaValle. *Planning Algorithms.* Cambridge University Press, 2006.
249. Jonathan R Lawton and Randal W Beard. Synchronized multiple spacecraft rotations. *Automatica*, 38(8):1359–1364, 2002.
250. Jonathan RT Lawton, Randal W Beard, and Brett J Young. A decentralized approach to formation maneuvers. *IEEE Transactions on Robotics and Automation*, 19(6):933–941, 2003.
251. Daero Lee. Nonlinear disturbance observer-based robust control for spacecraft formation flying. *Aerospace Science and Technology*, 76, 02 2018.
252. M Anthony Lewis and Kar-Han Tan. High precision formation control of mobile robots using virtual structures. *Autonomous Robots*, 4(4):387–403, 1997.
253. D Li, SS Ge, and TH Lee. Fixed-Time-Synchronized Consensus Control of Multi-Agent Systems. *IEEE Transactions on Control of Network Systems, in press, DOI: 10.1109/TCNS.2020.3034523*, 2020.
254. D Li, G Ma, Y Xu, W He, and SS Ge. Layered affine formation control of networked uncertain systems: a fully distributed approach over directed graphs. *IEEE Transactions on Cybernetics, in press, DOI: 10.1109/TCYB.2020.2965657*, 2020.
255. D Li, W Zhang, W He, C Li, and SS Ge. Two-layer distributed formation-containment control of multiple Euler-CLagrange systems by output feedback. *IEEE Transactions on Cybernetics*, 49(2):675–687, 2019.
256. Dongyu Li, Shuzhi Ge, Wei He, Guangfu Ma, and Lihua Xie. Multilayer formation control of multi-agent systems. *Automatica*, 109:108558, 2019.

257. Dongyu Li, Shuzhi Sam Ge, and Tong Heng Lee. Fixed-time-synchronized consensus control of multiagent systems. *IEEE Transactions on Control of Network Systems*, 8(1):89–98, 2020.
258. Dongyu Li, Shuzhi Sam Ge, and Tong Heng Lee. Simultaneous arrival to origin convergence: sliding-mode control through the norm-normalized sign function. *IEEE Transactions on Automatic Control*, 67(4):1966–1972, 2021.
259. Dongyu Li, Shuzhi Sam Ge, and Tong Heng Lee. *Time-Synchronized Control: Analysis and Design.* Springer, 2022.
260. Dongyu Li, Guangfu Ma, Wei He, Shuzhi Sam Ge, and Tong Heng Lee. Cooperative Circumnavigation Control of Networked Microsatellites. *IEEE Transactions on Cybernetics, in press, DOI: 10.1109/TCYB.2019.2923119*, 2019.
261. Dongyu Li, Guangfu Ma, Chuanjiang Li, Wei He, Jie Mei, and Shuzhi Ge. Distributed attitude coordinated control of multiple spacecraft with attitude constraints. *IEEE Transactions on Aerospace and Electronic Systems*, 1–1, 03 2018.
262. Dongyu Li, Guangfu Ma, Chuanjiang Li, Wei He, Jie Mei, and Shuzhi Sam Ge. Distributed attitude coordinated control of multiple spacecraft with attitude constraints. *IEEE Transactions on Aerospace and Electronic Systems*, 54(5):2233–2245, 2018.
263. Dongyu Li, Guangfu Ma, Yang Xu, Wei He, and Shuzhi Sam Ge. Layered affine formation control of networked uncertain systems: A fully distributed approach over directed graphs. *IEEE Transactions on Cybernetics*, 51(12):6119–6130, 2020.
264. Dongyu Li, Wei Zhang, Wei He, Chuanjiang Li, and Shuzhi Ge. Two-layer distributed formation-containment control of multiple euler-lagrange systems by output feedback. *IEEE Transactions on Cybernetics*, 1–13, 01 2018.
265. Hongjie Li, Chen Ming, Shigen Shen, and Wai Keung Wong. Event-triggered control for multi-agent systems with randomly occurring nonlinear dynamics and time-varying delay. *Journal of the Franklin Institute*, 351(5):2582–2599, 2014.
266. Huiming Li, Hao Chen, and Xiangke Wang. Affine formation tracking control of unmanned aerial vehicles. *Frontiers of Information Technology & Electronic Engineering*, 23(6):909–919, 2022.
267. Kewen Li and Shaocheng Tong. Observer-based finite-time fuzzy adaptive control for mimo non-strict feedback nonlinear systems with errors constraint. *Neurocomputing*, 341, 03 2019.
268. Lishuang Li, Wenting Fan, and Degen Huang. A two-phase bio-ner system based on integrated classifiers and multiagent strategy. *Computational Biology and Bioinformatics, IEEE/ACM Transactions on*, 10(4):897–904, 2013.
269. Ran Li and Yunhua Li. Localization of leader-follower formations using kinect and rtk-gps. In *Robotics and Biomimetics (ROBIO), 2014 IEEE International Conference on*, pages 908–913. IEEE, 2014.
270. Shihua Li, Haibo Du, and Xiangze Lin. Finite-time consensus algorithm for multi-agent systems with double-integrator dynamics. *Automatica*, 47(8):1706–1712, 2011.
271. Tao Li, Minyue Fu, Lihua Xie, and Ji-Feng Zhang. Distributed consensus with limited communication data rate. *IEEE Transactions on Automatic Control*, 56(2):279–292, 2010.

272. W Li. Notion of control-law module and modular framework of cooperative transportation using multiple nonholonomic robotic agents with physical rigid-formation-motion constraints. *IEEE Transactions on Cybernetics*, 46(5):1242, 2016.
273. Xiaohai Li, Jizhong Xiao, and Jindong Tan. Modeling and controller design for multiple mobile robots formation control. In *2004 IEEE International Conference on Robotics and Biomimetics*, pages 838–843. IEEE, 2004.
274. Xin Li and Daqi Zhu. An adaptive som neural network method for distributed formation control of a group of auvs. *IEEE Transactions on Industrial Electronics*, 65(10):8260–8270, 2018.
275. Xiuxian Li and Lihua Xie. Dynamic formation control over directed networks using graphical Laplacian approach. *IEEE Transactions on Automatic Control*, 9286(c):1–14, 2018.
276. Yahui Li, Sheng Qiang, Xianyi Zhuang, et al. Robust and adaptive backstepping control for nonlinear systems using rbf neural networks. *Neural Networks, IEEE Transactions on*, 15(3):693–701, 2004.
277. Yanan Li and Shuzhi Sam Ge. Impedance learning for robots interacting with unknown environments. *Control Systems Technology, IEEE Transactions on*, 22(4):1422–1432, 2013.
278. Z Li, J Deng, R Lu, X Yong, J Bai, and C Y Su. Trajectory-tracking control of mobile robot systems incorporating neural-dynamic optimized model predictive approach. *IEEE Transactions on Systems Man & Cybernetics Systems*, 46(6):740–749, 2016.
279. Z Li, B Huang, A Ajoudani, C Yang, C Su, and A Bicchi. Asymmetric bimanual control of dual-arm exoskeletons for human-cooperative manipulations. *IEEE Transactions on Robotics*, 34(1):264–271, 2018.
280. Zhijun Li, Chun-Yi Su, Liangyong Wang, Ziting Chen, and Tianyou Chai. Nonlinear disturbance observer-based control design for a robotic exoskeleton incorporating fuzzy approximation. *Industrial Electronics, IEEE Transactions on*, 62:5763–5775, 2015.
281. Zhongkui Li, Zhisheng Duan, Guanrong Chen, and Lin Huang. Consensus of multiagent systems and synchronization of complex networks: A unified viewpoint. *IEEE Transactions on Circuits and Systems I: Regular Papers*, 57(1):213–224, 2009.
282. Zhongkui Li, Wei Ren, Xiangdong Liu, and Mengyin Fu. Consensus of multi-agent systems with general linear and lipschitz nonlinear dynamics using distributed adaptive protocols. *IEEE Transactions on Automatic Control*, 58(7):1786–1791, 2012.
283. Zhongkui Li, Wei Ren, Xiangdong Liu, and Mengyin Fu. Distributed containment control of multi-agent systems with general linear dynamics in the presence of multiple leaders. *International Journal of Robust and Nonlinear Control*, 23(5):534–547, 2013.
284. X Liang, Y Fang, N Sun, and H Lin. A Novel Energy-Coupling-Based Hierarchical Control Approach for Unmanned Quadrotor Transportation Systems. *IEEE/ASME Transactions on Mechatronics*, 24(1):248–259, 2019.
285. Xiaoling Liang, Mingzhe Hou, and Guangren Duan. Adaptive dynamic surface control for integrated missile guidance and autopilot in the presence of input saturation. *Journal of Aerospace Engineering*, 28:04014121, 2014.

286. Xinwu Liang, Yun-Hui Liu, Hesheng Wang, Weidong Chen, Kexin Xing, and Tao Liu. Leader-following formation tracking control of mobile robots without direct position measurements. *IEEE Transactions on Automatic Control*, 61(12):4131–4137, 2016.
287. Xinwu Liang, Hesheng Wang, Yun Liu, Weidong Chen, and Tao Liu. Formation control of nonholonomic mobile robots without position and velocity measurements. *IEEE Transactions on Robotics*, 1–13, 12 2017.
288. Jie Lin, A Stephen Morse, and Brian DO Anderson. The multi-agent rendezvous problem. In *42nd IEEE International Conference on Decision and Control (IEEE cat. no. 03ch37475)*, volume 2, pages 1508–1513. IEEE, 2003.
289. Zhiyun Lin, Bruce Francis, and Manfredi Maggiore. Necessary and sufficient graphical conditions for formation control of unicycles. *IEEE Transactions on Automatic Control*, 50(1):121–127, 2005.
290. Zhiyun Lin, Lili Wang, Zhiyong Chen, Minyue Fu, and Zhimin Han. Necessary and sufficient graphical conditions for affine formation control. *IEEE Transactions on Automatic Control*, 61(10):2877–2891, 2015.
291. Zhiyun Lin, Lili Wang, Zhiyong Chen, Minyue Fu, and Zhimin Han. Necessary and sufficient graphical conditions for affine formation control. *IEEE Transactions on Automatic Control*, 61(10):2877–2891, 2016.
292. Zhiyun Lin, Lili Wang, Zhimin Han, and Minyue Fu. Distributed formation control of multi-agent systems using complex laplacian. *IEEE Transactions on Automatic Control*, 59(7):1765–1777, 2014.
293. Zhiyun Lin, Lili Wang, Zhimin Han, and Minyue Fu. Distributed formation control of multi-agent systems using complex laplacian. *IEEE Transactions on Automatic Control*, 59(7):1765–1777, 2014.
294. Bo Liu, Tianguang Chu, Long Wang, Zhiqiang Zuo, Guanrong Chen, and Housheng Su. Controllability of switching networks of multi-agent systems. *International Journal of Robust and Nonlinear Control*, 22(6):630–644, 2012.
295. Bo Liu, Housheng Su, Rong Li, Dehui Sun, and Weina Hu. Switching controllability of discrete-time multi-agent systems with multiple leaders and time-delays. *Applied Mathematics and Computation*, 228:571–588, 2014.
296. Bo Liu, Housheng Su, Licheng Wu, and Xixi Shen. Controllability for multi-agent systems with matrix-weight-based signed network. *Applied Mathematics and Computation*, 411:126520, 2021.
297. Dacai Liu, Zhi Liu, CL Philip Chen, and Yun Zhang. Finite-time distributed cooperative control for heterogeneous nonlinear multi-agent systems with unknown input constraints. *Neurocomputing*, 415:123–134, 2020.
298. Hui Liu, Zhiyun Lin, Ming Cao, Xiaoping Wang, and Jinhu Lü. Coordinate-free formation control of multi-agent systems using rooted graphs. *Systems & Control Letters*, 119:8–15, 2018.
299. Huiyang Liu, Long Cheng, Min Tan, and Zeng Guang Hou. Containment control of continuous-time linear multi-agent systems with aperiodic sampling. *Autom.*, 57:78–84, 2015.
300. Jun Liu, Hongbin Ma, Xuemei Ren, and Mengyin Fu. Optimal formation of robots by convex hull and particle swarm optimization. In *Computational Intelligence in Control and Automation (CICA), 2013 IEEE Symposium on*, pages 104–111. IEEE, 2013.

301. Lu Liu, Dan Wang, and Zhouhua Peng. Path following of marine surface vehicles with dynamical uncertainty and time-varying ocean disturbances. *Neurocomputing*, 173, 08 2015.
302. Lu Liu, Dan Wang, Zhouhua Peng, and Hugh Liu. Saturated coordinated control of multiple underactuated unmanned surface vehicles over a closed curve. *Science China Information Sciences*, 60, 07 2017.
303. Peng Liu and Tie-Hua Ma. Strong structural controllability of multi-agent systems with switching topologies. In *2018 10th International Conference on Modelling, Identification and Control (ICMIC)*, pages 1–6. IEEE, 2018.
304. Xiaomei Liu, Shuzhi Sam Ge, and Cher Hiang Goh. Neural-network-based switching formation tracking control of multiagents with uncertainties in constrained space. *IEEE Transactions on Systems, Man, and Cybernetics: Systems*, pages 1–10, 2017.
305. Xiaomei Liu, Shuzhi Sam Ge, Rui Jiang, and Cher-Hiang Goh. Intelligent speech control system for human-robot interaction. In *Control Conference (CCC), 2016 35th Chinese*, pages 6154–6159. TCCT, 2016.
306. Yan-Jun Liu, Mingzhe Gong, Shaocheng Tong, CL Philip Chen, and Dong-Juan Li. Adaptive fuzzy output feedback control for a class of nonlinear systems with full state constraints. *IEEE Transactions on Fuzzy Systems*, 26(5):2607–2617, 2018.
307. Yang Liu and Yingmin Jia. An iterative learning approach to formation control of multi-agent systems. *Systems & Control Letters*, 61(1):148–154, 2012.
308. Yingying Liu and Zhanshan Wang. Optimal output synchronization of heterogeneous multi-agent systems using measured input-output data. *Information Sciences*, 582:462–479, 2022.
309. Yongfang Liu and Zhiyong Geng. Finite-time optimal formation tracking control of vehicles in horizontal plane. *Nonlinear Dynamics*, 76:481–495, 04 2014.
310. Yonggui Liu, Bugong Xu, and Y. Ding. Convergence analysis of cooperative braking control for interconnected vehicle systems. *IEEE Transactions on Intelligent Transportation Systems*, PP:1–13, 10 2016.
311. Yuanchang Liu and Richard Bucknall. A survey of formation control and motion planning of multiple unmanned vehicles. *Robotica*, 36(7):1019–1047, 2018.
312. Zhixin Liu and Lei Guo. Synchronization of multi-agent systems without connectivity assumptions. *Automatica*, 45(12):2744–2753, 2009.
313. Zhixin Liu, Xiu You, Hongjiu Yang, and Ling Zhao. Leader-following consensus of heterogeneous multi-agent systems with packet dropout. *International Journal of Control, Automation and Systems*, 13(5):1067–1075, 2015.
314. Charles F Van Loan. The ubiquitous kronecker product. *Journal of Computational and Applied Mathematics*, 123(1):85–100, 2000. Numerical Analysis 2000. Vol. III: Linear Algebra.
315. T Lolla, PJ Haley Jr, and PFJ Lermusiaux. Path planning in multi-scale ocean flows: Coordination and dynamic obstacles. *Ocean Modelling*, 94:46–66, 2015.
316. A Lopez-Gonzalez, ED Ferreira, EG Hernandez-Martinez, José-Job Flores-Godoy, Guillermo Fernandez-Anaya, and P Paniagua-Contro. Multi-robot

formation control using distance and orientation. *Advanced Robotics*, 30(14):901–913, 2016.

317. Yao Lu, Yi Guo, and Zhaoyang Dong. Multiagent flocking with formation in a constrained environment. *Journal of Control Theory and Applications*, 8(2):151–159, 2010.
318. Yu Lu, Guoqing Zhang, Zhijian Sun, and Weidong Zhang. Robust adaptive formation control of underactuated autonomous surface vessels based on mlp and dob. *Nonlinear Dynamics*, 94, 10 2018.
319. Zehuan Lu, Zhiqiang Zhang, and Zhijian Ji. Strong targeted controllability of multi-agent systems with time-varying topologies over finite fields. *Automatica*, page 110404, 2022.
320. C Luo, X Li, Y Li, and Q Dai. Biomimetic Design for Unmanned Aerial Vehicle Safe Landing in Hazardous Terrain. *IEEE/ASME Transactions on Mechatronics*, 21(1):531–541, 2016.
321. Aleksandr Mikhailovich Lyapunov. The general problem of the stability of motion. *International Journal of Control*, 55(3):531–534, 1992.
322. Hongguang Lyu and Yong Yin. Colregs-constrained real-time path planning for autonomous ships using modified artificial potential fields. *The Journal of Navigation*, 72(3):588–608, 2019.
323. Yang Lyu, Quan Pan, Chunhui Zhao, Yizhai Zhang, and Jinwen Hu. Vision-based uav collision avoidance with 2d dynamic safety envelope. *IEEE Aerospace and Electronic Systems Magazine*, 31(7):16–26, 2016.
324. Lifeng Ma, Zidong Wang, Qing-Long Han, and Yurong Liu. Consensus control of stochastic multi-agent systems: a survey. *Science China Information Sciences*, 60, 12 2017.
325. Long Ma, Haibo Min, Shicheng Wang, Yuan Liu, and Shouyi Liao. An overview of research in distributed attitude coordination control. *IEEE/CAA Journal of Automatica Sinica*, 2(2):121–133, 2015.
326. Dana Mackenzie. A flapping of wings. *Science*, 335(6075):1430–1433, 2012.
327. Mohamed Maghenem, Antonio Loria, and Elena Panteley. A cascades approach to formation-tracking stabilization of force-controlled autonomous vehicles. *IEEE Transactions on Automatic Control*, 1–1, 11 2017.
328. Gian Luca Mariottini, Fabio Morbidi, Domenico Prattichizzo, Nicholas Vander Valk, Nathan Michael, George Pappas, and Kostas Daniilidis. Vision-based localization for leader–follower formation control. *IEEE Transactions on Robotics*, 25(6):1431–1438, 2009.
329. Joshua A Marshall, Mireille E Broucke, and Bruce A Francis. Pursuit formations of unicycles. *Automatica*, 42(1):3–12, 2006.
330. Joshua A Marshall, ME Brouke, and Bruce A Francis. A pursuit strategy for wheeled-vehicle formations. In *Decision and Control, 2003. Proceedings. 42nd IEEE Conference on*, volume 3, pages 2555–2560. IEEE, 2003.
331. Masoud and A Ahmad. Motion planning with gamma-harmonic potential fields. *Aerospace & Electronic Systems IEEE Transactions on*, 48(4):2786–2801, 2016.
332. Ahmad A Masoud and Samer A Masoud. Motion planning in the presence of directional and obstacle avoidance constraints using nonlinear, anisotropic, harmonic potential fields. In *Proceedings 2000 ICRA. Millennium Conference. IEEE International Conference on Robotics and Automation.*

Symposia Proceedings (Cat. No. 00CH37065), volume 3, pages 2944–2951. IEEE, 2000.

333. Silvia Mastellone, Juan S Mejía, Dušan M Stipanović, and Mark W Spong. Formation control and coordinated tracking via asymptotic decoupling for lagrangian multi-agent systems. *Automatica*, 47(11):2355–2363, 2011.
334. José Ramón Medina, Martin Lawitzky, Adam Molin, and Sandra Hirche. Dynamic strategy selection for physical robotic assistance in partially known tasks. In *Robotics and Automation (ICRA), 2013 IEEE International Conference on*, pages 1180–1186. IEEE, 2013.
335. José Ramón Medina, Dongheui Lee, and Sandra Hirche. Risk-sensitive optimal feedback control for haptic assistance. In *Robotics and Automation (ICRA), 2012 IEEE International Conference on*, pages 1025–1031. IEEE, 2012.
336. Farhad Mehdifar, Charalampos P Bechlioulis, Farzad Hashemzadeh, and Mahdi Baradarannia. Prescribed performance distance-based formation control of multi-agent systems. *Automatica*, 119:109086, 2020.
337. AR Mehrabian and K Khorasani. Constrained distributed cooperative synchronization and reconfigurable control of heterogeneous networked euler-lagrange multi-agent systems. *Information Sciences*, 370, 09 2015.
338. Ziyang Meng, Zongli Lin, and Wei Ren. Robust cooperative tracking for multiple non-identical second-order nonlinear systems. *Automatica*, 49(8):2363–2372, 2013.
339. Ziyang Meng, Wei Ren, and Zheng You. Distributed finite-time attitude containment control for multiple rigid bodies. *Automatica*, 46:2092–2099, 2010.
340. Ziyang Meng, Wei Ren, and Zheng You. Distributed finite-time attitude containment control for multiple rigid bodies. *Automatica*, 46(12):2092–2099, 2010.
341. Guoying Miao and Qian Ma. Group consensus of the first-order multi-agent systems with nonlinear input constraints. *Neurocomputing*, 161:113–119, 2015.
342. Zhiqiang Miao, Divya Thakur, R Scott Erwin, Jean Pierre, Yaonan Wang, and Rafael Fierro. Orthogonal vector field-based control for a multi-robot system circumnavigating a moving target in 3D. *2016 IEEE 55th Conference on Decision and Control, CDC 2016*, pages 6004–6009, 2016.
343. Mingming, Li, Yanan, Li, Shuzhi, Sam, Ge, Tong, Heng, and Lee. Adaptive control of robotic manipulators with unified motion constraints. *IEEE Transactions on Systems Man & Cybernetics Systems*, 2016.
344. Hamidreza Modares, Bahare Kiumarsi, Frank L Lewis, Frank Ferrese, and Ali Davoudi. Resilient and robust synchronization of multiagent systems under attacks on sensors and actuators. *IEEE Transactions on Cybernetics*, 50(3):1240–1250, 2019.
345. Sunghoon Mok, Yoonhyuk Choi, and Hyochoong Bang. Impulsive control of satellite formation flying using orbital period difference. In *Automatic Control in Aerospace*, volume 18, pages 368–373, 2010.
346. Eduardo Montijano, Eric Cristofalo, Dingjiang Zhou, Mac Schwager, and Carlos Sagüés. Vision-based distributed formation control without an external positioning system. *IEEE Transactions on Robotics*, 32(2):339–351, 2016.

347. Alexander Mörtl, Martin Lawitzky, Ayse Kucukyilmaz, Metin Sezgin, Cagatay Basdogan, and Sandra Hirche. The role of roles: Physical cooperation between humans and robots. *The International Journal of Robotics Research*, 31(13):1656–1674, 2012.
348. Nima Moshtagh, Nathan Michael, Ali Jadbabaie, and Kostas Daniilidis. Vision-based, distributed control laws for motion coordination of nonholonomic robots. *IEEE Transactions on Robotics*, 25(4):851–860, 2009.
349. Dongdong Mu, Guofeng Wang, Yunsheng Fan, Yiming Bai, and Yongsheng Zhao. Fuzzy-based optimal adaptive line-of-sight path following for underactuated unmanned surface vehicle with uncertainties and time-varying disturbances. *Mathematical Problems in Engineering*, 2018:1–12, 2018.
350. Dongdong Mu, Guofeng Wang, Yunsheng Fan, Xiaojie Sun, and Bingbing Qiu. Adaptive los path following for a podded propulsion unmanned surface vehicle with uncertainty of model and actuator saturation. *Applied Sciences*, 7:1232, 2017.
351. Indira Nagesh and Christopher Edwards. A multivariable super-twisting sliding mode approach. *Automatica*, 50(3):984–988, 2014.
352. Tiago Nascimento, Luis Costa, Andre Scolari Conceicao, and A. Moreira. Nonlinear model predictive formation control: An iterative weighted tuning approach. *Journal of Intelligent and Robotic Systems*, 80, 01 2015.
353. T Neubrunn. Quasi-continuity. *Real Analysis Exchange*, 14(2):259–306, 1988.
354. Helmut Neunzert. An introduction to the nonlinear boltzmann-vlasov equation. In *Kinetic Theories and the Boltzmann Equation*, pages 60–110. Springer, 1984.
355. Hanlin Niu, Al Savvaris, Antonios Tsourdos, and Ze Ji. Voronoi-visibility roadmap-based path planning algorithm for unmanned surface vehicles. *The Journal of Navigation*, 72(4):850–874, 2019.
356. Ali Noormohammadi Asl, Mohammad Bagher Menhaj, and Atena Sajedin. Control of leader–follower formation and path planning of mobile robots using asexual reproduction optimization (aro). *Applied Soft Computing*, 14:563–576, 2014.
357. G Notarstefano, M Egerstedt, and M Haque. Containment in leader-follower networks with switching communication topologies. *Automatica*, 47(5):1035–1040, 2011.
358. Petter Ogren, Magnus Egerstedt, and Xiaoming Hu. A control lyapunov function approach to multi-agent coordination. In *Decision and Control, 2001. Proceedings of the 40th IEEE Conference on*, volume 2, pages 1150–1155. IEEE, 2001.
359. Kwang-Kyo Oh, Myoung-Chul Park, and Hyo-Sung Ahn. A survey of multi-agent formation control. *Automatica*, 53, 10 2014.
360. Kwang Kyo Oh, Myoung Chul Park, and Hyo Sung Ahn. A survey of multi-agent formation control. *Automatica*, 53:424–440, 2015.
361. Ayano Okoso, Keisuke Otaki, and Tomoki Nishi. Multi-agent path finding with priority for cooperative automated valet parking. In *2019 IEEE Intelligent Transportation Systems Conference (ITSC)*, pages 2135–2140. IEEE, 2019.

362. R Olfati-Saber and RM Murray. Consensus problems in networks of agents with switching topology and time-delays. *IEEE Transactions on Automatic Control*, 49(9):1520–1533, 2004.
363. Reza Olfati-Saber. Flocking for multi-agent dynamic systems: Algorithms and theory. *Automatic Control, IEEE Transactions on*, 51(3):401–420, 2006.
364. Reza Olfati-Saber and Richard M Murray. Consensus problems in networks of agents with switching topology and time-delays. *Automatic Control, IEEE Transactions on*, 49(9):1520–1533, 2004.
365. Okechi Onuoha, Hilton Tnunay, Chunyan Wang, and Zhengtao Ding. Fully distributed affine formation control of general linear systems with uncertainty. *Journal of the Franklin Institute*, 357(17):12143–12162, 2020.
366. Hemanta Kumar Paikray, Pradipta Kumar Das, and Sucheta Panda. Optimal path planning of multi-robot in dynamic environment using hybridization of meta-heuristic algorithm. *International Journal of Intelligent Robotics and Applications*, 6(4):625–667, 2022.
367. Anil Kumar Pal, Shyam Kamal, Shyam Krishna Nagar, Bijnan Bandyopadhyay, and Leonid Fridman. Design of controllers with arbitrary convergence time. *Automatica*, 112:108710, 2020.
368. Zhenhua Pan, Chengxi Zhang, Yuanqing Xia, Hao Xiong, and Xiaodong Shao. An improved artificial potential field method for path planning and formation control of the multi-uav systems. *IEEE Transactions on Circuits and Systems II: Express Briefs*, 69(3):1129–1133, 2021.
369. Dimitra Panagou and Vijay Kumar. Cooperative visibility maintenance for leader–follower formations in obstacle environments. *IEEE Transactions on Robotics*, 30(4):831–844, 2014.
370. Bong Park. Adaptive formation control of underactuated autonomous underwater vehicles. *Ocean Engineering*, 96, 03 2015.
371. Bong Park and Sung Yoo. An error transformation approach for connectivity-preserving and collision-avoiding formation tracking of networked uncertain underactuated surface vessels. *IEEE Transactions on Cybernetics*, 1–12, 05 2018.
372. Bong Seok Park, Sung Jin Yoo, Jin Bae Park, and Yoon Ho Choi. A simple adaptive control approach for trajectory tracking of electrically driven nonholonomic mobile robots. *IEEE Transactions on Control Systems Technology*, 18(5):1199–1206, 2009.
373. Mirko Pastorelli, Riccardo Bevilacqua, and Stefano Pastorelli. Differential-drag-based roto-translational control for propellant-less spacecraft. *Acta Astronautica*, 114, 04 2015.
374. Stacy Patterson and Bassam Bamieh. Leader selection for optimal network coherence. In *49th IEEE Conference on Decision and Control (CDC)*, pages 2692–2697. IEEE, 2010.
375. Stacy Patterson, Neil McGlohon, and Kirill Dyagilev. Optimal k-leader selection for coherence and convergence rate in one-dimensional networks. *IEEE Transactions on Control of Network Systems*, 4(3):523–532, 2016.
376. Tobias Paul, Thomas R Krogstad, and Jan Tommy Gravdahl. Modelling of uav formation flight using 3d potential field. *Simulation Modelling Practice and Theory*, 16(9):1453–1462, 2008.

377. Ke Peng and Yupu Yang. Leader-following consensus problem with a varying-velocity leader and time-varying delays. *Physica A Statistical Mechanics & Its Applications*, 388(388):193–208, 2009.
378. X Peng, Z Sun, K Guo, and Z Geng. Mobile Formation Coordination and Tracking Control for Multiple Nonholonomic Vehicles. *IEEE/ASME Transactions on Mechatronics*, 25(3):1231–1242, 2020.
379. Z Peng, D Wang, Z Chen, X Hu, and W Lan. Adaptive dynamic surface control for formations of autonomous surface vehicles with uncertain dynamics. *IEEE Transactions on Control Systems Technology*, 21(2):513–520, 2013.
380. Zhouhua Peng, Dan Wang, Wei-yao Lan, and Gang Sun. Robust leader-follower formation tracking control of multiple underactuated surface vessels. *China Ocean Engineering*, 26, 09 2012.
381. Zhouhua Peng, Dan Wang, and Jun Wang. Cooperative dynamic positioning of multiple marine offshore vessels: A modular design. *IEEE/ASME Transactions on Mechatronics*, 21:1–1, 01 2015.
382. Zhouhua Peng, Dan Wang, and Wei Wang. Coordinated formation pattern control of multiple marine surface vehicles with model uncertainty and time-varying ocean currents. *Neural Computing and Applications*, 25:1771–1783, 12 2014.
383. Christopher Petersen and Rafael Fierro. Network-Lyapunov Technique For Spacecraft Formation Control. *AIAA Guidance, Navigation, and Control Conference*, pages 1–9, 2018.
384. Andrey Polyakov. Nonlinear feedback design for fixed-time stabilization of linear control systems. *IEEE Transactions on Automatic Control*, 57(8):2106–2110, 2012.
385. Andrey Polyakov. Nonlinear feedback design for fixed-time stabilization of linear control systems. *IEEE Transactions on Automatic Control*, 57(8):2106–2110, 2012.
386. MB Porter. Concerning absolutely continuous functions. *Bulletin of the American Mathematical Society*, 22(3):109–111, 1915.
387. Xue Qi and Zhi-jun Cai. Three-dimensional formation control based on nonlinear small gain method for multiple underactuated underwater vehicles. *Ocean Engineering*, 151:105–114, 2018.
388. Chunjiang Qian and Wei Lin. Non-Lipschitz continuous stabilizers for nonlinear systems with uncontrollable unstable linearization. *Systems and Control Letters*, 42(3):185–200, 2001.
389. Hongde Qin, Chengpeng Li, Yanchao Sun, Zhongchao Deng, and Yuhan Liu. Trajectory tracking control of unmanned surface vessels with input saturation and full-state constraints. *International Journal of Advanced Robotic Systems*, 15(5):1–9, 2018.
390. Bingbing Qiu, Guofeng Wang, Yunsheng Fan, Dongdong Mu, and Xiaojie Sun. Adaptive sliding mode trajectory tracking control for unmanned surface vehicle with modeling uncertainties and input saturation. *Applied Sciences*, 9(6), 2019.
391. Qian Qiu and Housheng Su. Distributed adaptive consensus of parabolic pde agents on switching graphs with relative output information. *IEEE Transactions on Industrial Informatics*, 18(1):297–304, 2021.

392. Hong Qu, Ke Xing, and Takacs Alexander. An improved genetic algorithm with co-evolutionary strategy for global path planning of multiple mobile robots. *Neurocomputing*, 120:509–517, 2013.
393. Amirreza Rahmani, Meng Ji, Mehran Mesbahi, and Magnus Egerstedt. Controllability of multi-agent systems from a graph-theoretic perspective. *SIAM Journal on Control and Optimization*, 48(1):162–186, 2009.
394. Sarvapali D Ramchurn, Maria Polukarov, Alessandro Farinelli, Cuong Truong, and Nicholas R Jennings. Coalition formation with spatial and temporal constraints. In *Proceedings of the 9th International Conference on Autonomous Agents and Multiagent Systems: volume 3-Volume 3*, pages 1181–1188. International Foundation for Autonomous Agents and Multiagent Systems, 2010.
395. D Ran, AHJ de Ruiter, W Yao, and X Chen. Distributed and reliable output feedback control of spacecraft formation with velocity constraints and time delays. *IEEE/ASME Transactions on Mechatronics*, 24(6):2541–2549, 2019.
396. Carmine Tommaso Recchiuto, Antonio Sgorbissa, and Renato Zaccaria. Visual feedback with multiple cameras in a uavs human–swarm interface. *Robotics and Autonomous Systems*, 80:43–54, 2016.
397. Beibei Ren, Shuzhi Sam Ge, Tong Heng Lee, and Miroslav Krstic. Region tracking control for multi-agent systems with high-order dynamics. In *American Control Conference (ACC), 2013*, pages 1266–1271. IEEE, 2013.
398. Wei Ren. Collective motion from consensus with cartesian coordinate coupling. *IEEE Transactions on Automatic Control*, 54(6):1330–1335, 2009.
399. Wei Ren and Ella Atkins. Distributed multi-vehicle coordinated control via local information exchange. *International Journal of Robust and Nonlinear Control*, 17(10-11):1002–1033, 2007.
400. Wei Ren and Randal Beard. Decentralized scheme for spacecraft formation flying via the virtual structure approach. *Journal of Guidance, Control, and Dynamics*, 27(1):73–82, 2004.
401. Wei Ren and Randal W Beard. Consensus seeking in multiagent systems under dynamically changing interaction topologies. *Automatic Control, IEEE Transactions on*, 50(5):655–661, 2005.
402. Wei Ren, Randal W Beard, and Ella M Atkins. Information consensus in multivehicle cooperative control. *Control Systems, IEEE*, 27(2):71–82, 2007.
403. Wei Ren, Randal W Beard, and Timothy W McLain. Coordination variables and consensus building in multiple vehicle systems. In *Cooperative Control*, pages 171–188. Springer, 2005.
404. Wei Ren and RW Beard. Consensus seeking in multiagent systems under dynamically changing interaction topologies. *IEEE Transactions on Automatic Control*, 50(5):655–661, 2005.
405. Craig W Reynolds. Flocks, herds and schools: A distributed behavioral model. In *ACM SIGGRAPH Computer Graphics*, volume 21, pages 25–34. ACM, 1987.
406. H Rezaee, F Abdollahi, and HA Talebi. Based motion synchronization in formation flight with delayed communications. *IEEE Transactions on Industrial Electronics*, 61(11):6175–6182, 2014.
407. Hamed Rezaee and Farzaneh Abdollahi. A decentralized cooperative control scheme with obstacle avoidance for a team of mobile robots. *IEEE Transactions on Industrial Electronics*, 61(1):347–354, 2013.

408. D Richert and J Cortés. Optimal leader allocation in uav formation pairs ensuring cooperation. *Automatica*, 49(11):3189–3198, 2013.
409. Kenneth Franklin Riley, Michael Paul Hobson, and Stephen John Bence. Mathematical methods for physics and engineering, *American Journal of Physics*, 67(2):165–169, 1999.
410. Alejandro Rodriguez-Angeles and Henk Nijmeijer. Cooperative synchronization of robots via estimated state feedback. In *Decision and Control, 2003. Proceedings. 42nd IEEE Conference on*, volume 2, pages 1514–1519. IEEE, 2003.
411. Dibyendu Roy, Madhubanti Maitra, and Samar Bhattacharya. Study of formation control and obstacle avoidance of swarm robots using evolutionary algorithms. In *2016 IEEE International Conference on Systems, Man, and Cybernetics (SMC)*, pages 003154–003159. IEEE, 2016.
412. Lorenzo Sabattini, Cristian Secchi, and Cesare Fantuzzi. Closed-curve path tracking for decentralized systems of multiple mobile robots. *Journal of Intelligent & Robotic Systems*, 71(1):109–123, 2013.
413. Chris Sabol, Rich Burns, and Craig A McLaughlin. Satellite formation flying design and evolution. *Journal of Spacecraft and Rockets*, 38(2):270–278, 2001.
414. Masao Saida, José Ramón Medina, and Sandra Hirche. Adaptive attitude design with risk-sensitive optimal feedback control in physical human-robot interaction. In *RO-MAN, 2012 IEEE*, pages 955–961. IEEE, 2012.
415. Stan Salvador and Philip Chan. Toward accurate dynamic time warping in linear time and space. *Intelligent Data Analysis*, 11(5):561–580, 2007.
416. Shuzhi Sam Ge, Jun Zhang, Xianbin Cao, and Xiaoming Sun. Region tracking control for high-order multi-agent systems in restricted space. *IET Control Theory & Applications*, 10(4):396–406, feb 2016.
417. Sanjay Sane and Michael Dickinson. The control of flight force by a flapping wing: Lift and drag production. *The Journal of Experimental Biology*, 204:2607–26, 2001.
418. Amit K Sanyal and Jan Bohn. Finite-time stabilisation of simple mechanical systems using continuous feedback. *International Journal of Control*, 88(4):783–791, 2015.
419. SS Sastry. Nonlinear systems: Analysis, stability, and control. 1999.
420. Luca Scardovi and Rodolphe Sepulchre. Synchronization in networks of identical linear systems. In *2008 47th IEEE Conference on Decision and Control*, pages 546–551. IEEE, 2008.
421. DP Scharf, FY Hadaegh, and SR Ploen. A survey of spacecraft formation flying guidance and control. part ii: control. In *Proceedings of the 2004 American Control Conference*, volume 4, pages 2976–2985, 2004.
422. Rastko R Šelmić and Frank L Lewis. Deadzone compensation in motion control systems using neural networks. *IEEE Transactions on Automatic Control*, 45(4):602–613, 2000.
423. Rastko R Selmic and Frank L Lewis. Neural-network approximation of piecewise continuous functions: Application to friction compensation. *IEEE Transactions on Neural Networks*, 13(3):745–751, 2002.
424. Rodolphe Sepulchre, Derek Paley, and Naomi Leonard. Collective motion and oscillator synchronization. In *Cooperative Control*, pages 189–205. Springer, 2005.

425. Rodolphe Sepulchre, Derek A Paley, and Naomi Ehrich Leonard. Stabilization of planar collective motion: All-to-all communication. *IEEE Transactions on Automatic Control*, 52(5):811–824, 2007.
426. Georg S Seyboth, Jingbo Wu, Jiahu Qin, Changbin Yu, and Frank Allgöwer. Collective circular motion of unicycle type vehicles with nonidentical constant velocities. *IEEE Transactions on Control of Network Systems*, 1(2):167–176, 2014.
427. Volkan Sezer and Metin Gokasan. A novel obstacle avoidance algorithm: Follow the gap method. *Robotics and Autonomous Systems*, 60(9):1123–1134, 2012.
428. Baike She, Siddhartha Mehta, Chau Ton, and Zhen Kan. Controllability ensured leader group selection on signed multiagent networks. *IEEE Transactions on Cybernetics*, 50(1):222–232, 2018.
429. Baike She, Siddhartha Mehta, Chau Ton, and Zhen Kan. Energy-related controllability of signed complex networks with laplacian dynamics. *IEEE Transactions on Automatic Control*, 66(7):3325–3330, 2020.
430. Guodong Shi, Karl Henrik Johansson, and Yiguang Hong. Reaching an optimal consensus: Dynamical systems that compute intersections of convex sets. *IEEE Transactions on Automatic Control*, 58(3):610–622, 2012.
431. Parisa Kaveh Shtessel and Yuri B. Attitude stabilization of rigid spacecraft with finite-time convergence. *International Journal of Robust and Nonlinear Control*, 18(October 2014):557–569, 2008.
432. Yuanchao Si and JinRong Wang. Relative controllability of multi-agent systems with input delay and switching topologies. *Systems & Control Letters*, 171:105432, 2023.
433. Yogang Singh. *Cooperative Swarm Optimisation of Unmanned Surface Vehicles*. PhD thesis, 04 2019.
434. J.-J. E. Slotine, W. Li, Applied Nonlinear Control (Prentice Hall, 1991).
435. Ge Song, Peng Shi, Shuoyu Wang, and Jeng-Shyang Pan. A new finite-time cooperative control algorithm for uncertain multi-agent systems. *International Journal of Systems Science*, 50(5):1006–1016, 2019.
436. Yongduan Song and Yujuan Wang. Cooperative control of nonlinear networked systems. *Cham: Springer-Verlag*, 2019.
437. Aakash Soni and Huosheng Hu. Formation control for a fleet of autonomous ground vehicles: A survey. *Robotics*, 7:67, 2018.
438. Eduardo D Sontag and Yuan Wang. On characterizations of the input-to-state stability property. *Systems & Control Letters*, 24(5):351–359, 1995.
439. Angel Soriano, Enrique J Bernabeu, Angel Valera, and Marina Vallés. Collision avoidance of mobile robots using multi-agent systems. In *Distributed Computing and Artificial Intelligence*, pages 429–437. Springer, 2013.
440. Andrew Sparks. Satellite formation keeping control in the presence of gravity perturbations. In *American Control Conference, 2000. Proceedings of the 2000*, volume 2, pages 844–848. IEEE, 2000.
441. Herbert Spohn. *Large Scale Dynamics of Interacting Particles*. Springer Science & Business Media, 2012.

442. Housheng Su, Xiaofan Wang, and Zongli Lin. Flocking of multi-agents with a virtual leader. *Automatic Control, IEEE Transactions on*, 54(2):293–307, 2009.
443. Shize Su, Zongli Lin, and Alfredo Garcia. Distributed synchronization control of multiagent systems with unknown nonlinearities. *IEEE Transactions on Cybernetics*, 46(1):325–338, 2015.
444. Tyler H Summers, Changbin Yu, Soura Dasgupta, and Brian DO Anderson. Control of minimally persistent leader-remote-follower and coleader formations in the plane. *IEEE Transactions on Automatic Control*, 56(12):2778–2792, 2011.
445. Bing Sun, Daqi Zhu, and Simon Yang. A bio-inspired cascaded approach for three-dimensional tracking control of unmanned underwater vehicles. *International Journal of Robotics and Automation*, 29, 01 2014.
446. Chao Sun, Guoqiang Hu, Lihua Xie, and Magnus Egerstedt. Robust finite-time connectivity preserving coordination of second-order multi-agent systems. *Automatica*, 89:21–27, 2018.
447. Haibin Sun and Choon Ki Ahn. Quantized decentralized adaptive neural network pi tracking control for uncertain interconnected nonlinear systems with dynamic uncertainties. *IEEE Transactions on Systems, Man, and Cybernetics: Systems*, 1–14, 06 2019.
448. Jiankun Sun, Jun Yang, Shihua Li, and Wei Xing Zheng. Estimate-based dynamic event-triggered output feedback control of networked nonlinear uncertain systems. *IEEE Transactions on Systems, Man, and Cybernetics: Systems*, 1–11, 06 2019.
449. Ran Sun, Jihe Wang, Dexin Zhang, Qingxian Jia, and Xiaowei Shao. Roto-translational spacecraft formation control using aerodynamic forces. *Journal of Guidance, Control, and Dynamics*, 40:1–13, 06 2017.
450. Tairen Sun, Frank Liu, Huan Pei, and Yuhong He. Brief paper-observer-based adaptive leader-following formation control for non-holonomic mobile robots. *Control Theory & Applications, IET*, 6(18):2835–2841, 2012.
451. Zhijian Sun, Guoqing Zhang, Yu Lu, and Weidong Zhang. Leader-follower formation control of underactuated surface vehicles based on sliding mode control and parameter estimation. *ISA Transactions*, 72, 11 2017.
452. Zhiyong Sun, Hector Garcia de Marina, Georg Seyboth, Brian Anderson, and Changbin Yu. Circular formation control of multiple unicycle-type agents with nonidentical constant speeds. *IEEE Transactions on Control Systems Technology*, 1–14, 01 2018.
453. Zhiyong Sun, Myoung-Chul Park, Brian DO Anderson, and Hyo-Sung Ahn. Distributed stabilization control of rigid formations with prescribed orientation. *Automatica*, 78:250–257, 2017.
454. CheeKuang Tam and Richard Bucknall. Cooperative path planning algorithm for marine surface vessels. *Ocean Engineering*, 57:25–33, 2013.
455. CheeKuang Tam, Richard Bucknall, and Alistair Greig. Review of collision avoidance and path planning methods for ships in close range encounters. *The Journal of Navigation*, 62(3):455–476, 2009.
456. Kar-Han Tan and M Anthony Lewis. Virtual structures for high-precision cooperative mobile robotic control. In *Proceedings of IEEE/RSJ International*

Conference on Intelligent Robots and Systems. IROS'96, volume 1, pages 132–139. IEEE, 1996.

457. Zhiqi Tang, Rita Cunha, Tarek Hamel, and Carlos Silvestre. Formation control of a leader–follower structure in three dimensional space using bearing measurements. *Automatica*, 128:109567, 2021.
458. Herbert Tanner and D. Christodoulakis. Decentralized cooperative control of heterogeneous vehicle groups. *Robotics and Autonomous Systems*, 55, 11 2007.
459. Herbert G Tanner, Ali Jadbabaie, and George J Pappas. Stable flocking of mobile agents, part i: Fixed topology. In *42nd IEEE International Conference on Decision and Control (IEEE Cat. No. 03CH37475)*, volume 2, pages 2010–2015. IEEE, 2003.
460. Herbert G Tanner, Ali Jadbabaie, and George J Pappas. Flocking in teams of nonholonomic agents. In *Cooperative Control*, pages 229–239. Springer, 2005.
461. Keng Peng Tee and Shuzhi Ge. Control of fully actuated ocean surface vessels using a class of feedforward approximators. *Control Systems Technology, IEEE Transactions on*, 16:750–756, 2006.
462. Keng Peng Tee, Shuzhi Ge, and Francis Tay. Barrier lyapunov functions for the control of output-constrained nonlinear systems. *Automatica*, 45:918–927, 04 2009.
463. John Toner and Yuhai Tu. Flocks, herds, and schools: A quantitative theory of flocking. *Physical Review E*, 58(4):4828, 1998.
464. Loring W Tu. *An Introduction to Manifolds.* Springer, 2008.
465. Vadim Utkin, Juergen Guldner, and Jingxin Shi. *Sliding Mode Control in Electro-Mechanical Systems.* CRC Press, 2009.
466. Umesh Vaidya, Prashant G Mehta, and Uday V Shanbhag. Nonlinear stabilization via control lyapunov measure. *IEEE Transactions on Automatic Control*, 55(6):1314–1328, 2010.
467. Patricio Vela, Amir Betser, James Malcolm, and Allen Tannenbaum. Vision-based range regulation of a leader-follower formation. *IEEE Transactions on Control Systems Technology*, 17(2):442–448, 2009.
468. Tamás Vicsek, András Czirók, Eshel Ben-Jacob, Inon Cohen, and Ofer Shochet. Novel type of phase transition in a system of self-driven particles. *Physical Review Letters*, 75(6):1226, 1995.
469. Rene Vidal, Omid Shakernia, and Shankar Sastry. Formation control of nonholonomic mobile robots with omnidirectional visual servoing and motion segmentation. In *2003 IEEE International Conference on Robotics and Automation (Cat. No. 03CH37422)*, volume 1, pages 584–589. IEEE, 2003.
470. Rong-Jong Wai and You-Wei Lin. Adaptive moving-target tracking control of a vision-based mobile robot via a dynamic petri recurrent fuzzy neural network. *IEEE Transactions on Fuzzy Systems*, 21(4):688–701, 2013.
471. B Wang, J Wang, B Zhang, and X Li. Global cooperative control framework for multiagent systems subject to actuator saturation with industrial applications. *IEEE Transactions on Systems, Man, and Cybernetics: Systems*, 47(7):1270–1283, 2016.
472. Chen Wang, Guangming Xie, and Ming Cao. Forming circle formations of anonymous mobile agents with order preservation. *IEEE Transactions on Automatic Control*, 58(12):3248–3254, 2013.

473. Chen Wang, Guangming Xie, and Ming Cao. Controlling anonymous mobile agents with unidirectional locomotion to form formations on a circle. *Automatica*, 50(4):1100–1108, 2014.
474. Fang Wang, Bing Chen, Chong Lin, Jing Zhang, and Xinzhu Meng. Adaptive neural network finite-time output feedback control of quantized nonlinear systems. *IEEE Transactions on Cybernetics*, 1–10, 06 2017.
475. Jia Wang, Xiaobei Wu, and Zhiliang Xu. Potential-based obstacle avoidance in formation control. *Journal of Control Theory and Applications*, 6(3):311–316, 2008.
476. Jianan Wang. Affine formation control for multi-agent systems with prescribed convergence time. *Journal of the Franklin Institute*, page 18, 2021.
477. Jianan Wang, Xiangjun Ding, Chunyan Wang, Zongyu Zuo, and Zhengtao Ding. Affine formation control of general linear multi-agent systems with delays. *Unmanned Systems*, 2022.
478. Jianan Wang and Ming Xin. Integrated optimal formation control of multiple unmanned aerial vehicles. *IEEE Transactions on Control Systems Technology*, 21(5):1731–1744, 2012.
479. Jianan Wang and Ming Xin. Integrated optimal formation control of multiple unmanned aerial vehicles. *Control Systems Technology, IEEE Transactions on*, 21(5):1731–1744, 2013.
480. Liangyong Wang, Tianyou Chai, and Lianfei Zhai. Neural-network-based terminal sliding-mode control of robotic manipulators including actuator dynamics. *IEEE Transactions on Industrial Electronics*, 56(9):3296–3304, 2009.
481. Lin Wang, Xiaofan Wang, and Xiaoming Hu. Connectivity preserving flocking without velocity measurement. *Asian Journal of Control*, 15(2):521–532, 2013.
482. Meiling Wang, Zhenchao Jia, Hongbin Ma, and Mengyin Fu. Three-robot minimum-time optimal line formation. In *Control and Automation (ICCA), 2011 9th IEEE International Conference on*, pages 1326–1331. IEEE, 2011.
483. Qiang Wang, Yuzhen Wang, and Huaxzhang Zhang. The formation control of multi-agent systems on a circle. *IEEE/CAA Journal of Automatica Sinica*, 1–7, 10 2016.
484. Rui Wang and Liu Jinkun. Adaptive formation control of quadrotor unmanned aerial vehicles with bounded control thrust. *Chinese Journal of Aeronautics*, 30, 02 2017.
485. Wei Wang, Dan Wang, Zhouhua Peng, and Tieshan Li. Prescribed performance consensus of uncertain nonlinear strict-feedback systems with unknown control directions. *IEEE Transactions on Systems, Man, and Cybernetics: Systems*, 46:1–8, 2015.
486. Xiaoli Wang and Yiguang Hong. Finite-time consensus for multi-agent networks with second-order agent dynamics. *IFAC Proceedings Volumes*, 41(2):15185–15190, 2008.
487. Y Wang, L Cheng, Zeng Guang Hou, J Yu, and M Tan. Optimal formation of multirobot systems based on a recurrent neural network. *IEEE Transactions on Neural Networks & Learning Systems*, 27(2):322–333, 2016.
488. Yan-Wu Wang, Xiao-Kang Liu, Jiang-Wen Xiao, and Yanjun Shen. Output formation-containment of interacted heterogeneous linear systems by distributed hybrid active control. *Automatica*, 93:26–32, 2018.

489. Yanwu Wang, Xiaokang Liu, and Jiangwen Xiao. Output formation-containment of coupled heterogeneous linear systems under intermittent communication. *Journal of the Franklin Institute*, 354(1):392–414, 2017.
490. Yujuan Wang, Yongduan Song, and Miroslav Krstic. Collectively Rotating Formation and Containment Deployment of Multiagent Systems: A Polar Coordinate-Based Finite Time Approach. *IEEE Transactions on Cybernetics*, 47(8):2161–2172, 2017.
491. Yujuan Wang, Yongduan Song, Miroslav Krstic, and Changyun Wen. Fault-tolerant finite time consensus for multiple uncertain nonlinear mechanical systems under single-way directed communication interactions and actuation failures. *Automatica*, 63:374–383, 2016.
492. Guanghui Wen, Zhisheng Duan, Wenwu Yu, and Guanrong Chen. Consensus in multi-agent systems with communication constraints. *International Journal of Robust and Nonlinear Control*, 22(2):170–182, 2012.
493. Guanghui Wen, Guoqiang Hu, Wenwu Yu, Jinde Cao, and Guanrong Chen. Consensus tracking for higher-order multi-agent systems with switching directed topologies and occasionally missing control inputs. *Systems & Control Letters*, 62(12):1151–1158, 2013.
494. Jiabao Wen, Jiachen Yang, Yang Li, Jingyi He, Zhengjian Li, and Houbing Song. Behavior-based formation control digital twin for multi-aug in edge computing. *IEEE Transactions on Network Science and Engineering*, 2022.
495. Michael Wooldridge. *An Introduction to Multiagent Systems*. John Wiley & Sons, 2009.
496. Guoqing Xia, Chuang Sun, Bo Zhao, and Jing Xue. Cooperative control of multiple dynamic positioning vessels with input saturation based on finite-time disturbance observer. *International Journal of Control, Automation and Systems*, 17, 01 2019.
497. Fan Xiao, Qingkai Yang, Xinyue Zhao, and Hao Fang. A framework for optimized topology design and leader selection in affine formation control. *IEEE Robotics and Automation Letters*, 7(4):8627–8634, 2022.
498. Feng Xiao and Long Wang. Asynchronous consensus in continuous-time multi-agent systems with switching topology and time-varying delays. *Automatic Control, IEEE Transactions on*, 53(8):1804–1816, 2008.
499. Wei Xiao, Jianglong Yu, Rui Wang, Xiwang Dong, Qingdong Li, and Zhang Ren. Time-varying formation control for time-delayed multi-agent systems with general linear dynamics and switching topologies. *Unmanned Systems*, 07, 10 2018.
500. H Xie, KH Low, and Z He. Adaptive Visual Servoing of Unmanned Aerial Vehicles in GPS-Denied Environments. *IEEE/ASME Transactions on Mechatronics*, 22(6):2554–2563, 2017.
501. Wenjing Xie, B Ma, Tyrone Fernando, and Herbert Iu. A new formation control of multiple underactuated surface vessels. *International Journal of Control*, 91:1–12, 2017.
502. Chengke Xiong, Danfeng Chen, Di Lu, Zheng Zeng, and Lian Lian. Path planning of multiple autonomous marine vehicles for adaptive sampling using voronoi-based ant colony optimization. *Robotics and Autonomous Systems*, 115:90–103, 2019.

503. Bin Xu and Fuchun Sun. Composite intelligent learning control of strict-feedback systems with disturbance. *IEEE Transactions on Cybernetics*, 1–12, 01 2017.
504. Chengjie Xu, Ying Zheng, Housheng Su, Changfan Zhang, and Michael ZQ Chen. Necessary and sufficient conditions for distributed containment control of multi-agent systems without velocity measurement. *IET Control Theory & Applications*, 8(16):1752–1759, 2014.
505. Yang Xu, Dongyu Li, Delin Luo, Yancheng You, and Haibin Duan. Two-layer distributed hybrid affine formation control of networked Euler–Lagrange systems. *Journal of the Franklin Institute*, 356(4):2172–2197, 2019.
506. Yang Xu, Delin Luo, Dongyu Li, Yancheng You, and Haibin Duan. Affine formation control for heterogeneous multi-agent systems with directed interaction networks. *Neurocomputing*, 330:104–115, 2019.
507. Yang Xu, Delin Luo, Dongyu Li, Yancheng You, and Haibin Duan. Target-enclosing affine formation control of two-layer networked spacecraft with collision avoidance. *Chinese Journal of Aeronautics*, 32(12):2679–2693, 2019.
508. Yang Xu, Delin Luo, Yancheng You, and Haibin Duan. Distributed Adaptive Affine Formation Control for Heterogeneous Linear Networked Systems. *IEEE Access*, 7:23354–23364, 2019.
509. Yang Xu, DeLin Luo, YanCheng You, and HaiBin Duan. Affine transformation based formation maneuvering for discrete-time directed networked systems. *Science China Technological Sciences*, 63(1):73–85, 2020.
510. Yang Xu, Shiyu Zhao, Delin Luo, and Yancheng You. Affine Formation Maneuver Control of Linear Multi-Agent Systems with Undirected Interaction Graphs. In *2018 IEEE Conference on Decision and Control (CDC)*, pages 502–507, Miami Beach, FL, December 2018. IEEE.
511. Yang Xu, Shiyu Zhao, Delin Luo, and Yancheng You. Affine Formation Maneuver Control of Linear Multi-Agent Systems with Undirected Interaction Graphs. In *2018 IEEE Conference on Decision and Control (CDC)*, pages 502–507. IEEE, 2018.
512. Yang Xu, Shiyu Zhao, Delin Luo, and Yancheng You. Affine formation maneuver control of multi-agent systems with directed interaction graphs. In *2018 37th Chinese Control Conference (CCC)*, pages 4563–4568, Wuhan, July 2018. IEEE.
513. Yang Xu, Shiyu Zhao, Delin Luo, and Yancheng You. Affine formation maneuver control of high-order multi-agent systems over directed networks. *Automatica*, 118:109004, 2020.
514. Yong Xu and Zheng-Guang Wu. Distributed adaptive event-triggered fault-tolerant synchronization for multiagent systems. *IEEE Transactions on Industrial Electronics*, 68(2):1537–1547, 2020.
515. Farnaz Adib Yaghmaie, Rong Su, Frank L Lewis, and Lihua Xie. Multiparty Consensus of Linear Heterogeneous Multiagent Systems. *IEEE Transactions on Automatic Control*, 62(11):5578–5589, 2017.
516. A Yang, W Naeem, GW Irwin, and K Li. Stability analysis and implementation of a decentralized formation control strategy for unmanned vehicles. *IEEE Transactions on Control Systems Technology*, 22(2):706–720, 2014.
517. Aolei Yang, Wasif Naeem, George W Irwin, and Kang Li. Stability analysis and implementation of a decentralized formation control strategy for

unmanned vehicles. *IEEE Transactions on Control Systems Technology*, 22(2):706–720, 2013.

518. E Yang and D Gu. Nonlinear formation-keeping and mooring control of multiple autonomous underwater vehicles. *IEEE/ASME Transactions on Mechatronics*, 12(2):164–178, 2007.
519. Jun Yang, Jiankun Sun, Wei Xing Zheng, and Shihua Li. Periodic event-triggered robust output feedback control for nonlinear uncertain systems with time-varying disturbance. *Automatica*, 94:324–333, 2018.
520. Junyi Yang, Feng Xiao, and Tongwen Chen. Formation Tracking of Nonholonomic Systems on the Special Euclidean Group under Fixed and Switching Topologies: An Affine Formation Strategy. *SIAM Journal on Control and Optimization*, 59(4):2850–2874, 2021.
521. Kwangjin Yang, Seng Keat Gan, and Salah Sukkarieh. An efficient path planning and control algorithm for ruav s in unknown and cluttered environments. *Journal of Intelligent and Robotic Systems*, 57(1-4):101–122, 2010.
522. Qingkai Yang, Hao Fang, Ming Cao, and Jie Chen. Planar Affine Formation Stabilization via Parameter Estimations. *IEEE Transactions on Cybernetics*, 52(6):5322–5332, June 2022.
523. Wen Yang, Ying Wang, Xiaofan Wang, and Hongbo Shi. Optimal controlled nodes selection for fast consensus. *Asian Journal of Control*, 18(3):932–944, 2016.
524. Yan Yang, Shuxin Wang, Zhiliang Wu, and Yanhui Wang. Motion planning for multi-hug formation in an environment with obstacles. *Ocean Engineering*, 38(17):2262–2269, 2011.
525. J Yeh. Real analysis: Theory of measure and integration, Singapore: World Scientific, 2006.
526. Namik Kemal Yilmaz, Constantinos Evangelinos, Pierre FJ Lermusiaux, and Nicholas M Patrikalakis. Path planning of autonomous underwater vehicles for adaptive sampling using mixed integer linear programming. *IEEE Journal of Oceanic Engineering*, 33(4):522–537, 2008.
527. Brett J Young, Randal W Beard, and Jed M Kelsey. A control scheme for improving multi-vehicle formation maneuvers. In *Proceedings of the 2001 American Control Conference.(Cat. No. 01CH37148)*, volume 2, pages 704–709. IEEE, 2001.
528. Changbin Yu, Brian DO Anderson, Soura Dasgupta, and Baris Fidan. Control of minimally persistent formations in the plane. *SIAM Journal on Control and Optimization*, 48(1):206–233, 2009.
529. P Yu, M Wu, J She, K Liu, and Y Nakanishi. Robust Tracking and Disturbance Rejection for Linear Uncertain System With Unknown State Delay and Disturbance. *IEEE/ASME Transactions on Mechatronics*, 23(3):1445–1455, 2018.
530. Wenwu Yu, Guanrong Chen, and Ming Cao. Some necessary and sufficient conditions for second-order consensus in multi-agent dynamical systems. *Automatica*, 46(6):1089–1095, 2010.
531. Wenwu Yu, Guanrong Chen, Ming Cao, and Jürgen Kurths. Second-order consensus for multiagent systems with directed topologies and nonlinear dynamics. *IEEE Transactions on Systems, Man, and Cybernetics, Part B (Cybernetics)*, 40(3):881–891, 2009.

532. Xiao Yu and Lu Liu. Distributed circular formation control of ring-networked nonholonomic vehicles. *Automatica*, 68:92–99, 2016.
533. Xiao Yu, Lu Liu, and Gang Feng. Distributed circular formation control of nonholonomic vehicles without direct distance measurements. *IEEE Transactions on Automatic Control*, 9286(1):1–8, 2018.
534. Michael M Zavlanos, Herbert G Tanner, Ali Jadbabaie, and George J Pappas. Hybrid control for connectivity preserving flocking. *IEEE Transactions on Automatic Control*, 54(12):2869–2875, 2009.
535. Daniel Zelazo, Paolo Robuffo Giordano, and Antonio Franchi. Bearing-only formation control using an se (2) rigidity theory. In *2015 54th IEEE Conference on Decision and Control (CDC)*, pages 6121–6126. IEEE, 2015.
536. DM Zhang, L Meng, XG Wang, and LL Ou. Linear quadratic regulator control of multi-agent systems. *Optimal Control Applications and Methods*, 36(1):45–59, 2015.
537. Hao Zhang, Bin Xin, Li-hua Dou, Jie Chen, and Kaoru Hirota. A review of cooperative path planning of an unmanned aerial vehicle group. *Frontiers of Information Technology & Electronic Engineering*, 21(12):1671–1694, 2020.
538. Sainan Zhang, Zhongliang Tang, Shuzhi Ge, and Wei He. Adaptive neural dynamic surface control of output constrained non-linear systems with unknown control direction. *IET Control Theory and Applications*, 11:2994–3003, 2017.
539. Shuang Zhang, Yiting Dong, Yuncheng Ouyang, Zhao Yin, and Kaixiang Peng. Adaptive neural control for robotic manipulators with output constraints and uncertainties. *IEEE Transactions on Neural Networks and Learning Systems*, 29(11):5554–5564, 2018.
540. Shuang Zhang, Pengxin Yang, Linghuan Kong, Wenshi Chen, Qiang Fu, and Kaixiang Peng. Neural Networks-Based Fault Tolerant Control of a Robot via Fast Terminal Sliding Mode. *IEEE Transactions on Systems, Man, and Cybernetics, in press, DOI: 10.1109/TSMC.2019.2933050*, 2019.
541. Wentao Zhang and Yang Liu. Distributed consensus for sampled-data control multi-agent systems with missing control inputs. *Applied Mathematics and Computation*, 240:348–357, 2014.
542. Yuwei Zhang, Wang Xingjian, Wang Shaoping, and Tian Xinyu. Distributed bearing-based formation control of unmanned aerial vehicle swarm via global orientation estimation. *Chinese Journal of Aeronautics*, 35(1):44–58, 2022.
543. Zheng Zhang, Shiming Chen, and Yuanshi Zheng. Fully distributed scaled consensus tracking of high-order multiagent systems with time delays and disturbances. *IEEE Transactions on Industrial Informatics*, 18(1):305–314, 2021.
544. Lin Zhao and Yingmin Jia. Neural network-based distributed adaptive attitude synchronization control of spacecraft formation under modified fast terminal sliding mode. *Neurocomputing*, 171, 07 2015.
545. Shiyu Zhao. Affine formation maneuver control of multi-agent systems. *IEEE Transactions on Automatic Control*, 1–1, 01 2018.
546. Shiyu Zhao. Affine formation maneuver control of multiagent systems. *IEEE Transactions on Automatic Control*, 63(12):4140–4155, 2018.
547. Shiyu Zhao, Feng Lin, Kemao Peng, Ben M Chen, and Tong H Lee. Finite-time stabilisation of cyclic formations using bearing-only measurements. *International Journal of Control*, 87(4):715–727, 2014.

548. Shiyu Zhao and Daniel Zelazo. Bearing rigidity and almost global bearing-only formation stabilization. *IEEE Transactions on Automatic Control*, 61(5):1255–1268, 2015.
549. Shiyu Zhao and Daniel Zelazo. Translational and scaling formation maneuver control via a bearing-based approach. *IEEE Transactions on Control of Network Systems*, 4(3):429–438, 2015.
550. Shiyu Zhao and Daniel Zelazo. Bearing rigidity and almost global bearing-only formation stabilization. *IEEE Transactions on Automatic Control*, 61(5):1255–1268, 2016.
551. Shiyu Zhao and Daniel Zelazo. Localizability and distributed protocols for bearing-based network localization in arbitrary dimensions. *Automatica*, 69:334–341, 2016.
552. Shiyu Zhao and Daniel Zelazo. Translational and scaling formation maneuver control via a bearing-based approach. *IEEE Transactions on Control of Network Systems*, 4(3):429–438, 2017.
553. Xudong Zhao, Haijiao Yang, and Guangdeng Zong. Adaptive neural hierarchical sliding mode control of nonstrict-feedback nonlinear systems and an application to electronic circuits. *IEEE Transactions on Systems, Man, and Cybernetics: Systems*, 47(7):1394–1404, 2016.
554. Yu Zhao, Guanghui Wen, Zhisheng Duan, Xiang Xu, and Guanrong Chen. A new observer-type consensus protocol for linear multi-agent dynamical systems. *Asian Journal of Control*, 15(2):571–582, 2013.
555. Z Zhao, W He, and SS Ge. Adaptive neural network control of a fully actuated marine surface vessel with multiple output constraints. *IEEE Transactions on Control Systems Technology*, 22(4):1536–1543, 2014.
556. Changwen Zheng, Mingyue Ding, Chengping Zhou, and Lei Li. Coevolving and cooperating path planner for multiple unmanned air vehicles. *Engineering Applications of Artificial Intelligence*, 17(8):887–896, 2004.
557. Ronghao Zheng, Yunhui Liu, and Dong Sun. Enclosing a target by nonholonomic mobile robots with bearing-only measurements. *Automatica*, 53:400–407, 2015.
558. Yuxin Zheng, Lei Zhang, Bing Huang, and Yumin Su. Distributed event-triggered affine formation control for multiple underactuated marine surface vehicles. *Ocean Engineering*, 265:112607, 2022.
559. Zhong Zheng and Shenmin Song. Autonomous attitude coordinated control for spacecraft formation with input constraint, model uncertainties, and external disturbances. *Chinese Journal of Aeronautics*, 27:602–612, 2014.
560. Hui Zhi, Liangming Chen, Chuanjiang Li, and Yanning Guo. Leader–follower affine formation control of second-order nonlinear uncertain multi-agent systems. *IEEE Transactions on Circuits and Systems II: Express Briefs*, 68(12):3547–3551, 2021.
561. Hui Zhi, Liangming Chen, Chuanjiang Li, and Yueyong Lv. Optimal leader-follower affine formation control of linear multi-agent systems. *Optimal Control Applications and Methods*, 43(1):304–320, 2022.
562. Hui Zhi, Liangming Chen, Chuanjiang Li, and Yueyong Lv. Optimal leader-follower affine formation control of linear multi-agent systems. *Optimal Control Applications and Methods*, 43(1):304–320, 2022.

563. Bin Zhou. Finite-time stability analysis and stabilization by bounded linear time-varying feedback. *Automatica*, 121:109191, 2020.
564. Bin Zhou and Zongli Lin. Consensus of high-order multi-agent systems with large input and communication delays. *Automatica*, 50, 06 2013.
565. Qidan Zhu, Junda Ma, Zhilin Liu, and Ke Liu. Containment control of underactuated ships with environment disturbances and parameter uncertainties. *Mathematical Problems in Engineering*, 2017:1–19, 2017.
566. Zheng Zhu, Yuanqing Xia, and Mengyin Fu. Attitude stabilization of rigid spacecraft with finite-time convergence. *International Journal of Robust and Nonlinear Control*, 21(6):686–702, 2011.
567. Yufei Zhuang, Haibin Huang, Sanjay Sharma, Dianguo Xu, and Qiang Zhang. Cooperative path planning of multiple autonomous underwater vehicles operating in dynamic ocean environment. *ISA Transactions*, 94:174–186, 2019.
568. An Min Zou and Krishna Dev Kumar. Neural network-based adaptive output feedback formation control for multi-agent systems. *Nonlinear Dynamics*, 70(2):1283–1296, 2012.
569. Y Zou, Z Zhou, X Dong, and Z Meng. Distributed formation control for multiple vertical takeoff and landing UAVs with switching topologies. *IEEE/ASME Transactions on Mechatronics*, 23(4):1750–1761, 2018.

Index

For Product Safety Concerns and Information please contact our EU representative GPSR@taylorandfrancis.com
Taylor & Francis Verlag GmbH, Kaufingerstraße 24, 80331 München, Germany

www.ingramcontent.com/pod-product-compliance
Lightning Source LLC
LaVergne TN
LVHW020600110826
845149LV00002B/332